Engineering and Humanities

Edited by

JAMES H. SCHAUB, Ph.D., P.E.
and
SHEILA K. DICKISON, Ph.D.
University of Florida

With

M. D. MORRIS, P.E.

1807 175 YEARS OF PUBLISHING 1982

A Wiley-Interscience Publication
JOHN WILEY & SONS
New York · Chichester · Brisbane · Toronto · Singapore

Library of Congress Cataloging in Publication Data:

Main entry under title:

Engineering and humanities.

"A Wiley-Interscience publication."
1. Engineering and humanities—Addresses, essays, lectures. I. Schaub, James Hamilton, 1925–
II. Dickison, Sheila K. (Sheila Kathryn), 1942–
III. Morris, M. D. (Morton Dan)
TA153.E55 001.3 81-23153
ISBN 0-471-08909-5 AACR2

Printed in the United States of America

10 9 8 7 6 5 4 3 2 1

This book is dedicated with gratitude to
the University of Florida and the National Endowment
for the Humanities, whose combined encouragement made its
conception and completion possible

Preface

As we write the preface to this book, we have been discussing two recent and significant publications that define our reasons for assembling this selection of readings far better than we can. *Newsweek* (October 13, 1980, p. 113), in "The Humanities Crisis," reports on the recently issued report of the Rockefeller Foundation, *The Humanities in American Life*. The article states, "The greatest challenge facing humanists . . . is . . . to demonstrate the importance of the humanities to education and to society."

Earlier this year the *Brookings Bulletin* (spring 1980) featured an article, "The Humanities in a Technological Society," by David W. Brennan. He states unequivocally:

> . . . the humanities are also a source of fundamental human skills much needed in a technological society. I refer to the skills of writing and analysis, to the appreciation of logic and reason, and to the possession of an historical sense, of cultural awareness, and of ability with other languages. Completed with a knowledge of mathematics and science, a person possessing such skills is well equipped to function effectively in a complex society.

We developed this book of readings for the purpose stated by Brennan and the Rockefeller Foundation report and directed it specifically toward the engineering profession with the intent that relating the career interests of engineers to the humanities would generate greater interest in the humanities and encourage readers to pursue them beyond the introduction this book provides.

The readings included in our book are not intended to provide a comprehensive coverage of the humanities. They were chosen to act as an introduction and, hopefully, to serve as an hors d'oeuvre for the intellectual feast the humanities offer. The references with individual articles and the selected additional readings at the end of each part are intended to provide some preliminary guidance for further reading. We are well aware that this effort to provide additional references is not complete but believe the suggestions given will be helpful to the interested reader.

The organization of the book requires brief explanation. The opening part develops the argument for the study of humanities by engineers. The next four parts deal directly with areas of study commonly accepted as humanities disciplines: literature, art, philosophy, and history. The last two parts intend to illustrate that many important engineering problems require a wide range of problem-solving skills, including those that a humanities perspective can provide.

We owe a debt of gratitude to many individuals and organizations without whose help this book could not have been completed. Unfortunately, we cannot acknowledge them all, but we would like to note particularly the help and encouragement of the National Endowment for the Humanities, the University of Florida; and Dr. Gareth Schmeling, principal investigator of the Humanities Perspectives on the Professions program. Credit also goes to those authors and publishers who have graciously permitted us to use their works. Specific acknowledgments follow this preface. Finally, we appreciate the patience of our students over several years of trial runs with the text material and the secretaries whose typing services were invaluable.

JAMES H. SCHAUB
SHEILA K. DICKISON

Gainesville, Florida
February 1982

Acknowledgments

We gratefully acknowledge the permissions of the authors and/or publishers to reprint this material:

PART 1. ENGINEERING AND THE HUMANITIES—WHY?

J. A. STRATTON, "The Humanities in Professional Education." Reprinted from *Science and The Educated Man* by J. A. Stratton with permission of The M.I.T. Press, Cambridge, Mass. Copyright © 1966 by The Massachusetts Institute of Technology.

S. C. FLORMAN, "The Civilized Engineer." Reprinted from *Engineering and the Liberal Arts*, by S. C. Florman. Copyright © 1968 by McGraw-Hill Book Company. Used with permission of the author and McGraw-Hill Book Company.

PART 2. LITERATURE AND ENGINEERING

R. SHECKLEY, "The Mnemone." Reprinted from *Can You Feel Anything When I Do This?*, Doubleday & Co., Inc. Copyright © 1971 by Robert Sheckley. Reprinted with permission of The Sterling Lord Agency, Inc.

R. A. BRYAN, "The Triumph of Science and the Loss of Innocence." An original essay (1970) printed with the permission of the author.

R. KIPLING, "The Sons of Martha," Copyright 1907 by Rudyard Kipling, from *Rudyard Kipling's Verse: Definitive Edition.* Reprinted by permission of the National Trust and Doubleday & Company, Inc.

K. ĈAPEK, "The Punishment of Prometheus." Reprinted from *Apocryphal Stories* by K. Ĉapek, translated by Dora Round, with permission of George Allen & Unwin (Publishers) Ltd. Copyright 1949 by George Allen & Unwin.

E. DICKINSON, "I Like To See It Lap the Miles." Reprinted by permission of the publishers and the Trustees of Amherst College from *The Poems of Emily Dickinson*, edited by Thomas H. Johnson, Cambridge, Mass: The Belknap Press of Harvard University Press. Copyright 1951, © 1955, 1979 by the President and Fellows of Harvard College.

H. D. THOREAU, "Sounds." Reprinted from *Walden* by H. D. Thoreau, with permission of Houghton Mifflin Company.

W. A. DROMGOOLE, "The Bridge Builder." Published in *Rare Old Chums* (1898) and reprinted with permission of *The Nashville Banner*.

W. CATHER, extract from *Alexander's Bridge* by Willa Cather. Copyright renewed 1950 by Edith Lewis. Reprinted with permission of Houghton Mifflin Company.

J. C. VITALE, "The Great Quebec Bridge Disaster." Reprinted from *Consulting Engineer*, ® February 1967, Volume 28, Number 2, pp. 92–95. Copyright © 1967 by Technical Publishing, A Division of Dun-Donnelley Publishing Corp., A Company of The Dun & Bradstreet Corporation—all rights reserved.

A. DEN DOOLAARD, "The Bridge Builder, 1938–1939," from *The Land Behind God's Back*, by A. Den Doolaard (C. Spoelstra). English translation, Copyright © 1958, by Mrs. H. C. Bruinwold Riedel. Reprinted with permission of Simon and Schuster, a Division of Gulf & Western Corporation.

W. HAWAII, extract from *The Robotosyncrasies*, by W. Hawaii. A Hearthstone Book, Carlton Press, Inc. Copyright © 1976 by W. Hawaii. Reprinted with permission of the author.

PART 3. ART AND ENGINEERING

G. SANDOR, "Engineering and Humanistic Creativity—Is There a Difference?" A paper presented at the 87th Annual Conference of the American Society for Engineering Education, June 1979. Reprinted with permission of the author.

A. TOCQUEVILLE, "Of the Spirt in Which the Americans Cultivate the Arts." Reprinted from *Democracy in America Vol. II* by A. Tocqueville (1840), with permission of Schocken Books, Inc. (1974 printing).

M. LEVY, "Structure and Sculpture," from Lecture on Civil Engineering Design, Department of Civil Engineering, University of Florida, 1980. Reprinted with permission of the author.

J. A. KOUWENHOVEN, "Two Traditions in Conflict." Reprinted from *The Arts in Modern American Civilization* by J. A. Kouwenhoven, published by Doubleday & Company, Inc., with permission of the author. Copyright © 1948 by John A. Kouwenhoven.

D. P. BILLINGTON, "Art in Engineering: The Need for a New Criticism." Reprinted with permission from *Engineering Issues*, October 1973, pp. 499–511. Copyright © 1973 American Society of Civil Engineers.

M. SALVADORI, "The Message of Structure," Crom Lecture on Civil Engineering Design, Department of Civil Engineering, University of Florida, 1978. Reprinted with permission of the author.

PART 4. PHILOSOPHY AND ENGINEERING

H. JONAS, "Technology and Responsibility: Reflections on the New Tasks of Ethics." Reprinted with permission of the author and The New School for Social Research from *Social Research*, Vol. 40, #1, spring 1973, pp. 31–54.

N. Balabanian, "Presumed Neutrality of Technology." Published with permission of the author and Transaction, Inc., from *Society*, Vol. 17, No. 3. Copyright © 1980 by Transaction, Inc.

K. Pavlovic, "Making, Doing, and Thinking." An original essay (1980) printed with permission of the author.

G. Hardin, "Living on a Lifeboat." Published with permission of *BioScience*, October 1974, Vol. 2, No. 10, pp. 561–568.

Engineers' Council for Professional Development, "Code of Ethics." Reprinted with permission of the Engineers' Council for Professional Development, Inc., as approved by the Board of Directors, October 1977.

A. Flores, "The Philosophical Basis of Engineering Codes of Ethics." An original essay (1980) printed with permission of the author.

PART 5. HISTORY AND ENGINEERING

P. F. Drucker, "The First Technological Revolution and Its Lessons." Reprinted from *Technology and Culture*, Vol. 7, No. 2, spring 1966, pp. 41–49, with permission of the author. Copyright © 1966 by Peter F. Drucker.

S. C. Gilfillan, "Lead Poisoning and the Fall of Rome." Reprinted with permission from the *Journal of Occupational Medicine*, Vol. 7, 1965, pp. 53–60.

L. White, Jr., "Technology Assessment from the Stance of a Medieval Historian." Reprinted with permission of the author from *American Historical Review*, Vol. 79, No. 1, February 1974, pp. 1–13. Copyright © 1974 by Lynn White, Jr.

T. C. Cochran, "The Social Impact of the Railroad." Reprinted from *The Railroad and the Space Program*, B. Mazlish (editor), pp. 163–181, as published by The M.I.T. Press, 1965, with permission of The American Academy of Arts and Sciences. Copyright © 1965 by The American Academy of Arts and Sciences.

J. E. Vollmar, Jr., "James B. Eads and the Great St. Louis Bridge," as published by The Engineers' Club of St. Louis, 1974. Reprinted with permission of the Board of Directors, The Engineers' Club of St. Louis.

D. McCullough, "The Plan." Reprinted with permission of the author from *The Great Bridge* by D. McCullough as published by Simon and Schuster, 1972. Copyright © 1972 by David McCullough.

M. C. Robinson, "Reclamation Origins." Reprinted with permission of the author from *Water for the West: The Bureau of Reclamation 1902–1977*, as published by the Public Works Historical Society, 1979. Copyright © 1979, by the Public Works Historical Society.

R. A. Kingery, R. D. Berg, and E. H. Schillinger, "Power by Persuasion." Reprinted from *Men and Ideas in Engineering*, by R. A. Kingery, R. D. Berg, and E. H. Schillinger, with permission of the University of Illinois Press, Urbana, Illinois. Copyright © 1967 by the Board of Trustees of the University of Illinois.

R. S. Cowan, "Women and Technology in American Life." Reprinted with permission of the author from *Technology at the Turning Point*, W. B. Pickett (editor), as published by the San Francisco Press, Inc. Copyright © 1977 by Ruth Schwartz Cowan.

PART 6. TECHNOLOGY AND DEVELOPING NATIONS

E. F. Schumacher, "Education for Good Work." Reprinted from *Good Work*, by E. F. Schumacher (New York: Harper & Row, London: Jonathan Cape, 1979), pp. 121–145. Copyright © 1979 by Verena Schumacher. Reprinted with permission of the publishers.

D. P. Mukerji, "Mahatma Gandhi's Views on Machines and Technology." Reprinted from *International Social Science*, Vol. 6, No. 3. Copyright © UNESCO 1954. Reproduced by permission of UNESCO.

D. S. Mansell, "Constraints on Engineering in Third World." Reprinted with permission from *Engineering Issues*, October 1978, pp. 281-292. Copyright © 1978 American Society of Civil Engineers.

A. Hailey, excerpt from *Wheels*. Copyright © 1971 by Arthur Hailey. Reprinted with permission of Doubleday & Company, Inc.

PART 7. THE TECHNO-HUMAN WORLD

M. Bates, "The Human Ecosystem." Reprinted with permission of the National Academy of Sciences from *Resources and Man*, as published by W. H. Freeman and Company, 1969, for the National Academy of Sciences. Copyright © 1969 by the National Academy of Sciences.

L. White jr., "The Historical Roots of Our Ecologic Crisis." Reprinted with permission of the author and *Science* from March 10, 1967 issue, Vol. 155, pp. 1203–1207. Copyright © 1967 by the American Association for the Advancement of Science.

I. Asimov, "The Polemics of Progress." Reprinted from *The Boston Globe*, February 23, 1975. Copyright 1975, Field Enterprises, Inc. Courtesy of Field Newspaper Syndicate.

J. Weizenbaum, "On the Impact of the Computer on Society." Reprinted with permission of the author and *Science* from May 12, 1972 issue, Vol. 176, pp. 609–614. Copyright 1972 by the American Association for the Advancement of Science.

C. A. Lindbergh, Personal letter to E. Q. Daddario, dated July 1, 1970, as published in *National Science and Technology Policy Issues, 1979, Part I*, pp. vii–ix. Reprinted with permission of the Committee on Science and Technology, U.S. House of Representatives and Emilio Q. Daddario.

A. J. Harrison, "Science, Technology and Public Benefits." Reprinted from *National Science and Technology Policy Issues, 1979, Part I*, pp. 39–41, with permission of the author and the Committee on Science and Technology, U.S. House of Representatives.

S. Ramo, "Science, Technology and the Economy." Reprinted from *National Science and Technology Policy Issues, 1979, Part I*, pp. 126–138, with permission of the author and the Committee on Science and Technology, U.S. House of Representatives.

M. Kranzberg, "The Social Ecology of Technology." Reprinted with permission of the author from *Technology and Society*, Oak Ridge Bicentennial Lectures, sponsored by Oak Ridge National Laboratory and published July 1977, pp. 1–14.

Contents

Engineering and Humanities

PART 1

Engineering and the Humanities—Why?

. . . we must remember that man is more than a scientist [engineer], more than a worker. He delights in painting, dancing and dreaming, and finds unspoken joy in the beauty of friendship, or in the smile of a child. He is an artist, a poet and a player.

Howard W. Johnson

There can be no adequate technical education which is not liberal, and no liberal education which is not technical: that is, no education which does not impart both technique and intellectual vision.

Alfred North Whitehead

The engineering profession has been concerned with the role of humanities in engineering education for many years. The question in its simplest form is, should the basic education for an engineering student be limited to the sciences, engineering sciences, and engineering studies necessary to prepare the student to function effectively as an engineer in today's world or should the curriculum prepare the student to perform effectively not only as an engineer but also as an informed citizen? There is little agreement among engineers [see Kent (1)] on this question and American Society of Civil Engineers (ASCE), for example, has been discussing it since at least 1874 when Thomas C. Clark (2) advocated the

---

Portions of this introduction were published in a slightly different form under the title "Humanities, Engineering and Education," J. H. Schaub, *Civil Engineering Education*, Vol. I, American Society of Civil Engineers, New York, 1979, pp. 644–650.

education of engineers to be strong in both the natural sciences and the humanities.

Jerome Kagan, a Harvard psychologist, provided a description of his educational aims during a conference. "The Educated Person in the Contemporary World," held in Aspen, Colorado, in 1974. These aims appear to be an excellent description of the educational needs of today's engineer (3):

> My list of educational aims—a sense of dignity, some understanding of the laws of nature and society, mastery of a set of technical skills that permit assumption of a vocation viewed as useful or creative, appreciation of the historical forces that sculpted society, the capacity to appreciate beauty and the motivation to participate in a creative enterprise, the capacity for serenity and honesty, charity and civility.

Kagan's list of aims include those that engineers recognize quickly as being of a technological nature—understanding the laws of nature, mastery of a set of vocational skills viewed as useful and creative, and concepts of ethical concern. The others in his list are not so much technically oriented as human oriented. And here lies the need for a strong and imaginative humanistic component in our engineering education.

The need for the nontechnical stem was recognized early by comments in the Grinter report (4) of 1955, in which the attributes of the well-educated engineer were defined:

> He must be not only a competent professional engineer, but also an informed and participating citizen, and a person whose living expresses high cultural values and moral standards. Thus, the competent engineer needs understanding and appreciation in the humanities and in the social sciences as well as in his own field of engineering. He needs to be able to deal with the economics, human and social factors of his professional problems. His facility with, and understanding of, ideas in the fields of humanities and social sciences not only provide an essential contribution to his professional engineering work, but also contribute to his success as a citizen and to the enrichment and meaning of his life as an individual.

Similar expressions of the need for the liberal education stem in engineering have been presented in the Olmsted report (5) and in the ASEE goals report (6). Numerous other papers since 1966, to consider only the relatively recent publications, illustrate the concerns with the liberal education of the engineering student.

If the concepts expressed by Kagan and by the Grinter report are accepted, the question still remains: Why the humanities? Kolenda (7) has stated unequivocally, "An education without humanities is a misnomer." In a like fashion, Hardy Cross declares in *Engineers and Ivory Towers* (8), "The purpose of education is to prepare the whole man to live a full life in a whole world." Much earlier, the Roman playwright Terence said, "Because I am a human being, I feel that nothing concerning a human being is foreign to me" (*Homo sum. Humani nihil a me alienum puto*). Engineers are human beings and citizens first, and their education must be more than training for the details of a technical job if they are to be equal participants in the full life.

The word *humanities* is derived from the Latin *humanitas;* which means the quality that makes a being "human." Although we speak of the humanities as if they were a grouping of a number of disciplines, they should be looked upon as a single discipline with all the components interrelated to a comprehensive whole. Engineering might well be considered within this whole. Alfred North Whitehead wrote in 1929 (9),

> There can be no adequate technical education which is not liberal, and no liberal education which is not technical: that is, no education which does not impart both technique and intellectual vision.

Unfortunately, the development of our university system has resulted in a degree of specialization in which each faculty member tends to teach a narrow field of interest rather than the unifying concept of *humanitas*.

It was at this point that the University of Florida found itself as it entered the 1970s. The general education requirement of the state university system included a block of courses entitled "The Humanities." These courses were taught as an integrated sequence intended to introduce and stimulate interest in that broad area of study. Conceptually, it was probably a good idea; however, it did not prove to be effective for engineering students. In fact, it often had opposite results. Two obvious reasons existed for this. First, the courses were offered in the first two years of the college experience. This is a time when students are least receptive to material they cannot relate to career goals. Second, the faculty teaching the humanities courses, although they were excellent teachers and well qualified in their fields of interest, were faced with the nearly impossible task of trying to be effective within a course concept that was so broad. Few, if any, mentioned the relationship between technology and the humanities in their presentations, and the results were what might be expected. The engineering students were not interested in the courses as presented, nor did they show any indication toward further work in the humanities as their education progressed.

The Florida experience is not unusual. The Olmsted report (5) quoted a study by Trent and Ruyle (10):

> . . . engineering is likely to attract the individual who seeks direct, sure lines to follow in life. This kind of individual tends to enter a profession where he need not be confused with the tensions and anxieties of conflicting, ambiguous, or novel ideas. . . .

Yet the confusion of ideas, the concern with ambiguity and complexity, and the lack of relevance is exactly what the humanities offer engineering students. No wonder they are confused and disenchanted. Yet the value of the humanities to the engineer is that they do not indoctrinate. They tend to encourage many conflicting points of view rather than the accepted solution. They do not provide ready answers but confuse issues in the same way that life and engineering are confused issues, because they are not the simple textbook problems with simple, straightforward answers as often presented in the classroom. The basic value of the humanities courses to the engineering student is that they present a way of approaching a question, a reading, or a problem that is different from the engineering disciplines. Rather, these courses cause the student to think and, in this way, they encourage mental growth and development.

Cross (11) recognized the value of the nontechnical courses.

> All of these disciplines [the liberal arts] appear when engineering principles are applied to the planning of engineering works. If they do so appear, the interest is created and the student may then or later seek out books, courses, men who can aid him in satisfying that interest. Time can be found in the undergraduate course to let the student begin this search, provided the courses in engineering create the interest.

It is likely that many, if not most, of the engineering faculty are lukewarm or less in their regard for the humanities [see Kent (1)]. It may be that a proper role for the humanities would be to define the relationship of the liberal arts to engineering to help create the interest to which Cross refers.

The two readings in this introductory part make an articulate case for viewing humanities as an integral part of professional education. Although J. A. Stratton, a physicist, is addressing the question from the point of view of an institute of technology, he makes a compelling case for a definition of the professional which includes a humanistic perspective on problem solving. The reading by S. C. Florman, who is both a practicing engineer and a humanist, takes Stratton's argument a step further and provides urgent reasons why engineers must study the liberal arts.

The problem is challenging: How to teach humanities to gain the good they offer to the education of engineering students without discouraging the student with the apparent conflict in approach between the humanities and the career? A partial solution might be the one suggested in the Florman chapter, which is developed in the remainder of his book *Engineering and the Liberal Arts.* Florman argues that between the liberal arts and engineering there are already existing bridges that only need to be discovered: the history of technology, for example, is the connecting link between history and engineering. The paths between engineering and other areas of the liberal arts may be more obscure. One of the aims of this anthology is to assist readers in making less obvious connections between the two disciplines and to begin a dialogue mutually beneficial to both humanities and engineering.

REFERENCES

1. Kent, James A., "The Role of the Humanities and Social Sciences in Engineering Education," *Journal of Engineering Education*, April 1978, pp. 725–729.
2. Clark, Thomas C., "The Education of Civil Engineers," *ASCE Transactions*, Vol. 3 1874, pp. 255–266.
3. *Harvard Magazine*, September 1974, p. 9.
4. Grinter, L. E., "Report on Evaluation of Engineering Education (1952–1955)," *Journal of Engineering Education*, September 1955, pp. 25–63.
5. "Liberal Learning for the Engineer," Report of the ASEE Humanistic Social Research Project, S. P. Olmsted, Director, *Journal of Engineering Education*, December 1968, pp. 303–342.
6. "Goals of Engineering Education," Final Report of the Goals Committee, ASEE, *Journal of Engineering Education*, January 1968, pp. 367–446.
7. Kolenda, Constantin, "The Role of Humanities in Engineering Education," *Journal of Engineering Education*, May 1966, pp. 329–331.
8. Cross, Hardy, *Engineers and Ivory Towers*, McGraw-Hill Book Company, New York, 1952, p. 29.

9. Whitehead, Alfred North, *The Aims of Education and Other Essays*, The Macmillan Company, New York, 1929, p. 74.

10. Trent, James W., and Janet H. Ruyle, "Practicality and Creativity in Engineering and the Liberal Arts," unpublished research report, 1968, p. 3 as quoted in Ref. 5.

11. Cross, Hardy, *Engineers and Ivory Towers*, McGraw-Hill Book Company, New York, 1952, p. 43.

1.1

The Humanities in Professional Education

JULIUS A. STRATTON

Some months ago, when I was invited to take part in the events surrounding this inauguration, the hope was expressed that I would comment tonight upon the relation of the humanities to professional education, particularly in the context of an institute of technology. I might have wished for an easier assignment, for the theme is hardly a new one. Since the founding of institutions such as ours fifty or a hundred years ago, the place of the humanities has been a lively subject of debate.

In many respects we were patterned after European models in the middle of the nineteenth century. But the idea that such subjects as English and history and modern languages should have a part in the professional education of engineers was distinctly an American innovation. It is significant that today, when nearly every country in Western Europe is engaged in a reassessment of higher education, they are examining closely the American experience and have shown a particular interest in the manner in which we have drawn the humanities into our programs in science and engineering.

The recent report of Lord Robbins, for example, proposes sweeping revisions in the plan of higher education in Britain. Several years ago both Dr. DuBridge and I were invited to appear in London before the Robbins Committee, and I recall that the questions that were pressed upon me most strongly dealt precisely with this issue of the humanities.

The German technical universities also are currently in a state of ferment and change, and they have taken major steps in the integration of liberal studies into the scientific and technical curricula.

Without any doubt, the humanities are a real presence on the campuses of such institutions as Case, Caltech, Carnegie, and M.I.T. We make much of them in our discourses to the public. The concept is fundamental, and the intent has been

---

An address given on the occasion of the inauguration of Dr. H. G. Stever as President of the Carnegie Institute of Technology on October 21, 1965.

sincere. Yet I think all of us must acknowledge that although a great deal has been accomplished and although the humanities have indeed had an impact upon our institutions, we still have far to go if they are to come finally into their own. The ideas that I present to you briefly this evening carry a certain sense of urgency. They express my own deep conviction that in the higher education of scientists and engineers, the humanities must be brought into full partnership.

I recall very vividly an incident that occurred several years ago at the Saturday Club, an old Boston institution with the traditions of Emerson, which describes itself as "a pleasant, utterly informal company of men, more or less eminent, having a long lunch together on the last Saturday of each month." The poet Robert Frost was there that day, and after a while the conversation turned to the teaching of humanities to scientists and engineers. Suddenly, to the dismay of almost everyone, Mr. Frost pounded the table and said to me with great bitterness: Let the scientists and technologists take care of their own affairs, which are going badly enough; let them play with their hardware, but let them leave the important problems of this world to us, to the poets and the philosophers.

Mr. Frost, of course, was not alone in his fear, and sometimes contempt, for the works of modern science. And it may very well be that he distrusted the scientists more than he did science itself. There are still those who believe in the desirability and even the feasibility of such a division of the world's labor. Yet this idea does injury to the cause of the humanist as well as to science and runs counter to reality. For whether we will or not, science cannot be isolated from the rest of human affairs. Nor can anyone understand the stream of art and literature and philosophy today without some perception, some awareness of the great currents of scientific thought. We are a single society, a single world, with a single complex of human problems. And science, in effect, is simply the search for comprehension, the attempt to unravel some part of the mysteries of the universe which we inhabit.

Contrary to a very common belief that the progress of science is leading us toward a fragmentation of knowledge and increasing specialization, I see clear evidence of a growing confluence of fields and disciplines. Today, for example, it is extremely difficult to draw a line of demarcation between what constitutes pure science and modern engineering. It is difficult to isolate physics from chemistry, or chemistry from biology. And engineering is itself a continuum, extending into economics, political science, and management. But this flowing over into broader areas goes much further. By the very nature of their wider concerns and the character of life as it is today, scientists and engineers are drawn deeper and deeper into issues and decisions whose import is far more than technical. And it is for such responsibilities, for this new order of citizenship, that we must prepare the graduates of our institutes of technology.

At the heart of this matter is the ideal of the true professional estate. Let us think for a moment not about scientists or engineers, or about lawyers or doctors, but about the qualities that should distinguish the professional man or woman.

First, of course, there is the essential requirement of complete competence, of a depth of understanding about some special field at a high intellectual level. Just the other day I came across as clear a statement of this requirement as one might wish. It appeared over ninety years ago in the annual report of the Head of the Department of English at M. I.T.

"The professional student," he said, "comes for a distinct purpose; he wishes, for instance, to be made an engineer, and he must be trained so thoroughly in engineering studies that his bridge will not break down through faults of construction. It would be small comfort to his employers if that should happen, that he could report in Addisonian English, and with unimpeachable logic, the precise reasons why it broke down."

This kind of mastery implies more than a command of current practice. It calls for such an understanding of fundamentals as to resist obsolescence and to instill a confidence in one's own power to maintain a position in the forefront. But beyond the requirement of commanding knowledge and adaptability to change and innovation, the professional estate is a way of life. It reflects a point of view—a concern for consequences beyond technical terms; a comprehension of the impact of one's works on the intellectual and spiritual life of the individual and of the society; a willingness to administer to the public welfare; a sense of civic responsibility; an understanding of our institutions that derives from a sense of history; a taste for excellence and a capacity to recognize and to appreciate the beauties of ideas, of art, and of literature. It calls for a commitment to the ethical principles on which decisions must rest. It is expressed through a reaching down for fundamentals with a desire to get to the essence of the matter, and, above all, to an understanding of one's own relation to what has gone before and what is to come—in sum, a human perspective. Admittedly, these are virtues that few men ever achieve completely, but ones to which we hope our graduates will aspire.

It would be unreasonable and ill-advised to assume that the faculties of English, of history, of philosophy, or of languages in our technical institutions can alone develop and instill these qualities in our students, or that the burden should be solely theirs. Yet from time immemorial, we have looked to the humanities for inspiration and leadership in just this realm of ideas. And if institutions such as Carnegie and M.I.T. are to fulfill their responsibility for professional education in this highest sense, the humanities must be more than peripheral. They must become central to our endeavor. It is in this perspective of a truly professional education that we must give them encouragement and support and that they, in turn, must respond to the challenge.

And now how are we going to bring this about? It will not be easy to achieve, nor is there in my judgment any unique way to augment the strength and consolidate the position of the humanities in our professional schools. We shall have to try many experiments and explore many roads.

While all these institutions have much in common, they also have different histories, different resources, and different manners of approach. Carnegie is fortunate, for example, in its College of Fine Arts, and it may capitalize on the extraordinary opportunities for the enjoyment of art and music offered by this city of Pittsburgh.

Of one thing I have become quite convinced: it will be difficult, if not pointless, to attempt to recreate a miniature of a traditional liberal arts college as an enclave in the midst of an institute of technology. The humanities must not be set apart. If they are to thrive, they must put down deep roots of their own for an authentic growth in our peculiar environment. What we seek is a style for the institution as a whole rather than the cultivation of a secluded little plot of ground.

This, in turn, implies new forms and demands new insights and imaginative solutions to new and old problems. I have no thought this evening of discussing details of administration or specific programs—the pros and cons of core curricula or course sequences—but I do want to emphasize the need for a revolutionary attack. My colleague Elting Morison has remarked that we must recognize that "unlike most universities, we are less custodians of what has been said and thought and more creators of the new. What goes on in science and engineering is a restless seeking for the undiscovered or untried. This spirit should be made to invest our work in the humanities."

At the same time, I must warn against what someone has called the "scientization" of the humanities. Linguistics, mathematical logic, and the quantitative applications of psychology are important in their own right and, by their very nature, are strongly linked to mathematics, to physics, to the communication sciences. The history and philosophy of science are also eminently germane to our interests. But these are not themselves humanities in the sense of which I speak, nor do they fulfill our need. We may fruitfully draw upon the modern world for examples and stress the society of our day rather than of antiquity. But the essence of humanistic studies is rooted in the timeless issues of mankind.

We must always face the question whether realistically these institutes of technology can vie successfully with the colleges and universities of the older liberal tradition in building strong faculties. I think they can, and I am confident that they will.

Of course the arts here will not be the same. They will develop their own form and character. It would be absurd to imagine that all areas of literature, for example, would find scholarly representation among us any more than one might expect to find every specialty of science in a liberal arts college. And yet I think we have much to offer to the scholar and the artist who will come to terms with science. There is a vitality in our institutions. There is a sense of involvement, of relevance to the world around us, to the problems of our time. There is a striking sense of forward movement. And there are students of superb intelligence, eager, extraordinarily receptive, open to ideas. This, I submit, is anything but sterile ground for humanistic studies.

Science and engineering today are the very matrix within which human affairs must be interpreted. As we speculate about the destiny of mankind, we draw upon what science teaches us of the origins of our earth and of the biological nature of life. As we strive to develop that destiny, we draw upon the power and implications of modern technology. As we seek to interpret that destiny, to move mankind forward in understanding and perception, we look with hope to the working partnership of the humanities and science in its broadest sense. As the greatness of the Renaissance lay in the meeting of the scientist and the humanist on common ground, so we must aspire to create together the man of the twenty-first century.

1.2

The Civilized Engineer

SAMUEL C. FLORMAN

Those of us who are engineers in the last third of the twentieth century are among the most fortunate of men.

In a time of widespread despair our constructive work gives us reason to be sanguine. In an age when most men are confused by the complexity of the scientific revolution, we are uniquely equipped to understand and enjoy the marvelous technological happenings all about us. It is said that the condition of man in our era is one of increasing alienation. But we engineers are needed by our fellow men; our place in society is secure; we feel at home in the world. Our work brings us comfortably in touch with the real world of "things"; our days are spiced with the tang of novelty and inventiveness. Financially, although we might not always consider ourselves adequately compensated for our efforts, we need never know want.

BUSY, SECURE, AND DISSATISFIED

Nevertheless, we are not content. We can see that the world is teeming with treasures of the arts about which we know little. We find ourselves somehow excluded from the intellectual and philosophical discourse in which the values and goals of our society are shaped. Not only are we not participating in the artistic and intellectual life of our time, but we find that our professional product, the technology of which we are so proud, is being misused and misdirected, dominated by forces beyond our control. We are unhappy about our "image" and about the fact that we are not receiving our proper share of respect and responsibility.

In sum, we may be busy and secure, but we are far from satisfied, either with our personal lives or with our role in society. Anyone familiar with the literature of our professional journals knows this to be the case.

Our dissatisfaction seems related to a certain flaw in our professional personality, a flaw which limits both our capacity to experience life to the full and our ability to play an important role in the political, cultural, and social developments of our time. This flaw stems from our failure to become "well rounded," a failure resulting from our admitted concentration on the technical aspect of our profession. To pinpoint it, the source of both our inadequacy and our discontent is rooted in our lack of a broadly based education, our lack of a civilizing education in the liberal arts.

According to Webster's dictionary, civilization consists of "progress in education, refinement of taste and feeling, and the arts that constitute culture." If we take this definition literally, the average engineer today is simply not civilized. It is paradoxical that without us civilization could not exist, yet we are somehow isolated from the civilizing influence of the culture of our time. This is a misfortune for us as individuals, for our profession, and for the world.

I am not talking about superficial refinement and the ability to sparkle at cocktail parties. I am talking about something more fundamental. It is not an exaggeration to say that liberal education for engineers could improve the quality of life for the average engineer, contribute to the sound development of the engineering profession, and help to preserve and enrich society as a whole. Let us consider some of the ways in which this is so.

INTELLECTUAL COMPETENCE AND IMAGINATION

First of all, a liberal education enlarges intellectual capacity, develops mental agility, improves our ability to think. As a noted educator has said, liberal education helps "to cultivate those skills and habits of reasoning which constitute intellectual competence, the capacity to think logically and clearly, the ability to organize one's thoughts on any subject on which essential facts are possessed or obtainable."[1]

If we need the liberal arts to maintain and improve our intellectual competence, we need them even more to develop imagination, for without imagination reason is not equal to even the minimum demands of our exploding technology. Lewis Mumford has warned us that a concentration on pure technical training

> ... might defeat even its immediate purposes by depriving original minds of the stimulus and enrichment of wider interests and activities.
>
> Has not Niels Bohr told us that he arrived at the doctrine of complementarity in physics by speculating on an ancient theological dilemma—the impossibility of reconciling perfect love with perfect justice? Men flew in dreams and communicated instantaneously in myths and fairy stories long before they achieved the technical apparatus for doing so. But would any chain of discoveries and inventions have produced a balloon or a telegraph if the dream had not first suggested these goals? Many significant inventions, from the helicopter to the motion picture, began as toys for amusing the young. Plainly the self-sufficiency of the specialist's world is a prisoner's illusion. It is time to open the gates.[2]

LIBERAL EDUCATION AND LEADERSHIP

As liberal education improves our intellectual competence and expands our imagination, it also develops those qualities of intellectual curiosity and general understanding, those traits of grace and wit and poise which characterize the leaders among men. Too often engineers are found lacking in these attributes. Scientifically made personality studies have revealed engineers to be "socially conforming, impersonal, introverted individuals."[3] In industry the effectiveness of engineers has been found to be limited by their lack of "people-wisdom," their reliance on "coldly rational judgment," and the recurrent appearance of a "noncommunicative syndrome."[4]

The president of a large corporation has said succinctly what many leaders of American industry have come to recognize as fact:

> The specialist cannot function effectively at the top level of management if all he brings to it is his specialty. At that level, the daily problems call for broad general knowledge, open-mindedness, an understanding of human nature, an insight into human frailties, a fairness of mind, a clarity of thought.... The qualifications needed for leadership in industry are developed largely through a liberal arts education.[5]

The proposition has scarcely changed in the nineteen centuries since Vitruvius, the great Roman engineer, wrote that those who have acquired

> ... skill without scholarship have never been able to reach a position of authority to correspond to their pains.... But those who have a thorough knowledge of both, like men armed at all points, have the sooner attained their object and carried authority with them.[6]

THE GOOD LIFE

In addition to helping each of us to do his job more effectively—and of even greater importance in the last analysis—liberal learning yields great riches to the individual in pursuit of the good life. Knowledge and understanding provide pleasure that needs no practical justification. Beauty evokes joyousness that is its own reward. The most precious treasure awaiting the engineer in the world of the liberal arts is enrichment of his personal life—enrichment and the tranquility of spirit which accompanies new insight.

We engineers pride ourselves on being members of a profession which engages our energies and challenges our capacities. We are usually too much absorbed in our interesting work to be overly bothered by the doubts and anxieties which plague many of our less fortunate brethren. This concentration on work is a blessing, but it contains a hidden flaw. Our questioning and doubting are liable to be postponed, only to emerge in later years, sometimes with disturbing effect. It is better surely to expose oneself early and often to the eternal problems of philosophy and art than to be awakened with a start in one's waning years by the sudden asking of the questions, "What is life all about? What have I been living for?" Socrates' admonition still rings true. "The unconsidered life is not worth living."

Mark Van Doren has spoken of the happiness won by the man who has sought inspiration and enlightenment in the "great tradition" of the liberal arts. "That happiness consists in the possession of his own powers, and in the sense that he has done all he could to avoid the bewilderment of one who suspects he has missed the main thing. There is no happiness like this."[7]

STATUS FOR THE PROFESSION

As the individual engineer profits from acquaintance with the liberal arts, so will the entire profession. Our lack of "status," our unsatisfactory "image"—these are concerns which gnaw away at our collective professional contentment. Only a vastly increased number of liberally educated engineers can remedy this situation. Self-praising pronouncements emanating from our professional organizations surely will not suffice.

The public relations problems of the engineering profession are nothing new. They already existed in the days of ancient Greece. Xenophon spoke for most of his fellow citizens when he said that "the mechanical arts carry a social stigma and are rightly dishonored in our cities." Technologists, he maintained, "simply have not got the time to perform the offices of friendship or citizenship. Consequently they are looked upon as bad friends and bad patriots . . . ."[8]

A hundred years ago Ralph Waldo Emerson looked at the technologists of his day and spoke sadly of "great arts and little men." "Look up the inventors," he wrote. "Each has his own knack; his genius is in veins and spots. But the great, equal symmetrical brain, fed from a great heart, you shall not find."[9] Even Thomas Henry Huxley, nineteenth-century advocate of science and technology, expressed concern about technologists becoming "lopsided men." "The value of a cargo," he pointed out, "does not compensate for a ship's being out of trim."[10]

In the early twentieth century, engineering achieved a certain level of prestige, although the profession was still regarded warily even by its greatest admirers. Thus Thorstein Veblen in 1917:

> Popular sentiment in this country will not tolerate the assumption of responsibility by the technicians, who are in the popular apprehension conceived to be a somewhat fantastic brotherhood of over-specialized cranks, not to be trusted out of sight except under the restraining hand of safe and sane businessmen. Nor are the technicians themselves in the habit of taking a greatly different view of their own case.[11]

Today, when in the midst of spectacular engineering achievements, this galling tradition persists. An American science editor informs us that "the image that has been projected of the engineering profession—and images are very hard to change—is of a prejudiced, conservative, non-involved group."[12] An English science editor comments that engineers, "the men who daily make history, are still not accepted as important citizens! . . . . Even the word engineer has connotations of a man in a boiler suit who is a kind of modern blacksmith."[13] A public relations man captures the essence of a prevalent attitude in a few deft sentences:

> I grew up in the tradition of the engineer being a man with no verbal skills whatsoever. He was one of these guys who if you gave him an applied kind of problem would go off in a corner and work it out for you but, God forbid, don't ask him to explain it. He's no good with the English language; don't expect him to articulate himself. He is this kind of faceless, anonymous character.[14]

Engineers are certainly not insensitive to public opinion, nor do they engage in self-deception. A study has shown that "engineers themselves are convinced that the general public does not hold them in as great esteem as other professions."[15]

Only liberally educated engineers can bring the profession the esteem it craves and, in so many important ways, deserves. For one thing, only liberally educated engineers will possess the eloquence with which to impress upon their fellow citizens the inherent worth of engineering and its importance to society. It has been charged, and rightly so, that the engineering profession "has not been in touch with the people and by default has permitted a working partner (science) to capture the imagination of the nation."[16]

But there is a more important goal than "telling the story" of engineering to the public. If engineers themselves, as individuals, become truly cultured—that is, become educated in the liberal arts—then the word will spread without a "good press." If we engineers do not want to be known as "bad friends and bad patriots," "little men," "lopsided men," "overspecialized cranks," if we resent being characterized as a "prejudiced, conservative, noninvolved group," if we cringe at being looked at as "a kind of modern blacksmith," as "this kind of faceless, anonymous character," then it is up to us to make sure we resemble these things as little as possible. If engineers become increasingly wise, sensitive, humane, and responsible, we will not need public-relations techniques to sell us to the public.

THE PUBLIC GOOD

And as the engineering profession gains prestige and authority, society as a whole will benefit. *For the world is desperately in need of the leadership that only engineers can give.* "The politicians, and even the statesmen," as James Reston of *The New York Times* has put it,

> ... are merely scrambling to deal with the revolutions in weapons, agriculture and industry created by the scientists and the engineers. The latter have transformed man's capacity to give life, to sustain and prolong life, and to take life; and the politicians no longer find that they can deal with all the new complexities and ambiguities....[17]

The world must listen to the engineer or it is doomed. Buckminster Fuller has stated the facts in the simplest terms:

> If humanity understood that the real world problem is that of upping the performances per pound of the world's metals and other resources, we might attempt to solve that problem deliberately, directly and efficiently.... But I find that approximately no one realizes what is going on. That is why we have been leaving it to the politician to make the world work. There is nothing political that the politician can do to make fewer resources do sixty percent more.[18]

We engineers already possess most of the technical knowledge required to provide food and shelter in abundance, restore purity to our air and water, heal the blight of our cities, untangle the snarl of traffic, harness our rivers, reap harvests from the oceans, husband our resources, and develop power from the sun and atom. We will—if the world will let us—subdue floods, minimize the danger from storms and earthquakes, and eventually control the weather. We will perform new miracles in the fields of medicine, communication, and transportation, and develop a continuous stream of marvelous fabrics and household appliances. We can—if called upon—contribute to the preservation of peace by assisting the underdeveloped nations and by devising improved means of arms control.

But unless we achieve a position of leadership, our talents will continue to be largely wasted and misdirected. The world will persist in demanding our blast furnaces but not our smoke-control devices, our highways but not our parks, our bombers but not our hospital ships. Running wild and out of control, technological progress will become a disease—a plague of asphalt and armaments, pollution and blight.

Admittedly, there is a school of thought which holds that the world's ills are not attributable to lack of leadership by the technologists, but rather to leadership heading in the wrong direction. "In every country in the world," according to George Orwell,

> . . . the large army of scientists and technicians, with the rest of us panting at their heels, are marching along the road of "progress" with the blind persistence of a column of ants. Comparatively few people want it to happen, plenty of people actively want it not to happen and yet it is happening. The process of mechanization has itself become a machine, a huge glittering vehicle whirling us we are not certain where, but probably toward the padded Wells-world and the brain in the bottle.[19]

"It is apparently our fate," echoes a French scholar, "to be facing a 'golden age' in the power of sorcerers who are totally blind to the meaning of the human adventure."[20]

Nor is this sentiment restricted to apprehensive artists and intellectuals. In his farewell address President Eisenhower warned the nation that its public policy might "become the captive of a scientific-technological elite." Senator Bartlett of Alaska has complained that "faceless technocrats in long, white coats are making decisions today which rightfully and by law should be made by the Congress."[21]

It is true that in government and industry engineers are numerically abundant and *potentially* powerful.[22]. But the extent of our *actual* influence in directing the course of our society is a moot question indeed. In either case—whether the world is racing toward disaster in spite of us or because of us—clearly what is needed is enlightened engineering leadership.

Someone must step forward to say, "We can afford to make that automobile a little safer," "Let us build a factory that is more attractive," "Let us consider the possible harmful effects of that insecticide before we market it," "Let us develop a plant process that will not pollute the water we use," "Let us make that machine a little quieter," "Let us not demolish that historically precious old building," "Let us locate that dam, not only where it will generate the most power, but also where it will serve the interests of the community—esthetically, politically, and socially," "Let us build a rapid transit system for this city rather than a freeway that will bring more cars into an area already choked with traffic."

This "someone" cannot be an ordinary citizen of good will. He must be able to bolster his arguments with facts—technical, scientific, and economic. Hunches and sentiments will not be sufficient. His recommendations, in order to be persuasive, must be founded in a knowledge of resources, materials, and energy conversion; statistics, probabilities, and decision theory; computers, controls, and systems engineering. Moreover, this "someone" must be concerned. He must be articulate. He must be esteemed. And he must have a highly developed moral and esthetic sensibility. In short, he must be a liberally educated engineer.

Liberal education will not make prophets and saints out of engineers, and even if it did, this would not necessarily bring about instant Utopia. But this prudent qualification should not discourage us from moving in what we know to be the right direction.

THE PUBLIC PHILOSOPHY

One thing further. A generation of liberally educated engineers would inevitably play an important role in the debates which are instrumental in shaping the public philosophy. Engineers are already developing techniques of decision making which will enable the leaders of society to choose rationally between alternative courses of action. When we are as wise as we are smart, we will be qualified to talk about new goals as well as methods, ultimate ends as well as means. We will have much of worth to contribute to the public philosophy, both in word and in deed.

Although world-wide conquest of poverty and need is far from being achieved, the more developed nations are already encountering an entirely new spectrum of problems born of affluence. Future generations, having won the battle for survival, will be confronted with the task of maintaining vitality and pride in our communities, conquering boredom, and renewing faith in the inherent value of creative labor.

The response of truly civilized engineers could lead to a renaissance of engineering that would dazzle and inspire the world. In every nation, in every town, there would be a flowering of towers, arches, and domes; concert halls, stadiums, theatres, museums, and libraries; universities and art centers; memorials, shrines, and churches; parks, gardens, fountains, pools, arcades, and promenades. Treasures of antiquity would be recovered, preserved, and reconstructed. Research centers and sanctuaries would flourish in the deserts, the jungles, and on the polar ice, under the sea and in outer space. In all of this, every line, every proportion of structure, vehicle, and machine would spring, not from greed or grim necessity, as so much of it does today, but from an inner human striving for beauty and excellence. Lewis Mumford has spoken hopefully of the eventual coming of "a new technology, one so finely adjusted, so delicately responsive, that it will meet all the needs and occasions of life at a minimum cost of human values."[23] Such a technology will not be beyond the reach of engineers who have been steeped in the liberal arts.

With the coming of a civilized technology, the world will discover that engineering is not merely a means to an end, but rather an inherently worthy way of life. A revitalized and enlightened engineering will someday achieve its rightful place as one of the sublime enterprises of mankind. As historian Lynn White, Jr., has said,

> Technology is a profoundly spiritual form of thought. It has flourished best in the context of the Judaeo-Christian presupposition that the physical universe was created for a good purpose, that it is not to be disregarded or transcended, but rather that, like the body itself, it is to be treasured and controlled as the necessary ground of psychic life.[24]

A RECAPITULATION OF REASONS

These, then, are some of the reasons that we engineers should study the liberal arts:

1. To improve our intellectual competence and expand our imaginative powers.
2. To develop those qualities of character and personality which make for leadership and successful careers.
3. To enrich our personal lives with new knowledge and insight, with a keener appreciation of beauty.
4. To elevate the standards of our profession and to gain it increased esteem in our society.
5. To contribute to the public good—first, by using our improved status to see to it that our sensible technological advice is taken; second, by applying our enlarged wisdom and sensitivity, not only to the solution of engineering problems but also to the selection of worthy goals for our society; and third, by demonstrating to the world that technology can be more than a tool, can in itself be a revitalizing and profoundly beneficial force in human affairs.

WHAT ARE THE LIBERAL ARTS?

In what has been said so far it has been assumed that we are all familiar with what is meant by a liberal education. And indeed we are, in a broad sense. Liberal learning, we know, concerns itself with the eternal quest for truth, goodness, and beauty. It grapples with such ultimate concerns as the meaning of justice, liberty, virtue, honor, love, and happiness. Its method is to study the works of scholars and artists, both past and present.

There are some who consider the ultimate purpose of liberal education to be the transmission of our cultural heritage and the creation of good citizens loyal to the ideals of our society. There are others who stress the development of the independent, sensitive, questioning mind, who see the liberal arts as having a critical, almost revolutionary, function. However, all agree that the ultimate goal of liberal learning is wisdom and a reverence for beauty.

Literally, arts that are "liberal" are those studies deemed fit for liberated, or free, men. In past ages only a few men were truly free; the vast majority lived as serfs, chained to a soil they did not own. So liberal education has an aristocratic tradition. Because it was restricted to the upper classes, it came to serve the purpose of artificially defining the upper classes. This "snobbish" side of liberal learning becomes an anachronism in a democratic society. An engineering educator has put it this way:

> The *Oxford Dictionary* defines liberal education as education fit for a gentleman. That is still an acceptable definition; it is the idea of a gentleman which has changed.... Modern gentlemen do not belong to the leisured class. Many of them work something like a seventy-hour week....[25]

In the universities of the Middle Ages there were established seven branches of liberal learning: grammar, logic, and rhetoric (in other words, literature and philosophy), and arithmetic, geometry, astronomy, and music (science and the fine arts). Through the years a liberal arts education has come to comprise all those studies which are not technical or vocational by nature, those which are not "useful arts." It is no longer said that the liberal arts are *for* free men. Instead one hears it said that the liberal arts *make* men free by liberating their minds from ignorance and dogma.

LIBERAL ARTS FOR THE ENGINEER

A truly liberal education includes the study of pure science and mathematics. In these fields the average engineer has received a considerable amount of instruction. Liberal learning also embraces the social sciences: sociology, anthropology, psychology, political science, economics, and the like. These are important and useful subjects, and no man can consider himself educated without some knowledge of them.

But when we say that engineers lack an adequate liberal arts education, we do not really mean that they need more training in the social sciences. For one thing, to the extent that engineers take liberal arts courses in college, they often select the social sciences, which appear to be more "useful" than such studies as, for example, literature. Also the social sciences, whatever their merits, have a certain "illiberal" quality about them. They are, after all, sciences, and their approach is essentially clinical. The social scientist is more a statistician than a philosopher. He is liable to make the engineer *more* "lopsided" rather than less.

We are left, then, with those subjects which constitute the true core of liberal learning: history, literature, philosophy, the fine arts, and music. These are the so-called *humanities* about which the average engineer has always known little and cared less. These are the subjects with which we must now try to gain some familiarity.

A WORLD TO BE WON

There is a world of wisdom and beauty to be won. Obviously, as busy engineers, we cannot explore more than the tiniest portion of it. But to explore even a tiny portion, to discover the coastal outlines and see the vistas, to touch in at port here and there is to change our lives for the good.

Our problem is not unlike that of the world traveler who is constantly torn between the impulse to see as much as he can and the desire to linger in one spot in order to know it well. How much shall we survey in breadth and how much can we plumb in depth?

Certainly we want the overall view, superficial as it may be. Perspective and broad understanding are necessary first objectives. But the overall view becomes meaningless unless we stop here and there to investigate in detail, unless we find something to linger over and make our own. Each of us must seek his own balance between the general and the particular.

The problem of selecting particular works of literature and art from the myriads which surround us in this age of mass production is extremely vexing, almost paralyzing. Jacques Barzun has stated the problem well:

> The very bulk of the output kills appetite. Symphonies in bars and cabs, classical drama on television any day of the week, highbrow paperbacks in mountainous profusion (easier to buy than to read), "art seminars in the home," capsule operas, "Chopin by Starlight," "The Sound of Wagner," the Best of World Literature: this cornucopia thrust at the inexperienced and pouring out its contents over us all deadens attention and keeps taste stillborn, like any form of gross feeding. Too much art in too many places means art robbed of its right associations, its exact forms, its concentrated power. We are grateful for the comprehensive repertoire which modern industry for the first time puts within our reach, but we turn sick at the aggressive temptation, like the novice in the sweetshop.[26]

If we live in a sweetshop, we must learn discrimination and self-restraint. We must learn to relish the individual morsel without bewailing the fact that we cannot gorge ourselves. A single work of art, studied with love and understood well, may contain in microcosm and by implication as much as a thousand other works together. At least this must be our hope and our expectation, since we are busy professional men, and cannot delude ourselves into thinking that our experience in the arts can ever be comprehensive.

A further dilemma involves the question of whether we should be guided by what is said to be "worthwhile" or follow our own inclinations. We must do both, of course. Acknowledged masterpieces should not be foolishly ignored; we cannot rely exclusively on our taste, which is, after all, also our ignorance and prejudice. But neither must we follow the herd. Our taste should be challenged, refined, and stretched, but never overwhelmed. Ultimately it must prevail. We are under no obligation to worship unthinkingly at old and musty shrines.

APOLOGETIC OR PROUD?

And this brings us to the question of how we, as engineers, should best approach the liberal arts. Should we come hat in hand, ashamed, apologetic, and willing to be melted down and molded anew? Or should we come proudly, willing to grow, but only with the understanding that we start from what we *are*? Eric Ashby, author of *Technology and the Academics*, leaves no doubt about the course to be taken. "The path to culture should be through a man's specialism, not by by-passing it. . . . The *sine qua non* for a man who desires to be cultured is a deep and enduring enthusiasm to do one thing excellently."[27]

Unless the liberal arts can be approached *through engineering* they will seem lifeless and frivolous to those of us who are professional engineers. Engineering educators are grappling with this problem now. They are agreed that liberal education for engineers is essential, but they are fearful of having it degenerate into mere "appreciation" or "ornamentation."

FIVE BRIDGES TO BE CROSSED

The approach in this book will be to seek a bridge between engineering and each of the five subjects to be considered. From such a bridge we can look back over our own profession and ahead to the new world to be explored, observing the similarities of terrain, and discerning the common ground on which all worthy human endeavor rests. There is much talk these days about building bridges between the two cultures. These bridges do not so much require building as discovering; they already exist.

One such bridge has been discovered only recently and it is being enthusiastically reconstructed at this time. It is at the frontier where engineering and history meet, and it is called "the history of technology."

A bridge between engineering and literature has long been sought, and I think rather fruitlessly, in such desert areas as "engineers as writers" and "report writing for engineers." Here an attempt will be made to explore a new and unmapped territory, "the engineer as a protagonist in fiction."

Across the gulf between engineering and philosophy is suspended "the truth of science," a fragile structure, scarcely known to most engineers, although it is well traveled by many scientists and philosophers.

Engineering and the fine arts are joined by the age-old link of utility and beauty, often crossed by architects and designers but all too seldom by engineers.

Between engineering and music we find "sound as environment," a span which lately is attracting attention from people in many disciplines.

Let us together venture upon these bridges, and then let us cross them into the alluring lands which lie beyond. We may enter these new lands as relative strangers, but we will know who we are, why we have come, and the route we have traveled. With this sense of confidence and purpose we may soon find ourselves very much at ease in our new surroundings. The liberal arts, after all, have nourished and delighted men since the beginning of civilization.

Notes

1. Earl J. McGrath, *Liberal Education in the Professions*, Teachers College, Columbia University, New York, 1959, p. 20.
2. Lewis Mumford, "From Erewhon to Nowhere," in Thomas Parke Hughes (ed.), *The Development of Western Technology Since 1500*, The Macmillan Company, New York, 1964, p. 22.
3. New England Consultants, Inc., *The Engineer Today*, prepared for Esso Research and Engineering Company, 1963, p. 10.
4. J. J. Dougherty, "The Engineer Among People," *American Engineer*, July, 1966.
5. Gilbert W. Chapman, "Specific Needs for Leadership in Management," in Robert A. Goldwin and Charles A. Nelson (eds.), *Toward the Liberally Educated Executive*, Mentor Books, New American Library of World Literature, Inc., New York, 1960, p. 10.

6. Marcus Vitruvius Pollio, "Selections from De Architectura," in Walter J. Miller and Leo E. A. Saidla (eds.), *Engineers as Writers*, D. Van Nostrand Co., Inc., Princeton, N. J., 1953, p. 14.
7. Mark Van Doren, *Liberal Education*, Beacon Press, Boston, 1959, p. 16.
8. Quoted in Benjamin Farrington, *Greek Science*, Penguin Books, Inc., Baltimore, 1944, pp. 28–29.
9. Ralph Waldo Emerson, "Works and Days," in Arthur O. Lewis, Jr. (ed.). *Of Men and Machines*, E. P. Dutton & Co., Inc., New York, 1963, p. 69.
10. Thomas H. Huxley, "Science and Culture," in John J. Cadden and Patrick R. Brostowin (eds.), *Science and Literature*, D. C. Heath and Co., Boston, 1964, p. 11.
11. Thorstein Veblen, *The Engineers and the Price System*, Harcourt, Brace & World, Inc., New York, 1963, pp. 139–140. Originally appeared as a series of essays in *The Dial* in 1919.
12. Earl Ubell, quoted in "Public Relations Seminar for the Engineering Profession," *American Engineer*, March, 1966.
13. Arthur Garratt (ed.), *Penguin Technology Survey 1966*, Penguin Books, Inc., Baltimore, 1966, p. 8.
14. William Ruder, "A Public Relations Executive Looks at the Engineer and His Role in Today's Society," *American Engineer*, March, 1966.
15. New England Consultants, Inc., *The Engineer Today*, prepared for Esso Research and Engineering Company, 1963, p. 35.
16. Richard D. Blanchard, quoted in "Public Relations Seminar for the Engineering Profession," *American Engineer*, March, 1966.
17. James Reston, *The New York Times*, December 13, 1964.
18. R. Buckminster Fuller, "Vision 65 Summary Lecture," *The American Scholar*, Spring, 1966.
19. George Orwell, "The Road to Wigan Pier," in Arthur O. Lewis, Jr. (ed.), *Of Men and Machines*, E. P. Dutton & Co., Inc., New York, 1963, p. 259.
20. Jacques Ellul, *The Technological Society*, trans. John Wilkinson, Alfred A. Knopf, Inc., New York, 1964. Passage was reprinted in *The Saturday Review*, February 6, 1965.
21. Quoted in Don K. Price, *The Scientific Estate*, Harvard University Press, Cambridge, 1965, p. 57.
22. See Don K. Price, *The Scientific Estate*, Harvard University Press, Cambridge, 1965, and Jay M. Gould, *The Technical Elite*, Augustus M. Kelley, New York, 1966.
23. Lewis Mumford, "From Erewhon to Nowhere," in Thomas Parke Hughes (ed.), *The Development of Western Technology Since 1500*, The Macmillan Company, New York, 1964, p. 26.
24. Lynn White, Jr., "Humanism and the Education of Engineers," *Studies of Courses and Sequences in Humanities, Fine Arts and Social Sciences for Engineering Students*, University of California, Department of Engineering, Los Angeles, 1963, p. 53.
25. Eric Ashby, *Technology and the Academics*, Macmillan & Co., Ltd., London, 1963, p. 81.
26. Jacques Barzun, *Classic, Romantic and Modern*, Anchor Books, Doubleday & Co., Inc., Garden City, N.Y., 1961, p. 148. Copyright 1943, © 1961 by Jacques Barzun, with permission of Atlantic Monthly Press, Little, Brown and Company, Boston.
27. Eric Ashby, *Technology and the Academics*, Macmillan & Co., Ltd., London, 1963, p. 81.

Recommended Reading

For the engineer who doubts the value of a liberal arts education, an eye-opening book is Robert A. Goldwin and Charles A. Nelson (eds.), *Toward the Liberally Educated Executive* (Mentor Books MD 299). This slim volume includes essays by prominent educators, businessmen, and creative artists, all directed toward the practical and personal benefits to be gained in a liberal education. Particularly interesting is a report on programs in liberal education for executives conducted by the Bell Telephone Company. An appendix lists institutions of higher education which conduct special programs for industry, and also provides book lists used in many of these programs.

A beautiful and inspiring work is Mark Van Doren, *Liberal Education* (Beacon Press BP 86).

From the point of view of the professional educator, an interesting and useful book is Edwin J. Holstein and Earl J. McGrath, *Liberal Education and Engineering* (Teachers College, Columbia University, 1960).

An excellent, informal history of engineering education, shedding much light on the sociological background of the profession, is Eric Ashby, *Technology and the Academics* (Macmillan Papermac 43).

A work of historic importance, a "must" for the reader interested in the interrelationship between technology and the liberal arts, is C. P. Snow, *The Two Cultures and a Second Look* (Mentor Books MP 557).

A unique and provocative anthology: W. H. Davenport and D. Rosenthal (eds.), *Engineering—Its Role and Function in Human Society* (EDP Report No. 3-66, University of California, Los Angeles).

Don K. Price, *The Scientific Estate* (Harvard University Press, 1965) is a fascinating book which tells of the engineer's place in the power structure of our nation. Jay M. Gould, *The Technical Elite* (Augustus M. Kelley, 1966) contains much enlightening statistical information about the rise of engineers in industry. Of historical interest is Thorstein Veblen, *The Engineers and the Price System* (Harbinger Books HO31).

Two statements of interest to the thoughtful citizen are Jacques Barzun, *The House of Intellect* (Harper Torchbooks TB 1051) and John W. Gardner, *Excellence* (Harper Colophon Books CN3H).

Selected Additional Readings

D. W. Breneman, "The Humanities in a Technological Society," *The Brookings Bulletin*, spring 1980, pp. 1–5.

W. H. Davenport and D. Rosenthal (editors), *Engineering: Its Role and Function in Human Society*, Pergamon Press, New York, 1967.

S. C. Florman, *The Existential Pleasures of Engineering*, St. Martin's Press, New York, 1976.

J. Gurland, "Engineering and Liberal Education," *Journal of Engineering Education*, February 1971, pp. 426–428.

Robert Hutchins, "Stamp Out Engineering Schools," *Engineer*, March–April 1968, pp. 17–19.

Henry Knepler, "The New Engineers," *Change*, June 1977, pp. 30–36.

Walter R. Lynn, "Engineering and Society Programs in Engineering Education," *Science*, January 1977, pp. 150–155.

R. Pirsig, *Zen and the Art of Motorcycle Maintenance*, William Morrow, New York, 1974.

C. P. Snow, *The Two Cultures and a Second Look*, Cambridge University Press, London, 1964.

P. E. Torgersen, "Engineering Education and the Second Obligation," *Engineering Education*, November 1979, pp. 169–174.

P. Woodring, "Finding a Place for Liberal Education," *The Higher Learning in America: A Reassessment*, McGraw-Hill Book Co., New York, 1968.

PART

2

Literature and Engineering

Great literature is simply language charged with meaning to the utmost possible degree.

Erza Pound, *How to Read*, Pt. II

One of the themes of this volume is the role of the engineer in our complex and changing world. In an article on engineering in the 1980s, E. Wenk eloquently depicts the future engineer as a "builder of bridges across the two cultures." He believes that from this interaction "engineers may begin to appreciate what other intellectual, social, and human considerations enter the policy process, to learn that life, unlike science, has perpetual ambiguities, and to find joys from wisdom, if not knowledge, of the nonengineer" (1). In this part we will begin that process by looking at a selection from several genres of literature (short stories, poems, essays, chapters from novels) that variously suggest the ambiguities and perplexities of human life. The criteria for selection are several but paramount is a capacity for providing pleasure to the reader. We made no attempt to limit the selection to great literature alone. Several authors have been included to introduce them to the reader (specifically Capek and Den Doolaard); many appropriate selections have had to be omitted.

Why literature? The humanities are sometimes said to provide "equipment or tools" for living. This is especially true of literature which at its best offers a look into the human condition: what it means to be a human being and how people react to each other and with their environment. On occasion literature can pose issues more effectively than nonliterature since it appeals to both the intellectual and emotional side of a person. "Those who devote their life to engineering are likely to find themselves in contact with almost every phase of human activity" (2).

for an engineer literature may be able to offer a new insight, another dimension, a different perspective in the search for the best solution to a problem.

The part opens with a provocative short story, "The Mnemone," which raises the question of why literature and history are essential to man's well-being. R. Sheckley (1928–), an award-winning writer of science fiction, has set his story in a future society, an antiutopia, where question asking is a dangerous activity. When the Mnemone arrives in the village, the inhabitants slowly and painfully learn the value of a fragment of poetry or a scrap of the past: "We are men: queer beasts with strange appetites." "Who," asks the narrator, "would have imagined us to possess a thirst for the ineffable?" (section 2.4).

Almost since its beginning Western literature has warned that people's thirst for knowledge might have a negative side to it. The power of technology and the technologist has also alternately fascinated and repelled the poet and writer. In the second selection R. A. Bryan, a scholar of English literature, examines aspects of man's ambiguity toward technology as "paradigms of human experience." His examples are drawn from the creation story in Genesis, the Prometheus myth, Milton's *Paradise Lost*, Wordsworth's "Michael," *I'll Take My Stand* (a collection of essays from the 1930s), and the movie *Butch Cassidy and the Sundance Kid*.

The next two selections illustrate the complexity of the relationship between technologists and the society they serve. Just how essential technology and engineering are to the working of this world is imaginatively expressed in Rudyard Kipling's (1865–1936) poem "The Sons of Martha." Under Kipling's pen the New Testament story of the pious and dutiful sisters Mary and Martha takes on a new significance: biblical cadences and religious imagery elevate the status of the "doers" of this world for "simple services simply given to his own kind in their common need." The powerful message of this high tribute to engineers is heightened by language which equates the accomplishments of the "doers" with the very work of the Lord.

The impact of the Kipling poem derives from a new reading of an old story; Čapek's "The Punishment of Prometheus" similarly draws our attention to the complex problem of technological progress by giving an odd twist to the famous myth. K. Čapek (1890–1938), an important Czech writer, is probably better known for his play *R.U.R.* (for which he invented the word "robot"), but *A Book of Apocryphal Tales*, from which the Prometheus story comes, will delight the reader. In this collection the author reworks characters and events from Greek mythology, history, and the Bible to point up a new truth or insight which may only be implied in the original story. Čapek's version of the Prometheus myth dispenses with the gods to concentrate on human reactions to the invention of fire. By using wit and humor to paint a negative portrait, Čapek forces the reader to raise questions about the important, beneficial side of technological innovation. His Prometheus anticipates the frequently maligned engineer of the last twenty years.

The next group of selections has as a theme some aspect of the railway. The fifty years between 1830–1880 mark the golden age of railroad development in the United States and with it enormous economic, political, and social changes (see for example, Cochran's article in the chapter on "History and Engineering"). Nor did the railroad go unnoticed by contemporary writers: the train is probably the most celebrated machine in literature of the period.

The locomotive in motion exercised a particular fascination on the mind of nineteenth-century poets: only two examples are included here. "I Like to See It Lap the Miles" by Emily Dickinson (1830–1886) uses a series of images and personification to capture the strength, power, and sound of a train as it rushes through valleys, around mountains, and finally home to the depot. In the poet's mind the train is an all-powerful monster, but despite its strength and physical power, it is "docile"—obedient to its creator, man.

In a more elaborate poem Walt Whitman (1819–1892) offers a hymn of praise to the locomotive for its beauty, strength, and reliability (3). To underscore that theme, Whitman chooses language derived from music (recitative, chant, trills, notes) and parallel sentence structure of almost epic effect, reminiscent of the King James Version of the Psalms. Repetitive language (especially at the beginning of the line) also suggests the beat and throb of the engine. In these different treatments of the same subject, the poet has elevated the machine to a thing of beauty and, in so doing, encourages the reader to see aspects of the machine which no technical handbook could offer.

The railroad has also provided the impetus for more extended contemplation. In July 1845 Henry David Thoreau (1817–1862) retired to Walden Pond, Concord, Massachusetts, to learn more about himself, humanity, and the world. But even in the solitude and isolation of Walden Pond, he could not escape the railroad. In his essay "Sounds," the fourth chapter of *Walden*, Thoreau elaborates the values of directly experiencing nature (in the previous chapter he had contemplated the importance of literature as a means to spiritual growth). A long, beautifully crafted sentence introduces the sound of the railroad cars, "like the beat of a partridge." But the locomotive also screams like a hawk and "penetrates my woods" with its mechanical sounds. In the remainder of the chapter Thoreau develops in language and imagery his ambivalent attitude toward the railroad: he could admire "this travelling demigod" for its strength and speed but, at the same time, regret its connection with the profit motive and base commercial interests (4). The modern reader may find Thoreau difficult; persistence, however, will gain a reward in the rich texture of his ideas.

Another work from the same period that features the railroad is Nathaniel Hawthorne's satire, "The Celestial Railroad," into which Hawthorne (1804–1864) has apparently incorporated his own experiences as a passenger on the Boston–Salem line (5). "The Celestial Railroad" is a nineteenth-century update of *Pilgrim's Progress.* Hawthorne's pilgrim follows the same itinerary as Bunyan's but goes by rail. For his trip to the Celestial City Christian has as his guide Mr. Smooth-It-Away, a railroad executive and stockholder. All the amenities of rail travel ease the rigors of the journey (compare the unpleasant trip of the two old-fashioned pilgrims who go on foot). The Slough of Despond has been bridged, a tunnel constructed through the Hill Difficulty, a causeway erected across the Valley of the Shadow of Death, now brightly lit by lamps fueled by the noxious gases there. A steam ferryboat now takes travelers across the last river to their final destination. The journey is no longer either difficult or dangerous but a tribute to American technological ingenuity (6).

By means of allegory and satire Hawthorne offers profound insight into the impact of technological progress on nineteenth-century life. Although Hawthorne's view is a balanced one (both advantages and disadvantages of the machine are implied), Hawthorne seems in the last analysis to suggest that the "road to hell is paved with good inventions" (7): technology may make the journey of life easier but it cannot replace individual effort and virtue.

The final selection in this group is a version of the most famous of the black American work songs of the last century—some scholars believe that John Henry worked on the Big Bend tunnel for the C&O Railroad in West Virginia (8). In the epic struggle between man and machine, John Henry triumphs but dies in the attempt. What does the song say that has made it so popular? It would seem that human beings are frequently gratified by the triumph of people over their inventions. But does the song tell the whole story about the impact of technology on the quality of the railworkers' lives?

There has always been an aura of magic attached to the bridge builder and the bridge: the word *pontiff*, for example, still bears traces of its original, double meaning of the priest who is also the bridge builder (Latin, *pontifex*). That special quality which attaches to the bridge and its maker is suggested in the simple poem "The Bridge Builder," by Will Allen Dromgoole (1860–1934).

It has been said that

> to know a bridge is to understand the civilization which produced it: it is to comprehend the scientific advance of a people; to learn their available natural resources; to become familiar with their technological advances; to judge of their artistic and aesthetic qualities; to study their economic status and their civil government; to become acquainted with their folklore and religion; to live with them their social life. (9)

To know a bridge is also to know its builder. The two substantial selections which follow will illustrate some of these perspectives for the reader.

The selection from *Alexander's Bridge* is the last chapter of Willa Cather's (1873–1947) earliest novel. Based in part on a true incident, the Quebec bridge disaster of 1907, the novel presents a sensitive portrait of the different kinds of conflicts in the life of a creative man, the bridge builder Alexander Bartley (the article by J. C. Vitale is included after the Cather chapter as background for the novel). Alexander Bartley, a prominent civil engineer, has everything that life could offer: economic success, the respect and admiration of the profession and fashionable society, a rich and talented Boston wife. But middle age is fast approaching (Alexander is forty-two) and with it a restlessness and longing for his lost youth. On a business trip to London Alexander rediscovers an earlier love of his life, Hilda Burgoyne, now a fascinating and famous actress. During the same period, he is building the Moorlock bridge in Canada, "the most important piece of bridge building going on in the world" (Cather, p. 47). But there are problems with the completion of the bridge (the longest single-span cantilever hitherto attempted) because of underbidding on the project and the use of lighter structural material than Alexander thought proper. Our chapter brings the story to a climax, as Alexander's double life and the unfinished bridge impact upon each other. The collapse of the bridge serves as the main symbol in *Alexander's Bridge*: it represents Alexander's own tragic flaw that causes the tragedy.

One of the themes of this finely crafted novel (although Cather herself expressed dissatisfaction with it) is the important question it raises about the relationship between a person's professional and private life, between a professional and the public he serves. Can a creative individual, for example,work in isolation from social responsibility and constraints? Earlier in the novel Alexander muses on what his life has become (Cather, pp. 48–49):

> His existence was becoming a network of great and little details. He had expected that success would bring him freedom and power; but it had brought only power that was in itself another kind of restraint.... He happened to be engaged in work of public utility, but he was not willing to become what is called a public man.... The one thing he had really wanted all his life was to be free; and there was still something unconquered in him, something besides the strong work-horse that his profession had made of him.

Although the novelist does not provide answers for these difficult questions, the problem is one all professionals ultimately have to face (10).

Den Doolaard's novel *The Land behind God's Back* uses the bridge to raise similar kinds of questions about an engineer's interaction with society. A. Den. Doolaard (1901–) is the pen name of the Dutch writer C. Spoelstra who lived for many years in the harsh mountain country he describes in the novel. *The Land behind God's Back* is set in pre-World War II Montenegro (now part of Yugoslavia), an underdeveloped, isolated, and poverty-stricken area of the world.

In growing up Wolf, the engineer-hero of the story, has been much influenced by the traditional values of his mountaineer grandfather, Sava, but he has also experienced another world during his engineering study with a master-builder in Italy. When chapter four opens, Wolf is returning to try for a bridge commission in the countryside where he grew up. The chapter centers on Wolf's struggle to design and build the bridge which means so much to him and his fellow-countrymen. The outside world, however, cannot be ignored, and by the end of the chapter a German and Italian invasion of Montenegro seems imminent. The bridge now becomes an instrument of invasion against his people.

The Land behind God's Back is in itself a moving story—Den Doolaard has few equals as a storyteller. But the novel is much more. It would be difficult to find a more inspiring depiction of the dedicated engineer or a more accurate representation of the struggles associated with the creative process. *The Land behind God's Back* also provides a sensitive examination of two important issues raised in this book: the relationship between aesthetics and engineering and the many dimensions of the impact of technology on a developing nation.

The last selection in the part brings us to the present and one of the major achievements of the twentieth century: computer technology. In *The Robotosyncrasies* ("fiction about actual science and technology," written by a professor of electrical engineering) the scientist-engineer hero creates a new kind of robot which embodies some of the qualities human beings ought to have: reliability, stability, agility, broadmindedness. In our selection (chapter six) their creator gives a keynote speech at a scientific conference and proves mathematically that his robots possess certain human qualities (11).

The hero's Human Intelligence Model (H.I.M.), on which his robot technology is based, raises serious and challenging questions about the essential nature of people: "I don't think that the human being is a supreme being with an innate complex mental mechanism. Rather, it is the simple concept of H.I.M., applied repetitively and continuously, that is responsible for many human qualities and wisdom" (*The Robotosyncrasies*, section 2.14). The premise about humans on which this novel is based takes us back again to the first selection of the chapter, "The Mnemone." One of the functions of the humanities (and of literature in particular) is to raise questions (perhaps insoluble) about the fundamental nature of people.

References

1. Wenk, E. "Engineering in the 80's," *The Bridge*, National Academy of Engineering, spring 1980, p. 21.
2. Cross, H. and R. C. Goodpasture, *Engineering and Ivory Towers*, reprinted in W. H. Davenport & D. Rosenthal (editors), *Engineering: Its Role and Function in Human Society*, Pergamon Press, New York, 1967, p. 81.
3. See Kasson, J. F., *Civilizing the Machine: Technology and Republican Values in America 1776–1900*, Grossman Publishers, Inc., New York, 1976, pp. 172–173.
4. On Thoreau, see Cronkhite, G. Ferris, "The Transcendental Railway," *New England Quarterly*, September 1951, pp. 315–321; Marx, L., *The Machine in the Garden: Technology and the Pastoral Ideal in America*, Oxford University Press, New York, 1964, pp. 242–265; also *Essays of E. B. White*, Harper & Row, New York, 1977, "The Railroad," pp. 208–221, and "A Slight Sound at Evening," pp. 234–242.
5. Cronkhite (1951), p. 322.
6. See Doubleday, N. F., "Hawthorne's Satirical Allegory," *College English*, January 1942, pp. 329–332.
7. Kasson (1976), p. 49.
8. On the John Henry tradition, see Courlander, H., *Negro Folk Music, USA*, Columbia University Press, New York, 1963, pp. 110–115.
9. Steinman, D. B., and S. R. Watson, *Bridges and Their Builders*, Dover, 1975, p. vii.
10. On Cather's view of creativity, see Randall, J. H., *The Landscape and the Looking Glass* Houghton Mifflin, Boston, 1960. pp. 21–57.
11. On the question of what the computer does to humanity's view of itself, see Simon, H. A. "What Computers Mean for Man and Society," *Science*, March 1977, pp. 1190–1191; and J. Weizenbaum, *Computer Power and Human Reason*, W. H. Freeman, San Francisco, 1976.

2.1

The Mnemone

ROBERT SHECKLEY (1961)

It was a great day for our village when the Mnemone arrived (1). But we did not know him at first, because he concealed his identity from us. He said that his name was Edgar Smith, and that he was a repairer of furniture. We accepted both statements at face value, as we receive all statements. Until then, we had never known anyone who had anything to conceal.

He came into our village on foot, carrying a knapsack and a battered suitcase. He looked at our stores and houses. He walked up to me and asked, "Where is the police station?"

"We have none," I told him.

"Indeed? Then where is the local constable or sheriff?"

"Luke Johnson was constable here for nineteen years," I told him. "But Luke died two years ago. We reported this to the county seat as the law requires. But no one has been sent yet to take his place."

"So you police yourselves?"

"We live quietly,," I said. "There's no crime in this village. Why do you ask?"

"Because I wanted to know," Smith said, not very helpfully. "A little knowledge is not as dangerous as a lot of ignorance, eh? Never mind, my blank-faced young friend. I like the look of your village. I like the wooden frame buildings and the stately elms. I like—"

"The stately what?" I asked him.

"Elms," he said, gesturing at the tall trees that lined Main Street. "Didn't you know their name?"

"It was forgotten," I said, embarrassed.

"No matter. Many things have been lost, and some have been hidden. Still, there's no harm in the name of a tree. Or is there?"

"No harm at all," I said. "Elm trees."

"Keep that to yourself," he said, winking. "It's only a morsel but there's no telling when it might prove useful. I shall stay for a time in this village."

"You are most welcome," I said. "Especially now, at harvest time."

Smith looked at me sharply. "I have nothing to do with that. Did you take me for an itinerant apple-picker?"

"I didn't think about it one way or another. What will you do here?"

"I repair furniture," Smith said.

"Not much call for that in a village this size," I told him.

"Then may be I'll find something else to turn my hand to." He grinned at me suddenly. "For the moment, however, I require lodgings."

I took him to the Widow Marsini's house, and there he rented her large back bedroom with porch and separate entrance. He arranged to take all of his meals there, too.

His arrival let loose a flood of gossip and speculation. Mrs. Marsini felt that Smith's questions about the police went to show that he himself was a policeman. "They work like that," she said. "Or they used to. Back fifty years ago, every third person you met was some kind of a policeman. Sometimes even your own children were policemen, and they'd be as quick to arrest you as they would a stranger. Quicker!"

But others pointed out that all of that had happened long ago, that life was quiet now, that policemen were rarely seen, even though they were still believed to exist.

But why had Smith come? Some felt that he was here to take something from us. "What other reason is there for a stranger to come to a village like this?" And others felt that he had come to give us something, citing the same argument.

But we didn't know. We simply had to wait until Smith chose to reveal himself.

He moved among us as other men do. He had knowledge of the outside world; he seemed to us a far-traveling man. And slowly, he began to give us clues as to his identity.

One day I took him to a rise which looks out over our valley. This was at midautumn, a pretty time. Smith looked out and declared it a fine sight. "It puts me in mind of that famous tag from William James," he said. "How does it go? 'Scenery seems to wear in one's consciousness better than any other element in life.' Eh? Apt, don't you think?"

"Who is or was this William James?" I asked.

Smith winked at me. "Did I mention that name? Slip of the tongue, my lad."

But that was not the last "slip of the tongue." A few days later I pointed out an ugly hillside covered with second growth pine, low coarse shrubbery, and weeds. "This burned five years ago," I told him. "Now it serves no purpose at all."

"Yes, I see," Smith said. "And yet—as Montaigne tells us—there is nothing useless in nature, not even uselessness itself."

And still later, walking through the village, he paused to admire Mrs. Vogel's late-blooming peonies. He said, "Flowers do indeed have the glances of children and the mouths of old men . . . Just as Chazal pointed out."

Toward the end of the week, a few of us got together in the back of Edmonds's Store and began to discuss Mr. Edgar Smith. I mentioned the things he had said to me. Bill Edmonds remembered that Smith had cited a man named Emerson, to the effect that solitude was impracticable, and society fatal. Billy Foreclough told us that Smith had quoted Ion of Chios to him: that Luck differs greatly from Art, yet creates many things that are like it. And Mrs. Gordon suddenly came up with the best of the lot; a statement Smith told her was made by the great Leonardo da Vinci: vows begin when hope dies.

We looked at each other and were silent. It was evident to everyone that Mr. Edgar Smith—or whatever his real name might be—was no simple repairer of furniture.

At last I put into words what we were all thinking. "Friends," I said, "this man appears to be a Mnemone."

Mnemones as a distinct class came into prominence during the last year of the War Which Ended All Wars. Their self-proclaimed function was to remember works of literature which were in danger of being lost, destroyed, or suppressed.

At first, the government welcomed their efforts, encouraged them, even rewarded them with pensions and grants. But when the war ended and the reign of the Police Presidents began, government policy changed. A general decision was made to jettison the unhappy past, to build a new world in and of the present. Disturbing influences were to be struck down without mercy.

Right-thinking men agreed that most literature was superfluous at best, subversive at worst. After all, was it necessary to preserve the mouthings of a thief like Villon, a homosexual like Genet, a schizophrenic like Kafka? Did we need to retain a thousand divergent opinions, and then to explain why they were false? Under such a bombardment of influences, how could anyone be expected to respond in an appropriate and approved manner? How would one ever get people to obey orders?

The government knew that if everyone obeyed orders, everything would be all right.

But to achieve this blessed state, divergent and ambiguous inputs had to be abolished. The biggest single source of confusing inputs came from historical and artistic verbiage. Therefore, history was to be rewritten, and literature was to be regularized, pruned, tamed, made orderly or abolished entirely.

The Mnemones were ordered to leave the past strictly alone. They objected to this most vehemently, of course. Discussions continued until the government lost patience. A final order was issued, with heavy penalties for those who would not comply.

Most of the Mnemones gave up their work. A few only pretended to, however. These few became an elusive, persecuted minority of itinerant teachers, endlessly on the move, selling their knowledge where and when they could.

We questioned the man who called himself Edgar Smith, and he revealed himself to us as a Mnemone. He gave immediate and lavish gifts to our village:

Two sonnets by William Shakespeare.

Job's Lament to God.

One entire act of a play by Aristophanes.

This done, he set himself up in business, offering his wares for sale to the villagers.

He drove a hard bargain with Mr. Ogden, forcing him to exchange an entire pig for two lines of Simonides.

Mr. Bellington, the recluse, gave up his gold watch for a saying by Heraclitus. He considered it a fair exchange.

Old Mrs. Heath exchanged a pound of goosefeathers for three stanzas from a poem entitled "Atalanta in Calydon," by a man named Swinburne.

Mr. Mervin, who owns the restaurant, purchased an entire short ode by Catullus, a description of Cicero by Tacitus, and ten lines from Homer's Catalogue of Ships. This cost his entire savings.

I had little in the way of money or property. But for services rendered, I received a paragraph of Montaigne, a saying ascribed to Socrates, and ten fragmentary lines by Anacreon.

An unexpected customer was Mr. Lind, who came stomping into the Mnemone's office one crisp winter morning. Mr. Lind was short, red-faced, and easily moved to anger. He was the most successful farmer in the area, a man of no nonsense who believed only in what he could see and touch. He was the last man whom you'd ever expect to buy the Mnemone's wares. Even a policeman would have been a more likely prospect.

"Well, well," Lind began, rubbing his hands briskly together. "I've heard about you and your invisible merchandise."

"And I've heard about you," the Mnemone said, with a touch of malice to his voice. "Do you have business with me?"

"Yes, by God, I do!" Lind cried. "I want to buy some of your fancy old words."

"I am genuinely surprised," the Mnemone said. "Who would ever have dreamed of finding a law-abiding citizen like yourself in a situation like this, buying goods which are not only invisible, but illegal as well!"

"It's not my choice," Lind said. "I have come here only to please my wife, who is not well these days."

"Not well? I'm not surprised," the Mnemone said. "An ox would sicken under the workload you give her."

"Man, that's no concern of yours!" Lind said furiously.

"But it is," the Mnemone said. "In my profession we do not give out words at random. We fit our lines to the recipient. Sometimes we find nothing appropriate, and therefore sell nothing at all."

"I thought you sold your wares to all buyers."

"You have been misinformed. I know a Pindaric ode I would not sell to you for any price."

"Man, you can't talk to me that way!"

"I speak as I please. You are free to take your business somewhere else."

Mr. Lind glowered and pouted and sulked, but there was nothing he could do. At last he said, "I didn't mean to lose my temper. Will you sell me something for my wife? Last week was her birthday, but I didn't remember it until just now."

"You are a pretty fellow," the Mnemone said. "As sentimental as a mink, and almost as loving as a shark! Why come to me for her present? Wouldn't a sturdy butter churn be more suitable?"

"No, not so," Lind said, his voice flat and quiet. "She lies in bed this past month and barely eats. I think she is dying."

"And she asked for words of mine?"

"She asked me to bring her something pretty."

The Mnemone nodded. "Dying! Well, I'll offer no condolences to the man who drove her to the grave, and I've not much sympathy for the woman who picked a creature like you. But I do have something she will like, a gaudy thing that will ease her passing. It'll cost you a mere thousand dollars."

"God in heaven, man! Have you nothing cheaper?"

"Of course I have," the Mnemone said. "I have a decent little comic poem in Scots dialect with the middle gone from it; yours for two hundred dollars. And I have one stanza of a commemorative ode to General Kitchener which you can have for ten dollars."

"Is there nothing else?"

"Not for you."

"Well . . . I'll take the thousand dollar item," Lind said. "Yes, by God, I will! Sara is worth every penny of it!"

"Handsomely said, albeit tardily. Now pay attention. Here it is."

The Mnemone leaned back, closed his eyes, and began to recite. Lind listened, his face tense with concentration. And I also listened, cursing my untrained memory and praying that I would not be ordered from the room.

It was a long poem, and very strange and beautiful. I still possess it all. But what comes most often to my mind are the lines

Charm'd magic casements, opening on the foam
Of perilous seas, in faery lands forlorn (2).

We are men: queer beasts with strange appetites. Who would have imagined us to possess a thirst for the ineffable? What was the hunger that could lead a man to exchange three bushels of corn for a single saying of the Gnostics? To feast on the spiritual—this seems to be what men must do; but who could have imagined it of *us*? Who would have thought us sufferers of malnutrition because we had no Plato? Can a man grow sickly from lack of Plutarch, or die from an Aristotle deficiency?

I cannot deny it. I myself have seen the results of abruptly withdrawing an addict from Strindberg.

Our past is a necessary part of us, and to take away that part is to mutilate us irreparably. I know a man who achieved courage only after he was told of Epaminondas, and a woman who became beautiful only after she heard of Aphrodite.

The Mnemone had a natural enemy in our schoolteacher, Mr. Vich, who taught the authorized version of all things. The Mnemone also had an enemy in Father Dulces, who ministered to our spiritual needs in the Universal Patriotic Church of America.

The Mnemone defied both of our authorities. He told us that many of the things they taught us were false, both in content and in ascription, or were perversions of famous sayings, rephrased to say the opposite of the original author's intention. The Mnemone struck at the very foundations of our civilization when he denied the validity of the following sayings:

—Most men lead lives of quiet aspiration.
—The unexamined life is most worth living.
—Know thyself within approved limits (3).

We listened to the Mnemone, we considered what he told us. Slowly, painfully, we began to think again, to reason, to examine things for ourselves. And when we did this, we also began to hope.

The neoclassical flowering of our village was brief, intense, sudden, and a delight to us all. Only one thing warned me that the end might be imminent. There was a day in early spring when I had been helping one of the neighbor's children with his lessons. He had a new edition of Dunster's *General History*, and I glanced through the section on the Silver Age of Rome. It took me a few minutes to realize that Cicero had been omitted. He wasn't even listed in the index, though many lesser poets and orators were. I wondered what retrospective crime he had been found guilty of.

And then one day, quite suddenly, the end came. Three men entered our village. They wore gray uniforms with brass insignia. Their faces were blank and broad, and they walked stiffly in heavy black boots. They went everywhere together, and they always stood very close to one another. They asked no questions. They spoke to no one. They knew exactly where the Mnemone lived, and they consulted a map and then walked directly there.

They were in Smith's room for perhaps ten minutes.

Then the three policemen came out again into the street, all three of them walking together like one man. Their eyes darted right and left; they seemed frightened. They left our village quickly.

We buried Smith on a rise of land overlooking the valley, near the place where he had first quoted William James, among late-blooming flowers which had the glances of children and the mouths of old men.

Mrs. Blake, in a most untypical gesture, has named her latest-born Cicero. Mr. Lind refers to his apple orchard as Xanadu. I myself have become an avowed Zoroastrian, entirely on faith, since I know nothing about that religion except that it directs a man to speak the truth and shoot the arrow straight.

But these are futile gestures. The truth is, we have lost Xanadu irretrievably, lost Cicero, lost Zoroaster. And what else have we lost? What great battles were fought, cities built, jungles conquered? What songs were sung, what dreams were dreamed? We see it now, too late, that our intelligence is a plant which must be rooted in the rich fields of the past.

In brief, our collective memories, the richest part of us, have been taken away, and we are poor indeed. In return for castles of the mind, our rulers have given us mud hovels palpable to the touch; a bad exchange for us.

The Mnemone, by official proclamation, never existed. By fiat he is ranked as an inexplicable dream or delusion—like Cicero. And I who write these lines, I too will soon cease to exist. Like Cicero and the Mnemone, my reality will also be proscribed.

Nothing will help me: the truth is too fragile, it shatters too easily in the iron hands of our rulers. I shall not be revenged. I shall not even be remembered. For if the great Zoroaster himself could be reduced to a single rememberer, and that one killed, then what hope is there for me?

Generation of cows! Sheep! Pigs! We have not even the spirit of a goat! If Epaminondas was a man, if Achilles was a man, if Socrates was a man, then are we also men?(4).

Annotations

1. A Mnemone is one who remembers, a rememberer. Sheckley has coined the word from the Greek root *mnemon*, "mindful, remembering."
2. Lines 69–70 from John Keats's (1795–1821) "Ode to a Nightingale."
3. The originals of these aphorisms are as follows: the mass of men lead lives of quiet desperation (Thoreau, "Economy," *Walden* (1854); the unexamined life is not worth living (Socrates in Plato, *Apology* 38a); know thyself (inscribed on the wall of the temple at Delphi, an important value of the ancient Greeks).
4. Any good dictionary or encyclopedia will provide help with the proper names in this story.

2.2

The Triumph of Science and the Loss of Innocence

(1970)

ROBERT A. BRYAN (1)

The impact of technology upon mankind is actually the history of mankind; any attempt to separate technology, to give it a special, highly localized place in history is naive. Technology *is* history and its impact begins with anthropology's hypothetical confrontation at Oldmaii between weapon-bearing apes and apes who had not learned to use sharp rocks as extensions of their limbs; technology's march continues with Apollo XI and the first man on the moon. Indeed technology—or man's history—cannot end unless multiple independent re-entry vehicles rain down upon the planet, bearing with them one of the greatest triumphs of technology, the harnessing of atomic fission. It should be clear to you from what I have already said that the terms, "science" and "technology" are, for my purposes, synonymous. To be sure, "science" is the art of knowing and technology is the application of scientific knowledge, but in the historical looking glass I wish to use, knowing and doing become one.

How is a humanist affected by technology? We could discuss any number of obvious relationships: the invention of the press in the fifteenth century, a technological contribution that began the humanities industry; the invention of a functional camera in the nineteenth century; the invention of paperback books in the 1940s, a technological achievement that has made the entire spectrum of learning available to all but the poorest; the computer, which affords dazzling opportunities in textual analysis. Yet these finally seem trivial compared to the larger question: how has technology affected not merely the humanist, but the human? And in asking this question we ask, really, another question: what has been the price of knowledge? Or put yet a third way, what has supplanted our loss of innocence?

It seems to me that we have available to us works of art which tell us vividly how technology or science or scientia or knowledge has affected the human race. We have in these works of art paradigms of human experience, each in its own way telling us the same story, teaching us a variant of the same lesson. Let us look at five or six such works: the chapter in the Old Testament in the Bible known as Genesis, written in about the fourth century B.C.; some Greek mythology which comes to us from before the ninth century B.C.; the great English poem by John Milton, *Paradise Lost*, written at about the midpoint of the seventeenth century; *Michael*, a poem by Wordsworth written at the beginning of the nineteenth century; *I'll Take My Stand*, a book of essays written in the 1930s by a group of southern writers at Vanderbilt University; and a recent movie, "Butch Cassidy and the Sundance Kid."

It seems only appropriate to begin with Genesis. According to the story, the Lord constructed a paradise within a garden and put Adam and Eve there with only one restriction: they could not eat of the fruit of the tree of knowledge. That they ate is known to us all, and the consequence of their disobedience is our mortality, the human condition. The Genesis myth can be interpreted in a number of different ways, but using our particular historical looking glass, it is obvious that it represents the conflict between science and religion as they contended for control of the human mind. The story is an early version of a familiar pattern: the mystics in the hill country versus the technocrats on the plains. And every time the plainsmen have prevailed. In the Garden of Eden the economy is agrarian. Do not eat of the tree of knowledge meant: do not convert your agrarian economy to a scientific, technological one. Yet Adam and Eve tried to do just that; they were not satisfied with the conditions of their existence. The tragedy of their fall is bound up in paradox: as man falls he rises to power. Each step of the way has been costly and each step has led him further from his agrarian origins. Each step has widened the separation of man and nature. As we shall note in passing, the literature of western civilization again and again reflects the longing of man for his innocent, agrarian origins: witness the popularity of the pastoral motif in literature, the popularity of Platonic doctrine, the Romantic movement of the early nineteenth century, and the re-emergence of this life style in our present culture, a re-emergence that resurrects in an uncanny way the very details of early nineteenth century romanticism: beards, bells, flowing clothes, fascination with things oriental, worship of nature, a fondness for mysticism and esoteric religions, and, most importantly, a reaction against technology and technocrats.

The Genesis myth tries to preach the doctrine of innocence: there are limits to the human understanding. To eat the fruit, to seek too much knowledge is to adopt a secular, technocratic attitude; to disobey God is to embrace science. To eat the fruit is to prey on nature, to use nature, to put nature to unnatural uses, to become, in short, a technocrat. Later in Genesis we are told about Cain and Abel. Cain became a planter; he broke from his family's traditional occupation, sheepherding, and tried to make scientific his pastoral livelihood. God rejected his offering; God must reject the offerings of all technology, and, in despair, Cain slew Abel. God could not let Cain/science triumph; He had to flood the earth.

If we are not convinced by the Old Testament that the human condition is constantly a falling from innocence into knowledge, from mysticism into technology, from happiness into despair, let us consider an even older group of myths, those furnished by the ancient Greeks. Reflect upon the story of Prometheus. He brought fire (knowledge, science, technology) from the heavens to man. This precious gift freed man from his innocence and plunged him into knowing. The knowledge Prometheus brings frees man from his tyrant-god, Zeus. Yet Prometheus is made to suffer: chained to a rock, vultures eternally tear and pluck at his liver. He is crucified and punished eternally. The cost of scientific knowledge is tremendous. This is the lesson both Greek and Hebrew understand; this is the lesson they continually preach.

Let us jump forward in time about 2000 years and consider the central doctrine of the greatest poem in the English language, *Paradise Lost.* The poem is a retelling, in twelve books, of events in the Old Testament, but it is mainly concerned with Adam's fall from grace and the possibility of eventual salvation. Glory, ruin, and restoration—these are the great subjects of *Paradise Lost.* We note, however, that Milton does not tamper with the lessons of Greek and Hebrew mythology: the fall of man is tragic and costly. In the poem Satan is a fully developed person. Known originally as Lucifer (the Promethean light-bringer) he became Satan after his fall from grace. He becomes the voice of the new science; the contemporary enlightenment of the seventeenth century. He is conversant with all the latest scientific discoveries, Galileo's telescope, Kepler's astronomy, and Tycho Brahe's new mathematics. A demigod himself in the poem, he urges man to become God. Listen to him as he persuades Eve to eat of the fruit of the tree of knowledge; addressing the Tree in the presence of Eve, he says,

> O Sacred, Wise, and Wisdom-giving Plant,
> Mother of Science, Now I feel thy Power
> Within me cleere, not only to discerene
> Things in thir Causes, but to trace the wayes
> Of highest Agents, deemd however, wise.
> Queen of this Universe, doe not believe
> Those rigid threats of Death; ye shall not Die:
>
> How should ye? by the Fruit? it gives you Life
> To Knowledge: By the Threatner? look on mee,
> Mee who have toucht and tasted, yet both live,
> And life more perfet have attaind than Fate
> Meant me, by ventring higher than my Lot.
>
> Why then was this forbid? Why but to awe,
> Why but to keep ye low and ignorant,
> His worshippers; he knows that in the day
> Ye Eate thereof, your Eyes that seem so cleere,
> Yet are but dim, shall perfetly be then
> Op'nd and cleerd, and yee shall be as Gods,
> Knowing both Good and Evil as they know.
>
> What can your knowledge hurt him, or this Tree
> Impart against his will if all be his?
> Or is it envie, and can envie swell
> In heav'nly brests? these, these and many more
> Causes import your need of this fair Fruit.
> Goddess humane, reach then, and freely tast.

Eve is overwhelmed and eats the fatal apple and, in so doing, gives nature a wound from which we have never recovered:

> So saying her rash hand in evil hour
> Forth reaching to the Fruit, she pluckd, she eat:
> Earth felt the wound, and Nature from her seat
> Sighing through all her Works gave signs of woe,
> That all was lost.

Adam, in a gesture both foolish and grand, both simple and heroic, in a gesture, in short, that is *human*, eats the apple as soon as he learns of Eve's transgression and the two of them, doomed to die, are banished from the Garden of Eden. Hence science and technology again have triumphed and man again must pay dearly for that triumph. Prelapsarian Adam and Eve enjoyed the world as the work of God; nature is, to the innocent, the metaphor of God, and the innocent man attempts to enjoy the whole of nature. To the knowledgeable man, the man of science, the technocrat, nature is an object to be exploited. The technocrat deforms it to serve his own ends; science is willing to sacrifice the whole for the part. The apple, for example, is an esthetic whole to be appreciated so long as we do not partake; but plucked and eaten and thus *known*, it fills the stomach and is useful—and is lost. Nowhere in the poem is this lesson so bitterly taught as in the depiction of the marriage relationship of the fallen Adam and Eve. After the fall, their love turns to lust, for after all, lust is the science of love. Where once in mutual respect and entirely nude they walked through the Garden in connubial bliss, now they

> Soon found thir Eyes how op'nd, and thir minds
> How dark'nd; Innocence, that as a veile
> Had shadowd them from knowing ill, was gon,
> Just Confidence, and native Righteousness,
> And Honour from about them, naked left
> To Guiltie Shame: hee coverd, but his Robe
> Uncovered more. So rose the Danite strong
> Herculean Samson from the Harlot-lap
> Of Philistean Dalilah, and wak'd
> Shorn of his strength, They destitute and bare
> Of all thir vertue; silent, and in face
> Confounded long they sate, as strook'n mute . . .

Let us examine next a classic example of the anti-scientific thinking of the Romantic movement, the poem, "Michael," published by William Wordsworth in 1800. The poem tells the story of an old sheepherder and his wife and their only child, a boy named Luke. The old man and his family live in a hidden valley and seldom come down from the mountains. Their food, shelter, and clothing were grown and made by hand in their mountain valley; all their wants were supplied by nature. All was idyllic and edenic; all was pastoral and innocent. Then came the bad news from the world of commerce on the plains: the old man had years before countersigned a note for his brother who went to the city on the plain to make his fortune. He had failed, and the old sheepherder had to make good the note. Rather than sell his patrimonial lands, he decided to send his son Luke to the city to make enough money to pay off the note. Then he would return to the mountain valley. As a pact, father and son agreed to begin work on a sheepfold; the day that Luke left for the city he laid the first stone on the sheepfold. Then Luke left for the city. The resolution of this story is best told by Wordsworth:

> Of Luke and his well-doing: and the Boy
> Wrote loving letters, full of wondrous news,
> Which, as the Housewife phrased it, were throughout
> "The prettiest letters that were ever seen."
> Both parents read them with rejoicing hearts.
> So many months passed on: and once again
> The Shepherd went about his daily work
> With confident and cheerful thoughts; and now
> Sometimes when he could find a leisure hour
> He to that valley took his way, and there
> Wrought at the Sheepfold. Meantime Luke began
> To slacken in his duty; and at length,
> He in the dissolute city gave himself
> To evil courses: ignominy and shame
> Fell on him, so that he was driven at last
> To seek a hiding-place beyond the seas.
> There is a comfort in the strength of love;
> 'T will make a thing endurable, which else
> Would overset the brain, or break the heart:
> I have conversed with more than one who well
> Remember the old Man, and what he was
> Years after he had heard this heavy news.
> His bodily frame had been from youth to age
> Of an unusual strength. Among the rocks
> He went, and still looked up to sun and cloud,
> And listened to the wind; and as before,
> Performed all kinds of labour for his sheep,
> And for the land, his small inheritance.
> And to that hollow dell from time to time
> Did he repair, to build the Fold of which
> His flock had need. 'T is not forgotten yet
> The pity which was then in every heart
> For the old Man—and 't is believed by all
> That many and many a day he thither went,
> And never lifted up a single stone.

The moral is obvious. Commerce and technology are the enemies of nature and innocence. And science will triumph over innocence, and great will be the fall thereof.

As noted earlier, this nineteenth century romanticism is prevalent today in the neoromantic subculture of America's youth. But it has always been a part of America, for the land itself was—and perhaps still is—the world's last frontier, the world's last Garden of Eden, the world's last Shangri-la. Even so eminent a group of writers as the fugitive poets, a group active at Vanderbilt University in the 1930's, could fall prey to the most simple-minded versions of the revolt against technology. In a book called *I'll Take My Stand*, these men argued against the industrialization of the South, paid homage to a South that had no industry and no science. Clothes were to be made on looms; food was to be grown in everyone's backyard; craftsmanship in woodcarving and stonecutting would again flourish, and manners and civilized discourse would again run throughout the land. In short, the writers of *I'll Take My Stand* were unwilling to pay the price of knowledge. They wished to remain innocent—and ignorant.

The old tug towards innocence and Eden still remains in these shattered times, and there seems no better testament to its continual lure than the famous movie, "Butch Cassidy and the Sundance Kid." Out on the frontier of America, Butch and the Sundance Kid playfully rob trains, and seem not so much interested in the money they get from these exploits as the discomfort they give to those agents of progress and technology, the railroad managers and the owner, Mr. Harriman. All is lighthearted and innocent until one day Mr. Harriman decides to track them down using all the latest devices that technology could afford him. Suddenly Eden becomes a nightmare ("Who are *those* guys?" Butch and the Kid continually ask themselves about their pursuers), and after a daring and preposterous escape from Harriman's technocrats, the two bandits and their lovely Eve leave for a greener, less technological, Shangri-la, Bolivia. Even their escape from Harriman's minions is instructive: the long, long, long leap from a cliff into a mountain stream reminds us all of the Saturdays of our childhood innocence when in movie serial after movie serial we would watch in fascinated horror as our hero fell off a cliff at the end of the reel—and cheered the next Saturday when we watched him fall into a pool of water. Bolivia finally also becomes a nightmare for Butch and the Kid. Their girl goes home, and, trapped by an Army battalion, Butch and the Kid are killed and frozen into a daguerrotype of themselves. Technology killed them; civilization closed in upon them and, just before they become martyrs to their own lost innocence, they gasp out a plan to move on to Australia, a new land where surely there would be no soldiers—and no technocracy.

But does all this mean that the loss of innocence is an unmitigated tragedy? That science is an unmixed evil? If we can return to *Paradise Lost* for a moment, it may be possible to understand that our fall from innocence is a *felix culpa*—a fortunate fall. Milton, at least, saw it this way and, while lamenting our grandparents' fall from grace, he noted that Adam and Eve, while banished from Eden, had, on the other hand, the *world* in which to live. No longer children, Adam and Eve now had a chance to live as adults; to live, in other words, with science; to make a wild and untamed and desolate world a civilization that would be a monument to man—and to God. Consider carefully the last lines of *Paradise Lost* which describe Adam and Eve, now reconciled to one another, banished from the Garden, facing, hand-in-hand their world:

> . . . whereat
> In either hand the hastning Angel caught
> Our lingring Parents, and to th' Eastern Gate
> Led them direct, and down the Cliff as fast

To the subjected Plaine; then disappeered.
They looking back, all th' Eastern side beheld
Of Paradise, so late thir happie seat,
Wav'd over by that flaming Brand, the Gate
With dreadful Faces throngd and fierie Armes:
Som natural tears they dropd, but wip'd them soon;
The World was all before them, where to choose
Thir place of rest, and Providence thir guide:
They hand in hand with wandering steps and slow,
Through Eden took thir solitarie way.

Annotations

1. R. A. Bryan is currently professor of English and vice-president for academic affairs at the University of Florida, Gainesville.

2.3

The Sons of Martha[1]

(1907)

RUDYARD KIPLING

The Sons of Mary seldom bother, for they have
inherited that good part;
But the Sons of Martha favour their Mother of the
careful soul and the troubled heart.
And because she lost her temper once, and because
she was rude to the Lord her Guest,
Her Sons must wait upon Mary's Sons, world with-
out end, reprieve, or rest.

It is their care in all the ages to take the buffet
and cushion the shock.
It is their care that the gear engages; it is their
care that the switches lock.
It is their care that the wheels run truly; it is their
care to embark and entrain,
Tally, transport, and deliver duly the Sons of Mary
by land and main.

They say to mountains, 'Be ye removèd.' They
say to the lesser floods 'Be dry.'
Under their rods are the rocks reprovèd—they are
not afraid of that which is high.
Then do the hill-tops shake to the summit—then
is the bed of the deep laid bare,
That the Sons of Mary may overcome it, pleasantly
sleeping and unaware.

They finger death at their gloves' end where they
piece and repiece the living wires.

He rears against the gates they tend: they feed him hungry
  behind their fires.
Early at dawn, ere men see clear, they stumble into
  his terrible stall,
And hale him forth like a haltered steer, and goad
  and turn him till evenfall.

To these from birth is Belief forbidden; from these
  till death is Relief afar.
They are concerned with matters hidden—under
  the earth-line their altars are:
The secret fountains to follow up, waters withdrawn
  to restore to the mouth,
And gather the floods as in a cup, and pour them
  again at a city's drouth.

They do not preach that their God will rouse them
  a little before the nuts work loose.
They do not teach that His Pity allows them to
  leave their work when they damn-well choose.
As in the thronged and the lighted ways, so in the
  dark and the desert they stand,
Wary and watchful all their days that their breth-
  ren's days may be long in the land.

Raise ye the stone or cleave the wood to make a
  path more fair or flat;
Lo, it is black already with blood some Son of
  Martha spilled for that!
Not as a ladder from earth to Heaven, not as a wit-
  ness to any creed,
But simple service simply given to his own kind in
  their common need.

And the Sons of Mary smile and are blessèd—they
  know the angels are on their side.
They know in them is the Grace confessèd, and for
  them are the Mercies multiplied.
They sit at the Feet—they hear the Word—they
  see how truly the Promise runs:
They have cast their burden upon the Lord, and—
  the Lord He lays it on Martha's Sons!

Annotations

1. For the original version of the Martha and Mary story see Luke 10:38–42 (King James Version). Note the echoes of biblical language in the Kipling poem.

2.4

The Punishment of Prometheus

(1932)

KAREL ČAPEK

With much coughing and clearing of throats and after prolonged proceedings for the collection of evidence, the extraordinary Senate betook itself to the meeting, which was held in the shade of the sacred olive grove (1).

"Well, gentlemen," yawned Hypometheus (2), the president of the Senate, "what a confoundedly long time all this has taken! I think I hardly need to sum up; however, to obviate any formal objections—The accused, Prometheus, a citizen of this place, being summoned before the court on a charge of inventing fire and thereby—hm, hm—upsetting the existing order of things, confessed, firstly, that he actually invented fire; further, that he is able to produce the same at any time by the action called kindling; thirdly, that he did not keep this mystery—this shocking phenomenon—secret or report it to the appropriate authority but deliberately revealed it, in fact handed it over to the use of unauthorized persons, as is proved by the evidence of the persons concerned, whom we have just interrogated. I think that is all and that we can proceed at once to declare him guilty and to pronounce the sentence."

"Excuse me, Mr. President," objected the lay magistrate Apometheus, "but I consider that in view of the importance of this extraordinary tribunal it would have been more suitable if we proceeded to pronounce judgment after deliberation and, so to speak, general discussion."

"As you please, gentlemen," assented the conciliatory Hypometheus. "The case is perfectly clear, but if any of you wish to make a remark, please do so."

"I would venture to point out," said Ametheus, a member of the tribunal, coughing primly, "that in my opinion one aspect of the whole matter should be specially emphasized. Gentlemen, I am thinking of the religious aspect. I ask you, what is this fire? What is this kindled spark? As Prometheus himself admits, it is nothing other than lightning, and lightning, as we all know, is the expression of the extraordinary power of Zeus the Thunderer. Will you explain to me, gentlemen, how an ordinary fellow like Prometheus had access to this divine fire? By what right did he seize it? Where did he get it from? Prometheus has tried to persuade us that he simply invented it; but that is a silly excuse—if it were as simple and innocent as all that, why should not one of us have invented fire? I am convinced, gentlemen, that Prometheus simply stole this fire from our gods. His denial and prevarication do not mislead us. I would describe his crime as common theft on the one hand, and as the crime of blasphemy and sacrilege on the other. We are here to punish with the utmost severity this impious presumptuousness and to protect the sacred property of our national gods. That is all I wished to say," concluded Ametheus, and blew his nose energetically on the corner of his chlamys.

"Well said," agreed Hypometheus. "Does anyone else wish to make a remark?"

"I ask your indulgence," said Apometheus, "but I cannot agree with the argument of my esteemed colleague. I watched how the said Prometheus kindled this fire; and I tell you frankly, gentlemen, that—between ourselves—there's absolutely nothing in it. The discovery of fire could have been made by any idler, loafer or goatherd; we ourselves did not happen upon it simply because a serious man hasn't the time and doesn't dream of playing about with stones to make fire. I assure my colleague Ametheus that these are quite ordinary natural forces which it is beneath the dignity of a thinking man, much less a god, to occupy himself with. In my opinion fire is too insignificant a phenomenon to affect in any way matters which are sacred to us all. But the case has another aspect, to which I must call the attention of my distinguished colleagues. It appears that fire is a very dangerous element, in fact even harmful. You have heard a number of witnesses who have deposed that in trying Prometheus's boyish invention they have suffered serious burns and in some cases even damage to property. Gentlemen, if through the fault of Prometheus the use of fire becomes general, which now, unfortunately, seems impossible to prevent, neither the property nor even the lives of any of us will be safe; and that, gentlemen, may mean the end of all civilization. It needs only the least carelessness—and at what will this mischief-working element stop? Prometheus, gentlemen, has committed an act of criminal irresponsibility in bringing into the world so harmful a thing. I should describe his crime as the causing of grievous bodily harm and endangering the public safety. In view of this I am in favour of a life sentence of imprisonment with a hard pallet and manacles. I have finished, Mr. President."

"You are perfectly right, sir," grunted Hypometheus. "And I would just like to say, gentlemen, what did we need this fire for, anyway? Did our forefathers use fire? To invent such a thing is simply disrespect to the inherited order, it is—hm, merely revolutionary activity. Playing about with fire, who ever heard of such a thing? And consider, gentlemen, what it leads to: people will relax by the fire, they will wallow in warmth and comfort instead of—well, instead of fighting and things like that. It can only lead to effeminacy, the degeneration of morals and—hm, general disorder and so on. In a word, something must be done against such unhealthy signs. The times are grave, very grave. That is all I wished to point out."

"Most rightly said," declared Antimetheus. "We all certainly agree with our President that Prometheus's fire may have unforseeable consequences. Gentlemen, do not let us conceal the fact from ourselves, it is a tremendous thing. To have fire in one's power—what new possibilities open out before one! I will only mention a few of them haphazard: to burn the enemy's crops, set fire to his olive groves and so on. In fire, gentlemen, our people have been given a new force and a new weapon; through fire we shall become almost equal to the gods," whispered Antimetheus, and suddenly exploded fiercely, "I accuse Prometheus of having entrusted this divine and irresistible element of fire to shepherds and slaves, to the first comer; I accuse him of not giving it up into authorized hands which would have guarded it as a treasure of the state and governed by its means. I accuse Prometheus of thus being a dishonest trustee of the discovery of fire, which should be a secret of the priesthood. I accuse Prometheus," shouted Antimetheus, carried away by emotion, "of teaching even foreigners how to kindle fire! of not concealing it even from our enemies! Prometheus stole fire from us by giving it to everyone! I accuse Prometheus of high treason! I accuse him of conspiracy against the community!" His voice rose to a scream and he broke off, coughing. "I propose the death sentence," he managed to get out.

"Well, gentlemen," said Hypometheus, "does anyone else wish to speak? Then in the opinion of the court the accused Prometheus is found guilty firstly of the crime of blasphemy and sacrilege, secondly of the crime of inflicting grievous bodily harm, causing damage to the property of others and endangering the public safety, and thirdly of the crime of high treason. Gentlemen, I propose to pass sentence upon him either of life imprisonment, rendered more rigorous by hard pallet and manacles, or sentence of death. Ahm."

"Or both," said Ametheus thoughtfully, "so as to comply with both proposals."

"How do you mean, both sentences?" asked the president.

"I've just been thinking it over," grunted Ametheus. "Perhaps we could manage it like this . . . condemn Prometheus to be chained to a rock for the rest of his life . . . perhaps let vultures peck out his godless liver, does Your Excellency understand?"

"That would do," said Hypometheus placidly. "Gentlemen, that would be an exemplary punishment for such a—ahem—criminal eccentricity, wouldn't it? Has anyone any objection? Then the session is closed."

° ° °

"And why did you condemn Prometheus to death, Daddy?" Hypometheus's son, Epimetheus, asked him at supper.

"You wouldn't understand," grunted Hypometheus, gnawing a leg of mutton as he spoke. "Upon my word, this mutton tastes better roasted than raw; so you see, this fire is some use after all. It was for reasons of public interest, do you see? Where should we be if anyone who liked was allowed to come along with great, new inventions unpunished? See what I mean? But there's something this meat still needs—Ah, I've got it!" he exclaimed delightedly. "Roast mutton ought to be salted and rubbed with garlic! That's the right way to do it! Now that is a real discovery! You know, a fellow like Prometheus would never have thought of that!"

Annotations

1. Different phases of the struggles between Zeus and mankind are told in Hesiod's *Theogony* and *Works and Days* (ca. 750B.C.). In the *Theogony* (a family tree of the Greek gods) Prometheus ("forethinker or he who thinks ahead") twice tricks Zeus on behalf of mankind: once he gets Zeus to accept the bony part of an ox-sacrifice, saving the edible portion for mankind; a second time he

steals fire for mankind from Zeus who had hidden it. In anger Zeus exacted a terrible penalty from both mankind and Prometheus. To mankind he sent Pandora, a beautiful woman, as bringer of evil and affliction to the world—Prometheus's slow-witted brother Epimetheus ("afterthinker") accepted her as a gift from the gods despite his brother's warning. Prometheus received an excruciating punishment: he was chained to a rock and Zeus sent an eagle to gnaw on his liver which was restored each night. Other versions of the Prometheus story represent him not only as the bringer of fire to mankind but also as the teacher of the arts and sciences and bringer of civilization to a brutish world (see Aeschylus's tragedy *Prometheus Bound*, 11. 440–508). Because of his suffering on behalf of mankind, Prometheus has also become in literature a symbol of man's indomitable struggle against a tyrannical oppressor.

2. Čapek has made up names for other characters in his story by combining the Greek root for "thinker" with different Greek prefixes.

2.5

I Like to See It Lap the Miles

(c. 1862)

EMILY DICKINSON

I like to see it lap the Miles—
And lick the Valleys up—
And stop to feed itself at Tanks—
And then—prodigious step

Around a Pile of Mountains—
And supercilious peer
In Shanties—by the sides of Roads—
And then a Quarry pare

To fit its Ribs
And crawl between
Complaining all the while
In horrid—hooting stanza—
Then chase itself down Hill—

And neigh like Boanerges(1)—
Then—punctual as a Star
Stop—docile and omnipotent
At its own stable door—

Annotations

1. See Mark 3:17 where Christ calls his disciples John and James *Boanerges* ("sons of thunder"). The word is now used to describe a loud-mouthed preacher or orator (*American Heritage Dictionary*).

2.6

To a Locomotive in Winter

(1876)

WALT WHITMAN

Thee for my recitative (1).
Thee in the driving storm even as now, the snow, the winter-
day declining,
Thee in thy panoply (2), thy measur'd dual throbbing and thy
beat convulsive,
Thy black cylindric body, golden brass and silvery steel,
Thy ponderous side-bars, parallel and connecting rods, gyrating,
shuttling at thy sides,
Thy metrical, now swelling pant and roar, now tapering in
the distance,
Thy great protruding head-light fix'd in front,
Thy long, pale, floating vapor-pennants, tinged with delicate
purple,
The dense and murky clouds out-belching from thy smokestack,
Thy knitted frame, thy springs and valves, the tremulous
twinkle of thy wheels,
Thy train of cars, behind, obedient, merrily following,
Through gale or calm, now swift, now slack, yet steadily careering;
Type of the modern—emblem of motion and power—pulse of
the continent,
For once come serve the Muse and merge in verse, even as
here I see thee,
With storm and buffeting gusts of wind and falling snow,
By day thy warning ringing bell to sound its notes,
By night thy silent signal lamps to swing,

Fierce-throated beauty!
Roll through my chant with all thy lawless music, thy swinging
lamps at night,

Thy madly-whistled laughter, echoing, rumbling like an earthquake,
  rousing all,
Law of thyself complete, thine own track firmly holding,
(No sweetness debonair of tearful harp or glib piano thine,)
Thy trills of shrieks by rock and hills return'd,
Launch'd o'er the prairies wide, across the lakes,
To the free skies unpent (3) and glad and strong.

Annotations

1. Recitative is a musical style used in opera, which imitates the rhythm of natural speech.
2. A panoply is a suit of armor.
3. Unpent means not shut in, not closely confined.

2.7

Sounds

(1854)

HENRY DAVID THOREAU

But while we are confined to books, though the most select and classic, and read only particular written languages, which are themselves but dialects and provincial, we are in danger of forgetting the language which all things and events speak without metaphor, which alone is copious and standard. Much is published, but little printed. The rays which stream through the shutter will be no longer remembered when the shutter is wholly removed. No method nor discipline can supersede the necessity of being for ever on the alert. What is a course of history or philosophy, or poetry, no matter how well selected, or the best society, or the most admirable routine of life, compared with the discipline of looking always at what is to be seen? Will you be a reader, a student merely, or a seer? Read your fate, see what is before you, and walk on into futurity.

I did not read books the first summer; I hoed beans. Nay, I often did better than this. There were times when I could not afford to sacrifice the bloom of the present moment to any work, whether of the head or hands. I love a broad margin to my life. Sometimes, in a summer morning, having taken my accustomed bath, I sat in my sunny doorway from sunrise till noon, rapt in a reverie, amidst the pines and hickories and sumachs, in undisturbed solitude and stillness, while the birds sang around or flitted noiseless through the house, until by the sun falling in at my west window, or the noise of some traveller's waggon on the distant highway, I was reminded of the lapse of time. I grew in those seasons like corn in the night, and they were far better than any work of the hands would have been. They were not time subtracted from my life, but so much over and above my usual allowance. I realised what the Orientals mean by contemplation and the forsaking of works. For the most part, I minded not how the hours went. The day advanced as if to light some work of mine; it was morning, and lo! now it is evening, and nothing memorable is accomplished. Instead of singing like the birds, I silently smiled at my incessant good fortune. As the sparrow had its trill, sitting on the hickory before my door, so had I my chuckle or suppressed warble which he might hear out of my nest. My days were not days of the week, bearing the stamp of any heathen deity, nor were they minced into hours and fretted by the ticking of a clock; for I lived like the Puri Indians, of whom it is said that "for yesterday, to-day, and to-morrow they have only one word, and they express the variety of meaning by pointing backward for yesterday, forward for to-morrow, and overhead for the passing day." This was sheer idleness to my fellow-townsmen, no doubt; but if the birds and flowers had tried me by their standard, I should not have been found wanting. A man must find his occasions in himself, it is true. The natural day is very calm, and will hardly reprove his indolence.

I had this advantage, at least, in my mode of life, over those who were obliged to look abroad for amusement, to society and the theatre, that my life itself was become my amusement and never ceased to be novel. It was a drama of many scenes and without an end. If we were always indeed getting our living, and regulating our lives according to the last and best mode we had learned, we should never be troubled with ennui. Follow your genius closely enough, and it will not fail to show you a fresh prospect every hour. Housework was a pleasant pastime. When my floor was dirty, I rose early, and, setting all my furniture out of doors on the grass, bed and bedstead making but one budget, dashed water on the floor, and sprinkled white sand from the pond on it, and then with a broom scrubbed it clean and white; and by the time the villagers had broken their fast the morning sun had dried my house sufficiently to allow me to move in again, and my meditations were almost uninterrupted. It was pleasant to see my whole household effects on the grass, making a little pile like a gipsy's pack, and my three-legged table, from which I did not remove the books and pen and ink, standing amid the pines and hickories. They seemed glad to get out themselves, and as if unwilling to be brought in. I was sometimes tempted to stretch an awning over them and take my seat there. It was worth the while to see the sun shine on these things, and hear the free wind blow on them; so much more interesting most familiar objects look out of doors than in the house. A bird sits on the next bough, life-everlasting grows under the table, and blackberry vines run round its legs; pine cones, chestnut burs, and strawberry leaves are strewn about. It looked as if this was the way these forms came to be transferred to our furniture, to tables, chairs, and bedsteads—because they once stood in their midst.

My house was on the side of a hill, immediately on the edge of the larger wood, in the midst of a young forest of pitch pines and hickories, and half a dozen rods from the pond, to which a narrow footpath led down the hill. In my front yard grew the strawberry, blackberry, and life-everlasting, johnswort and goldenrod, shrub oaks and sand-cherry, blueberry and ground-nut. Near the end of May, the sand-cherry (*cerasus pumila*), adorned the sides of the path with its delicate flowers arranged in umbels cylindrically about its short stems, which last, in the fall, weighed down with good-sized and handsome cherries, fell over in wreaths like rays on every side. I tasted them out of compliment to Nature, though they were scarcely palatable. The sumach (*rhus glabra*) grew luxuriantly about the house, pushing up through the embankment which I had made, and growing five or six feet the first season. Its broad pinnate tropical leaf was pleasant though strange to look on. The large buds, suddenly pushing out late in the spring from dry sticks which had seemed to be dead, developed themselves as by magic into graceful green and tender boughs, an inch in diameter; and sometimes, as I sat at my window, so heedlessly did they grow and tax their weak joints, I heard a fresh and tender bough suddenly fall like a fan to the ground, when there was not a breath of air stirring, broken off by its own weight. In August, the large masses of berries, which, when in flower, had attracted many wild bees, gradually assumed their bright velvety crimson hue, and by their weight again bent down and broke the tender limbs.

As I sit at my window this summer afternoon, hawks are circling about my clearing; the tantivy of wild pigeons, flying by twos and threes athwart my view, or perching restless on the white pine boughs behind my house, gives a voice to the air; a fish-hawk dimples the glassy surface of the pond and brings up a fish; a mink steals out of the marsh before my door and seizes a frog by the shore; the sedge is bending under the weight of the reed-birds flitting hither and thither; and for the last half hour I have heard the rattle of railroad cars, now dying away and then reviving like the beat of a partridge, conveying travellers from Boston to the country. For I did not live so out of the world as that boy, who, as I hear, was put out to a farmer in the east part of the town, but ere long ran away and came home again, quite down at the heel and home-sick. He had never seen such a dull and out-of-the-way place; the folks were all gone off somewhere; why, you couldn't even hear the whistle! I doubt if there is such a place in Massachusetts now:—

> "In truth, our village has become a butt
> For one of those fleet railroad shafts, and o'er
> Our peaceful plain its soothing sound is—Concord."

The Fitchburg Railroad touches the pond about a hundred rods south of where I dwell. I usually go to the village along its causeway, and am, as it were, related to society by this link. The men on the freight trains, who go over the whole length of the road, bow to me as to an old acquaintance, they pass me so often, and apparently they take me for an employee: and so I am. I too would fain be a track-repairer somewhere in the orbit of the earth.

The whistle of the locomotive penetrates my woods summer and winter, sounding like the scream of a hawk sailing over some farmer's yard, informing me that many restless city merchants are arriving within the circle of the town, or adventurous country traders from the other side. As they come under one horizon, they shout their warning to get off the track to the other, heard sometimes through the circles of two towns. Here come your groceries, country; your rations, countrymen! Nor is there any man so independent on his farm that he can say them nay. And here's your pay for them! screams the countryman's whistle; timber like long-battering rams going twenty miles an hour against the city walls, and chairs enough to seat all the weary and heavy-laden that dwell within them. With such huge and lumbering civility the country hands a chair to the city. All the Indian huckleberry hills are stripped, all the cranberry meadows are raked into the city. Up comes the cotton, down goes the woven cloth; up comes the silk, down goes the woollen; up come the books, but down goes the wit that writes them.

When I meet the engine with its train of cars moving off with planetary motion—or, rather, like a comet, for the beholder knows not if with that velocity and with that direction it will ever revisit this system, since its orbit does not look like a returning curve—with its steam-cloud like a banner streaming behind in golden and silver wreaths, like many a downy cloud which I have seen, high in the heavens, unfolding its masses to the light—as if this travelling demigod, this cloud-compeller, would ere long take the sunset sky for the livery of his train; when I hear the iron horse make the hills echo with his snort like thunder, shaking the earth with his feet, and breathing fire and smoke from his nostrils (what kind of winged horse or fiery dragon they will put into the new Mythology I don't know), it seems as if the earth had got a race now worthy to inhabit it. If all were as it seems, and men made the elements their servants for noble ends! If the cloud that hangs over the engine were the perspiration of heroic deeds, or as beneficent as that which floats over the farmer's fields, then the elements and Nature herself would cheerfully accompany men on their errands, and be their escort.

I watch the passage of the morning cars with the same feeling that I do the rising of the sun, which is hardly more regular. Their train of clouds stretching far behind and rising higher and higher, going to heaven while the cars are going to Boston, conceals the sun for a minute and casts my distant field into the shade, a celestial train beside which the petty train of cars which hugs the earth is but the barb of the spear. The stabler of the iron horse was up early this winter morning by the light of the stars amid the mountains, to fodder and harness his steed. Fire, too, was awakened thus early to put the vital heat in him and get him off. If the enterprise were as innocent as it is early! If the snow lies deep, they strap on his snow-shoes, and with the giant plough plough a furrow from the mountains to the seaboard, in which the cars, like a following drill-barrow, sprinkle all the restless men and floating merchandise in the country for seed. All day the fire-steed flies over the country, stopping only that his master may rest, and I am awakened by his tramp and defiant snort at midnight, when in some remote glen in the woods he fronts the elements incased in ice and snow; and he will reach his stall only with the morning star, to start once more on his travels without rest or slumber. Or perchance, at evening, I hear him in his stable blowing off the superfluous energy of the day, that he may calm his nerves and cool his liver and brain for a few hours of iron slumber. If the enterprise were as heroic and commanding as it is protracted and unwearied!

Far through unfrequented woods on the confines of towns, where once only the hunter penetrated by day, in the darkest night dart these bright saloons without the knowledge of their inhabitants; this moment stopping at some brilliant station-house in town or city, where a social crowd is gathered, the next in the Dismal Swamp, scaring the owl and fox. The startings and arrivals of the cars are now the epochs in the village day. They go and come with such regularity and precision, and their whistle can be heard so far, that the farmers set their clocks by them, and thus one well-conducted institution regulates a whole country. Have not men improved somewhat in punctuality since the railroad was invented? Do they not talk and think faster in the depôt than they did in the stage-office? There is something electrifying in the atmosphere of the former place. I have been astonished at the miracles it has wrought; that some of my neighbours, who, I should have prophesied, once for all, would never get to Boston by so prompt a conveyance, are on hand when the bell rings. To do things "railroad fashion" is now the by-word; and it is worth the while to be warned so often and so sincerely by any power to get off its track. There is no stopping to read the riot act, no firing over the heads of the mob, in this case. We have constructed a fate, an *Atropos* (1), that never turns aside. (Let that be the name of your engine.) Men are advertised that at a certain hour and minute these bolts will be shot toward particular points of the compass; yet it interferes with no man's business, and the children go to school on the other track. We live the steadier for it. We are all educated thus to be sons of Tell (2). The air is full of invisible bolts. Every path but your own is the path of fate. Keep on your own track, then.

What recommends commerce to me is its enterprise and bravery. It does not clasp its hands and pray to Jupiter. I see these men every day go about their business with more or less courage and content, doing more even than they suspect, and perchance better employed than they could have consciously devised. I am less affected by their heroism who stood up for half an hour in the front line at Buena Vista (3), than by the steady and cheerful valour of the men who inhabit the snow-plough for their winter quarters; who have not merely the three o'clock in the morning courage, which Bonaparte thought was the rarest (4), but whose courage does not go to rest so early, who go to sleep only when the storm sleeps or the sinews of their iron steed are frozen. On this morning of the Great Snow, perchance, which is still raging and chilling men's blood, I hear the muffled tone of their engine bell from out the fog-bank of their chilled breath, which announces that the cars *are coming*, without long delay, notwithstanding the veto of a New England northeast snow-storm, and I behold the ploughmen covered with snow and rime, their heads peering above the mould-board which is turning down other than daisies and the nests of field-mice, like boulders of the Sierra Nevada, that occupy an outside place in the universe.

Commerce is unexpectedly confident and serene, alert, adventurous, and unwearied. It is very natural in its methods, withal, far more so than many fantastic enterprises and sentimental experiments, and hence its singular success. I am refreshed and expanded when the freight train rattles past me, and I smell the stores which go dispensing their odours all the way from Long Wharf to Lake Champlain, reminding me of foreign parts, of coral reefs, and Indian oceans, and tropical climes, and the extent of the globe. I feel more like a citizen of the world at the sight of the palm-leaf, which will cover so many flaxen New England heads the next summer, the Manilla hemp and cocoa-nut husks, the old junk, gunny bags, scrap iron, and rusty nails. This car-load of torn sails is more legible and interesting now than if they should be wrought into paper and printed books. Who can write so graphically the history of the storms they have weathered as these rents have done? They are proof-sheets which need no correction. Here goes lumber from the Maine woods, which did not go out to sea in the last freshet, risen four dollars on the thousand because of what did go out or was split up; pine, spruce, cedar—first, second, third, and fourth qualities, so lately all of one quality, to wave over the bear, and moose, and caribou. Next rolls Thomaston lime, a prime lot, which will get far among the hills before it gets slacked. These rags in bales, of all hues and qualities, the lowest condition to which cotton and linen descend, the final result of dress—of patterns which are now no longer cried up, unless it be in Milwaukee, as those splendid articles, English, French, or American prints, ginghams, muslins, etc.—gathered from all quarters both of fashion and poverty, going to become paper of one colour or a few shades only, on which, forsooth, will be written tales of real life, high and low, and founded on fact! This closed car smells of salt fish, the strong New England and commercial scent, reminding me of the Grand Banks and the fisheries. Who has not seen a salt fish, thoroughly cured for this world, so that nothing can spoil it, and putting the perseverance of the saints to the blush? with which you may sweep or pave the streets, and split your kindlings, and the teamster shelter himself and his lading against sun, wind, and rain behind it—and the trader, as a Concord trader once did, hang it up by his door for a sign when he commences business, until at last his oldest customer cannot tell surely whether it be animal, vegetable, or mineral, and yet it shall be as pure as a snowflake, and if it be put into a pot and boiled, will come out an excellent dun fish for a Saturday's dinner. Next, Spanish hides, with the tails still preserving their twist and the angle of elevation they had when the oxen that wore them were careering over the pampas of the Spanish main—a type of obstinacy, and evincing how almost hopeless and incurable are all constitutional vices. I confess, that practically speaking, when I have learned a man's real disposition, I have no hopes of changing it for the better or worse in this state of existence. As the Orientals say, "A cur's tail may be warmed, and pressed, and bound round with ligatures, and after a twelve years' labour bestowed upon it, still it will retain its natural form." The only effectual cure for such inveteracies as these tails exhibit is to make glue of them, which I believe is what is usually done with them, and then they will stay put and stick. Here is a hogshead of molasses or of brandy, directed to John Smith, Cuttingsville, Vermont, some trader among the Green Mountains, who imports for the farmers near his clearing, and now perchance stands over his bulk-head and thinks of the last arrivals on the coast, how they may affect the price for him, telling his customers this moment, as he has told them twenty times before this morning, that he expects some by the next train of prime quality. It is advertised in the *Cuttingsville Times.*

While these things go up other things come down. Warned by the whizzing sound, I look up from my book and see some tall pine, hewn on far northern hills, which has winged its way over the Green Mountains and the Connecticut, shot like an arrow through the township within ten minutes, and scarce another eye beholds it; going

> "to be the mast
> Of some great ammiral (5)."

And hark! here comes the cattle train bearing the cattle of a thousand hills, sheepcots, stables, and cowyards in the air, drovers with their sticks, and shepherd boys in the midst of their flocks, all but the mountain pastures, whirled along like leaves blown from the mountains by the September gales. The air is filled with the bleating of calves and sheep, and the hustling of oxen, as if a pastoral valley was going by. When the old bell-wether at the head rattles his bell, the mountains do indeed skip like rams and the little hills like lambs. A car-load of drovers, too, in the midst, on a level with their droves now, their vocation gone, but still clinging to their useless sticks as their badge of office. But their dogs, where are they? It is a stampede to them; they are quite thrown out; they have lost the scent. Methinks I hear them barking behind the Peterboro' Hills, or panting up the western slope of the Green Mountains. They will not be in at the death. Their vocation, too, is gone. Their fidelity and sagacity are below par now. They will slink back to their kennels in disgrace, or perchance run wild and strike a league with the wolf and the fox. So is your pastoral life whirled past and away. But the bell rings, and I must get off the track and let the cars go by—

What's the railroad to me?
I never go to see
Where it ends.
It fills a few hollows,
And makes banks for the swallows,
It sets the sand a-blowing,
And the blackberries a-growing,

but I cross it like a cart-path in the woods. I will not have my eyes put out and my ears spoiled by its smoke, and steam, and hissing.

Now that the cars are gone by and all the restless world with them, and the fishes in the pond no longer feel their rumbling, I am more alone than ever. For the rest of the long afternoon, perhaps, my meditations are interrupted only by the faint rattle of a carriage or team along the distant highway.

Sometimes, on Sundays, I heard the bells, the Lincoln, Acton, Bedford, or Concord bell, when the wind was favourable, a faint, sweet, and, as it were, natural melody, worth importing into the wilderness. At a sufficient distance over the woods this sound acquires a certain vibratory hum, as if the pine needles in the horizon were the strings of a harp which it swept. All sound heard at the greatest possible distance produces one and the same effect, a vibration of the universal lyre, just as the intervening atmosphere makes a distant ridge of earth interesting to our eyes by the azure tint it imparts to it. There came to me in this case a melody which the air had strained, and which had conversed with every leaf and needle of the wood, that portion of the sound which the elements had taken up and modulated and echoed from vale to vale. The echo is, to some extent, an original sound, and therein is the magic and charm of it. It is not merely a repetition of what was worth repeating in the bell, but partly the voice of the wood; the same trivial words and notes sung by a wood-nymph.

At evening, the distant lowing of some cow in the horizon beyond the woods sounded sweet and melodious, and at first I would mistake it for the voices of certain minstrels by whom I was sometimes serenaded, who might be straying over the hill and dale; but soon I was not unpleasantly disappointed when it was prolonged into the cheap and natural music of the cow. I do not mean to be satirical, but to express my appreciation of those youths' singing, when I state that I perceived clearly that it was akin to the music of the cow, and they were at length one articulation of Nature.

Regularly at half past seven, in one part of the summer, after the evening train had gone by, the whippoorwills chanted their vespers for half an hour, sitting on a stump by my door, or upon the ridge-pole of the house. They would begin to sing almost with as much precision as a clock, within five minutes of a particular time, referred to the setting of the sun, every evening. I had a rare opportunity to become acquainted with their habits. Sometimes I heard four or five at once in different parts of the wood, by accident one a bar behind another, and so near me that I distinguished not only the cluck after each note, but often that singular buzzing sound like a fly in a spider's web, only proportionally louder. Sometimes one would circle round and round me in the woods a few feet distant as if tethered by a string, when probably I was near its eggs. They sang at intervals throughout the night, and were again as musical as ever just before and about dawn.

When other birds are still the screech owls take up the strain, like mourning women their ancient u-lu-lu. Their dismal scream is truly Ben Jonsonian (6). Wise midnight hags! It is no honest and blunt to-whit tu-who of the poets, but, without jesting, a most solemn graveyard ditty, the mutual consolations of suicide lovers remembering the pangs and the delights of supernal love in the infernal groves. Yet I love to hear their wailing, their doleful responses, trilled along the woodside, reminding me sometimes of music and singing birds; as if it were the dark and tearful side of music, the regrets and sighs that would fain be sung. They are the spirits, the low spirits and melancholy forebodings, of fallen souls that once in human shape nightly walked the earth and did the deeds of darkness, now expiating their sins with their wailing hymns or threnodies in the scenery of their transgressions. They give me a new sense of the variety and capacity of that nature which is our common dwelling. *Oh-o-o-o-o that I never had been bor-r-r-r-n!* sighs one on this side of the pond, and circles with the restlessness of despair to some new perch on the grey oaks. Then—*That I never had been bor-r-r-r-n!* echoes another on the farther side with tremulous sincerity, and—*bor-r-r-n!* comes faintly from far in the Lincoln woods.

I was also serenaded by a hooting owl. Near at hand you could fancy it the most melancholy sound in Nature, as if she meant by this to stereotype and make permanent in her choir the dying moans of a human being—some poor weak relic of mortality who has left hope behind, and howls like an animal, yet with human sobs, on entering the dark valley, made more awful by a certain gurgling melodiousness—I find myself beginning with the letters gl when I try to imitate it—expressive of a mind which has reached the gelatinous mildewy stage in the mortification of all healthy and courageous thought. It reminded me of ghouls and idiots and insane howlings. But now one answers from far woods in a strain made really melodious by distance—*Hoo hoo hoo, hoorer hoo*; and indeed for the most part it suggested only pleasing associations, whether heard by day or night, summer or winter.

I rejoice that there are owls. Let them do the idiotic and maniacal hooting for men. It is a sound admirably suited to swamps and twilight woods which no day illustrates, suggesting a vast and undeveloped nature which men have not recognised. They represent the stark twilight and unsatisfied thoughts which all have. All day the sun has shone on the surface of some savage swamp, where the single spruce stands hung with usnea lichens, and small hawks circulate above, and the chickadee lisps amid the evergreens, and the partridge and rabbit skulk beneath; but now a more dismal and fitting day dawns, and a different race of creatures awakes to express the meaning of Nature there.

Late in the evening I heard the distant rumbling of waggons over bridges—a sound heard farther than almost any other at night—the baying of dogs, and sometimes again the lowing of some disconsolate cow in a distant barnyard. In the meanwhile all the shore rang with the trump of bull-frogs, the sturdy spirits of ancient wine-bibbers and wassailers, still unrepentant, trying to sing a catch in their Stygian lake (7)—if the Walden nymphs will pardon the comparison, for though there are almost no weeds, there are frogs there—who would fain keep up the hilarious rules of their old festal tables, though their voices have waxed hoarse and solemnly grave, mocking at mirth, and the wine has lost its flavour, and become only liquor to distend their paunches, and sweet intoxication never comes to drown the memory of the past, but mere saturation and waterloggedness and distention. The most aldermanic, with his chin upon a heart-leaf, which serves for a napkin to his drooling chaps, under this northern shore quaffs a deep draught of the once scorned water, and passes round a cup with the ejaculation *tr-r-r-oonk, tr-r-r-oonk, tr-r-r-oonk!* and straightway comes over the water from some distant cove the same password repeated, where the next in seniority and girth has gulped down to his mark; and when this observance has made the circuit of the shores, then ejaculates the master of ceremonies, with satisfaction, *tr-r-r-oonk!* and each in his turn repeats the same down to the least distended, leakiest, and flabbiest paunched, that there be no mistake; and then the bowl goes round again and again, until the sun disperses the morning mist, and only the patriarch is not under the pond, but vainly bellowing *troonk* from time to time, and pausing for a reply.

I am not sure that ever I heard the sound of cock-crowing from my clearing, and I thought that it might be worth the while to keep a cockerel for his music merely, as a singing bird. The note of this once wild Indian pheasant is certainly the most remarkable of any bird's, and if they could be naturalised without being domesticated, it would soon become the most famous sound in our woods, surpassing the clangour of the goose and the hooting of the owl; and then imagine the cackling of the hens to fill the pauses when their lords' clarions rested! No wonder that man added this bird to his tame stock—to say nothing of the eggs and drumsticks. To walk in a winter morning in a wood where these birds abounded, their native woods, and hear the wild cockerels crow on the trees, clear and shrill for miles over the resounding earth, drowning the feebler notes of other birds—think of it! It would put nations on the alert. Who would not be early to rise, and rise earlier and earlier every successive day of his life, till he became unspeakably healthy, wealthy and wise? This foreign bird's note is celebrated by the poets of all countries along with the notes of their native songsters. All climates agree with brave Chanticleer (8). He is more indigenous even than the natives. His health is ever good, his lungs are sound, his spirits never flag. Even the sailor on the Atlantic and Pacific is awakened by his voice; but its shrill sound never roused me from my slumbers. I kept neither dog, cat, cow, pig, nor hens, so that you would have said there was a deficiency of domestic sounds; neither the churn, nor the spinning-wheel, nor even the singing of the kettle, nor the hissing of the urn, nor children crying, to comfort one. An old-fashioned man would have lost his senses or died of ennui before this. Not even rats in the wall, for they were starved out, or rather were never baited in—only squirrels on the roof and under the floor, a whippoorwill on the ridge-pole, a blue-jay screaming beneath the window, a hare or woodchuck under the house, a screech-owl or a cat-owl behind it, a flock of wild geese or a laughing loon on the pond, and a fox to bark in the night. Not even a lark or an oriole, those mild plantation birds, ever visited my clearing. No cockerels to crow now nor hens to cackle in the yard. No yard! but unfenced Nature reaching up to your very sills. A young forest growing up under your windows, and wild sumachs and blackberry vines breaking through into your cellar; sturdy pitch-pines rubbing and creaking against the shingles for want of room, their roots reaching quite under the house. Instead of a scuttle or a blind blown off in the gale—a pine tree snapped off or torn up by the roots behind your house for fuel. Instead of no path to the front yard gate in the Great Snow—no gate—no front yard—and no path to the civilised world!

Annotations

1. In Greek mythology Atropos is one of the three Fates who preside over human destiny. The root word means "not turned aside."
2. According to Swiss legend, William Tell was required to shoot an apple from his son's head with a crossbow.
3. Buena Vista is a battle in the Mexican-American war in which the U.S. forces under Zachary Taylor defeated the Mexicans (1847).
4. Napoleon Bonaparte (1769–1821), French general and emperor.
5. From John Milton (1608–1674), *Paradise Lost* I. 293–294.
6. Ben Jonson (1573–1637), English poet and dramatist.
7. In Greek mythology the Styx is one of the rivers which surrounds Hades, land of the dead.
8. Chanticleer is a rooster (from the Old French, "he who sings clearly").

2.8

The Celestial Railroad

(1843)

NATHANIEL HAWTHORNE

Not a great while ago, passing through the gate of dreams, I visited that region of the earth in which lies the famous City of Destruction (1). It interested me much to learn that by the public spirit of some of the inhabitants a railroad has recently been established between this populous and flourishing town and the Celestial City. Having a little time upon my hands, I resolved to gratify a liberal curiosity by making a trip thither. Accordingly, one fine morning after paying my bill at the hotel, and directing the porter to stow my luggage behind a coach, I took my seat in the vehicle and set out for the station-house. It was my good fortune to enjoy the company of a gentleman—one Mr. Smooth-it-away—who, though he had never actually visited the Celestial City, yet seemed as well acquainted with its laws, customs, policy, and statistics, as with those of the City of Destruction, of which he was a native townsman. Being, moreover, a director of the railroad corporation and one of its largest stockholders, he had it in his power to give me all desirable information respecting that praiseworthy enterprise.

Our coach rattled out of the city, and at a short distance from its outskirts passed over a bridge of elegant construction, but somewhat too slight, as I imagined, to sustain any considerable weight. On both sides lay an extensive quagmire, which could not have been more disagreeable, either to sight or smell, had all the kennels of the earth emptied their pollution there.

"This," remarked Mr. Smooth-it-away, "is the famous Slough of Despond—a disgrace to all the neighborhood; and the greater that it might so easily be converted into firm ground."

"I have understood," said I, "that efforts have been made for that purpose from time immemorial. Bunyan mentions that above twenty thousand cartloads of wholesome instructions had been thrown in here without effect."

"Very probably! And what effect could be anticipated from such unsubstantial stuff?" cried Mr. Smooth-it-away. "You observe this convenient bridge. We obtained a sufficient foundation for it by throwing into the slough some editions of books of morality; volumes of French philosophy and German rationalism; tracts, sermons, and essays of modern clergymen; extracts from Plato, Confucius, and various Hindoo sages, together with a few ingenious commentaries upon texts of Scripture,—all of which by some scientific process, have been converted into a mass like granite. The whole bog might be filled up with similar matter."

It really seemed to me, however, that the bridge vibrated and heaved up and down in a very formidable manner; and, spite of Mr. Smooth-it-away's testimony to the solidity of its foundation, I should be loath to cross it in a crowded omnibus, especially if each passenger were encumbered with as heavy luggage as that gentleman and myself. Nevertheless we got over without accident, and soon found ourselves at the stationhouse. This very neat and spacious edifice is erected on the site of the little wicket gate, which formerly, as all old pilgrims will recollect, stood directly across the highway, and by its inconvenient narrowness, was a great obstruction to the traveller of liberal mind and expansive stomach. The reader of John Bunyan will be glad to know that Christian's old friend Evangelist, who was accustomed to supply each pilgrim with a mystic roll, now presides at the ticket office. Some malicious persons it is true deny the identity of this reputable character with the Evangelist of old times, and even pretend to bring competent evidence of an imposture. Without involving myself in a dispute I shall merely observe that, so far as my experience goes, the square pieces of pasteboard now delivered to passengers are much more convenient and useful along the road than the antique roll of parchment. Whether they will be as readily received at the gate of the Celestial City I decline giving an opinion.

A large number of passengers were already at the station-house awaiting the departure of the cars. By the aspect and demeanor of these persons it was easy to judge that the feelings of the community had undergone a very favorable change in reference to the celestial pilgrimage. It would have done Bunyan's heart good to see it. Instead of a lonely and ragged man with a huge burden on his back, plodding along sorrowfully on foot while the whole city hooted after him, here were parties of the first gentry and most respectable people in the neighborhood setting forth towards the Celestial City as cheerfully as if the pilgrimage were merely a summer tour. Among the gentlemen were characters of deserved eminence—magistrates, politicians, and men of wealth, by whose example religion could not but be greatly recommended to their meaner brethren. In the ladies' apartment, too, I rejoiced to distinguish some of those flowers of fashionable society who are so well fitted to adorn the most elevated circles of the Celestial City. There was much pleasant conversation about the news of the day, topics of business and politics, or the lighter matters of amusement; while religion, though indubitably the main thing at heart, was thrown tastefully into the background. Even an infidel would have heard little or nothing to shock his sensibility.

One great convenience of the new method of going on pilgrimage I must not forget to mention. Our enormous burdens, instead of being carried on our shoulders as had been the custom of old, were all snugly deposited in the baggage car, and, as I was assured, would be delivered to their respective owners at the journey's end. Another thing, likewise, the benevolent reader will be delighted to understand. It may be remembered that there was an ancient feud between Prince Beelzebub and the keeper of the wicket gate, and that the adherents of the former distinguished personage were accustomed to shoot deadly arrows at honest pilgrims while knocking at the door. This dispute, much to the credit as well of the illustrious potentate above mentioned as of the worthy and enlightened directors of the railroad, has been pacifically arranged on the principle of mutual compromise. The prince's subjects are now pretty numerously employed about the station-house, some in taking care of the baggage, others in collecting fuel, feeding the engines, and such congenial occupations; and I can conscientiously affirm that persons more attentive to their business, more willing to accommodate, or more generally agreeable to the passengers, are not to be found on any railroad. Every good heart must surely exult at so satisfactory an arrangement of an immemorial difficulty.

"Where is Mr. Greatheart?" inquired I. "Beyond a doubt the directors have engaged that famous old champion to be chief conductor on the railroad?"

"Why, no," said Mr. Smooth-it-away, with a dry cough. "He was offered the situation of brakeman; but, to tell you the truth, our friend Greatheart has grown preposterously stiff and narrow in his old age. He has so often guided pilgrims over the road on foot that he considers it a sin to travel in any other fashion. Besides, the old fellow had entered so heartily into the ancient feud with Prince Beelzebub that he would have been perpetually at blows or ill language with some of the prince's subjects, and thus have embroiled us anew. So, on the whole, we were not sorry when honest Greatheart went off to the Celestial City in a huff and left us at liberty to choose a more suitable and accommodating man. Yonder comes the engineer of the train. You will probably recognize him at once."

The engine at this moment took its station in advance of the cars, looking, I must confess, much more like a sort of mechanical demon that would hurry us to the infernal regions than a laudable contrivance for smoothing our way to the Celestial City. On its top sat a personage almost enveloped in smoke and flame, which, not to startle the reader, appeared to gush from his own mouth and stomach as well as from the engine's brazen abdomen.

"Do my eyes deceive me?" cried I. "What on earth is this! A living creature? If so, he is own brother to the engine he rides upon!"

"Poh, poh, you are obtuse!" said Mr. Smooth-it-away, with a hearty laugh. "Don't you know Apollyon, Christian's old enemy, with whom he fought so fierce a battle in the Valley of Humiliation? He was the very fellow to manage the engine; and so we have reconciled him to the custom of going on pilgrimage, and engaged him as chief engineer.'

"Bravo, bravo!" exclaimed I, with irrepressible enthusiasm; "this shows the liberality of the age; this proves, if anything can, that all musty prejudices are in a fair way to be obliterated. And how will Christian rejoice to hear of this happy transformation of his old antagonist! I promise myself great pleasure in informing him of it when we reach the Celestial City."

The passengers being all comfortably seated, we now rattled away merrily, accomplishing a greater distance in ten minutes than Christian probably trudged over in a day. It was laughable, while we glanced along, as it were, at the tail of a thunderbolt, to observe two dusty foot travellers in the old pilgrim guise, with cockle shell and staff, their mystic rolls of parchment in their hands and their intolerable burdens on their backs. The preposterous obstinacy of these honest people in persisting to groan and stumble along the difficult pathway rather than take advantage of modern improvements, excited great mirth among our wiser brotherhood. We greeted the two pilgrims with many pleasant gibes and a roar of laughter; whereupon they gazed at us with such woful and absurdly compassionate visages that our merriment grew tenfold more obstreperous. Apollyon also entered heartily into the fun, and contrived to flirt the smoke and flame of the engine, or of his own breath, into their faces, and envelop them in an atmosphere of scalding steam. These little practical jokes amused us mightily, and doubtless afforded the pilgrims the gratification of considering themselves martyrs.

At some distance from the railroad Mr. Smooth-it-away pointed to a large, antique edifice, which, he observed, was a tavern of long standing, and had formerly been a noted stopping-place for pilgrims. In Bunyan's road-book it is mentioned as the Interpreter's House.

"I have long had a curiosity to visit that old mansion," remarked I.

"It is not one of our stations, as you perceive," said my companion. "The keeper was violently opposed to the railroad; and well he might be, as the track left his house of entertainment on one side, and thus was pretty certain to deprive him of all his reputable customers. But the footpath still passes his door, and the old gentleman now and then receives a call from some simple traveller, and entertains him with fare as old-fashioned as himself."

Before our talk on this subject came to a conclusion we were rushing by the place where Christian's burden fell from his shoulders at the sight of the Cross. This served as a theme for Mr. Smooth-it-away, Mr. Live-for-the-world, Mr. Hide-sin-in-the-heart, Mr. Scaly-conscience, and a knot of gentlemen from the town of Shun-repentance, to descant upon the inestimable advantages resulting from the safety of our baggage. Myself, and all the passengers indeed, joined with great unanimity in this view of the matter; for our burdens were rich in many things esteemed precious throughout the world; and, especially, we each of us possessed a great variety of favorite Habits, which we trusted would not be out of fashion even in the polite circles of the Celestial City. It would have been a sad spectacle to see such an assortment of valuable articles tumbling into the sepulchre. Thus pleasantly conversing on the favorable circumstances of our position as compared with those of past pilgrims and of narrow-minded ones at the present day, we soon found ourselves at the foot of the Hill of Difficulty. Through the very heart of this rocky mountain a tunnel has been constructed of most admirable architecture, with a lofty arch and a spacious double track, so that, unless the earth and rocks should chance to crumble down, it will remain an eternal monument of the builder's skill and enterprise. It is a great though incidental advantage that the materials from the heart of the Hill Difficulty have been employed in filling up the Valley of Humiliation, thus obviating the necessity of descending into that disagreeable and unwholesome hollow.

"This is a wonderful improvement, indeed," said I. "Yet I should have been glad of an opportunity to visit the Palace Beautiful and be introduced to the charming young ladies—Miss Prudence, Miss Piety, Miss Charity, and the rest—who have the kindness to entertain pilgrims there."

"Young ladies!" cried Mr. Smooth-it-away, as soon as he could speak for laughing. "And charming young ladies! Why, my dear fellow, they are old maids, every soul of them—prim, starched, dry, and angular; and not one of them, I will venture to say, has altered so much as the fashion of her gown since the days of Christian's pilgrimage."

"Ah, well," said I, much comforted, "then I can very readily dispense with their acquaintance."

The respectable Apollyon was now putting on the steam at a prodigious rate, anxious, perhaps, to get rid of the unpleasant reminiscences connected with the spot where he had so disastrously encountered Christian. Consulting Mr. Bunyan's road-book, I perceived that we must now be within a few miles of the Valley of the Shadow of Death, into which doleful region, at our present speed, we should plunge much sooner than seemed at all desirable. In truth, I expected nothing better than to find myself in the ditch on one side or the quag on the other; but on communicating my apprehensions to Mr. Smooth-it-away, he assured me that the difficulties of this passage, even in its worst condition, had been vastly exaggerated, and that, in its present state of improvement, I might consider myself as safe as on any railroad in Christendom.

Even while we were speaking the train shot into the entrance of this dreaded Valley. Though I plead guilty to some foolish palpitations of the heart during our headlong rush over the causeway here constructed, yet it were unjust to withhold the highest encomiums on the boldness of its original conception and the ingenuity of those who executed it. It was gratifying, likewise, to observe how much care had been taken to dispel the everlasting gloom and supply the defect of cheerful sunshine, not a ray of which has ever penetrated among these awful shadows. For this purpose, the inflammable gas which exudes plentifully from the soil is collected by means of pipes, and thence communicated to a quadruple row of lamps along the whole extent of the passage. Thus a radiance has been created even out of the fiery and sulphurous curse that rests forever upon the valley—a radiance hurtful, however, to the eyes, and somewhat bewildering, as I discovered by the changes which it wrought in the visages of my companions. In this respect, as compared with natural daylight, there is the same difference as between truth and falsehood; but if the reader have ever travelled through the dark Valley, he will have learned to be thankful for any light that he could get—if not from the sky above, then from the blasted soil beneath. Such was the red brilliancy of these lamps that they appeared to build walls of fire on both sides of the track, between which we held our course at lightning speed, while a reverberating thunder filled the Valley with its echoes. Had the engine run off the track,—a catastrophe, it is whispered, by no means unprecedented,—the bottomless pit, if there be any such place, would undoubtedly have received us. Just as some dismal fooleries of this nature had made my heart quake there came a tremendous shriek, careering along the valley as if a thousand devils had burst their lungs to utter it, but which proved to be merely the whistle of the engine on arriving at a stopping-place.

The spot where we had now paused is the same that our friend Bunyan—a truthful man, but infected with many fantastic notions—has designated, in terms plainer than I like to repeat, as the mouth of the infernal region. This, however, must be a mistake, inasmuch as Mr. Smooth-it-away, while we remained in the smoky and lurid cavern, took occasion to prove that Tophet has not even a metaphorical existence (2). The place, he assured us, is no other than the crater of a half-extinct volcano, in which the directors had caused forges to be set up for the manufacture of railroad iron. Hence, also, is obtained a plentiful supply of fuel for the use of the engines. Whoever had gazed into the dismal obscurity of the broad cavern mouth, whence ever and anon darted huge tongues of dusky flame, and had seen the strange, half-shaped monsters, and visions of faces horribly grotesque, into which the smoke seemed to wreathe itself, and had heard the awful murmurs, and shrieks, and deep shuddering whispers of the blast, sometimes forming themselves into words almost articulate, would have seized upon Mr. Smooth-it-away's comfortable explanation as greedily as we did. The inhabitants of the cavern, moreover, were unlovely personages, dark, smoke-begrimed, generally deformed, with misshapen feet, and a glow of dusky redness in their eyes as if their hearts had caught fire and were blazing out of the upper windows. It struck me as a peculiarity that the laborers at the forge and those who brought fuel to the engine, when they began to draw short breath, positively emitted smoke from their mouth and nostrils.

Among the idlers about the train, most of whom were puffing cigars which they had lighted at the flame of the crater, I was perplexed to notice several who, to my certain knowledge, had heretofore set forth by railroad for the Celestial City. They looked dark, wild, and smoky, with a singular resemblance, indeed, to the native inhabitants, like whom, also, they had a disagreeable propensity to ill-natured gibes and sneers, the habit of which had wrought a settled contortion of their visages. Having been on speaking terms with one of these persons,—an indolent, good-for-nothing fellow, who went by the name of Take-it-easy,—I called him, and inquired what was his business there.

"Did you not start," said I, "for the Celestial City?"

"That's a fact," said Mr. Take-it-easy, carelessly puffing some smoke into my eyes. "But I heard such bad accounts that I never took pains to climb the hill on which the city stands. No business doing, no fun going on, nothing to drink, and no smoking allowed, and a thrumming of church music from morning till night. I would not stay in such a place if they offered me house room and living free."

"But, my good Mr. Take-it-easy," cried I, "why take up your residence here, of all places in the world?"

"Oh," said the loafer, with a grin, "it is very warm hereabouts, and I meet with plenty of old acquaintances, and altogether the place suits me. I hope to see you back again some day soon. A pleasant journey to you."

While he was speaking the bell of the engine rang, and we dashed away after dropping a few passengers, but receiving no new ones. Rattling onward through the Valley, we were dazzled with the fiercely gleaming gas lamps, as before. But sometimes, in the dark of intense brightness, grim faces, that bore the aspect and expression of individual sins, or evil passions, seemed to thrust themselves through the veil of light, glaring upon us, and stretching forth a great, dusky hand, as if to impede our progress. I almost thought that they were my own sins that appalled me there. These were freaks of imagination—nothing more, certainly—mere delusions, which I ought to be heartily ashamed of; but all through the Dark Valley I was tormented, and pestered, and dolefully bewildered with the same kind of waking dreams. The mephitic gases of that region intoxicate the brain. As the light of natural day, however, began to struggle with the glow of the lanterns, these vain imaginations lost their vividness, and finally vanished with the first ray of sunshine that greeted our escape from the Valley of the Shadow of Death. Ere we had gone a mile beyond it I could well-nigh have taken my oath that this gloomy passage was a dream.

At the end of the valley, as John Bunyan mentions, is a cavern, where, in his days, dwelt two cruel giants, Pope and Pagan, who had strown the ground about their residence with the bones of slaughtered pilgrims. These vile old troglodytes are no longer there; but into their deserted cave another terrible giant has thrust himself, and makes it his business to seize upon honest travellers and fatten them for his table with plentiful meals of smoke, mist, moonshine, raw potatoes, and sawdust. He is a German by birth, and is called Giant Transcendentalist; but as to his form, his features, his substance, and his nature generally, it is the chief peculiarity of this huge miscreant that neither he for himself, nor anybody for him, has ever been able to describe them. As we rushed by the cavern's mouth we caught a hasty glimpse of him, looking somewhat like an ill-proportioned figure, but considerably more like a heap of fog and duskiness. He shouted after us, but in so strange a phraseology that we knew not what he meant, nor whether to be encouraged or affrighted.

It was late in the day when the train thundered into the ancient city of Vanity, where Vanity Fair is still at the height of prosperity, and exhibits an epitome of whatever is brilliant, gay, and fascinating beneath the sun. As I purposed to make a considerable stay here, it gratified me to learn that there is no longer the want of harmony between the town's-people and pilgrims, which impelled the former to such lamentably mistaken measures as the persecution of Christian and the fiery martyrdom of Faithful. On the contrary, as the new railroad brings with it great trade and a constant influx of strangers, the lord of Vanity Fair is its chief patron, and the capitalists of the city are among the largest stockholders. Many passengers stop to take their pleasure or make their profit in the Fair, instead of going onward to the Celestial City. Indeed, such are the charms of the place that people often affirm it to be the true and only heaven; stoutly contending that there is no other, that those who seek further are mere dreamers, and that, if the fabled brightness of the Celestial City lay but a bare mile beyond the gates of Vanity, they would not be fools enough to go thither. Without subscribing to these perhaps exaggerated encomiums, I can truly say that my abode in the city was mainly agreeable, and my intercourse with the inhabitants productive of much amusement and instruction.

Being naturally of a serious turn, my attention was directed to the solid advantages derivable from a residence here, rather than to the effervescent pleasures which are the grand object with too many visitants. The Christian reader, if he have had no accounts of the city later than Bunyan's time, will be surprised to hear that almost every street has its church, and that the reverend clergy are nowhere held in higher respect than at Vanity Fair. And well do they deserve such honorable estimation; for the maxims of wisdom and virtue which fall from their lips comes from as deep a spiritual source, and tend to as lofty a religious aim, as those of the sagest philosophers of old. In justification of this high praise I need only mention the names of the Rev. Mr. Shallow-deep, the Rev. Mr. Stumble-at-truth, that fine old clerical character the Rev. Mr. This-to-day, who expects shortly to resign his pulpit to the Rev. Mr. That-to-morrow; together with the Rev. Mr. Bewilderment, the Rev. Mr. Clog-the-spirit, and, last and greatest, the Rev. Dr. Wind-of-doctrine. The labors of these eminent divines are aided by those of innumerable lecturers, who diffuse such a various profundity, in all subjects of human or celestial science, that any man may acquire an omnigenous erudition without the trouble of even learning to read. Thus literature is etherealized by assuming for its medium the human voice; and knowledge, depositing all its heavier particles, except, doubtless, its gold, becomes exhaled into a sound, which forthwith steals into the ever-open ear of the community. These ingenious methods constitute a sort of machinery, by which thought and study are done to every person's hand without his putting himself to the slightest inconvenience in the matter. There is another species of machine for the wholesale manufacture of individual morality. This excellent result is effected by societies for all manner of virtuous purposes, with which a man has merely to connect himself, throwing, as it were, his quota of virtue into the common stock, and the president and directors will take care that the aggregate amount be well applied. All these, and other wonderful improvements in ethics, religion, and literature, being made plain to my comprehension by the ingenious Mr. Smooth-it-away, inspired me with a vast admiration of Vanity Fair.

It would fill a volume, in an age of pamphlets, were I to record all my observations in this great capital of human business and pleasure. There was an unlimited range of society—the powerful, the wise, the witty, and the famous in every walk of life; princes, presidents, poets, generals, artists, actors, and philanthropists,—all making their own market at the fair, and deeming no price too exorbitant for such commodities as hit their fancy. It was well worth one's while, even if he had no idea of buying or selling, to loiter through the bazaars and observe the various sorts of traffic that were going forward.

Some of the purchasers, I thought, made very foolish bargains. For instance, a young man having inherited a splendid fortune, laid out a considerable portion of it in the purchase of diseases, and finally spent all the rest for a heavy lot of repentance and a suit of rags. A very pretty girl bartered a heart as clear as crystal, and which seemed her most valuable possession, for another jewel of the same kind, but so worn and defaced as to be utterly worthless. In one shop there were a great many crowns of laurel and myrtle, which soldiers, authors, statesmen and various other people pressed eagerly to buy; some purchased these paltry wreaths with their lives, other by a toilsome servitude of years, and many sacrificed whatever was most valuable, yet finally slunk away without the crown. There was a sort of stock or scrip, called Conscience, which seemed to be in great demand, and would purchase almost anything. Indeed, few rich commodities were to be obtained without paying a heavy sum in this particular stock, and a man's business was seldom very lucrative unless he knew precisely when and how to throw his hoard of conscience into the market. Yet as this stock was the only thing of permanent value, whoever parted with it was sure to find himself a loser in the long run. Several of the speculations were of a questionable character. Occasionally a member of Congress recruited his pocket by the sale of his constituents; and I was assured that public officers have often sold their country at very moderate prices. Thousands sold their happiness for a whim. Gilded chains were in great demand, and purchased with almost any sacrifice. In truth, those who desired, according to the old adage, to sell anything valuable for a song, might find customers all over the Fair; and there were innumerable messes of pottage, piping hot, for such as chose to buy them with their birthrights. A few articles, however, could not be found genuine at Vanity Fair. If a customer wished to renew his stock of youth the dealers offered him a set of false teeth and an auburn wig; if he demanded peace of mind, they recommended opium or a brandy bottle.

Tracts of land and golden mansions, situated in the Celestial City, were often exchanged, at very disadvantageous rates, for a few years' lease of small, dismal, inconvenient tenements in Vanity Fair. Prince Beelzebub himself took great interest in this sort of traffic, and sometimes condescended to meddle with smaller matters. I once had the pleasure to see him bargaining with a miser for his soul, which, after much ingenious skirmishing on both sides, his highness succeeded in obtaining at about the value of six-pence. The prince remarked with a smile, that he was a loser by the transaction.

Day after day, as I walked the streets of Vanity, my manners and deportment became more and more like those of the inhabitants. The place began to seem like home; the idea of pursuing my travels to the Celestial City was almost obliterated from my mind. I was reminded of it, however, by the sight of the same pair of simple pilgrims at whom we had laughed so heartily when Apollyon puffed smoke and steam into their faces at the commencement of our journey. There they stood amidst the densest bustle of Vanity; the dealers offering them their purple and fine linen and jewels, the men of wit and humor gibing at them, a pair of buxom ladies ogling them askance, while the benevolent Mr. Smooth-it-away whispered some of his wisdom at their elbows, and pointed to a newly-erected temple; but there were these worthy simpletons, making the scene look wild and monstrous, merely by their sturdy repudiation of all part in its business or pleasures.

One of them—his name was Stick-to-the-right—perceived in my face, I suppose, a species of sympathy and almost admiration, which, to my own great surprise, I could not help feeling for this pragmatic couple. It prompted him to address me.

"Sir," inquired he, with a sad, yet mild and kindly voice, "do you call yourself a pilgrim?"

"Yes," I replied, "my right to that appellation is indubitable. I am merely a sojourner here in Vanity Fair, being bound to the Celestial City by the new railroad."

"Alas, friend," rejoined Mr. Stick-to-the-right, "I do assure you, and beseech you to receive the truth of my words, that that whole concern is a bubble. You may travel on it all your lifetime, were you to live thousands of years, and yet never get beyond the limits of Vanity Fair. Yea, though you should deem yourself entering the gates of the blessed city, it will be nothing but a miserable delusion."

"The Lord of the Celestial City," began the other pilgrim, whose name was Mr. Foot-it-to-heaven, "has refused, and will ever refuse, to grant an act of incorporation for this railroad; and unless that be obtained, no passenger can ever hope to enter his dominions. Wherefore every man who buys a ticket must lay his account with losing the purchase money, which is the value of his own soul."

"Poh, nonsense!" said Mr. Smooth-it-away, taking my arm and leading me off, "these fellows ought to be indicted for a libel. If the law stood as it once did in Vanity Fair we should see them grinning through the iron bars of the prison window."

This incident made a considerable impression on my mind, and contributed with other circumstances to indispose me to a permanent residence in the city of Vanity; although, of course, I was not simple enough to give up my original plan of gliding along easily and commodiously by railroad. Still, I grew anxious to be gone. There was one strange thing that troubled me. Amid the occupations or amusements of the Fair, nothing was more common than for a person—whether at feast, theatre, or church, or trafficking for wealth and honors, or whatever he might be doing, and however unseasonable the interruption—suddenly to vanish like a soap bubble, and be never more seen of his fellows; and so accustomed were the latter to such little accidents that they went on with their business as quietly as if nothing had happened. But it was otherwise with me.

Finally, after a pretty long residence at the Fair, I resumed my journey towards the Celestial City, still with Mr. Smooth-it-away at my side. At a short distance beyond the suburbs of Vanity we passed the ancient silver mine, of which Demas was the first discoverer, and which is now wrought to great advantage, supplying nearly all the coined currency of the world. A little further onward was the spot where Lot's wife had stood forever under the semblance of a pillar of salt. Curious travellers have long since carried it away piecemeal. Had all regrets been punished as rigorously as this poor dame's were, my yearning for the relinquished delights of Vanity Fair might have produced a similar change in my own corporeal substance, and left me a warning to future pilgrims.

The next remarkable object was a large edifice, constructed of mossgrown stone, but in a modern and airy style of architecture. The engine came to a pause in its vicinity, with the usual tremendous shriek.

"This was formerly the castle of the redoubted giant Despair," observed Mr. Smooth-it-away; "but since his death Mr. Flimsy-faith has repaired it, and keeps an excellent house of entertainment here. It is one of our stopping-places."

"It seems but slightly put together," remarked I, looking at the frail yet ponderous walls. "I do not envy Mr. Flimsy-faith his habitation. Some day it will thunder down upon the heads of the occupants."

"We shall escape at all events," said Mr. Smooth-it-away, "for Apollyon is putting on the steam again."

The road now plunged into a gorge of the Delectable Mountains, and traversed the field where in former ages the blind men wandered and stumbled among the tombs. One of these ancient tombstones had been thrust across the track by some malicious person, and gave the train of cars a terrible jolt. Far up the rugged side of a mountain I perceived a rusty iron door, half overgrown with bushes and creeping plants, but with smoke issuing from its crevices.

"Is that," inquired I, "the very door in the hill-side which the shepherds assured Christian was a by-way to hell?"

"That was a joke on the part of the shepherds," said Mr. Smooth-it-away, with a smile. "It is neither more nor less than the door of a cavern which they use as a smoke-house for the preparation of mutton hams."

My recollections of the journey are now, for a little space, dim and confused, inasmuch as a singular drowsiness here overcame me, owing to the fact that we were passing over the enchanted ground, the air of which encourages a disposition to sleep. I awoke, however, as soon as we crossed the borders of the pleasant land of Beulah. All the passengers were rubbing their eyes, comparing watches, and congratulating one another on the prospect of arriving so seasonably at the journey's end. The sweet breezes of this happy clime came refreshingly to our nostrils; we beheld the glimmering gush of silver fountains, overhung by trees of beautiful foliage and delicious fruit, which were propagated by grafts from the celestial gardens. Once, as we dashed onward like a hurricane, there was a flutter of wings and the bright appearance of an angel in the air, speeding forth on some heavenly mission. The engine now announced the close vicinity of the final station-house by one last and horrible scream, in which there seemed to be distinguishable every kind of wailing and woe, and bitter fierceness of wrath, all mixed up with the wild laughter of a devil or a madman. Throughout our journey, at every stopping-place, Apollyon had exercised his ingenuity in screwing the most abominable sounds out of the whistle of the steam-engine; but in this closing effort he outdid himself and created an infernal uproar, which, besides disturbing the peaceful inhabitants of Beulah, must have sent its discord even through the celestial gates.

While the horrid clamor was still ringing in our ears we heard an exulting strain, as if a thousand instruments of music, with height and depth and sweetness in their tones, at once tender and triumphant, were struck in unison, to greet the approach of some illustrious hero, who had fought the good fight and won a glorious victory, and was come to lay aside his battered arms forever. Looking to ascertain what might be the occasion of this glad harmony, I perceived, on alighting from the cars, that a multitude of shining ones had assembled on the other side of the river, to welcome two poor pilgrims, who were just emerging from its depths. They were the same whom Apollyon and ourselves had persecuted with taunts, and gibes, and scalding steam, at the commencement of our journey--the same whose unworldly aspect and impressive words had stirred my conscience amid the wild revelers of Vanity Fair.

"How amazingly well those men have got on," cried I to Mr. Smooth-it-away. "I wish we were secure of as good a reception."

"Never fear, never fear!" answered my friend. "Come, make haste; the ferry boat will be off directly, and in three minutes you will be on the other side of the river. No doubt you will find coaches to carry you up to the city gates."

A steam ferry boat, the last improvement on this important route, lay at the river side, puffing, snorting, and emitting all those other disagreeable utterances which betoken the departure to be immediate. I hurried on board with the rest of the passengers, most of whom were in great perturbation; some bawling out for their baggage; some tearing their hair and exclaiming that the boat would explode or sink; some already pale with the heaving of the stream; some gazing affrighted at the ugly aspect of the steersman; and some still dizzy with the slumberous influences of the Enchanted Ground. Looking back to the shore, I was amazed to discern Mr. Smooth-it-away waving his hand in token of farewell.

"Oh, no!" answered he with a queer smile, and that same disagreeable contortion of visage which I had remarked in the inhabitants of the Dark Valley. "Oh, no! I have come thus far only for the sake of your pleasant company. Good-by! We shall meet again."

And then did my excellent friend Mr. Smooth-it-away laugh outright, in the midst of which cachinnation a smoke-wreath issued from his mouth and nostrils, while a twinkle of lurid flame darted out of either eye, proving indubitably that his heart was all of a red blaze. The impudent fiend! To deny the existence of Tophet, when he felt its fiery tortures raging within his breast. I rushed to the side of the boat, intending to fling myself on shore; but the wheels, as they began their revolutions, threw a dash of spray over me so cold—so deadly cold, with the chill that will never leave those waters until Death be drowned in his own river--that with a shiver and a heartquake I awoke. Thank Heaven it was a Dream!

Annotations

1. In order to appreciate fully Hawthorne's "The Celestial Railroad" the reader is urged to review the principal scenes and characters in John Bunyan's *The Pilgrim's Progress* (1678).
2. Tophet (Jer. 19:6) is a place where heinous things take place.

2.9

John Henry

(c. 1870)

ANONYMOUS

When John Henry was a little baby,
Sittin' on his daddy's knee,
He grabbed himself a hammer and a piece of steel,
Said, "This hammer'll be the death of me, Lawd, Lawd,
This hammer'll be the death of me."

Now the captain said to John Henry,
"Gonna bring that steam drill round,
Gonna take that steam drill out on the job,
Gonna whop that steel on down, Lawd, Lawd,
Gonna whop that steel on down."

John Henry told his captain,
"A man ain't nuthin' but a man,
But before that steam drill beat me down,
I'll die with my hammer in my hand, Lawd, Lawd,
I'll die with my hammer in my hand."

John Henry said to his shaker,
"Shaker, why don't you sing?
'Cause I'm throwin' twelve pounds from my hips on down,
Just listen to that cold steel ring, Lawd, Lawd,
Just listen to that cold steel ring."

The man that invented the steam drill,
He thought he was mighty fine,
But John Henry he made sixteen feet,
And the steam drill only made nine, Lawd, Lawd,
And the steam drill only made nine.

John Henry hammered on the mountain
Till his hammer was strikin' fire.
He drove so hard he broke his pore heart,
Then he laid down his hammer and died, Lawd, Lawd,
He laid down his hammer and died.

They took John Henry to the graveyard,
And they buried him in the sand,
And every locomotive comes roarin' by,
Says, "There lies a steel-drivin' man, Lawd, Lawd,
There lies a steel-drivin' man."

2.10

The Bridge Builder

(1898)

WILL ALLEN DROMGOOLE [1]

An old man, going a lone
highway,
Came at the evening, cold and
gray,
To a chasm, vast and deep and
wide,
Through which was flowing a
sullen tide.
The old man crossed in the
twilight dim;
That sullen stream had no fears
for him;
But he turned, when he reached
the other side,
And built a bridge to span the
tide.
"Old man," said a fellow pilgrim
near,
"You are wasting strength in
building here;
Your journey will end with the
ending day;
You never again must pass this
way;
You have crossed the chasm,
deep and wide—
Why build you the bridge at the
eventide?"

The builder lifted his old gray
  head:
"Good friend, in the path I have
  come," he said
"There followeth after me today
A youth, whose feet must pass this
  way.
This chasm, that has been naught
  to me,
To that fair-haired youth may a
  pitfall be.
He, too, must cross in the twilight
  dim;
Good friend, I am building the
  bridge for him."

Annotations

1. Miss Will Allen Dromgoole was for many years the literary editor of the *Nashville Daily Banner*. "The Bridge Builder" was published in the volume *Rare Old Chums*.

2.11

Alexander's Bridge

(1912)

WILLA S. CATHER

On Tuesday afternoon a Boston lawyer, who had been trying a case in Vermont, was standing on the siding at White River Junction when the Canadian Express pulled by on its northward journey. As the day-coaches at the rear end of the long train swept by him, the lawyer noticed at one of the windows a man's head, with thick rumpled hair. "Curious," he thought; "that looked like Alexander, but what would he be doing back there in the day-coaches?"

It was, indeed, Alexander.

That morning a telegram from Moorlock had reached him, telling him that there was serious trouble with the bridge and that he was needed there at once, so he had caught the first train out of New York. He had taken a seat in a day-coach to avoid the risk of meeting any one he knew, and because he did not wish to be comfortable. When the telegram arrived, Alexander was at his rooms on Tenth Street, packing his bag to go to Boston. On Monday night he had written a long letter to his wife, but when morning came he was afraid to send it, and the letter was still in his pocket. Winifred was not a woman who could bear disappointment. She demanded a great deal of herself and of the people she loved; and she never failed herself. If he told her now, he knew, it would be irretrievable. There would be no going back. He would lose the thing he valued most in the world; he would be destroying himself and his own happiness. There would be nothing for him afterward. He seemed to see himself dragging out a restless existence on the Continent—Cannes, Hyères, Algiers, Cairo—among smartly dressed, disabled men of every nationality; forever going on journeys that led nowhere; hurrying to catch trains that he might just as well miss; getting up in the morning with a great bustle and splashing of water, to begin a day that had no purpose and no meaning; dining late to shorten the night, sleeping late to shorten the day.

And for what? For a mere folly, a masquerade, a little thing that he could not let go. *And he could even let it go,* he told himself. But he had promised to be in London at midsummer, and he knew that he would go . . . . It was impossible to live like this any longer.

And this, then, was to be the disaster that his old professor had foreseen for him: the crack in the wall, the crash, the cloud of dust. And he could not understand how it had come about. He felt that he himself was unchanged, that he was still there, the same man he had been five years ago, and that he was sitting stupidly by and letting some resolute offshoot of himself spoil his life for him. This new force was not he, it was but a part of him. He would not even admit that it was stronger than he; but it was more active. It was by its energy that this new feeling got the better of him. His wife was the woman who had made his life, gratified his pride, given direction to his tastes and habits. The life they led together seemed to him beautiful. Winifred still was, as she had always been, Romance for him, and whenever he was deeply stirred he turned to her. When the grandeur and beauty of the world challenged him—as it challenges even the most self-absorbed people—he always answered with her name. That was his reply to the question put by the mountains and the stars; to all the spiritual aspects of life. In his feeling for his wife there was all the tenderness, all the pride, all the devotion of which he was capable. There was everything but energy; the energy of youth which must register itself and cut its name before it passes. This new feeling was so fresh, so unsatisfied and light of foot. It ran and was not wearied, anticipated him everywhere. It put a girdle round the earth while he was going from New York to Moorlock. At this moment, it was tingling through him, exultant, and live as quicksilver, whispering, "In July you will be in England."

Already he dreaded the long empty days at sea, the monotonous Irish coast, the sluggish passage up the Mersey, the flash of the boat train through the summer country. He closed his eyes and gave himself up to the feeling of rapid motion and to swift, terrifying thoughts. He was sitting so, his face shaded by his hand, when the Boston lawyer saw him from the siding at White River Junction.

When at last Alexander roused himself, the afternoon had waned to sunset. The train was passing through a gray country and the sky overhead was flushed with a wide flood of clear color. There was a rose-colored light over the gray rocks and hills and meadows. Off to the left, under the approach of a weather-stained wooden bridge, a group of boys were sitting around a little fire. The smell of the wood smoke blew in at the window. Except for an old farmer, jogging along the highroad in his box-wagon, there was not another living creature to be seen. Alexander looked back wistfully at the boys, camped on the edge of a little marsh, crouching under their shelter and looking gravely at their fire. They took his mind back a long way, to a campfire on a sandbar in a Western river, and he wished he could go back and sit down with them. He could remember exactly how the world had looked then.

It was quite dark and Alexander was still thinking of the boys, when it occurred to him that the train must be nearing Allway (1). In going to his new bridge at Moorlock he had always to pass through Allway. The train stopped at Allway Mills, then wound two miles up the river, and then the hollow sound under his feet told Bartley that he was on his first bridge again. The bridge seemed longer than it had ever seemed before, and he was glad when he felt the beat of the wheels on the solid roadbed again. He did not like coming and going across that bridge, or remembering the man who built it. And was he, indeed, the same man who used to walk that bridge at night, promising such things to himself and to the stars? And yet, he could remember it all so well: the quiet hills sleeping in the moonlight, the slender skeleton of the bridge reaching out into the river, and up yonder, alone on the hill, the big white house; upstairs in Winifred's window, the light that told him she was still awake and still thinking of him. And after the light went out he walked alone, taking the heavens into his confidence, unable to tear himself away from the white magic of the night, unwilling to sleep because longing was so sweet to him, and because, for the first time since first the hills were hung with moonlight, there was a lover in the world. And always there was the sound of rushing water underneath, the sound which, more than anything else, meant death; the wearing away of things under the impact of physical forces which men could direct but never circumvent or diminish. Then, in the exaltation of love, more than ever it seemed to him to mean death, the only other thing as strong as love. Under the moon, under the cold, splendid stars, there were only those two things awake and sleepless; death and love, the rushing river and his burning heart.

Alexander sat up and looked about him. The train was tearing on through the darkness. All his companions in the day-coach were either dozing or sleeping heavily, and the murky lamps were turned low. How came he here among all these dirty people? Why was he going to London? What did it mean—what was the answer? How could this happen to a man who had lived through that magical spring and summer, and who had felt that the stars themselves were but flaming particles in the far-away infinitudes of his love?

What had he done to lose it? How could he endure the baseness of life without it? And with every revolution of the wheels beneath him, the unquiet quicksilver in his breast told him that at midsummer he would be in London. He remembered his last night there: the red foggy darkness, the hungry crowds before the theatres, the hand-organs, the feverish rhythm of the blurred, crowded streets, and the feeling of letting himself go with the crowd. He shuddered and looked about him at the poor unconscious companions of his journey, unkempt and travel-stained, now doubled in unlovely attitudes, who had come to stand to him for the ugliness he had brought into the world.

And those boys back there, beginning it all just as he had begun it; he wished he could promise them better luck, if one could assure a single human being of happiness! He had thought he could do so, once; and it was thinking of that that at last he fell asleep. In his sleep, as if it had nothing fresher to work upon, his mind went back and tortured itself with something years and years away, an old, long-forgotten sorrow of his childhood.

When Alexander awoke in the morning, the sun was just rising through pale golden ripples of cloud, and the fresh yellow light was vibrating through the pine woods. The white birches, with their little unfolding leaves, gleamed in the lowlands, and the marsh meadows were already coming to life with their first green, a thin, bright color which had run over them like fire. As the train rushed along the trestles, thousands of wild birds rose screaming into the light. The sky was already a pale blue and of the clearness of crystal. Bartley caught up his bag and hurried through the Pullman coaches until he found the conductor. There was a stateroom unoccupied, and he took it and set about changing his clothes. Last night he would not have believed that anything could be so pleasant as the cold water he dashed over his head and shoulders and the freshness of clean linen on his body.

After he had dressed, Alexander sat down at the window and drew into his lungs deep breaths of the pine-scented air. He had awakened with all his old sense of power. He could not believe that things were as bad with him as they had seemed last night, that there was no way to set them entirely right. Even if he went to London at midsummer, what would that mean except that he was a fool? And he had been a fool before. That was not the reality of his life. Yet he knew that he would go to London.

Half an hour later the train stopped at Moorlock. Alexander sprang to the platform and hurried up the siding, waving to Phillip Horton, one of his assistants, who was anxiously looking up at the windows of the coaches. Bartley took his arm and they went together into the station buffet.

"I'll have my coffee first, Philip. Have you had yours? And now, what seems to be the matter up here?"

The young man, in a hurried, nervous way, began his explanation.

But Alexander cut him short. "When did you stop work?" he asked sharply.

The young engineer looked confused. "I have n't stopped work yet, Mr. Alexander. I did n't feel that I could go so far without definite authorization from you."

"Then why did n't you say in your telegram exactly what you thought, and ask for your authorization? You'd have got it quick enough."

"Well, really, Mr. Alexander, I could n't be absolutely sure, you know, and I did n't like to take the responsibility of making it public."

Alexander pushed back his chair and rose. "Anything I do can be made public, Phil. You say that you believe the lower chords are showing strain, and that even the workmen have been talking about it, and yet you've gone on adding weight."

"I'm sorry, Mr. Alexander, but I had counted on your getting here yesterday. My first telegram missed you somehow. I sent one Sunday evening, to the same address, but it was returned to me."

Have you a carriage out there? I must stop to send a wire."

Alexander went up to the telegraph-desk and penciled the following message to his wife:—

I may have to be here for some time. Can you come up at once? Urgent!

BARTLEY.

The Moorlock Bridge lay three miles above the town. When they were seated in the carriage, Alexander began to question his assistant further. If it were true that the compression members showed strain, with the bridge only two thirds done, then there was nothing to do but pull the whole structure down and begin over again. Horton kept repeating that he was sure there could be nothing wrong with the estimates.

Alexander grew impatient. "That's all true, Phil, but we never were justified in assuming that a scale that was perfectly safe for an ordinary bridge would work with anything of such length. It's all very well on paper, but it remains to be seen whether it can be done in practice. I should have thrown up the job when they crowded me. It's all nonsense to try to do what other engineers are doing when you know they're not sound."

"But just now, when there is such competition," the younger man demurred. "And certainly that's the new line of development."

Alexander shrugged his shoulders and made no reply.

When they reached the bridge works, Alexander began his examination immediately. An hour later he sent for the superintendent. "I think you had better stop work out there at once, Dan. I should say that the lower chord here might buckle at any moment. I told the Commission that we were using higher unit stresses than any practice has established, and we've put the dead load at a low estimate. Theoretically it worked out well enough, but it had never actually been tried." Alexander put on his overcoat and took the superintendent by the arm. Don't look so chopfallen, Dan. It's a jolt, but we've got to face it. It is n't the end of the world, you know. Now we'll go out and call the men off quietly. They're already nervous, Horton tells me, and there's no use alarming them. I'll go with you, and we'll send the end riveters in first."

Alexander and the superintendent picked their way out slowly over the long span. They went deliberately, stopping to see what each gang was doing, as if they were on an ordinary round of inspection. When they reached the end of the river span, Alexander nodded to the superintendent, who quietly gave an order to the foreman. The men in the end gang picked up their tools and, glancing curiously at each other, started back across the bridge toward the river-bank. Alexander himself remained standing where they had been working, looking about him. It was hard to believe, as he looked back over it, that the whole great span was incurably disabled, was already as good as condemned, because something was out of line in the lower chord of the cantilever arm.

The end riveters had reached the bank and were dispersing among the tool-houses, and the second gang had picked up their tools and were starting toward the shore. Alexander, still standing at the end of the river span, saw the lower chord of the cantilever arm give a little, like an elbow bending. He shouted and ran after the second gang, but by this time everyone knew that the big river span was slowly settling. There was a burst of shouting that was immediately drowned by the scream and cracking of tearing iron, as all the tension work began to pull asunder. Once the chords began to buckle, there were thousands of tons of iron-work, all riveted together and lying in midair without support (2). It tore itself to pieces with roaring and grinding and noises that were like the shrieks of a steam whistle. There was no shock of any kind; the bridge had no impetus except from its own weight. It lurched neither to right nor left, but sank almost in a vertical line, snapping and breaking and tearing as it went, because no integral part could bear for an instant the enormous strain loosed upon it. Some of the men jumped and some ran, trying to make the shore.

At the first shriek of the tearing iron, Alexander jumped from the downstream side of the bridge. He struck the water without injury and disappeared. He was under the river a long time and had great difficulty in holding his breath. When it seemed impossible, and his chest was about to heave, he thought he heard his wife telling him that he could hold out a little longer. An instant later his face cleared the water. For a moment, in the depths of the river, he had realized what it would mean to die a hypocrite, and to lie dead under the last abandonment of her tenderness. But once in the light and air, he knew he should live to tell her and to recover all he had lost. Now, at last, he felt sure of himself. He was not startled. It seemed to him that he had been through something of this sort before. There was nothing horrible about it. This, too, was life, and life was activity, just as it was in Boston or in London. He was himself, and there was something to be done; everything seemed perfectly natural. Alexander was a strong swimmer, but he had gone scarcely a dozen strokes when the bridge itself, which had been settling faster and faster, crashed into the water behind him. Immediately the river was full of drowning men. A gang of French Canadians fell almost on top of him. He thought he had cleared them, when they began coming up all around him, clutching at him and at each other. Some of them could swim, but they were either hurt or crazed with fright. Alexander tried to beat them off, but there were too many of them. One caught him about the neck, another gripped him about the middle, and they went down together. When he sank, his wife seemed to be there in the water beside him, telling him to keep his head, that if he could hold out the men would drown and release him. There was something he wanted to tell his wife, but he could not think clearly for the roaring in his ears. Suddenly he remembered what it was. He caught his breath, and then she let him go (3).

The work of recovering the dead went on all day and all the following night. By the next morning forty-eight bodies had been taken out of the river, but there were still twenty missing. Many of the men had fallen with the bridge and were held down under the débris. Early on the morning of the second day a closed carriage was driven slowly along the river-bank and stopped a little below the works, where the river boiled and churned about the great iron carcass which lay in a straight line two thirds across it. The carriage stood there hour after hour, and word soon spread among the crowds on the shore that its occupant was the wife of the Chief Engineer; his body had not yet been found. The widows of the lost workmen, moving up and down the bank with shawls over their heads, some of them carrying babies, looked at the rusty hired hack many times that morning. They drew near it and walked about it, but none of them ventured to peer within. Even half-indifferent sight-seers dropped their voices as they told a newcomer: "You see that carriage over there? That's Mrs. Alexander. They have n't found him yet. She got off the train this morning. Horton met her. She heard it in Boston yesterday—heard the newsboys crying it in the street."

At noon Philip Horton made his way through the crowd with a tray and a tin coffee-pot from the camp kitchen. When he reached the carriage he found Mrs. Alexander just as he had left her in the early morning, leaning forward a little, with her hand on the lowered window, looking at the river. Hour after hour she had been watching the water, the lonely, useless stone towers, and the convulsed mass of iron wreckage over which the angry river continually spat up its yellow foam.

"Those poor women out there, do they blame him very much?" she asked, as she handed the coffee-cup back to Horton.

"Nobody blames him, Mrs. Alexander. If any one is to blame, I'm afraid it's I. I should have stopped work before he came. He said so as soon as I met him. I tried to get him here a day earlier, but my telegram missed him, somehow. He did n't have time really to explain to me. If he'd got here Monday, he'd have had all the men off at once. But, you see, Mrs. Alexander, such a thing never happened before. According to all human calculations, it simply could n't happen."

Horton leaned wearily against the front wheel of the cab. He had not had his clothes off for thirty hours, and the stimulus of violent excitement was beginning to wear off.

"Don't be afraid to tell me the worst, Mr. Horton, Don't leave me to the dread of finding out things that people may be saying. If he is blamed, if he needs any one to speak for him,"—for the first time her voice broke and a flush of life, tearful, painful, and confused, swept over her rigid pallor,—"if he needs any one, tell me, show me what to do." She began to sob, and Horton hurried away.

When he came back at four o'clock in the afternoon he was carrying his hat in his hand, and Winifred knew as soon as she saw him that they had found Bartley. She opened the carriage door before he reached her and stepped to the ground.

Horton put out his hand as if to hold her back and spoke pleadingly: "Won't you drive up to my house, Mrs. Alexander? They will take him up there."

"Take me to him now, please. I shall not make any trouble."

The group of men down under the river-bank fell back when they saw a woman coming, and one of them threw a tarpaulin over the stretcher. They took off their hats and caps as Winifred approached, and although she had pulled her veil down over her face they did not look up at her. She was taller than Horton, and some of the men thought she was the tallest women they had ever seen. "As tall as himself," some one whispered. Horton motioned to the men, and six of them lifted the stretcher and began to carry it up the embankment. Winifred followed them the half-mile to Horton's house. She walked quietly, without once breaking or stumbling. When the bearers put the stretcher down in Horton's spare bedroom, she thanked them and gave her hand to each in turn. The men went out of the house and through the yard with their caps in their hands. They were too much confused to say anything as they went down the hill.

Horton himself was almost as deeply perplexed. "Mamie," he said to his wife, when he came out of the spare room half an hour later, "will you take Mrs. Alexander the things she needs? She is going to do everything herself. Just stay about where you can hear her and go in if she wants you."

Everything happened as Alexander had foreseen in that moment of prescience under the river. With her own hands she washed him clean of every mark of disaster. All night he was alone with her in the still house, his great head lying deep in the pillow. In the pocket of his coat Winifred found the letter that he had written her the night before he left New York, water-soaked and illegible, but because of its length, she knew it had been meant for her.

For Alexander death was an easy creditor. Fortune, which had smiled upon him consistently all his life, did not desert him in the end. His harshest critics did not doubt that, had he lived, he would have retrieved himself. Even Lucius Wilson did not see in this accident the disaster he had once foretold.

When a great man dies in his prime there is no surgeon who can say whether he did well; whether or not the future was his, as it seemed to be. The mind that society had come to regard as a powerful and reliable machine, dedicated to its service, may for a long time have been sick within itself and bent upon its own destruction.

Annotations

1. Allway is the site of Alexander's first suspension bridge and also the place where he met his wife, Winifred.
2. Compare Vitale's description (section 2.12) of the actual fall of the Quebec bridge.

3. At the beginning of the novel his old professor says to Alexander: "I always used to feel that there was a weak spot where someday strain would tell. . . . The more dazzling the front you presented, the higher your facade rose, the more I expected to see a big crack zigzagging from top to bottom . . . then a crash and clouds of dust (Cather, p. 15).

2.12

The Great Quebec Bridge Disaster

(1967)

JOHN C. VITALE

Sixty years ago, during the era of rapid advances in bridge engineering that followed quickly upon the heels of the new 20th century, the greatest single feat of bridge construction in North America was centered in the building of the Quebec Bridge—a huge and magnificent cantilever structure spanning the St. Lawrence River at Quebec. Undoubtedly to the American engineer the building of the Quebec Bridge, whose original and daring design specified an 1800′ center span, the most massive single span yet attempted in the world, was a monumental step forward in engineering. With its great center span cutting the skyline with graceful curves of laced steel, the Quebec Bridge became at once the symbol of progressive engineering. It captured the imagination of the world, almost seeming in a sense to forecast the great technological wonders that were about to unfold upon the new century.

PRAISE TURNS TO CRITICISM

So close was the tie between the building of the Quebec Bridge and the field of civil engineering that *Scientific American* again and again reminded the public of the technological association. One article, appearing in June 1907, stated, "If we were to select out of the many fields of activity which are covered by modern civil engineering some particular one in which the American engineer has displayed most signally his originality and freedom from tradition, we would choose that of bridge engineering; and if we had been called upon to name some one particular structure which stood as the highest exemplification of his skill in the particular branch of his profession, we would have selected the great cantilever bridge across the St. Lawrence River at Quebec."

Yet, at the very moment the general public was awed by the engineering daring of the bridge builders, unexpectedly and with a suddenness that shocked the world, the Quebec Bridge collapsed. It carried 81 workmen to their death in the wake of an enormous tangle of twisted and broken steelwork. Consequently on August 29, 1907—the infamous date of the great disaster—the entire profession of civil engineering went on trial.

Spurred on by the great loss of lives, *Scientific American*, whose pages had once been filled with nothing but praise for the enormous project, now echoed the remarks of every newspaper in North America and questioned the theories of bridge building. "The tremendous significance of this disaster," it stated shortly after the collapse, "lies in the suspicion, which today is staring every engineer coldly in the face that there is something wrong with our theories of bridge building, at least as applied to a structure of the size of the Quebec Bridge."

Later, during the period of confusion which followed the disaster, while engineers across the country themselves paused momentarily to contemplate the significance of the collapsed bridge, an entire series of questions were raised. "Are we to conclude, then," *Scientific American* asked, "that those theories, those formulae, upon which we have been building our bridges so successfully during the past quarter of a century, are inapplicable when the structure exceeds a certain magnitude? Can it be that for some unsuspected reason a stress per square inch which is perfectly safe in the end-post of a 500′ railroad truss becomes perilous when used in the bottom chord of an 1800′ cantilever? As far as our engineering knowledge goes, there is no reason whatever this disparity should exist. But if not, why is the Quebec Bridge now lying at the bottom of the St. Lawrence River?"

To find the answers to these pressing questions, every detail of the giant structure was reviewed, studied, and then reviewed again. Within a few short weeks, the history of the Quebec Bridge became common knowledge, and discussed as any general topic of the period. Those interested soon discovered that the fallen structure, which was designed to cross the St. Lawrence River a few miles above Quebec, consisted basically of two deck spans, each 210′ long, reaching from the abutment to the anchor piers, and a huge cantilever construction, with a total length of 2800′ between the anchor piers. Each cantilever consisted of an anchor or shore arm 500′ long and a cantilever arm 562½′ long. Suspended between the ends of the cantilever arm was a 675′ truss span, the longest simple truss span ever attempted at the time.

Initially, as the fact finders found, the building of the steelwork for the bridge was begun in 1902, some five years before the collapse, when the 210′ deck trusses were erected. Work on the main cantilever construction began in July 1905, when steel falsework of a special design was erected beneath the south cantilever arm to carry the weight of that arm during its erection. This falsework alone weighed 1000 tons. Upon this, during the construction season, were erected six panels of the south anchor arm, weighing altogether 5346 tons.

During the season of 1906, the south anchor arm was completed, and the south cantilever arm erected. Up to the time of the sudden disaster, about one-third of the central suspended truss had been built out by overhanging beyond the south cantilever arm. When the bridge collapsed, the steelwork projected about 800′ out over the river from the main pier.

One of the main points of interest, both to professional civil engineers and the man in the street, centered around *stress*. Concerning this point, the Phoenix

Bridge Company, whose engineers kept records of the deflections and movements of the bridge under the changing loads as the work advanced, reported that at the time of the failure of the bridge the maximum compressive stresses were about 16,000 psi in the lower chord of the anchor arm near the tower pier, or only 45 percent of the maximum safe stress to which these chords might be theoretically subjected without risk of failure. Consequently, the Phoenix Bridge Company concluded the failure came totally without warning.

However, despite the first reports by the company, the disaster was *not* totally without warning. As later reports indicated, two days before the fatal accident it had been observed that the bottom chord of the anchor arm of the cantilever showed incipient signs of yielding; it was springing from 1½″–2″ out of line, the deflection being toward the inside of the truss. Instead of immediately suspending work on the structure once this had been observed, an engineer was dispatched to New York to see the consulting engineer, while at the same time a second engineer was sent to Phoenixville to the works of the bridge company. Ironically at about the very hour when instructions were wired from New York to suspend work, the bridge fell.

Speculation on exactly what had happened to the bridge eventually made several facts clear. One, supported by the testimony of eyewitnesses who survived the accident, was that the bridge did not fall over sideway. Instead, buckling under strain, it settled vertically upon itself, slowly at first and then with a rush as the entire length of the structure gave way. Another observation which supported the theory that the bridge did not fall by lateral distortion was that the lateral bracing between the top chords laid symmetrically in the water, though, of course, twisted and broken in places by the tremendous impact of its fall. Furthermore, had the bridge failed through the rupture of the tension members, that is, of the top chords or the diagonal ties, the parting of the metal would have been instantaneous and accompanied by a loud report of colossal magnitude. At the time, however, witnesses agreed that although there were subsequent thunderous crashes while the bridge was falling, the *commencement* of its settlement was not marked by any loud report.

When the final evidence was assembled and subsequently revealed, it was proven that the failure occurred in the lower chord of the anchor arm, in the second panel out from the foot of the tower. To ascertain this, the Royal Commission of Inquiry met for a period of five months of continuous investigation. Then, on March 9, 1908, a report was released. The document was one of the most voluminous and valuable accounts ever presented in the history of civil engineering. In part, the Commission found:

> The collapse of the Quebec Bridge resulted from the failure of the lower chords in the anchor arm near the main pier. The failure of these chords was due to their defective design.
>
> We do not consider that the specifications for the work were satisfactory or sufficient, the unit stresses in particular being higher than any established by past practices. The specifications were accepted without protest by all interested.
>
> A grave error was made in assuming the dead load of the calculations at too low a value and not afterward revising this assumption. This error was of sufficient magnitude to have required the condemnation of the bridge even if the details of the

> lower chords had been of sufficient strength because, if the bridge had been completed as designed, the actual stresses would have been considered greater than those permitted by the specifications.

Thus, by releasing this report to the public, the general theories of engineering then being practiced in bridge building were completely exonerated, and the entire blame laid at the feet of *human error.* Looking back from the '60s, however, the Quebec Bridge disaster stands today as a reminder to all who practice the profession of engineering—a grim and terrible reminder that human error can breed a disaster. More important, it can cast a dark shadow of suspicion on the very profession of engineering. Fortunately, history has a long memory; the Quebec Bridge disaster will not soon be forgotten.

2.13

The Bridge Builder 1938–1939

(1956)

A. DEN DOOLAARD

The taxi drove to the station through muddy streets. He didn't see the town; he saw Giulia and through her the phantom of a graceful silhouette—his bridge. But the farther the train pulled away from Florence the fainter her image grew and the clearer that of the bridge. At first Wolf's fellow passengers cast occasional questioning glances at the solitary man in the corner of the compartment, covering page after page of a large sketchbook with thin curved lines. Then they forgot him, for he kept apart from the newly formed small community of talkers, eaters and card players. He worked on steadily with the feverish feeling of someone on the track of something good. At last he stopped. He was thirsty, and he walked to the refreshment car in search of a bottle of beer.

At the table with its neat white cloth covering, he looked over his sketches. The first one, the one made at the Tara itself, and the last one, done only five minutes ago—they were no more alike than an elephant and a deer. He now understood why Radich had shaken his head over the first sketches. "It can be done better, Wolf. All this is sound in construction, but something's missing. I won't say more, it wouldn't be fair to your competitors."

He realized his mistake. The strength was there, but grace was lacking. At Christmas, when he had set out for the Tara in his haste to be the first, he had misunderstood the landscape to which the bridge must be suited. Winter with its ice and snow had made the world bleak. But there was summer as well, when under the high sun the Tara Valley was enchanting. The gorge itself remained majestic and stern, but above it the umbrella pines with their sinuous branches moved like dancers with outstretched arms and flowing hair. Something of this he had captured in his new design; what rose before him on high, slender pillars, with arches almost Gothic in their delicacy and supports thin as matchsticks, was a bridge both strong and graceful; a bridge—if such a thing were possible—such as a man and woman might build together.

He sat up, dazed at the thought. Was it possible that her beauty had penetrated his strength to such an extent that his drawing pencil now executed with joyful ease lines which he formerly wouldn't have dared attempt? Lines playful as the smile on her lips, arcs which seemed the embodiment of the curves he had traced with his mouth night after night? Or was it just his imagination because she was still in his blood, beating in his veins?

He broke suddenly into a sweat. Could he be sure that all this beauty and grace was mechanically sound? Forces of pressure and stress, of waves and wind and amount of friction—under their impact everything literally stood or fell. A bridge is not a woman kept as a luxury; it must be capable of carrying loads. He began to calculate with furious haste, in a desperate attempt to save the design with which he was now almost in love. Almost?

The whole time he was haunted by doubt. It was not the right method; to search for an æsthetic solution first and afterward test if it was mechanically sound was scientifically unjustifiable. But why? Were this paper bridge to tumble down, nothing would be lost but an illusion and he could start all over again in a new sketchbook.

"Pardon, signore, I must lay the table for supper . . . "

He looked up, saw he was alone with the waiter in the deserted car, furiously slammed his book shut and walked through the swaying corridors. He smoked a cigarette and calmed himself. Perhaps the man had arrived just in time. There was no point in losing oneself in a dream removed from reality. He'd better go back to the Tara first and look it over with new eyes, eyes which in Italy had seen that the useful can be beautiful as well—as beautiful as a flower, or as a woman.

"If you won't extend hospitality to me in this office where I've worked so long I'll regard it as an insult," growled Wolf.

Radich scratched his gray head. "It's against the rules, my dear fellow, to come and work out your plans in the office of one of the judges. It's true you have your preliminary reappointment here, but even that hasn't been officially confirmed yet."

"Let me work here after office hours, then," Wolf continued. "After three o'clock there's no one here anyway. I'll see that everything is cleared away by morning. I'll be finished in a week."

That might be possible. There's only one thing I have against night work. The night is instinct, the day reason. In the dark one tends to lose one's critical sense. Women take advantage of that," he concluded with a grin.

"You sound like an Italian," Wolf said woodenly.

"Only on that point! Individually I sometimes really like Italians, but in the mass I don't trust them. In politics they're like whores; they sell themselves to the highest bidder. After 1918 it was touch and go whether they'd grab the whole of Dalmatia, and they've still got an eye on it even though they've behaved for the past few years as though butter wouldn't melt in their mouths. The worst of it is that everybody forgives them everything because they happen to inhabit the world's largest museum."

"That proves how great is the influence of beauty."

"Which you won't have denied yourself over there, I trust. . . . Now don't put on that wooden look again." Radich shook his head. "Man, what can be more pleasurable than to drink a glass with a good friend and to pass in review a series of beds?"

"In that case I'm afraid you'll have to do all the talking."

"Don't be so modest! But really, I'm a little annoyed with you. I can see in your face that something happened and you won't tell me about it."

"What is it you see in my face, then?"

"Something you yourself wouldn't be aware of. In the old days when you wanted to say or ask something you always had to overcome some slight insecurity. That's gone now. I notice it perhaps because I take a fatherly interest in you. I can see you have had experiences which few of us escape at a certain age, and a good thing too. You've gone through heaven and hell, and now at last your feet are firmly planted on the ground. Do you know what was the matter with you? Belated puberty. Now that you've got over that once and for all you'll find you'll be able to work better too. Why do you think I maneuvered to get you sent to Italy? That was one of the reasons, my boy. Abroad, and especially in Italy, one does things one wouldn't dare do at home—except myself, of course. And now if you're determined to work here you'd better get down to it. You know where you sleep. Here's the key."

After Radich had left, Wolf fell to thinking. Life seemed a conspiracy, all the more mysterious since the conspirators were ignorant of one another's intentions. Radich had managed to get him sent to Italy. Tomassio had brought him to Giulia. Now he was going to build a bridge. Later—if it worked out, at any rate—people would say it was his bridge, the bridge of Wolf Crnojevich. But that would be only partly true. It would be Radich's sound knowledge, Tomassio's sharp criticism, Giulia's elegance—all would have gone into its building. Especially Giulia would have helped, despite her complete ignorance of technical matters. The bridge must span the ravine with the same easy grace as Giulia's when she reached across the table for her gloves. At the same time it must have the strength of his stern grandfather, Sava. It would be more than a sober engineering construction: it would be born out of Wolf's deepest experience, out of all that life had hitherto taught him.

He began to draw and worked till his eyelids were glued together. The next day he found an error. Radich was right; it was better to stop earlier; frantic work was lost effort. After eight long afternoons and evenings the topographical sketches were finished, the sections, the specifications. Something seemed wrong; in comparison with bridges of similar construction which he had studied in the textbooks, the total quantity of concrete appeared to be much too small. He checked his additions again. The large span over the gorge where the roadway would be suspended more than two hundred yards above the water; the four arches spanning the steep incline of the southern bank, based on pillars of decreasing height; and finally, the subsidiary bridge to the abutment—he could find no error, and yet the total volume seemed small for a bridge some four hundred yards in length.

When he held his sketches beside photographs of other bridges the difference became clear to him. Suddenly he remembered one of Tomassio's clever comparisons—his bridge differed from the others as a foil differs from a saber. It was all slenderness and grace. Weightless on tenuous arches, it seemed to float, not stand. It spanned the valley just as Giulia walked, effortlessly, swinging her legs from the hips.

Some capture love and a sense of beauty in a poem, others in a painting, or in a piece of music. He had tried to do it in a bridge, an object for use, the construction of which was governed by strict disciplines. But didn't that make the triumph of beauty over matter of even greater value? Wolf felt he had created something and now he was tired. However, a deep-rooted distrust of himself made him go through his calculations once more. He fell asleep stretched over the drawing table. In the middle of the night he wakened shivering. The fire had gone out. Outside the stars looked big in the sky. Suddenly resolved, he rolled his plans up and put them in a cardboard cylinder, addressed it according to the instructions and added a few lines to Radich:

"I'm going home for a few days, if you don't mind. So far I haven't been officially assigned to my new job, and besides, I have another ten days' vacation due me. I'd be on tenterhooks here anyway. I'll walk. The weather is settled, so don't worry. Thanks for everything. W."

An hour later, skis over his shoulder, he was slushing through the half-frozen mud puddles of dirty Plevlje. The moon was in its second quarter and he knew the way; by daybreak he could reach the Tara, the following evening Jezera, the third evening home.

Wolf sat at his desk writing a report on the urgent need for improving the road between the Tara and Niksic. The window made a square of sunlight, and on the sill the shadows of flowers danced. Shouting stridently, a black-veiled woman rode by on an donkey that picked its way so carefully over the pointed cobbles that it seemed afraid of breaking them. Over the rooftops a muezzin's prayer call fluttered like the screech of a bird.

Old, half-blind Zharko Djurich, who had squandered on *rakija* the savings from his years in America, returned from his first trip to the café. He stopped in front of Wolf's window, thumped with his thorn stick on the stones, and gazed at Wolf with red-rimmed eyes. Wolf looked up, greeted him briefly, and wrote on. The life of Plevlje, the untidy little frontier town between the Orthodox and Islamic worlds, hardly touched him. Work was his only reality, and that he tackled systematically. Before he began, he arranged his drawing instruments and other gear within reach, somewhat like a general getting his troops into battle formation. In front of him the box of compasses, the inkstands, pencil trays; on the left, the calendar, rolls of maps, reference books; to the right, scissors, ruler, set of squares, slide rule and magnifying glass.

"You are right," Radich had mocked only yesterday. "A systematically arranged table saves several seconds on each hand movement, and in the course of a month several hours, and in a lifetime several years. The only thing is, and that's what is important, how do you spend the years you've economized, apart from lamenting the inevitable arteriosclerosis?"

The old postman trudged past. Perhaps he'd bring another letter from Jelena. There was one still waiting to be answered. It was too soon for a letter from his mother; the road past Zabljak wasn't usually free of snow until the first of May.

"Wolf! Wolf! Wolf!" It was Radich's voice, growing louder so quickly that he must surely be chasing through the office at full gallop. The doors swung open, Radich stumbled over the threshold, crashed against the desk and sent compasses, pencils, rulers and squares flying in all directions. He scrambled to his feet, waved a letter and cried, "Numskull, blockhead, failure, jobbing engineer—you've won it!" He hoisted himself up on his toes, kissed Wolf on both cheeks, clapped him on the shoulder and held out his hand. "For a start, a hundred dinars for wine and *rakija!* No more work is going to be done in this office today!"

Wolf stood stiffly beside his chair. "I had a premonition of this," he said, getting bright red in the face.

"Why do you stand there like a statue?" Radich shouted in despair. "Why don't you throw inkpots up to the ceiling, jump over your desk, embrace me? Why don't you say, 'Dear Uncle Radich, I'm grateful you managed to secure the award for me by banging together the heads of the political schemers in Belgrade'? Why don't you? In the first place because you're a stiff mountain Montenegrin, a piece of stone from the Durmitor; secondly, you haven't the faintest notion of what's been going on behind the scenes. But I'll tell you all that tonight. First, we're going to celebrate! Selim! Selim! Where is that lazy man of mine? . . . There he is, at last. Selim, go to the tavern, fetch *rakija*, wine, onions, sausage and tell them to spit a suckling pig right away."

And so it wasn't until dark that a hiccupping Radich, his fleshy hands stroking his belly in deep contentment, told Wolf the story. "As soon as I saw your design, boy, I knew you deserved the prize. I also got quite a shock, for such a thing had never been seen in this country before—a bridge which didn't go straight but in a curve. . . . How did you come to do that, after the first design?"

"Which you rightly didn't think much of," Wolf confessed. "At Christmas, I overlooked one thing in my haste. The course of the large span was fixed once and for all. One pier had to be based on the peak of rock which nature had so helpfully placed on the left bank; the other end of the big arch had to get its support from the vertical rock wall. From there I wanted to carry on in a straight line, for traditionally such a line happens to be the shortest distance between two points. But on my second visit I immediately saw that the river curves right from the spot where the large span would come, and with it the whole wall of the valley on the opposite bank. The other side swung away from the bridge, as it were. If I built in a straight line, the bridge would be longer than it had to be. I didn't want to give it a sharp twist in the middle; that would be, first, too awkward for traffic, and second, displeasing to the eye. So I made the four spans I had designed each in turn swing slightly to the right. That way I could keep my bridge well over thirty yards shorter. But once I started, in complete defiance of tradition, to build a bridge with a curve, I felt I would have to do it boldly, without attempting to disguise it; there is no real beauty without sincerity."

"Except in the case of women," grunted Radich.

"That's why I made the footpaths alongside the roadway protrude and designed the balustrade in a clear, pure curve."

Radich grinned. "Sound and simple as a recipe for roast suckling pig, but the construction involved mathematical problems which only someone like you could solve in such a short time. And do you know what else weighted the scales in your favor, apart from the great saving of materials, labor and wages achieved in your design? The fact that you had made the arches of your spans so slender that they could be built in turn, one after the other, because they didn't have to keep each other in balance. What gave you that brilliant idea?"

"Nothing brilliant about it." Wolf reddened again. "I must make a confession. I drew the arches like that out of my imagination first and didn't start to calculate until afterward. It didn't dawn on me until I was doing this how many practical advantages such a method of construction offered. In this region no contractor would be able to get a sufficient number of workers together to build the four arches simultaneously. There'll be quite a few Montenegrins among the workers and I know them. They're lazy by nature and they like variety. When they've worked a month they go home and they come back two weeks later—maybe. Whereas now each arch can be finished in turn and the mold for the concrete need be made only once, since it can be used four times in succession. It occurred to me that that would save a lot of money..."

"Do you know what the cost differences were between your design and that of your most dangerous rival? It's only a rough estimate so far, but I make it over thirty per cent."

"And who was the third competitor? Or aren't you permitted to tell?"

Radich chuckled, "You nearly broke your neck on him. Not because of his design—an elephant of a bridge, fat, lazy, clumsy and ugly." He leaned toward Wolf and hiccupped. "But he's the son-in-law of the present Minister of Works."

"Is that so?" Wolf said softly. "And, unseen, that made him number one."

"Exactly. But I have a few connections in the capital myself, and what's more, I happened to know something about that superior of ours which was not known to a lot of influential people who are waiting for an opportunity to oust him. Now don't look so upset, greenhorn. Better get used to the idea that in a system like ours, in spite of—or rather, because of all your merits—you'll find yourself under a steamroller unless you're able to defend yourself by blackmail. Here, take a drink to get over the shock and forget it. I'll see that the intriguers don't get near you."

"I knew Belgrade was a nest of vipers, but..."

"You don't know the half of it," Radich raged. "Today's departmental head sits in the porter's lodge tomorrow, but when the wind shifts again, he'll be secretary general." He heaved his bulky figure from his chair and looked about him. "No spies around? Anyway, tonight I don't give a damn. Let me tell you that..."

Wolf put his hand on Radich's shoulder. "Let's not wallow in the filth of monarcho-fascist politics tonight, Uncle Radich. Instead, tell me when we can start, and tell me—an old hand like you knows more about these things than I do—how many months you think the work will take."

Radich snorted "Months? We can start as soon as I get over tonight's hangover. But no one will walk across your bridge this year. If we finish the piers, the subsidiary bridge, and the foundations for the large span this year, I'll give a keg of *rakija*. Don't forget, either, how long it will take before we have detailed blueprints at our disposal. You'll get help with those, of course. I've seen to that already. But even then we'll have nothing but paper..."

"And there are the hutments—" Wolf counted on his fingers—"the concrete, the quarry, tipcarts for getting the stone to the crusher, transport for the sand, concrete mixers, firs for the scaffolding, carpentry for the molds and casts, sawing machines, moving steel for the reinforced concrete, and—"

"And liquor glasses for the canteen girls, and the pretty nurse who will bandage the fellows who hammer their own thumbs or each other's noses," Radich jeered. "You can see you alone haven't hands enough. Better leave all that to me and the contractor whose bill comes out of the hat when the work is tendered. Largely politics again, so leave that to me too. There'll be graft in any case, but at least I'll see to it that the job isn't messed up. What are you going to do now—why are you getting up?"

"There's so much to be done," murmered Wolf. "I'm going back to my drawing table."

"Tonight you're going to be a normal human being. For once you're going to get good and tight." Radich admonished Wolf with an uncertain finger. "Remember I've got you in my power. If you don't do as I say I'll saddle you with a contractor who'll build your bridge upside down, piers in the air!"

Radich was right. Tomorrow the dream would be over—the fever of creation, the excitement about new conceptions. Tomorrow reality began. The dust clouds of blasting, the dizzy climbing exploits along high scaffoldings. The fear about what would happen to the thin arches once their casings were dismantled and they were suspended in the sky. The never-abating anxiety about the quality of the concrete. All that was for tomorrow; today not the bridge but the bottle stood at the center of the swaying world. Things swirled before Wolf's eyes and he drank and Radich poured him another glass and he drank again and ate fresh sheep's cheese, the greasy crumbs staining his coat lapels without his noticing it, though normally he brushed off every speck of dust. He began to sing, and even the musical Radich didn't care how much out of tune he was, for Radich, too, was drunk; and they started to quarrel about the moon—whether it was still waxing or already waning—and made up over another drink. Until a policeman, who had listened patiently to their bragging and granted them right in all things, took them to Radich's house, where Wolf stumbled on the doorstep and in the early morning wakened on the front porch, chilled and clammy, with stonecold feet and a pain in his neck. He lay with his head on his shoes, into which he had, from habit, pushed his socks.

"Now everything will be all right," Radich said as they looked into each other's bleary eyes later that morning over a third cup of steaming Turkish coffee. "In Italy you went to bed with the signorinas all by yourself, and last night in my care you drank yourself into a stupor. Now finally I can be sure that I have a full-grown human being as a first assistant. Shall we start work now? Let's begin at the beginning. We'll have to work out carefully where the piers are to be set, the abutment on the southern bank and the niche which will have to be blasted out of the rock on the northern wall to brace the arch of the large span—"

"Since you'll be the big chief and I only the Jack of all trades I want to ask you about that," interrupted Wolf. "I'll have more to do with the workers than you. On the few big projects I've worked on, including the mountain road in Tuscany, I've learned something. You can get the best workers to do absolutely anything as long as they respect you. My way of winning their respect is to share both their efforts and their hazards, not at all times, but at critical moments. So it would be a good thing for me to take an active part in making the niche. To excavate it, holes will have to be drilled for the blasting charges from a platform overhanging a gorge about two hundred yards deep. Once the workers see that the engineer with whom they're in daily contact doesn't think himself above perching on a joist dangling on a couple of steel wires suspended over an abyss—"

"Stop," pleaded Radich. "You make me dizzy and my eyes aren't too steady today anyway. I think it's the wrong method; an engineer should never have to dirty his hands on a job. But I know you. If I forbid you to risk your neck I'm afraid you'll rebel and I don't want that either. So you can go ahead, but only as an exception and in this particular case. You're getting a first rate driller to whom you can easily leave things like that. An expert, but an adventurer; you'll often find that with fellows of his type. A friend from Belgrade recommended him; for very good reasons he walked out of a copper mine and is now waiting for other work. Between ourselves, he fought in Spain too, so you can guess his political leanings and you'd better forget them right away. As long as he works well I don't care if he worships the devil. I've made a note of his name—Ante, he's called, surname Djulich or something like that. Now let's have another look at the maps."

Six weeks later Wolf stood on the spot where the previous Christmas he had pushed through thick brushwood to the edge of the rock wall, to throw a stone at the peak on the other side of the gorge. Now the shrubs had been cut and from the bare rock there rose the first tier of a steel construction tower. The piers climbed vividly into the blue sky, their red-lead coating barely dry. From two protruding beams hung the lift chair which would presently be lowered down the perpendicular wall with the aid of a winch.

From his vantage point, Wolf overlooked the whole project. He felt proud. The rock wall reflected the bright glitter of metal and stone. On the peak three men were hewing at the weathered layers to find a firm base for the pier which was to carry the other end of the large arch. At four places along the sloping incline of the valley pickaxes rose and fell, laying bare the foundations for the other piers. From higher up fierce hammering sounded. The third hutment was being knocked together; the canteen had already been finished and its chimney was smoking. The contractors' shed was ready and the rafters of the engineers' and foremen's shack were just being put into position. Wolf's plan had ended the green peace of the valley; from wide patches of red earth rose the clamor of voices and tools. Only the Tara remained as it was; in eternal hurry and eternal repose the water foamed around the smoothed rocks, and the stripped trunks of spruce trees twirled lazily with the mountain stream. Viewed from the rock wall they looked like matchsticks; but out of these matchsticks the huge scaffolding for the arches was to be built. Downstream, at the pebble beach just in front of the old bridge, two Montenegrins caught the beams with boat hooks. The red and gold of their caps sparkled against the green bank. Suddenly a picture in the Uffizi came to his mind: Botticelli's "Goldsmith"—the same red, gold and green. It seemed years ago since he had seen it, together with her; the red gold of the large coin in the picture was the color of her hair. . . .

Someone tugged at his sleeve just in time. He looked into the laughing blue eyes of a small man. A red scar ran across his sallow forehead and like a prickly wreath the reddish stubble of his beard surrounded his sallow chin. On a coiled rope slung across the shoulders of his coarse-woven jacket hung a little wooden box. "I'm Ante Djulich," a high voice said. By way of further introduction he lifted the lid of the little box. Explosives.

"Ah, you're the driller."

"Right, Mister Engineer. Shall we go?" He pointed to the rope ladder suspended from the crossbeam and reaching down to the platform which had already been lowered by the man at the winch.

"But wouldn't it be better to have the box lowered on a rope afterward, so that you could keep your hands free?"

Ante caressed the lid lovingly and swung the box to his back. "No, Mister Engineer."

"I'd prefer you to call me Mister Wolf. Crnojevich is my name."

"No one touches the box but myself, Mister Wolf. That box means to me what Our Lord means to the priest."

A smooth customer, Wolf thought. Presently I'll put him through his paces. He stepped forward, but Ante stopped him.

"No, Mister. I go first. I'm the driller, you're the engineer." Quick as a monkey he lowered himself along the swinging ladder and held it tight so that Wolf reached the platform without difficulty. "Now the drills. What the hell is keeping the drills? I like to get on with things. So do you, I can see that from your face, but those Montenegrins here are a lazy as bats in winter."

"I'm a Montenegrin too."

"Well, then, you're a favorable exception. But I really like it here, you know. A little fresh air at last instead of that damned mine, though the pay was better . . . "

"What made you leave it, then?"

Ante gave him a wink. "A widow, if you please. A big, fat widow a head taller than I. I couldn't show my face aboveground or there she was, waiting for me like a dog in front of a rabbit hole. Widows are my fate, you know. This was the third, and each time I've had to take to my heels or there'd have been the devil to pay."

Good God, thought Wolf, must everybody I come across be a woman chaser? He lowered his voice. "Did one of those widows live in Spain, by any chance?"

Wolf would never have asked the question if he hadn't been tempted by the familiarity and the comical chatter of the little driller, to whom he had immediately felt drawn. The attraction must surely be mutual, because, after the first brief exchange, Ante looked up at him as a dog at his master—a dog, that is to say, like the big sheep dogs from the mountains, who don't surrender their independence even to their masters and who show their affection only by wagging their tails slowly and with dignity. That's why the sudden change of expression on Ante's face now gave Wolf a shock. The pale forehead turned a sallow red and he bared his crumbling teeth.

"I'm sorry if I framed my question badly," Wolf said, embarrassed. "I didn't mean it personally so much as—er—how shall I say it—politically." He clenched his fist and put it on Ante's shoulder.

The driller gave him a searching look. "Oh well, that's different," he said carefully. "You couldn't know, but you touched a raw spot just now. A fellow's corns are sometimes not on his toes alone. Watch out, here come the drills!"

Two black rubber hoses came winding down alongside the red rock. Ante unfastened the rope tied round the detonation chamber and looked at the cold, formidable piece of metal as affectionately as if he were holding a woman in his arms.

Levertara, May. In haste.

DEAR JELENA,

Haven't heard from you for a long time, probably because of your approaching exam, but tonight I must write to you, for I'm excited and that's not often the case. It has really begun now; today the new driller and I drilled the first holes in the steep rock wall where the arch of the large span will be based. It's an exciting, somewhat dangerous job, and the driller, a little Croatian, has given me a good lesson. He explained that I applied too much pressure and therefore didn't get as good results as he. It was a solemn moment for me when the first explosions went off less than a minute after we'd been pulled up again, platform and all. Immediately following the two tremendous blasts, stones were hurled into the water; one piece even, landed on the opposite bank, right on top of the peak—which I had had evacuated of course. That bit of stone now rests on my drawing table as a paperweight. A nice memento.

Those explosions sounded like music to me. They proclaimed the opening of a new era for our neglected country, which has lain for so long behind God's back only because no one dared look the future in the eye. You never heard our grandfather, Sava, play the gusla. When he began a new song, voice and string together sounded like the roll of thunder. I'm only a prosaic engineer who can neither sing nor play. Nevertheless, today I began a new song, with modern orchestration. The pneumatic drill is my gusla and the dynamite my voice. My song doesn't tell of heroic deeds of long ago, but of the great peaceful work of today. So far only the first few lines have been written, but it promises to be a long song, a song not in words but in stone. That song will be followed by others, for there is still so much to be done. The future, as I see it, still lies hidden behind the mountain ranges, blocked not by snow and storms but by stupidity and sluggishness and traditions wrongly understood. Toward that far future I'm building a rainbow. Do you remember last December, in the middle of a sleepless night, I asked you: "Little sister, what does man do with his life?" Everything was different then. I was being torn between passion and duty as if tied between two horses turned tail to tail. That is all over now. I know the answer to my question: BUILD. For that reason, too, I must leave you now. It is midnight and at five the canteen keeper brings me a cup of black coffee in bed.

When you travel home after your exams, for which I send you my best wishes, do take the route through Plevlje. You know where the office is; my old boss Radich will undoubtedly see to it that you get a lift on some truck or other. Then I can show you some of the work and I'll see to it that you get home from here.

Your loving brother,
WOLF

P.S. Sorry! There is a letter of yours after all. I didn't get down to answering it in all this bustle. Anyway, it's largely about things which I'd rather discuss with you in person.

Jelena was angry, as angry as she could be with a brother she admired and loved; but for that very reason she was angry. He had disappointed her. She didn't dare write him this because she knew the effect words can have when dissociated from the voice that speaks them, from the gesture that clarifies, from the eyes that may lovingly forgive while the mouth is still thin with reproach. Words on paper are like the thorny stem of a rose which is wilted by the time the letter arrives. So the sad reflections she could not keep to herself she confided to her diary alone.

MAY 17, 1938.

A letter from Wolf in reply to mine, or rather, without a reply. Apparently he is completely absorbed in his work and forgets the world around him, the terrible world of today. He is building a bridge across a mountain stream, but he doesn't see the stream of history which might well sweep his bridge away long before it is finished. When he was here with me at Christmas he looked so unhappy that afterward I couldn't put it into words, though I would have liked to write to him straight out: Apparently, my brother, you are unhappy over some love affair, with an Italian at that, and a fascist, most likely. You didn't once ask about your college friends, though you had spent years with them. But I, your young sister, know what has happened to some of them. Dragutin is fighting in Spain; there has been news that Zharko is seriously wounded, and Milutin is missing. He won't be the only one, for all our best boys are going to Spain. While your whole earth and sky consist of a beautiful Italian, I suffer on account of friends I got to know at meetings.

Later on I did write you, in cautious words: how our people are being persecuted here; that our country is being sold and betrayed by its rulers. Our Prime Minister gives our old allies, the Czechs, a kick in the pants and travels to Berlin and Rome. Austria has been occupied by the Germans and it is exactly as our people's manifesto so clearly put it: Today they're still at our frontiers, but tomorrow, without warning, they'll cross them. I wept with rage a few weeks ago, the day their air force flew over Belgrade as a demonstration of so-called friendship, the friendship of the wolf for the goat. And in the meantime you, brother, are building a bridge to a better future. If only your eyes could see, you would realize the terrible irony of it all; but you are blinded by your bridge.

BJELO, JULY 10

I had to take a day's rest before being able to write. I was tired before I started; the exam was torture—my thoughts were occupied with other things. However, thank goodness another year has passed. The journey was exhausting, all the more so because I always get into conversations with people, and their poverty makes me feel poor, their dejection makes me feel dejected. Everyone fell silent when the gendarme in his fine uniform came to scrutinize the papers of the people in their worn clothing, clothing that is symbolic of our whole social system—worn to the last thread. A small bourgeoisie growing rich on corruption and treason; the broad mass of the people poor, disunited, incited against each other—Serbians hating Croatians, Croatians contemptuously looking down on the Moslem Bosnians; Macedonians treated as inferiors by everybody, and the proud Montenegrins behaving as if they were alone in the world. Now I've put it too gloomily; among themselves the poorest people are all brothers, at least the peasants and the conscious workers. But the shadow is broader than the light.

Mr Radich saw to it that I was given a lift by a truck. I didn't recognize the Tara Gorge. A furious din arose from it, which frightened me at first.

We were nearly there when the truck stopped with such a jerk that I was thrown against the windshield. A half-naked fellow with a red cloth around his dusty hair jumped suddenly to the middle of the road and waved his arms. A moment later we heard the explosion. A white cloud of rubble hid the sun. The driver, who was carrying sacks of cement, explained that the rubble was put into mixers with cement, sand and water, and this mixture is called concrete. But the first I saw of the bridge was nothing but wood. From the high bank I was able to look across to the other side and at five or six places I saw wooden scaffoldings over which the workers crawled like ants. Above their heads floated some sort of cableway with gray buckets. I began to understand it more or less. They grabbed the buckets, emptied them into the wooden molds, and then tamped the mixture down. In this way, layer by layer, the piers were being built up. Because the driver knew I was Wolf's sister, he treated me with much more deference than a man normally shows a woman in these parts. All the workers, except the idlers, were very fond of Wolf. I had heard that almost before we left Plevlje.

A foreman told us that Wolf should be somewhere on the other side of the river. I went down but didn't hear the water until I was walking across the wooden emergency bridge. Every now and then a thundering sound came from the stone quarry, the tipcarts screeched along the creaking rails, and the machines which swallowed stones in order to chew them to bits made a hellish din. It took me nearly an hour to get to the huts, and there I was greeted by the whine of power saws in a large wooden shed. I fled to the canteen but came out on the wrong side. On the grass, right out in the open, a huge man with shoulders like a bull was slaughtering an ox. With one hammer blow he felled the big brown animal, which had been looking at him good-naturedly. Then, the animal's open eyes still watching patiently, he slit its throat. The blood spouted into a tub and a wide spurt stained the grass dark red. There and then I decided to live in the future as a vegetarian, but in the next moment I knew I was being hypocritical, since I always eat meat when I can afford it.

I had already asked a dozen people where Wolf was and they had all seen him three or five or ten minutes before in different places. But the butcher knew where he was. Raising an arm dripping with blood, he pointed up toward a rock high above the encampment and said, "He's sitting up there; he often sits there."

Here I had expected to find my brother at his drawing table in an office, or on top of the bridge issuing orders like a *vojvoda!* But he was sitting on a rock among the bramble bushes, all by himself. On the knee of his breeches lay a pocket chess set which he was studying with a frown of concentration, as if nothing existed in the world but that piece of leather with its little carved figures. He didn't really come to until my shadow fell on him. I was moved, for it was different this time from Christmas. This time he didn't embrace me with his arms only, but with his heart as well.

He will always remain a solitary person, I'm afraid, for there is something within him which he can't share with anyone. It's something like a stone, something which has no voice, something—I believe—which no one can touch or destroy.

"I was faced with a difficulty, you know," he said slowly. "And at the office, people keep walking in and out. At times all those people tire me and I suddenly feel like being alone, as alone as I used to be when I tended the cows on Krnovo. It's an important problem of construction, and to solve it I first have to try and make my dusty brain clear and cool, like a stone lying in the middle of the Tara. That's why I do something completely different first: I solve a chess problem!" He laughed, and only then I saw that he had dark rings under his eyes. It was as I had thought. How many hours a night do you sleep, brother? As many as are left over. . . .

He pulled a writing pad from under his shirt and began to draw diagrams, forgetting that I was there. All at once he said, "I've got it at last. Let's go and have a drink in the canteen. Then I'll show you your bed. You're sleeping with the nurse. No, I have time for you now. I'll work out these details tonight."

He showed me over the project and when I returned, confused and benumbed because it was all too big and too complicated and too noisy, I partly understood his not speaking about the present-day world. The bridge is still far from complete but is constantly before his eyes; the arches he drew against the sky are to him tangible and visible reality, far more so than I am. The bridge, the seething mass of people, the chaos of materials, the droning and screeching and hammering and screaming are a world in themselves, a world of which he is the center. Somewhere on a different planet there flows a river across which he is building a bridge. Occasionally he takes his spyglass and peers at the earth. When he sees that it still exists he is satisfied. That is enough for him.

Yet it is wrong; for he lives just like the old monks in Orthodox monasteries. They mutter their prayers and sing their masses as if this were the age of the holy Sava and not the twentieth century. Many of them live in a monastery hidden in a mountain gorge, far away from everything and everybody. Wolf lives like that; only his monastery is noisier. I didn't have the courage to say anything about it until that evening, in the canteen, when he introduced me to a little fellow with a roguish look and crooked smile—Ante, his driller. "With him you can talk politics," Wolf whispered suddenly. So we did, and Wolf listened. Ante is more or less his eyes and ears, for the workers don't dare talk freely with an engineer. The food is good, but of course there are complaints about wages. Wolf can't do anything about that, but I got the impression that he is continuously at loggerheads with the contractor, who wants to grab as much as he can out of the project. And then there are the usual tensions. Of course the Montenegrins and the Moslem Bosnians don't get along together. What's more, the foremen are nearly all Serbians.

"There's only one way to put an end to all that," Ante said suddenly. "No more monarchy with a Serbian king and a Serbian clique lording it over everybody else, but a republic with equal rights for all our five peoples."

"That's all right with me," Wolf said patiently. "Go ahead, once the bridge is finished, try and fix it." Then, in spite of my protests, he went to his office, and Ante and I talked on for hours. He is a fine comrade even though he doesn't know a lot of theory. But he sees things Wolf can't or won't see.

The next morning I accompanied two old peasants who had to go to Zabljak. I spent the night there and made the rest of the journey on my own. It was a long walk—nine hours—and my feet were sore by the time I arrived. Town life makes one soft.

BELGRADE, SEPTEMBER 29

Father took me to the Tara on horseback. We rode in turns. Good old Father; a great many wrinkles and the young, innocent heart of a childlike man who has never been able to cope with the intricacies of life.

On the way he told me something I had to promise not to mention to Wolf, as he has enough worries already. Dushan Mirkovich, the son of old Bajo who was killed in the World War, has made himself rich on land speculations. Just after Christmas Wolf had trekked home on skis and in his innocence had told about the bridge and the road-development scheme. Thereupon the scoundrel began to buy up plots of land in the area about to be requisitioned. Previously he used to speculate in bonds for war invalids, Father said; first they had been issued, then canceled, then declared valid again. We must be careful with him, for he is hand in glove with the Governor of Cetinje; he is a government spy and suspects anyone who can read and write better than himself of being a Communist.

Father wanted to see the work without bothering Wolf. The whole summer Wolf has been home only once, although he now has a motorcycle. Mother sighs over his being so near and yet so far, but Father says, "Better leave him to make sure the job's well done. Imagine if something went wrong on account of his being absent! I worked myself to the bone to send him to college. His whole life depends on this job, so we'd better let him go on in his own way." Father of course is as proud as a peacock of his son. Wolf has realized what Father himself always wanted but was unable to accomplish.

By the time Father took me back there the project had made very good progress. Five piers were completed and between the first and the second pier a tremendous scaffolding of pine was being erected. When one of the beams was hoisted into position and nearly slipped out of the strap, Father immediately pitched in and worked till he sweated. Wolf didn't dare say anything, for Father remains boss over his three learned sons. The carpenters were busy making large wooden arches, leaving them open at the top. Wolf explained that the concrete will be poured into these. They hope to complete the first arch before the winter, for as soon as the frost starts it is impossible to pour concrete. In one way I wish it would begin to freeze; early in the morning, when we had just got up, Wolf looked as white as chalk. All I could do was shake my head. He is working himself to death on that bridge while the world is about to explode.

October 10

Now Wolf has had a shock too. The news that the British and French have sold out our Czech brothers to the German fascists has been greeted with fury by our people. Students marched to the Czech Embassy to enlist as volunteers, and so did hundreds of workers. The gendarmes closed off the streets around the embassy and it ended in a pitched battle.

The following day Ante turned up at my place, his face smeared with blood. The canteen keeper at the bridge had heard the news of the Munich betrayal over the radio. A quarter of an hour later the whole project had stopped work. Ante climbed onto the casing of the first arch and blew the trumpet with which the oldest timekeeper calls the workers to meals. The concrete mixers halted, the quarry fell silent, the power saws stopped, the tipcarts stood still—and suddenly the rustling of the river could be heard again. Ante made a speech, or rather he told what had happened and ended up with the question, "What now, men—what now?" An old Montenegrin who works in the quarry fired his revolver into the sky. That was the answer. Without knowing what was happening in Belgrade the men formed themselves into two volunteer detachments. Only the Moslems didn't join it. Then a collection was taken and Ante was deputed to go to Belgrade on behalf of the men on the project, in order to present the list of names to the Czech ambassador. Wolf lent him his motorcycle—a good sign. Wolf's name headed the list, as well.

But Ante didn't have a chance. It's all over now. The Czechs gave in without a shot being fired, and Ante and I both wept, he with rage and regret, I with despair. If the Czechs, who have a strong, well-equipped army, don't dare fight, what's going to happen when our turn comes—divided as we are between royalist Serbians and republican Croats, with leaders largely on the wrong side? We must overthrow this Government before it is too late. But only politically conscious workers and intellectuals realize that; the mass of the people act from instinct. And they have been suppressed for too long. All the same I'm proud of our people

Everywhere in Europe there has been rejoicing because peace has been bought with shame. Only we have spat in the faces of the traitors and tyrants. Are we such a special people then? It would seem so. We may be poor and backward but at least we're not afraid to die.

April 25, 1939.

I had wanted to stay in Belgrade over Easter because my exam is getting near. But Mother was ill and I went home. Also I was really rather curious to see how the work was getting along, for Wolf has infected me with his excitement. And his letters too—he writes in such a way that you see it all happen. They have been in luck; a mild winter with less snow than usual. Another arch has been finished; a miracle how a thing which looks so fragile can remain suspended in the air apparently without support.

On the first arch props are being built which from a distance look like matchsticks pointing upward. The roadway will be laid on top of them; incredible that such little spikes can carry such a load. Wolf laughed and offered to show me how to work it out, but I've never had a head for figures.

Mother's hands were swollen with rheumatism; there's been a lot of damp fog this winter. I kept house, but what good were those few days to her? And I left her in despair. On Easter Sunday, when we were testing the eggs, the news came over our little crystal set that the Italians had occupied Albania. Father dug out the old school atlas, and his stiff forefinger, which will remain the color of earth until he dies, traced the frontiers. "It's like the Turkish times all over again," he said with a sigh. "We're surrounded on all sides. Italy, Germany, Austria, Hungary, Bulgaria—all enemies, and it's no good relying on the Greeks."

Afterward he saddled the horse and rode to Shavnik to talk it over with his friends. Men are like that, but we women stay at home. Wolf didn't turn up, though he had promised; his work again, of course. But in Mother's eyes he was there just the same. She's very fond of Stojan and Mirko, but somehow it's Wolf who is her favorite. Our men are not tender as a rule, but he is with her, always. He strokes her hand and looks at her as if they were lovers. And on bad days like this, her first thought is for him. When she started to cry, I knew the reason; but she confused me when she whispered suddenly, with profound sadness, the name of Sava. Through her tears she told me how that old gamecock, our Grandfather Sava, had held his hand, bigger than the child's head, over Wolf's cradle; how in the prophetic language natural to old *guslars* like him, he had predicted that Wolf would become a warrior whether he wished to or not—his mother should consider herself lucky that one day she would have a hero to mourn. It is now twenty-three years since Wolf buried the old man in the mountains above the village, but Mother is still scared of Grandfather Sava's eyes and it worries her that Wolf, when he is working, has the same expression. The first time Father went to the project he came back and said, "Mother, you wouldn't have known your son, standing on top of the wooden arch and shouting orders. His jaws really looked like those of a wolf and his eyes were so hard that no one dared contradict him. He was as like his Grandfather Sava as two peas in a pod; only the mustache is missing."

July 1939

I sleep and sleep. Father sits there smiling happily all the time. Three of his four children are now university graduates—though I'm only a girl and hardly count here. The men who visit us give me a wide berth and watch me shyly through half-closed eyes as though I were a strange animal. (Yet I look like any other Montenegrin woman when I'm here. I like walking barefoot; it makes me feel closer to the earth.) Nevertheless, my success has enhanced Father's prestige both here and in Shavnik. The men listen to him, something they never used to do. They always outshouted him because of his mildness. The only person who cold-shoulders him is Dushan Mirkovich, the speculator. Sheer envy, of course. In the public elections Dushan was the only one in the village who voted for the Government candidate. Everyone else voted for the opposition candidate, quite openly, for of course the elections were not secret. Apart from falsifying the figures, the Government didn't dare to do a thing about it.

Yesterday Wolf came, with Ante on the pillion. Three flat tires on the way. No wonder; how they managed to get over that little path through the mountains, heaven knows. But suddenly we heard something no one in the village had ever heard before: the pop-popping of a motorcycle. And there came Wolf, lurching and swaying, his feet sliding along the ground, careering up the slope between the rocks. He was excited and his eyes shone with satisfaction, but his mouth was hard. Two days ago the casting of the large arch had been completed. That was the last difficult problem. Later, when Ante explained to me that it had nearly gone wrong, I understood why Wolf had looked so grim. At a critical point, just where the roadway starts to "catch on," as he called it, the concrete mixture had been meddled with. Wolf discovered it only when the concrete was already hardening. He nearly strangled the foreman who was responsible; when Wolf released the man he fell backward and would have plunged down two hundred yards into the river if his clothes hadn't caught on a wire. It's time the work were finished; Wolf is on the verge of a nervous breakdown. All he managed to eat was *grushavina* the way Mother makes it. I can also guess what made him come home, quite apart from wanting to tell about his triumph. He was still filled with cold rage that someone should have wanted to destroy his bridge. And he was furious with himself for having almost committed a murder in a fit of temper. But as soon as Mother lays her old, cold hand on his hot anger he quiets down.

August 29, 1939

Yesterday was the great day—Wolf's bridge was inaugurated. Wolf himself says, with a smile which isn't a hundred per cent modest, that it's Radich's bridge, for he was the engineer in charge. But Radich gives all honor to Wolf and it's funny to see those two compete in generosity. After all, it was Wolf's design, even though Radich introduced some practical alterations.

When I awoke cold clouds hung down from the Vojnik; but before we reached the bridge the sun was shining. Wolf himself came to meet us in Shavnik with a car!

He said triumphantly, "Presently the Governor of Cetinje may open the bridge, but I have already inaugurated it. Yesterday I came from Plevlje and this is the first car to cross the bridge."

But for me the most moving moment had taken place before we met him. On the way I told Wolf about it (I sat in front, next to him). Mother was riding on the horse when we went to Shavnik. I supported her and Father walked in front, the reins around his wrist. When we came past the last fall of stone and saw Shavnik lying down below, he suddenly stopped and said to Mother, "Do you remember how I took you to Shavnik in November of the year 1905? In those days there was nothing yet—only your wish for a daughter—yes, I knew it. And now? The same stones, the same river, the same Shavnik. But now it lies on a wide road, and that road is no longer the dead end it used to be but it leaps over the Tara. Our son has done that for us." Mother only nodded and her smile hid a sob. I've never before seen her look so young.

That was how I told it to Wolf while the car jolted through the holes the rain had made in the road. To my surprise he suddenly stopped the car, got out, and embraced Mother in the back seat. Then he drove on as if nothing had happened. I looked at him quizzically, but only two turns farther along he said in a stern voice, "Now I have inaugurated the bridge. Last night also, over a bottle of *rakija*, Radich and I inaugurated it, in a different fashion. The rest is all nonsense—I'd rather do without it."

"In that case why did you put on such a neat new suit?" I teased. It was a blue suit with a thin white stripe; he wore blue silk socks—still from Italy, I suppose—and he had even plastered down his unruly hair.

"My new suit?" he laughed. "That's part of the official disguise. I can't very well leave Radich standing alone in his striped trousers, can I? I loathe all the fuss, but I've made up my mind to play along with it. Our time hasn't come yet, we have to howl with the wolves. But don't let Father and Mother notice how much I hate all this; for them, of course, it will be a great day."

On the way we caught up with lots of people. Our neighbors from Bjelo had already started out the previous evening, on horseback. How marvelous our country is when the sun shines! The high, grassy plateaus covering a vast expanse until suddenly, past Zabljak, the high mountains loom up. The Durmitor with its three forbidding peaks, the Medjede, the Savina Kuk; and, above, the sky full of hurrying white clouds. Far below this infinity walk the little horses with their riders—on the way, eternally on the way, from nowhere to nowhere. This day, however, they knew where they were going—to a feast, an interruption in the monotony of their existence.

When the hairpin bends started, on the descent to the Tara, we made Mother sit in front. She was afraid the continuous turning and twisting and twisting and turning would make her sick. All of a sudden, close by a high umbrella pine, Wolf clamped on the brakes and pointed. Deep down lay the bridge, shining white between the red earth and green slopes. Mother took his hand and Father patted his shoulder; I was nearly in tears and Wolf gave a sigh. "I feel terribly tired now," he said. At each new turn we saw the bridge again, getting bigger and bigger. Now I understand completely what has so possessed him, for this bridge is no ordinary bridge between two banks but has some of the surprise of a painting in a new style. It looks so strong and peaceful, so proud and full of grace. It moves one the way a work of art moves.

We halted before the bridge. Right across the road stretched red-white-and-blue ribbon. There were at least twelve cars, and everywhere on the slopes people were sitting beside their grazing horses. Our neighbors, too, were there already. They had spread out their linen cloth and were eating bread with sheep's cheese and sausage. But Wolf couldn't drink *rakija* with them, for the ceremony was about to begin and Radich, in a ridiculously smart getup, dragged him by the arm, saying "Last night you had more than enough, old boozer." A good thing Mother didn't hear it.

A gentleman in black, holding a top hat in his hand, got out of a beflagged car: the Governor. Radich introduced Wolf to him and to my horror I saw Wolf shaking hands and even bowing. Is my brother a rebel in word only, after all? The people came flocking in, the Governor mounted a wooden platform and read a short speech from a piece of paper. The paper fluttered in the wind so much that he had to stop every now and then; so finally he put it inside his top hat. An old peasant guffawed loudly, but the Governor pretended not to notice and quickly scrambled through the rest of his address.

The contractor handed him silver scissors. The two men, accompanied by Radich and Wolf, walked up to the ribbon. I secretly hoped he would fail to cut it, for men can sometimes be so deliciously clumsy. However, Radich and the contractor kept it taut, so he couldn't go wrong.

The Governor waved his top hat and shouted, "In the name of our beloved and revered," etcetera etcetera, "I declare the bridge open!" He called for three cheers for Their Royal Highnesses. Some of the crowd joined in, and I saw Wolf raise his arm as well. Bah! Flanked by Radich, the Governor next set his patent-leather shoes in motion—and at last Wolf did something nice. Quick as lightning he turned around, stepped back, took Mother by the arm and signaled Father to take the other. After a moment's hesitation Father did so, and then they strolled slowly across the white concrete structure, behind the three black-suited men who were walking, carefully in step, toward the other side as if measuring the length of the bridge with their strides.

First they walked in dead silence, but then from the other side, where all the workers stood assembled, rose a wild and confused cheering. I suppose the Governor thought it was meant for him, for I saw him solemnly waving his top hat, by way of thanks. But it was Wolf they cheered! Because they liked him, and because he had had the courage to show everybody—at least so they assumed—for whom the bridge had been built. Not for the Governor or for the little King, but for his father and mother, for the people. Had this really been Wolf's intention? When I asked him about it later he wouldn't say. After a cup of honor had been emptied in the management shed, the Governor fortunately took himself off.

Mother was tired, so Wolf took us back in the car. I could have stayed behind and gone on to Belgrade, but I still have two weeks before starting as a teacher at the high school there, and the longer I can give Mother a hand the better. So I had to forgo the feast, unfortunately. Ante told me later what it was like. Fires everywhere, with sheep roasting on the spit, and a lot of *rakija*—too much of it!" In the middle of the night Wolf suddenly got up from the company of people who sat drinking in the shed—Radich, the contractor, the foremen, and Ante of course. He was a bit drunk and said, "And now you can all go to blazes. I want to be by myself for a while." When Ante left the shed a few moments later he found Wolf right in the center of the bridge. He was stamping on it the way a horse likes to stamp; then he walked a little farther along, stamped again and finally hung across the balustrade watching the water of the Tara below. Ante tried to coax him away, but Wolf shook himself free and said, "Leave me alone." At the end of the bridge a gendarmery post has been built; the Governor litters the whole country with those dog kennels, to keep us down. Ante knew one of the three gendarmes who already manned the post—a Croat from the parts where he was born. This man told him the following morning how throughout the night he had heard Wolf walking to and fro. Toward dawn he must have gone to sleep somewhere among the bushes, for later he turned up in the canteen quite crumpled, just as Ante was trying to buck himself up with black coffee.

That same afternoon they were back here. Of course I didn't mention anything to Wolf; he wouldn't talk about such things anyway. He went to sleep right away, in the grass.

Ante told me something which made me all the more sorry I didn't stay. During the feast, by the campfire, an old gusla player sang a song about Wolf, the Tara, and the bridge. It just poured out of him like water from a well, Ante said. Our people have an incurable desire for hero worship; and to the people in the Tara Valley Wolf is not a hero. Unfortunately Ante could remember the words only vaguely, and the accompaniment not at all. He is about as musical as a cat, and what's more he regards the gusla as a barbarous instrument. "I'd rather hear explosions," he said. "That's my music." When I asked Wolf about the song he beat about the bush, and when Father kept after him he got angry. "If they make a hero of me it's because of their antiquated stupidity," he said angrily. "All that hero business is Montenegro's destruction. Please let Grandfather Sava be the last hero in our family; he at least deserved it, for he paid for it with his life. All I've done is pocket my salary, and at most get my hands dirty with work." Mother got up to kiss him, but he gave her such a look that she sat down again. Sometimes Wolf gazes down on us as coldly and inexorably as the mountains.

September 2, 1939

At last—the war has broken out. The Soviet Union keeps neutral, on account of that incomprehensible nonaggression pact between it and Germany. That sticks in our throat. Party discipline is fine, but people stop at swallowing this. For we always want to understand, and this just can't be understood.

Before I left for Belgrade I had a talk with Wolf. He came to get me on his motorcycle.

"We're still neutral—" I began.

He interrupted, "Only as long as it pleases the Germans."

"And what next?" I asked. "What will you do?"

"I'm a reserve officer," he answered. "I'll have to await orders."

"And if those orders clash with your conscience?"

"I'll see about that when it happens," he said shortly.

"Why don't you join the Party?" I asked. "It is guided by people who can judge the situation better than a single individual; that's why they're able to give leadership."

He put his fist on his heart and looked at me with his bulldog face. "For the time being my party is here."

"But surely you don't exist for yourself alone, Wolf!" I chided.

"One should always make important decisions all by oneself," he said seriously.

"Wolf, why don't you try to be realistic for once?" I pleaded with him. "Suppose we are drawn into the war. Suppose the enemy wants to invade Montenegro, across your bridge. What will you do then?"

"Defend the bridge, of course—if I happen to be near."

"And if, as so often happens in war, it were necessary—"

I saw him clench his fists. "You mean that the bridge might have to be blown up? Someone else can do it in that case; I'll be damned if I'll ever do that. You might as well ask a mother to kill her child." He looked at me with a frown. "What makes you torture me with such questions? For the moment the fighting is a long way off, in Poland. For the moment the war has nothing to do with my bridge."

"Yes, for the moment," I whispered. "That's just it. We live for the moment, we work for the moment. And for the moment we don't know if we'll ever see each other again alive."

It was a sad farewell in Plevlje. When we passed over the bridge he looked neither up nor sideways; he just drove across it as if it were any other bridge—a piece of stone between two banks. But that was purely for my benefit, of course. I'm afraid he really does cling to that bridge as a mother to her child. We are strange people, we Montenegrins. We are peaceful in time of peace; but in war terrible instincts seem to be awakened. I haven't experienced it so far, but I know it from Mother's tales. In the Turkish wars whole villages used to make for the mountains, otherwise the Turks exterminated every living soul. Then it sometimes happened that mothers who could no longer carry their infants left them among the rocks. For the older ones had to be saved, the boys in the first place; and to that end the smaller children were sacrificed. So our mothers sometimes left their young ones to die for the sake of the people; and the rest of their lives they dragged that sorrow along with them, like a stone inside their bodies.

There's something both great and terrible about our people, just like the sky we live under, just like the rocks that enclose us. The war will come to us, too, everybody knows that. But I'm afraid of what will happen then, to Wolf even more than to others.

2.14

The Robotosyncrasies

(1976)

WAYNE HAWAII

The annual meeting of the Federation of Engineering and Scientific Societies (FESS) of the USSR began today and had its keynote session scheduled at 2 P.M. in the 6,000-seat Conference Hall of the Palace of Congresses under the watchful eyes of Lenin, whose likeness adorned the giant curtain.

Rob sat nervously on the stage along with other dignitaries. Dr. Nicholas Krylov, president of the FESS, delivered the welcoming address and expressed his distinct pleasure to have a most distinguished American engineer-scientist, Dr. Jack Rob Tains, as their keynote speaker. And a glowing introduction of Dr. Tains, delivered by Professor Avenir A. Serov, director of the Institute of Automatic and Telecommunications, followed, with this concluding statement: "Today, we have with us the creator of the 'robot,' which is also known as the 'Human Intelligence Model' and which is inconceivably used as the workhorse in almost all electronic, communication, computer, and control systems now in use throughout the world. And his invention-and-theory and his present responsibility may yet usher in a new era for all of us, an era of space exploration with human moon landings merely as the first step."

The audience listened in silence and awe. So did Rob on the stage and Helga sitting among the audience.

After a thunderous ovation, Rob rose from the seat and walked to the rostrum. He thanked his Russian hosts profusely, "I am grateful for this invitation and for this opportunity to meet with you. Professor Serov was very kind to give me a great deal of credit for my accomplishments. But the frontiers of science and technology have really been established collectively by scientists and engineers throughout the world. And let me use this opportunity to thank all of you and your professional colleagues throughout the Soviet Union for your outstanding contributions." And another round of applause ensued.

Now, Rob proceeded with his keynote speech on "The Robot and Its Robotosyncrasies." He had been forewarned that, since the audience was to be a heterogeneous group of engineers and scientists of many different disciplines, the speech should be of popular nature and cater to the laymen rather than the specialists. An abbreviated version of Rob's speech[1] is given with captions and in first-person expressions as follows:

The Crude Robot

For a *command* or *input* x in terms of a signal, we want a robot to have a *response* or *effect* y (which is often called a "*control variable*" in technical terms) to obey and follow the command or input.

First let's consider a "robot without intelligence" or a "crude robot" as shown in Fig. 1 of the slide projection on the screen (which is also Fig. 1 of this book). A crude robot, often called a *controller*, is designed with many system components (e.g., transistors, diodes, resistors or capacitors if the crude robot is an amplifier; amplifers, motors, networks and/or other components if the crude robot is a certain type of control system) to do a specific job. It has a *gain* G relating the output y to input x with the relation:

$$y = Gx$$

or

$$G = \frac{y}{x}$$

The Bandwidth: The Breadth of a Robot's Mind

The input to a robot is usually a signal.

We are all familiar what is a signal. When we hit a piano key, we have an *elementary signal* of a certain frequency f.

As an example, we may have elementary signals of these frequencies:

$f_1 = 10$ cps (or 10 hertz)
$f_2 = 35$ cps
$f_3 = 10{,}000$ cps
$f_4 = 1{,}000{,}000$ cps

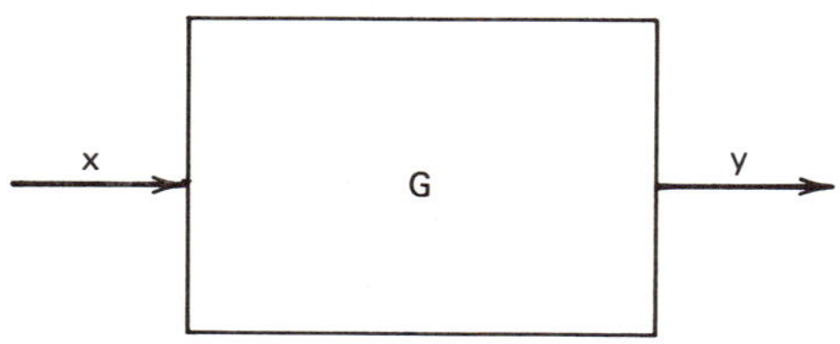

Figure 1

[1]This speech is scientifically factual and requires only simple algebra to understand.

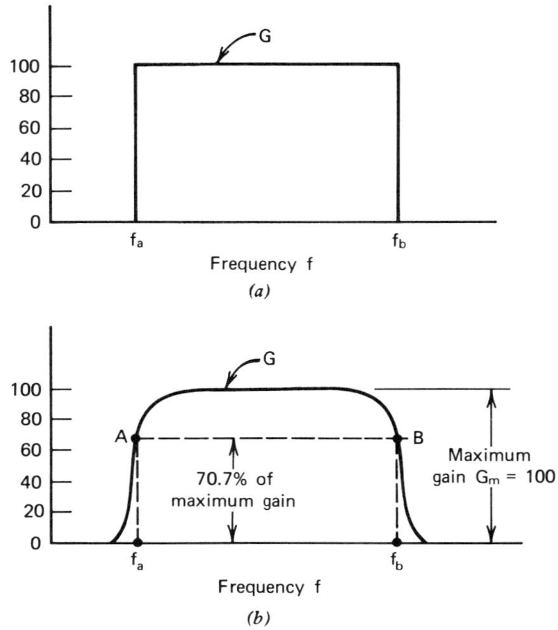

Figure 2

The *frequency* f of an elementary signal is the number of vibrations (or cycles) per second with a unit in cps (or cycles per second) or hertz (meaning cps).

An elementary signal has a *frequency* f in cps as well as a *magnitude* x (or y), say, in volts or other units.

For a controller (i.e., a crude robot) or a robot (which has yet to be introduced), we may feed in an input x at a certain frequency f, measure its output y, and compute its ratio or gain $G = y/x$ at this frequency. By repeating this step for different frequencies, we may obtain the values of G at different frequencies and may plot a *gain characteristic* G as shown in Fig. 2*b*.

As can be shown mathematically, when the gain characteristic G of a controller (or robot) is a *constant*, say, within certain reasonable frequency limits f_a and f_b as shown in Fig. 2*a*, the output y is almost identical in form to the input x. This is the most desirable situation, as the output or response y follows the input x faithfully just as the controller (or robot) is required to do.

Note that, for any frequency f between f_a and f_b in the example of Fig. 2*a*, if the input is $x = 1$, the output is $y = Gx = 100$. On the other hand, for any frequency outside these limits f_a and f_b, e.g., $f < f_a$ (or $f > f_b$), if the input is $x = 1$, the output is $y = Gx = 0$. It is easy to define the *bandwidth* of the controller or robot of this "ideal case" with a constant gain to be:

$$W = f_b - f_a$$

For humans, we have our range of emotion, and for robots, they have their range of frequency. The bandwidth is, therefore, the breadth of a robot's mind to accommodate signals or commands.

However, the gain characteristic G in Fig. 2*a* with sharp cutoff frequencies f_a and f_b for its passband is an *ideal* case, and exists only mathematically.

In the physical world, the gain characteristic G is a curve and is therefore called a function of the frequency f (namely, the value of G is different at different frequencies), as is shown in Fig. 2*b* and note that it tapers off at two ends gradually rather than abruptly. Then, how do we define its bandwidth? In these physical cases, we define f_a and f_b to be the frequencies at which the gain G drops to 70.7% of the maximum gain (and f_a and f_b are called "half-power frequencies," at which the output y has ½ of the maximum power and has $1/\sqrt{2} = 0.707$ or 70.7% of the maximum magnitude). The bandwidth, as the breadth of the robot's mind, is still defined to be:

$$W = f_b - f_a$$

The Robot as a Human Intelligence Model (H.I.M.)

Many years ago, an idea struck me when I tried to park my automobile in a space along the curb of a busy street in a Long Island township. At each moment, I compared my "objective," i.e., the final desired parked position of the car, with my "observation" of the current status, i.e., the current position of the car. In this process the *actual input* w to my mind was therefore the difference ($w = x - y$) of the *reference input* x (namely, my objective) and *observed output* y (namely, my observation); and this process is a continuous one. This is the concept of my Human Intelligence Model (H.I.M.). I don't think that the human being is a supreme being with an innate complex mental mechanism. Rather, it is the simple concept of the H.I.M., applied repetitively and continuously, that is responsible for many human qualities and wisdom.

Then, why can't we apply the concept of the H.I.M. to the crude robot, making it a *robot* with human intelligence? This is how I introduced the system model for a robot in Fig. 3*a*. The circular device in the schematic is a sensing device or comparator, taking two inputs x and z and offering its difference $w = x - z$ as output. Note here in Fig. 3*a* that, for a reference input x and an observed output y, their difference $w = x - z = x - y$ is fed into the controller for processing.

Later, I introduced an additional degree of freedom in the system model of robot in Fig. 3*b* when a "crude robot" with a gain H is inserted in the feedback loop. The simpler model in Fig. 3*a* is, therefore, its special case with $H = 1$.

THE FUNDAMENTAL EQUATION

For the system model in Fig. 3*b* we want to obtain its input-output relation, namely, an equation relating x to y.

From Fig. 3*b*, we have these relations:

$$z = Hy \tag{1}$$

$$w = x - z \tag{2}$$

$$y = Gw \tag{3}$$

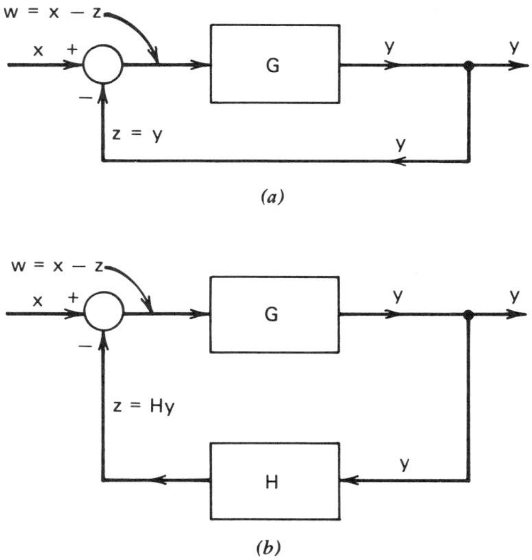

Figure 3

From these three simultaneous equations, we can readily eliminate two variables w and z, and obtain Eq. (A) below relating x to y.

For completeness, we shall carry out this process of eliminating w and z [which the reader may *omit* and proceed to Eq. (A) directly.]

First substituting (1) into (2), we obtain:

$$w = x - Hy \tag{4}$$

We then substitute (4) into (3):

$$y = G\ (x - Hy) = Gx - GHy$$

which may be written in the form of:

$$y + GHy = Gx$$

or

$$y(1 + GH) = Gx \tag{5}$$

Changing the form of the above, we have the *fundamental equation* of the robot:

$$y = \frac{G}{1 + GH}\, x \tag{A}$$

Defining the *system gain* $G^* = y/x$ as in Fig. 4, the fundamental equation may also be written in this form:

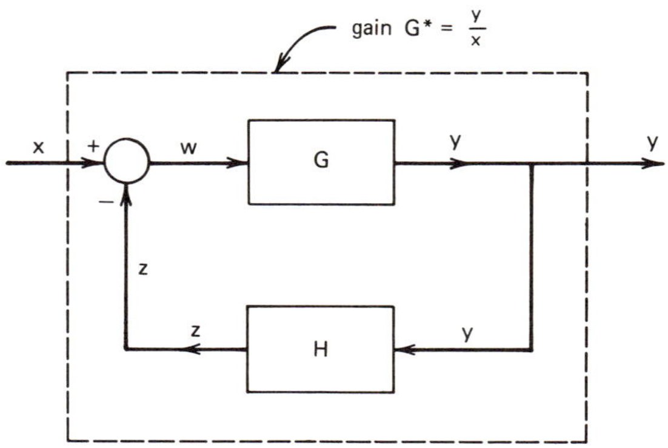

Figure 4

$$G^* = \frac{G}{1 + GH} \qquad \text{(B)}$$

The fundamental equation (A) or (B) will be found to be a powerful tool to establish the various robotosyncrasies, as we shall do in the remaining portion of this lecture.

Robotosyncrasy 1: Broadminded

We can readily show that a robot is much more broadminded than the crude robot or controller.

We shall postpone our illustrative example to establish both idiosyncrasies 1 and 2 together.

Robotosyncrasy 2: Even-Tempered

We shall also show that a robot is much more even-tempered than the crude robot or controller.

ILLUSTRATIVE EXAMPLE

Let's consider the system model in Fig. 3*b* with $H = 0.09$ (which is typical for a feedback-amplifier network arrangement).

Let's assume the gain characteristic G of the controller be represented by this table:

Frequency (in cps)	Gain G
f_1 = 10	15
f_2 = 100	35
f_3 = 1,000	55
f_4 = 10,000	100
f_5 = 100,000	20
f_6 = 1,000,000	10

Table A

or by the graph in Fig. 5*a*.

(Note that, for $f_1 = 10$; $f_2 = 100$; $f_3 = 1{,}000, \ldots,$ the points are equal-spaced along the frequency axis in Fig. 5*a*. This is because the frequency axis is on a *logarithmic scale.* It is a standard practice to use a specially prepared graph paper with a "logarithmic scale" for the frequency axis in plotting gain characteristics.)

Note that with the gain characteristic in Table A or Fig. 5*a* for a constant input $x = 1$, the output will be $y = 35$ at the frequency f_2 and $y = 100$ at the frequency

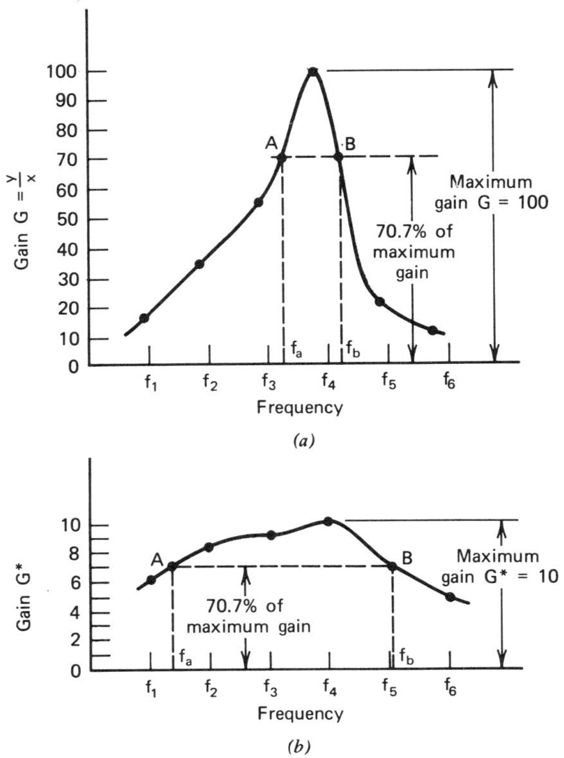

Figure 5

f_4. The outputs $y = 35$ and $y = 100$ are radically different for the same input $x = 1$. The controller or crude robot is very uneven-tempered and is temperamental in the sense that it will not obey commands uniformly at different frequencies. In other words, it is a disreputable (crude) robot.

In general, we choose the controller or crude robot with a gain characteristic much better and more even-tempered than the one in Fig. 5*a*. However, we deliberately choose a bad one for illustration.

From Fig. 5*a* (where $f_1, f_2 \ldots$ are defined in Table A), we may readily determine:

$$f_a = 3{,}000 \text{ cps}$$
$$f_b = 30{,}000 \text{ cps}$$

and the bandwidth of the controller or crude robot is, therefore,

$$W = f_b - f_a = 27{,}000 \text{ cps}$$

which is a rather narrow bandwidth. And the controller or crude robot is therefore rather "narrow-minded."

We shall now investigate the robot with the system mode in Fig. 3*b* and $H = 0.90$. Using the values of G in Table A in the fundamental equation (B) we may compute, for example at $f_1 = 10$,

$$G^* = \frac{15}{1 + 15 \times 0.09} = \frac{15}{1 + 1.35} = \frac{15}{2.35} = 6.4 \qquad (\text{for } f_1 = 10)$$

Repeating for other frequencies $f_2, f_3 \ldots$ we now have

Frequency (in cps)	Gain G^*
$F_1 = 10$	6.4
$f_2 = 100$	8.4
$f_3 = 1{,}000$	9.3
$f_4 = 10{,}000$	10
$f_5 = 100{,}000$	7.2
$f_6 = 1{,}000{,}000$	5.0

Table B

And these data are now plotted in Fig. 5*b*.

From Fig. 5*b* (where $f_1, f_2 \ldots$ are defined in Table B), we may readily determine:

$$f_a = 20 \text{ cps}$$
$$f_b = 100{,}000 \text{ cps}$$

and the bandwidth of the robot with the system model in Fig. 3*b* is now:

$$W = f_b - f_a = 100{,}000 - 20 = 99{,}980 \text{ cps}$$

which is much more *broad-minded* than that (27,000 cps) of the controller or crude robot.

By comparing Figs. 5*b* and 5*a* it is easy to see that the robot is *more even-tempered* than the controller or crude robot.

Robotosyncrasy 3: Agile

Assuming a step-function input x as command, the "ideal" output y as effect or response is exactly the *same* as the input in form, as is shown in Fig. 6. However, this is an impossibility for a physical system.

Rob then proceeded to derive with mathematical techniques that for an input x as shown in Fig. 7a, its output y is shown in Fig. 7b with a small time delay t_d and a buildup time t_B, where

$$t_B = \frac{1}{2f_c} = \frac{1}{W}$$

and W is the bandwidth) and f_c is the cutoff frequency of a low-pass controller or robot).[2]

It is now easy to see that for zero time delay (i.e., $t_d = 0$) and an infinitely large bandwidth (i.e., $W = \infty$, and $t_B = 0$), the output y in Fig. 7*b* will be almost *identical* to the input x in Fig. 7*a* in form.

Ignoring the small delay t_d, we note that the broader the bandwidth W, the smaller the buildup time t_B and the more agile is the crude robot or the robot in responding to the command.

In the above *Illustrative Example*, the controller or crude robot has a bandwidth $W = 27{,}000$ or a buildup time $t_B = 1/W = 37 \times 10^{-6}$ secs; and the robot has a bandwidth of $W = 99{,}980$ or a buildup time $t_B = 1/W = 10 \times 10^{-6}$ secs. We can readily note that the robot is much *more agile* (3.7 times faster for this case) than the crude robot in responding to the command.

Robotosyncrasy 4: Stable and Exercising Moderation

Let's now consider the system model in Fig. 8*a* in which the reference input x is directly compared to the observed output y, where their difference or "error" $w = x - y$ is fed into the process. The reference input or command x is assumed to be a step function as shown in Fig. 8*b*.

A "robot" system operation could be stable or unstable. There are criteria for detecting stability which I shall also discuss in this lecture. Using these stability

---

[2]See W. H. Chen, *The Analysis of Linear Systems*, McGraw-Hill Book Co., New York, 1963, pp. 200, 209–210, 213–220.

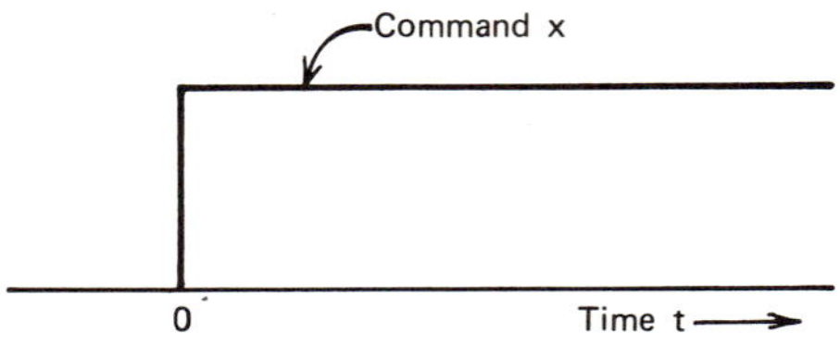

(a)

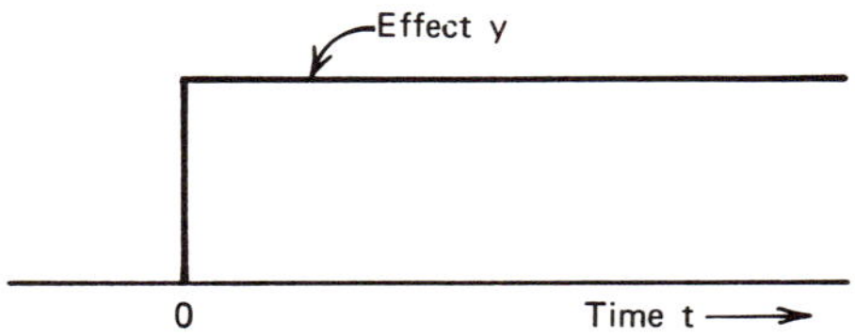

(b)

Figure 6

criteria and compensation design techniques, "robot" systems are always designed to be stable.

Now, let's briefly discuss stability vs. instability.

STABLE OPERATION

In Fig. 8*c* we sketch the input or command x in dotted form, as a reference for comparison with the output or effect y. And we note:

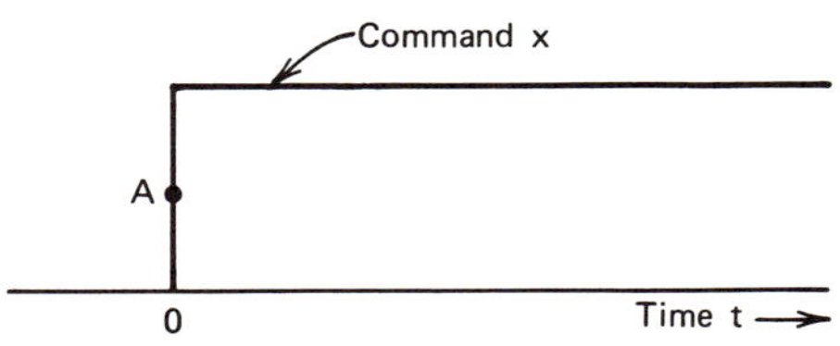

(a)

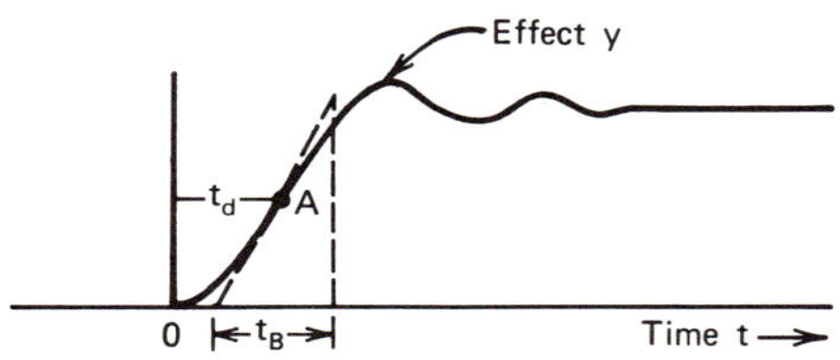

(b)

Figure 7

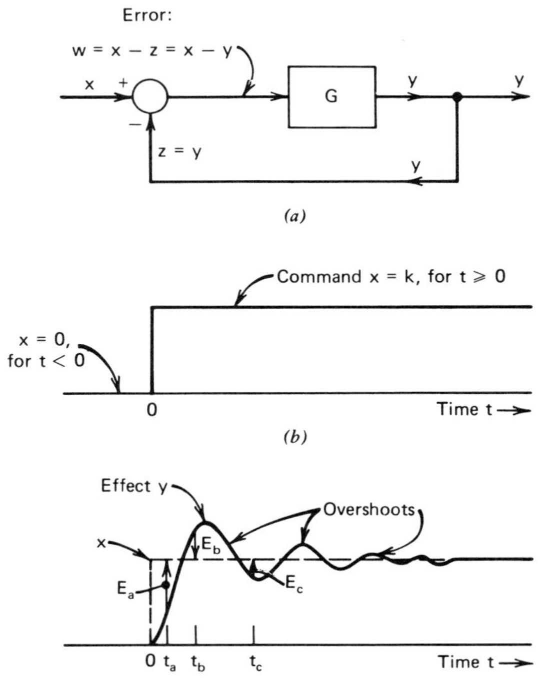

Figure 8 (a–c)

1. At the instant $t = t_a$, x is larger than y in magnitude, and their difference:

$$x - y = E_a$$

is positive. In Fig. 8*c* we use an "upward" arrowhead to indicate E_a as a positive quantity. As E_a is *positive*, the controller tends to *increase* or boost the output magnitude y.
2. As the output magnitude y increases, y becomes greater than x.
3. Now at the instant $t = t_b$, y is greater than x, and their difference

$$x - y = E_b$$

is negative. In Fig. 8*c*, we use a "downward" arrowhead to indicate E_b is a negative quantity. As E_b is *negative*, the controller tends to *decrease* or pull back the output y.
4. As the output magnitude y decreases, y gradually becomes smaller than x.
5. Now at the instant $t = t_c$, y is smaller than x and their difference

$$x - y = E_c$$

is positive again. As E_c is *positive*, the controller tends to *increase* or boost the output y.

6. Repeat this process many times until the output y settles down to a constant value as the input x. Here we have the output y (solid curve) as shown in Fig. 8*c*, which is a reasonable approximation of the input x in Fig. 8*b*.

In this operation, the controller *reacts* with positive or negative error (i.e., E_a, E_b, E_c . . .), and *compensates* the output y by increasing or decreasing it.

We are very fortunate that a well-designed robot neither *overreacts* nor *overcompensates*. As a result, we have stable operations, as just described, with decreasing overshoots and with the output or effect y settling to a constant value as prescribed.

UNSTABLE OPERATION

On the other hand, an unstable operation is described in Fig. 8*d*. The above detailed steps describing a stable operation may be used to describe an unstable operation with a slight modification: the robot *overreacts* and *overcompensates* with the result that the overshoots grow with each cycle as shown in Fig. 8*d*.

With the growing overshoots, where does it lead to?

Theoretically (namely, mathematically), it will lead to overshoots of infinite magnitude, and the overshoots will perpetually grow in magnitude. In technical terms, the output or effect y is called a *runaway transient* response.

The principles of physical science dictate that a phenomenon (say, an output response) of an infinite magnitude requires an infinite amount of energy. And no creature or device or system has an infinite amount of energy, not even a rich robot. Therefore, the unstable response y in Fig. 8*d* cannot grow indefinitely. Instead, it will grow into a *sustained oscillation*, limited by the energy content of the robot, as shown in Fig. 8*e* (where Fig. 8*d* has been shrunk in scale and extended).

Now, let's see what happens with an unstable operation. For a given step-function input or command x as in Fig. 8*b* we wish to obtain an output or effect y identical to x in form. Instead, we obtain the runaway transient y in Fig. 8*d* or the sustained oscillation y in Fig. 8*e*. Of course these responses are not what we intend to have. The robot is functioning unintelligently in a retarded fashion, or is running wild or mad in an insane mood.

In the human world, we put the retarded and the insane in institutions or asylums, and they do live a useful life in a perhaps limited sense. What about robots? Fortunately, the oscillatory robots are useful, in a more limited sense than the (respectable) robots, as *oscillators*. For the use of oscillators, particularly, in electronic circuits or systems, ask any electrical engineer.

REMARKS

It is gratifying to note that even a robot believes in and exercises *moderation*—not to overreact, and not to overcompensate. It is "moderation" that is responsible for the successful operation of the robot in the stable mood.

However, this "moderation" does not come by accident. It is carefully designed

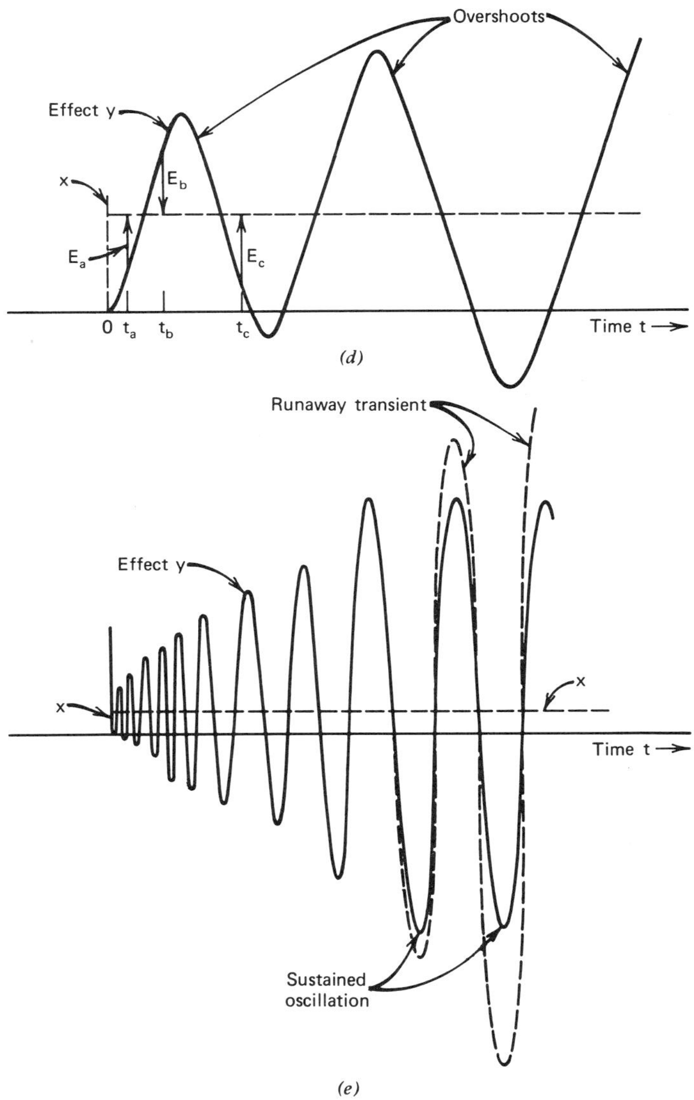

Figure 8 (d–e)

by men of ingenuity. And I shall later discuss these design considerations under the captions of stability criteria and compensation techniques.

Robotosyncrasy 5: Visionary

As human beings, our mental process is actuated by signals in terms of intelligence or emotion. A signal is often accompanied by a noise (i.e., disturbance or doubt) and is corrupted by this noise. If the noise is too large, the signal becomes non-decipherable and meaningless. This is also true for the robot.

We shall consider the signal-and-noise relations of both 1) a controller or crude robot and 2) a robot, where the noise comes from the internal sources, and compare these two situations.

In order to obtain quantitative measures, we shall define the *signal/noise ratio* as:

$$\frac{\text{Signal}}{\text{Noise}} = \frac{x}{n} \left(\text{or} = \frac{y}{m} \right)$$

where a signal x (or y) is accomplished by a noise n (or m) and where x and n (or y and m) are magnitudes in the same unit (e.g., volt). In general, we shall find that *the higher the signal-noise ratio, the more desirable is the situation* in the sense that the signal is less corrupted by the noise.

First, let's consider the controller or crude robot with a gain G. In order to accommodate an internal noise source n, we shall use the schematic in Fig. 9*a* where the controller is split into two components with gains G_1 and G_2, and $G = G_1G_2$. The output of G_1 and the noise n add together, forming the input to G_2.

We then consider the robot with an internal noise source n and a system model as shown in Fig. 9*b*.

Assuming some typical operating values of the various parameters and holding the "same" noise input n for both cases, Rob was able to obtain a) a signal/noise ratio, y/m = 10, for the output of the controller or crude robot in Fig. 9a and b) a signal/noise ratio, y/m = 100, for the output of the robot in Fig. 9b.[3]

It is now obvious that the robot, with a much higher signal/noise in its output, has its output or effect much less corrupted by its internal noise in terms of internal disturbances and self-doubt. In other words, the robot is more *visionary* than a controller or crude robot.

Robotosyncrasy 6: Resolute and Reliable

A controller or crude robot will have dozens, hundreds, thousands (or even millions) of components. To operate satisfactorily, it must have a constant gain G, say G = 500, for its operating frequency range (and will allow the gain characteristic to drop outside this range). Suppose that one or several of these dozens or hundreds or thousands of components "age" or become "faulty," the gain could drop to G = 450 and the result could be catastrophic. (Consider this controller is in the control system of a spaceship!) But, to make all these dozens, hundreds or thousands of components "precision" and "fault-proof " components is both enormously costly and almost impossible. Then, what shall we do?

The robot with its system model actually saved us. This is perhaps the most important of all robotosyncrasies and is based upon a very simple principle.

---

[3]See W. H. Chen, *The Analysis of Linear Systems*, McGraw-Hill Book Co., New York, 1963, pp. 537–545.

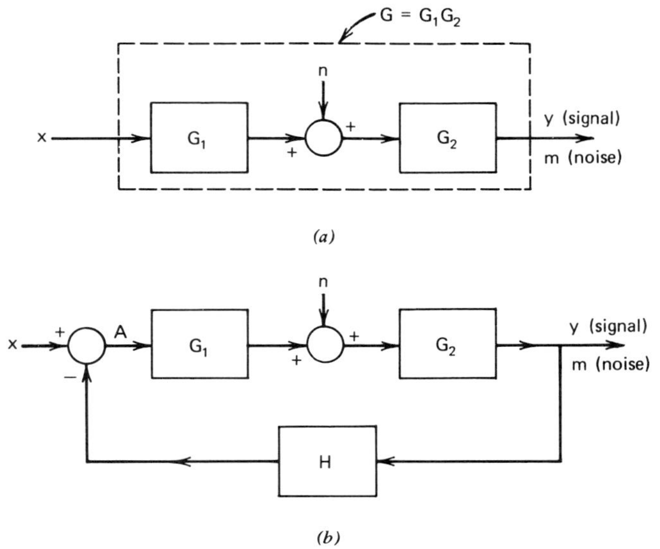

Figure 9

Let's consider the fundamental equation (B), which has been earlier derived for the system model in Fig. 4:

$$G^* = \frac{G}{1 + GH}$$

Suppose that we divide both the numerator and denominator on the right side of the equation by G, thus obtaining:

$$G^* = \frac{G/G}{(1 + GH)/G} \quad \text{or} \quad G^* = \frac{1}{(1/G) + H} \tag{C}$$

When 1/G is very small (and it does matter whether $G = 450$ or $G = 500$) as compared to H, Equation (C) reduces to the following approximation:

$$G^* \cong \frac{1}{H} \tag{D}$$

Even when several components in the controller failed and its gain dropped from $G = 500$ to $G = 450$, the system gain G^* is unchanged and is dependent upon the value of H alone. And we need only to make H reliable to have a reliable robot system.

For some systems, e.g., transistor feedback amplifiers, the H box in Fig. 4 is merely a single resistor. By using a single expensive reliable "precision" resistor for the H box (and using numerous cheap components for the G box, i.e., the controller or transistor amplifiers), a reliable system is created.

We have now shown that a robot is far more *resolute* and *reliable*, with amazing results, than a controller or crude robot.

Criteria for Stability

Rob then went on to discuss the system function, characteristic equation, and open-loop transfer function in terms of stability (Chen[4], pp. 474–478), as well as the Routh's Criterion for Stability (loc. cit., pp. 478–485), Nyquit Criterion for Stability (loc. cit., pp. 489–514), and Root-Locus Method (loc. cit., pp. 487–489).

He also remarked that we must obtain the system function, characteristic equation or open-loop transfer function before we can apply any of these criteria for stability, and that we may use signal-flow techniques (loc. cit., pp. 387–465), to obtain these functions or equations.

Compensation for Improving Stability and Performance

With the criteria for determining stability now at our disposal, Rob continued his discussion with the methods of compensation for improving stability and performance for robot design (loc. cit., pp. 549–569).

Concluding Remarks

Ladies and gentlemen, I have just concluded my technical presentation. I do, however, have some general concluding remarks to make.

Again, I wish to thank you for this significant honor of inviting me to present this keynote speech.

Our two nations, the two superpowers of our time, have been at odds on many issues. However, I am extremely pleased that we now have a common language, the Robotese, that has brought us together today.

We have now learned that the robot is *broad-minded, even-tempered, agile, stable, visionary, resolute* and *reliable*. Shall we then learn from him, the robot of our creation, these virtues?

We have also learned that a wise robot resists self-doubt, exercises moderation, and never overreacts. And now shall we follow his example to rediscover these cardinal rules of nature for the salvation of our own world?

When a standing ovation ebbed, President M. V. Keldysh of the Academy of Sciences of the USSR came forward to the rostrum, presiding over the presentation ceremony. Professor Serov read Dr. Jack Rob Tains' biography and presented Rob to President Keldysh for the conferring of a honorary membership of the Soviet Academy of Sciences with the citation "For extraordinary contributions and services to mankind."

Academician Rob Tains took a deep bow as the finale of this historical event.

---

[4]W. H. Chen, *The Analysis of Linear Systems*, McGraw-Hill, 1963, *or* other appropriate books on closed-loop control systems or linear systems.

Annotations:

1. Wayne Hawaii, *The Robotosyncrasies*, Carlton Press, 1976, is currently out of print. The chapter which is reprinted in our anthology will appear as part II of a new book under the author's real name: Wayne Chen, *The Year of the Robot*, Dilithium Press, 1981. Dr. Chen is dean of the College of Engineering at the University of Florida.

Suggested Additional Readings

To make a brief list of appropriate and enjoyable literature would be difficult. Thus we urge readers to consult these sources for suggestions and guidance.

S. C. Florman, *Engineering and the Liberal Arts: A Technologist's Guide to History, Literature, Philosophy, Art and Music*, McGraw-Hill Book Co., New York, 1968, Ch. 1 (recommended reading), ch. 4.

S. C. Florman, *The Existential Pleasures of Engineering*, St. Martin's, New York, 1976, Chs. 8–11.

A. O. Lewis (editor), *Of Men and Machines*, Dutton, New York, 1963.

J. Winton, "Hey Mac!—the Engineer in Modern English Fiction," *The Chartered Mechanical Engineer*, June 1970, pp. 252–254.

PART 3

Art and Engineering

In art, it is our experience of the process of construction that counts, not the finished product.

Victor Shklovsky

Everything beautiful and noble is the result of reason and calculation.

Baudelaire

There are many definitions of art, and few, if any, recognize engineering as one with painting, sculpture, music, and other more generally accepted artistic endeavors. But at least one concept of art, that of the use of human ingenuity and intelligence to bring something into being, would clearly include engineering. Bronowski (1) expressed this concept very clearly when he said, "There is a likeness between the creative acts of the mind in art and in science [engineering]." This part discusses three aspects of the interaction, commonality, and relationship of art and engineering: creativity; engineering works as subject matter for the artist; and engineering as art.

Note Bronowski's use of the word *creative* in the preceding quotation. Certainly, art in all its many forms is creative; in a like fashion, true engineering is creative for it brings variety together to form a new unity. This production of something original from the raw materials of human thought and activity is creativity. Theodore von Karman (2) extended the argument further when he defined the engineer as one who "creates what has never been." Creativity, then, provides one connection between art and engineering and is a characteristic of equal importance to the mature artist and the accomplished engineer. Sandor pursues this common characteristic in his article "Engineering and Humanistic Creativity—Is There a Difference?" A successful and accomplished mechanical engineer, George Sandor, denies the usual conception of the engineer. A poet, musician, connoisseur of music and ballet, his interests permit him to discuss the bond of creativity between the engineer and the artist from personal knowledge and experience. His article provides an appropriate opening for this chapter.

The interaction of art and engineering extends beyond expression in different forms by creative individuals. "Art reflects society—its aspirations, confusions, and inheritance." (3) Artists—be they painters or poets, sculptors or musicians, architects or engineers—reflect the culture of the era in which they work and the materials, techniques, and knowledge available to them. This may be seen readily in the graphic arts with the introduction of perspective in the early Renaissance, the developing theories of color that were applied by the impressionists, and the pop art of the mid-twentieth century. Note again the architectural and engineering changes that have occurred throughout history, with the use of new materials and new outlooks on what constitutes aesthetic achievement in structure. Of equal significance, "art reflects society" through artists' choices of engineering topics as the subject for many of their works. In still other ways, engineering relates to art as it makes possible the artist's expression and as it is art in itself through its contributions to the aesthetic quality of the artificial landscape. These relationships are introduced here and are illustrated more fully by the selections chosen for this part.

A great deal may be said about the graphic artists' choices of engineering topics for subject matter. Billington (4) has several excellent examples in his article "Engineering Education and the Art Museum," and the book *The Art of the Industrial Revolution* by Francis D. Klingender (5) is devoted to the "area where art and technology meet and interpenetrate." From the works of sixteenth-century Flemish artists who have provided us with excellent illustrations of the machinery and architecture of the era (Pieter Bruegel the Elder, for example) to those of the landscape artists of the nineteenth century, engineering works are pictured frequently. In some instances, the artist reflects the belief of his or her era that technology is the grand achievement of the age, while in other cases the beginnings of a concern of the effect of engineering on nature and people is reflected. The futurists, precisionists, and, more recently, the pop-art advocates of the twentieth century have both glorified and spoofed technology and the accomplishments of engineering through their art. These reactions continue to demonstrate that "art reflects society." Joseph Wright of Derby, J. C. Bourne, J. M. W. Turner, Robert Delaunay, John F. Weir, John Kane, Joseph Stella, and Piet Mondrian are a few of the artists whose works serve as illustration of this facet of the art and engineering relationship.

Unfortunately, the limitations of this book prevent a detailed investigation of the fascinating interplay between the works of graphic artists and those of the engineers. The selected additional readings at the end of the part will provide a starting point for those who wish to pursue this subject further.

Alexis de Tocqueville provides his impressions of our developing nation and its art in the extract from *Democracy in America*, first published in 1840, that is included in this part. The fine arts and the industrial arts in America were not well accepted by most of the world in the nineteenth century. In particular, industrial art was in the period of rapid expansion of the American system of manufacturers—a system that shocked Europe with its many triumphs at the Crystal Palace exhibition in London in 1851. The "Official Description and Illustrated Catalog of the Great Exhibition" described the American exhibit in these words (6):

> . . . impart to the productions of American industry a character distinct from that of many other countries. The expenditure of months or years of labour upon a single article, not to increase its intrinsic value, but solely to augment its cost or its estimation as an object of 'virtu,' is not common in the United States. On the contrary, both manual and mechanical labour are applied with direct reference to increasing the number or the quantity of articles suited to the wants of a whole people. . . .

De Tocqueville's discussion anticipated the Crystal Palace description. He recognized that the social and economic framework of the newly emerging American nation was creating a new attitude toward the arts.

The cooperation between the sculptor as an artist and the engineer was a part of the new attitude toward art. Matthys Levy has drawn attention to the sculptor–engineer relationship in art in his article "Structure and Sculpture." The monumental sculptures represented by the Statue of Liberty by Bartholdi and the modern works of Noguchi, Newman, and Picasso are possible only by the ingenious uses of new materials and the application of engineering mechanics to produce engineered frameworks. Levy also points out several illustrations of the engineer's work that may be viewed in the role of sculpture and, through these illustrations, brings us to an aspect of art that is closest to engineering in its development—architecture.

Architects and engineers have always directed their efforts toward making structures attractive. The way in which this has been accomplished has varied in time depending on materials, needs, and upon society's evaluations. Here again "art reflects society." In the nineteenth and early twentieth centuries, American structural aesthetics were achieved, for the most part, by adding decoration to the finished structure, be it an architectural or engineering product. This is well illustrated for engineering by the massive allegorical figures of the portal sculptures of the Manchester Bridge built in 1874 over the Alleghany River at Pittsburgh. Although add-on decoration produced the pleasing affect that was desired for that period, it did not stand the test of time. "A beautiful structure does not mean an ordinary structure (whatever that may be) with applied ornament added to it" (7). The change in American architecture and the associated engineering concern with aesthetics are the subject of the remaining articles in this part.

The growth of architecture and engineering aesthetics in this country is traced by Kouwenhoven in the article "Two Traditions in Conflict" taken from his classic book, *The Arts in Modern American Civilization.* The development of these two closely related aspects of art reflects the influence of the physical and social characteristics of the United States, developments in materials, construction techniques, and the philosophy of design that grew under the influence of Greenough, Jenney, and Sullivan. The growth of constructed forms suitable to the developing U.S. culture is emphasized. Note that other chapters in Kouwenhoven's book consider similar developments in other aspects of technology and art. We recommend them for further study of this subject.

Engineering construction can achieve aesthetic value through its design. This concept is defined in the selection "Art in Engineering—The Need for a New Criticism" by David P. Billington, which was directed to engineers when it was published by the American Society of Civil Engineers. The author emphasizes a need for aesthetic criticism of design by knowledgeable designers to pursue function and beauty as integral parts of the design process.

The final article in this part was prepared by Mario Salvadori, an eminent engineer with extensive experience in the application of his art of structural engineering as supportive of architectural projects. Salvadori sees a structure as providing a message to its viewers. The interpretation of the message of structure will be influenced by the individual experiences and cultural background of each viewer. The message of structure can be made more acceptable aesthetically by judicious design. Salvadori's thesis and that of Levy both present structure as art, and, as any art form does, the structure provides a message—a message that results from design and that may be greatly different for each viewer.

An engineer's struggles as he attempts to merge his engineering knowledge and innate feeling for beauty in an effort to build a bridge that is a useful, yet aesthetic, addition to the environment of a wild river gorge is described in "The Bridge Builder," an earlier reading in this book (see section 2.13). The result is a work of art developed from human imagination and experience and created with the designer's application of engineering fundamentals. This story, though drawn from fiction, exemplifies numerous real-life examples of the creative act with aesthetic value that is produced by the accomplished engineer.

There are many paths you may take to connect art in its various forms with engineering in its variety of applications. Lewis Mumford (8) has stated in *The Myth of the Machine*: "The purpose of art has never been labor-saving, but labor-loving, a deliberate elaboration of function, form and symbolic ornament to enhance the interest of life itself." Certainly this is the case with the great works of engineering.

References

1. Bronowski, Jacob, *Science and Human Values*, Julian Messner, New York, 1956, p. 14.
2. *The Function of the Engineer & the Scientist*. National Society of Professional Engineers, Washington, D.C., 1962, cover.
3. Plumb, J. H., *The Penguin Book of the Renaissance*, Penguin Books, Ltd., Harmondsworth, England 1964, p. 9.
4. Billington, D.P., "Engineering Education and the Art Museum," *The Eads Bridge*, The Art Museum, Princeton University, Princeton, N.J., 1974, pp. 12–16.
5. Klingender, F. D., *Art and the Industrial Revolution*, edited by A. Elton, Evelyn, Adams & Mackay Ltd., Chatham, England, 1968.
6. "Official Description and Illustrated Catalogue of the Great Exhibition of the Works of Industry of all Nations," as quoted by Nathan Rosenberg, *The American Systems of Manufacturers,* University Press, Edinburgh, 1969, p. 5
7. Faber, Oscar, "The Aesthetic Aspect of Civil Engineering Design," *The Aesthetic Aspect of Civil Engineering Design*, Institution of Civil Engineers, London, 1945, p.3.
8. Mumford, L., *The Myth of the Machine: The Pentagon of Power* (Vol. 2), Harcourt Brace Jovanovich, New York, 1964, p. 137.

3.1

Engineering and Humanistic Creativity—Is There a Difference?[1]

GEORGE N. SANDOR[2]

I. DEFINITION OF ENGINEERS AND HUMANISTS

Reading the title of this paper, you will doubtless ask: What do you mean by "humanities" and by "engineering"?

I am an engineer, so I think I know what engineering is. For the purpose of this comparison of humanistic and engineering creativity, allow me to define engineering this way: *Engineering fills mankind's needs, wants and desires creatively.*

But—you will continue—what about the butcher, the baker, the candlestick maker, don't they fill mankind's needs, wants and desires, too?

Of course they do. But they fill it routinely: day by day, they reproduce much-needed, wanted and desired products that have existed for centuries, using long-standing proven methods. But the engineer worth his salt *creates new* things that did not exist before and new methods to bring them into being. Engineers fill our needs *creatively*.

Now take the humanities. I think I have heard a distinguished humanist colleague say that a poet is not a humanist! Surprised? Of course I was, just as you are now! He said a poet, a fiction writer, a sculptor or a painter is an *artist* not a humanist.

Well, who *is* a humanist?

Historians, linguists, anthropologists, sociologists, archeologists, they are the real

[1]This article was presented as a paper at the Annual Conference of the American Society for Engineering Education, held at Louisina State University, June 24–28, 1979.

[2]Research Professor and Director, Mechanical Engineering Design and Rotordynamics Laboratories, University of Florida; Life Fellow, ASME.

humanists—he said—because they search for the understanding of human traits, human deeds and human events—. On the other hand, other humanist colleagues do include among the humanists the poets, the fiction writers, sculptors and painters, as well as historians, archeologists, anthropologists and sociologists, and many, many more.

Let me adopt this latter view and offer the following definition for a humanist: *Humanists fill mankind's needs, wants and desires creatively.*

Yes, that's right. It is the same definition as that of the engineer.

Honest, I did try to make distinctions:—"engineers apply the physical sciences"—I said—"and humanists . . . "—and here I got stuck, because archeologists, anthropologists and even historians have been known to apply physical sciences!

Here is another try—"Engineers apply economics"—I continued. Yes, but so do sociologists, historians, and all the rest!

Of course there *are* differences. You would not expect a historian to design an automobile. But neither would the engineer who builds bridges attempt to design a jet engine! So, differences prevail between engineers and engineers, as well as between humanists and humanists. And perhaps, if we really tried, by writing long, elaborate, painstakingly circumscribed definitions, we could find some differences between *all* humanists on one hand and *all* engineers on the other.

But my aim in this paper is to try to show the *similarities*, the common goals, the mutual interests, and even the essentially *identical methods* of engineers and humanists. So, please do bear with me, and just for today accept the foregoing two identical definitions for both humanists and engineers.

II. ENGINEERING AND HUMANISTIC TRAITS

Let me start my attempt to show the similarities by paraphrasing the title question of this paper:

What, if any, are the differences and/or similarities between the methods of creative humanists and creative engineers?

To explore this, let us step back one notch and ask a dual pair of questions:

Do Engineers need Humanistic Traits?
Do Humanists need Technical Knowledge?

If so, for what? There are reasons on both sides. Some of these are:

Engineers need humanistic traits

- To discern what will *appeal*, as well as satisfy a *need*;
- To know not only what *can* be done, but also what *should* be done [5].

On the other hand, humanists need technical knowledge

- To *write convincingly* in today's "age of technology";

- To ensure that paintings don't *fade* away or darken into a gravy-brown gloom;
- To discover *new media*;
- To ascertain that sculptures are *safe structurally and mechanically.*

Here are a few examples.

Take the longest surviving automobile of all time: Porsche's original 1933 Volkswagen design, the "Beetle." Not only did it fill a genuine perennial need for economical, efficient, personal transportation, but it also presented, and still does present today, an acceptable, or even pleasing appearance after 46 years! In contrast to this, how many other 1933 models (discounting antique curiosities) are in use today? But count the Beetles passing by!

And another thing about the Beetle. Porsche *could* design large "gas-guzzlers" too, but he knew they *should not be built!*

So, the engineer who created the most durable car-design on record not only filled a *need* with an *appealing* design, but also chose correctly what *should* be done.

How about a dual example from the humanities?

Here is one. What was the secret of the movie scriptwriters for *China Syndrome*, or for the original *Airport* film? Here is what: they both *wrote convincingly*, because they had, or were clever enough to obtain from experts, the necessary level of technical knowledge in nuclear power generation and in air traffic control, respectively.

These are positive examples. I can also quote a negative one, where the lack of technical knowledge of an artist well-nigh led to a disaster. A few years back, a sculptor of large stand-alone mobiles loaned one of his works to the institution where I was teaching. It was a beautiful, huge, upright, stainless steel creation erected in open space. Its tall stanchion supported a freely rotating topmast with a crosswise axle on which two symmetrically juxtaposed large flat parallelepipeds were rotatably mounted. Their horizontal axis of rotation was off their geometric centerline, but they were statically balanced by lead ballast on the short side. In still air they made an impressive sight, having stopped in random out-of-phase positions, gleaming in the sun. When the wind came up, they faced it with uneven surfaces above and below their axes. Hence, in a slight breeze they lazily turned about their horizontal axle and around the vertical axis of the topmast. I often rested my eyes on their fascinating never-repeated random motion. The stronger the breeze, the livelier they became. In a gale or squall they would automatically "feather" and ride out the blow in a weather-cocked position.

That is, they rode it out until one fateful night. Next morning we found one of the "wings" on the ground, sadly crumpled. The sculptor asked me why it came down, and how could this be prevented in the future. His concern was heightened by the fact that the identical twin of this sculpture was installed in a West European city, as he described it, among tall buildings in the middle of a locally narrowing busy thoroughfare. What a Venturi channel to accelerate the wind!

Examining our damaged mobile, I found the horizontal axle across the topmast intact. However, examination of the bearing mounting on the horizontal axle revealed that there was no endwise retention provided and that the rotating "wing" just backed off the axle unhindered!

The "fix" of the mounting was simple, and, to prevent a similar accident overseas, the new design, showing proper endwise retention, was airmailed to Europe. All this could have been avoided by the application of some minimal *technical "know-how"* in the first place!

These examples clearly show that *engineers* do *need humanistic traits* and that *technical knowledge is indispensable* in some *humanistic endeavors.*

III. THE UNIVERSAL CREATIVE PROCESS

And now for a trio of related questions on the engineer's and the humanist's respective methods of creativity:

- Are there similarities?
- Would humanists benefit from adapting the engineer's methods?
- Can engineers use engineering methods to produce a humanistic work?

A brief review of engineers' and humanists' method of creation may throw some light on these. So, here we go.

The engineer solves problems by *design.* One way to picture the creative design process is the practice- and classroom-proven sequence of Fig. 1:

Here's an example how an engineer may follow this method:

Stage One: *Confrontation*	Manufacturer to engineer: "Improve this workplace, the labor cost is too high." Engineer: Collects data on operations, there and elsewhere.
Stage Two: *Formulation*	Engineer finds real need: overall cost reduction. Defines the task: Automate!
Stage Three: *Conceptualization*	Engineer prepares plant layout, conceptual drawings of automatic production machines.
Stage Four: *Synthesis*	Engineer designs plant and machines in every detail.
Stage Five: *Modeling*	Engineer prepares production flowcharts, mathematical and simplified physical models of machine units, etc.
Stage Six: *Analysis*	Engineer performs tests, calculations, defines optimization criteria, optimizes the design.
Stage Seven: *Presentation*	Engineer discusses plans with contractors, machine builders, then presents it to his client, the Manufacturer.

This is an ultra-brief thumbnail sketch of the engineer's method. Iterations between any two stages are likely to occur before a final plan is developed.

And now for the humanist's method. How does the *humanist create* an "opus" (a novel, sculpture, painting, play, concerto, poem, etc.)? The elements of his methods are similar:

"The Seven Stages of Engineering Design" [1]

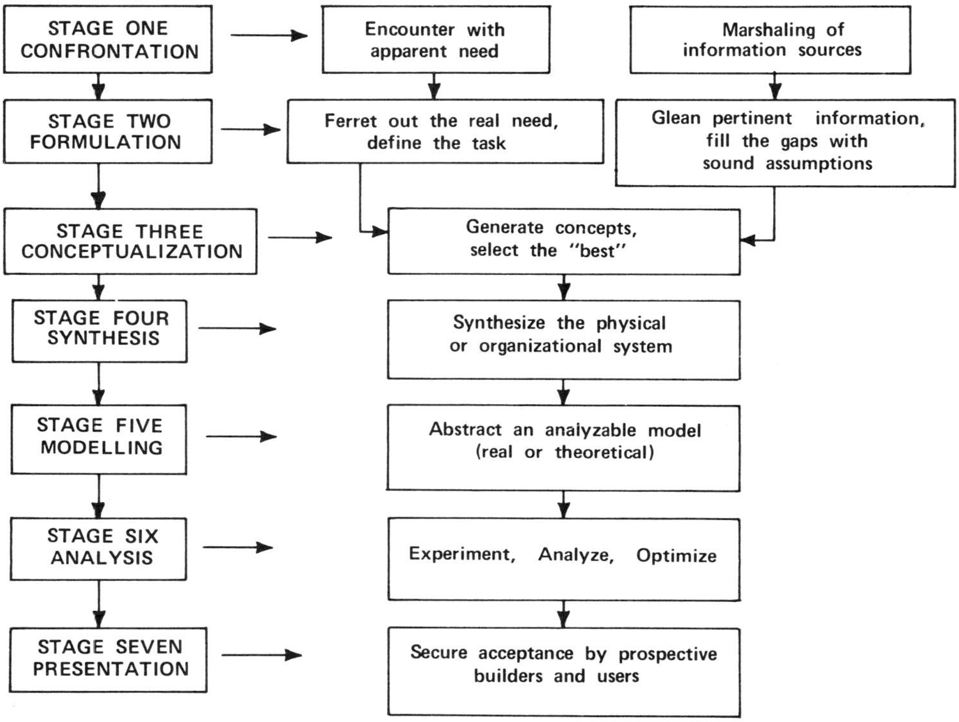

[1] Numbers in brackets identify references at end of paper.

Figure 1

Stage One: *Confrontation* — Let a painter, composer, dancer, poet, writer, or sculptor walk through an automobile assembly plant. What may each "see"?

Painter — A picture of men struggling with machines;

Composer — Clanking din of machinery with a monotonous under-theme;

Poet — Repetitive, plaintive laments of a bored worker;

Writer: — Sad story of a depressed worker and his family life;

Choreographer: — The beauty of coordinated rhythmic movements of men and machines;

Sculptor: — The grim faces and straining muscles of crowded figures.

Information gathering — Each will seek additional *sources of information*—visit workers, consult foremen, go to managers, read up in library, etc. Then each will *select* what interests him most, use imagination to fill the gaps with *assumptions* (What did that silent workman "think"? Why was the foreman "cross"?).

Stage Two: *Formulation*	Define the intended work:
Painter:	Sketch? Etching? Lithograph? Mural?
Composer:	Concerto? Song? Rhapsody? Musical comedy?
Poet	Sonnet? Limerick? Metered verse?
Writer:	Short story? Novel? Stage play?
Choreographer:	Ballet? Pantomime? Solo? Couple? Group?
Stage Three: *Conceptualization:*	
Painter:	Central figure in foreground, others in perspective? Stylized forms?
Sculptor:	Face only? Bust? Full figure?
Composer:	Andante? Scherzo? Allegro? Vivace? Staccato? Main theme, auxiliary themes.
Stage Four: *Synthesis*	
Painter:	Soften and select brushes, mix paints, spread and prepare canvas, make first picture:
Sculptor:	Fashion a wire frame, mix clay, form first version;
Composer:	Sit down to piano, write first version, orchestrate;
Poet:	Write first draft;
Writer:	Detail out the plot, characters, chapter plan; write first draft;
Choreographer:	Detail out motion sequence, plot plan, timing correlation between dancers.
Stage Five: *Modeling*	
Painter:	To see a "model" of the painting, step back, squint to get light values, get overall impression;
Sculptor:	Walk around, project shadows to see silhouette, move light source around to see "self-shadows" on form (shadow of nose on face, of arm on torso, etc.);
Composer:	Play the work with a few representative musicians;
Poet:	Read through fast to get overall impression;
Writer:	Skim through to get overview;
Choreographer:	Go through movements with small group, just a piano (no orchestra).

Stage Six: *Analysis*		Each will "experiment," "analyze," and "optimize" in his own way:
	Painter:	Change this color, that contour, step back and contemplate, try to get "best" effect;
	Sculptor:	Scrape off here, add clay there, reshape the nose, bend the arm, spray with white or bronze, change texture;
	Composer:	Add or omit percussions, change emphasis from strings to woodwinds, tone down the brasses, decide on repeating passages;
	Poet:	Insert or rearrange stanzas, replace a word or phrase, improve or eliminate a rhyme, alter the rhythm;
	Writer:	Add a flashback, replace a narrative with a dialogue, rearrange, rewrite;
	Choreographer:	Speed up or slow down here or there, add or cut out a move, recast the roles.
Stage Seven: *Presentation*		Each will seek to secure acceptance by two categories of people:
	Painter:	1. Sponsor or art dealer; 2. Peers and critics, the public;
	Sculptor:	Ditto;
	Composer:	1. Conductor and concert master; 2. Audience;
	Poet:	1. Publisher (who will want to make money); 2. Listener (who will want to be moved or amused);
	Writer:	1. Publisher (reviewers); 2. Readers.

This completes the paralleling of engineering and humanities creativity. I believe it shows that:

1. There is *no difference in basic elements* of creativity;
2. There are *differences in detail* between various engineering and humanistic procedures, and even between various engineering disciplines as well as between different art forms.

IV. CAN THE ENGINEER'S METHODS HELP HIM CREATE HUMANISTIC WORKS?

The question of whether engineers can use the principles of creativity from their own field to create artistic works, can be treated in the classroom by a demonstration, involving the students themselves. For example, one can form small groups (3 to 5 students), and ask each to prepare a skit or "miniplay" to fill a 90 second "TV spot." The topic can be left to the students, with the hint that it is best to choose one that they can all be excited about: ecology, conservation, hobbies, sports, living habits, nutrition, etc. are typical examples. The objective should be to urge the audience (the rest of the class and the faculty) to take some action in favor of the selected topic: contribute to "save the whales," take up hang-gliding, etc. I usually give them a few hints about organizing the script of their little play:

Stage 1: *Confrontation*	Get attention by a surprise, a shocking statement, or startling action;
Stage 2: *Formulation:*	State your point early and clearly so everyone knows right from the start what you are driving at;
Stage 3: *Concepts*	Give reasons why the topic is important and why the audience should adopt your viewpoint;
Stage 4: *Synthesis*	Develop your subject fully, in as much detail as time permits;
Stage 5: *Modeling*	Give a convincing example, something the audience can identify with or recognize among their own experiences;
Stage 6: *Analysis:*	Compare the example with your subject, show relevance.
Stage 7: *Presentation*	Call for some specific action by the audience.

Next, we take a specific time period, say 30–40 minutes, for each group to prepare the script and "cast" the play. This is followed by a series of presentations or "performances," one by each group, with subsequent critique by peers and faculty.

The result of the entire experience is often a lasting impression on both the engineering and the humanist students that their respective methods of "creation" are not all that different as they might have assumed before. Their perspective on each other's field is opened up somewhat by their participation in this class work.

Here is another example of an engineer's humanistic work: *The Robotosynchrasies* by Wayne Hawaii [2] (real name: Dr. Wayne H. Chen, Dean of Engineering, University of Florida). It is the first work of fiction about real science, not science fiction, an idyllic love story between a newspaper woman and the fictitious inventor of modern feedback control, the basic "robot" and its "robotosynchrasies," which he parallels with human "idiosynchrasies," such as, for example:

- Wide-band frequency response with broadmindedness,
- Uniform gain over response band with even temper,
- Speed of response with agility,
- Stability with moderation,
- Instability with overreaction,
- Selectivity with vision and judgement, and so on.

And in a modest way, even my own review of this book, published in *Mechanical Engineering* [3] and other journals and newspapers, can perhaps be counted among humanistic works by an engineer.

Still another example is a delightful, as yet unpublished small book of poetry [4] written by an outstanding materials engineer, Dr. Larry L. Hench, professor at the University of Florida.

I am sure there are many more examples; perhaps you know of a few yourself. Most of us who, like myself, like to dabble in these areas, will probably never produce any outstanding humanistic works, but we can surely have a lot of fun and enjoyment trying!

V. SUMMARY

An attempt has been made to:

1. Demonstrate that humanists and engineers need acquaintance with each other's fields.
2. Show that their respective methods of "creation" are founded on similar basic elements.
3. Show that engineers can use their own systematic methods to achieve some limited "humanistic" accomplishments.

It is hoped that you, the reader, if you haven't already done so, may now be encouraged to try your hand at it.

References

1. Sandor, George N., "The Seven Stages of Engineering Design," *Mechanical Engineering*, Vol. 86, April 1964, pp. 21–25.
2. Hawaii, Wayne, *The Robotosyncrasies*, Carlton Press, New York, 1976.
3. Sandor, George N., "Science in Fiction but Not Science Fiction," review of Wayne Hawaii's book, *The Robotosyncrasies*, *Mechanical Engineering*, August 1977, pp. 111–112.
4. Hench, Larry L., Unpublished poems, University of Florida, Gainesville, Florida.
5. Schaub, James H., Lecture Notes, University of Florida.

3.2

Of the Spirit in Which the Americans Cultivate the Arts

A. TOCQUEVILLE

It would be to waste the time of my readers and my own, if I strove to demonstrate how the general mediocrity of fortunes, the absence of superfluous wealth, the universal desire of comfort, and the constant efforts by which every one attempts to procure it, make the taste for the useful predominate over the love of the beautiful in the heart of man. Democratic nations, amongst which all these things exist, will therefore cultivate the arts which serve to render life easy, in preference to those whose object is to adorn it. They will habitually prefer the useful to the beautiful, and they will require that the beautiful should be useful.

But I propose to go further; and after having pointed out this first feature, to sketch several others.

It commonly happens that in the ages of privilege the practice of almost all the arts becomes a privilege; and that every profession is a separate walk, upon which not allowable for every one to enter. Even when productive industry is free, the fixed character which belongs to aristocratic nations gradually segregates all the persons who practise the same art, till they form a distinct class, always composed of the same families, whose members are all known to each other, and amongst whom a public opinion of their own, and a species of corporate pride soon spring up. In a class or guild of this kind, each artisan has not only his fortune to make, but his reputation to preserve. He is not exclusively swayed by his own interest, or even by that of his customer, but by that of the body to which he belongs; and the interest of that body is, that each artisan should produce the best possible workmanship. In aristocratic ages, the object of the arts is therefore to manufacture as well as possible,—not with the greatest dispatch, or at the lowest rate.

When, on the contrary, every profession is open to all,—when a multitude of persons are constantly embracing and abandoning it,—and when its several members are strangers to each other, indifferent and from their numbers hardly seen amongst themselves, the social tie is destroyed, and each workman, standing alone, endeavours simply to gain the greatest possible quantity of money at the least possible cost. The will of the customer is then his only limit. But at the same time a corresponding revolution takes place in the customer also. In countries in which riches as well as power are concentrated and retained in the hands of the few, the use of the greater part of this world's goods belongs to a small number of individuals, who are always the same. Necessity, public opinion, or moderate desires exclude all others from the enjoyment of them. As this aristocratic class remains fixed at the pinnacle of greatness on which it stands, without diminution or increase, it is always acted upon by the same wants and affected by them in the same manner. The men of whom it is composed naturally derive from their superior and hereditary position a taste for what is extremely well-made and lasting. This affects the general way of thinking of the nation in relation to the arts. It often occurs, among such a people, that even the peasant will rather go without the objects he covets, than procure them in a state of imperfection. In aristocracies, then, the handicraftsmen work for only a limited number of very fastidious customers: the profit they hope to make depends principally on the perfection of their workmanship.

Such is no longer the case when, all privileges being abolished, ranks are intermingled, and men are for ever rising or sinking upon the ladder of society. Amongst a democratic people a number of citizens always exist whose patrimony is divided and decreasing. They have contracted, under more prosperous circumstances, certain wants, which remain after the means of satisfying such wants are gone; and they are anxiously looking out for some surreptitious method of providing for them. On the other hand, there are always in democracies a large number of men whose fortune is upon the increase, but whose desires grow much faster than their fortunes: and who gloat upon the gifts of wealth in anticipation, long before they have means to command them. Such men are eager to find some short cut to these gratifications, already almost within their reach. From the combination of these two causes the result is, that in democracies there are always a multitude of individuals whose wants are above their means, and who are very willing to take up with imperfect satisfaction, rather than abandon the object of their desires.

The artisan readily understands these passions, for he himself partakes in them: in an aristocracy he would seek to sell his workmanship at a high price to the few; he now conceives that the more expeditious way of getting rich is to sell them at a low price to all. But there are only two ways of lowering the price of commodities. The first is to discover some better, shorter, and more ingenious method of producing them; the second is to manufacture a larger quantity of goods, nearly similar, but of less value. Amongst a democratic population, all the intellectual faculties of the workman are directed to these two objects: he strives to invent methods which may enable him not only to work better, but quicker and cheaper; or, if he cannot succeed in that, to diminish the intrinsic qualities of the thing he makes, without rendering it wholly unfit for the use for which it is intended. When none but the wealthy had watches, they were almost all very good ones; few are now made which are worth much, but everybody has one in his pocket. Thus the democratic principle not only tends to direct the human mind to the useful arts, but it induces the artisan to produce with great rapidity a quantity of imperfect commodities, and the consumer to content himself with these commodities.

Not that in democracies the arts are incapable of producing very commendable works, if such be required. This may occasionally be the case, if customers appear who are ready to pay for time and trouble. In this rivalry of every kind of industry,—in the midst of this immense competition and these countless experiments, some excellent workmen are formed who reach the utmost limits of their craft. But they have rarely an opportunity of displaying what they can do; they are scrupulously sparing of their powers; they remain in a state of accomplished mediocrity, which condemns itself, and, though it be very well able to shoot beyond the mark before it, aims only at what it hits. In aristocracies, on the contrary, workmen always do all they can; and when they stop, it is because they have reached the limit of their attainments.

When I arrive in a country where I find some of the finest productions of the arts, I learn from this fact nothing of the social condition or of the political constitution of the country. But if I perceive that the productions of the arts are generally of an inferior quality, very abundant and very cheap, I am convinced that, amongst the people where this occurs, privilege is on the decline, and that ranks are beginning to intermingle, and will soon be confounded together.

The handicraftsmen of democratic ages endeavour not only to bring their useful productions within the reach of the whole community, but they strive to give to all their commodities attractive qualities which they do not in reality possess. In the confusion of all ranks every one hopes to appear what he is not, and makes great exertions to succeed in this object. This sentiment indeed, which is but too natural to the heart of man, does not originate in the democratic principle; but that principle applies it to material objects. To mimic virtue is of every age; but the hypocrisy of luxury belongs more particularly to the ages of democracy.

To satisfy these new cravings of human vanity, the arts have recourse to every species of imposture: and these devices sometimes go so far as to defeat their own purpose. Imitation diamonds are now made which may be easily mistaken for real ones; as soon as the art of fabricating false diamonds shall have reached so high a degree of perfection that they cannot be distinguished from real ones, it is probable that both one and the other will be abandoned, and become mere pebbles again.

This leads me to speak of those arts which are called the fine arts, by way of distinction. I do not believe that it is a necessary effect of a democratic social condition and of democratic institutions to diminish the number of men who cultivate the fine arts; but these causes exert a very powerful influence on the manner in which these arts are cultivated. Many of those who had already contracted a taste for the fine arts are impoverished: on the other hand, many of those who are not yet rich begin to conceive that taste, at least by imitation; and the number of consumers increases, but opulent and fastidious consumers become more scarce. Something analogous to what I have already pointed out in the useful arts then takes place in the fine arts; the productions of artists are more numerous, but the merit of each production is diminished. No longer able to soar to what is great, they cultivate what is pretty and elegant; and appearance is more attended to than reality.

In aristocracies a few great pictures are produced; in democratic countries, a vast number of insignificant ones. In the former, statues are raised of bronze; in the latter, they are modelled in plaster.

When I arrived for the first time at New York, by that part of the Atlantic Ocean which is called the Narrows, I was surprised to perceive along the shore, at some distance from the city, a considerable number of little palaces of white marble, several of which were built after the models of ancient architecture. When I went the next day to inspect more closely the building which had particularly attracted my notice, I found that its walls were of whitewashed brick, and its columns of painted wood. All the edifices which I had admired the night before were of the same kind.

The social condition and the institutions of democracy impart, moreover, certain peculiar tendencies to all the imitative arts which it is easy to point out. They frequently withdraw them from the delineation of the soul to fix them exclusively on that of the body: and they substitute the representation of motion and sensation for that of sentiment and thought: in a word, they put the Real in the place of the Ideal.

I doubt whether Raphael studied the minutest intricacies of the mechanism of the human body as thoroughly as the draftsmen of our own time.[1] He did not attach the same importance to rigorous accuracy on this point as they do, because he aspired to surpass nature. He sought to make of man something which should be superior to man, and to embellish beauty's self. David and his scholars were, on the contrary, as good anatomists as they were good painters. They wonderfully depicted the models which they had before their eyes, but they rarely imagined anything beyond them: they followed nature with fidelity: whilst Raphael sought for something better than nature. They have left us an exact portraiture of man; but he discloses in his works a glimpse of the Divinity.

This remark as to the manner of treating a subject is no less applicable to the choice of it. The painters of the Middle Ages generally sought far above themselves, and away from their own time, for mighty subjects, which left to their imagination an unbounded range. Our painters frequently employ their talents in the exact imitation of the details of private life, which they have always before their eyes; and they are for ever copying trivial objects, the originals of which are only too abundant in nature.

Notes

1. [I trust I may be allowed to enter a protest against an opinion which some readers will perhaps agree with me in thinking a mistaken one. Whoever has examined the accurate study, the close fidelity to nature, and the profound scientific knowledge of Raphael and of almost all the great masters of art, in their drawings, (where the means by which their perfect works were achieved are more easily discernible than amidst the total splendour of those works themselves,) will, I think, prefer them to the crude design or the pedantic modelling of more recent times. Nor has the industry of the elder artists in these details ever been surpassed. For instance, in the series of sketches for Raphael's finest works, one represents the skeletons of the whole group, in perfect and not unmeaning osteology; a second reproduces the same group *fleshed* with complete anatomical science; in a third, the artist proceeds to put drapery on his figures. To compare the drawing of Raphael with that of David, is to compare the science of a surgeon with that of a butcher. The former penetrated by his art into the hidden beauty and truth of nature: the latter dragged nature to the easel, and deprived her at once of life, truth, and freedom.

   I would be understood to confine this remark to the illustration here, as I think, inappropriately selected; for of the general influence of democracy on the choice of subjects, on the range of art, on the spirit, faith, and education of artists, and, last though not least, on the sympathy of the public, the passage above gives a very true analysis. The pictures of Ostade, Teniers, and Paul Potter belong as truly to the burghers of Holland, as those of Raphael and Michael Angelo to the aristocracy of the Church, or those of Titian and Velasquez, Rubens and Vandyke to the courts and palaces of Europe,—*Translator's Note.*]

3.3

Structure and Sculpture

MATTHYS P. LEVY

This talk has as its origin two events. The first has to do with my involvement over the years with the design of numerous sculptures. The second concerns the movement of architecture called Post-Modernism for which Robert Venturi was the theoretician and which was legitimized by Philip Johnson. This architectural trend reintroduces decoration or sculptural facades to buildings. In the 1930's, the American roadside was dotted with restaurants in the shape of giant ducks and gas stations in the shape of large shells. Now we have the Chippendale broken pediment at the top of the ATT Building currently under construction in New York, as the "crowning" achievement of Philip Johnson. Whatever the reasons for this trend, and I can think of many . . . a reaction against sameness, or the antiseptic look of the international style, or a desire for humanism in architecture . . . the trend is nevertheless with us and means that the engineer is once more involved in architectural sculpture. A similar trend can be seen in art as super-realism, in which we see a photo-like multiple portrait of Marilyn Monroe in multi-colors. It seems we want to recognize what we are looking at. Perhaps we'll now see gas stations responding to the energy situation: in the shape of gushing geysers on even days and sputtering, sagging towers on odd days.

I believe it is important to place this movement in perspective by looking at how engineers have handled this involvement in sculpture in the past.

In 1881 Frederic August Bartholdi designed a statue, Liberty Enlightening the World (Fig. 1). The plaster model showed a feminine figure with raised torch and spiked crown which was to be translated to a statue 151 feet tall, with a head 17 feet from chin to cranium and an index finger eight feet long. Although it was originally designed to be a gift to the United States for the 1876 Centennial, problems of fund raising and other delays resulted in delivery of the work to its harbor location at the entrance to the New York Harbor in 1885.

This was to be the greatest statue ever built, towering over the 60-foot high Sphinx of Giza and the 100-foot-high Colossus of Rhodes of ancient times and the

Figure 1

fourteenth century bronze Buddha in Kamakura. Bartholdi, unsure of how to translate his plaster model into the modern colossus that was to dominate New York Harbor, approached his friend Gustave Eiffel, an engineer renowned for his steel bridge designs. Eiffel proposed a skeletal framework consisting of four major cross-braced columns to be anchored to a granite pedestal which the Americans were to provide on Bedloe's Island (Fig. 2). From this column structure, a trussed frame projects to support the copper sheathing which is the visible surface of the sculpture. The arm was built of a similarly cross-braced box frame cantilevering from the column structure. The total structure, weighing 81 tons, supports the 100-ton weight of hammered copper and provides the necessary rigidity to sustain the ever present wind force.

Eiffel, who was fifty-five years old at the time, was still to produce his own outstanding achievement for the 1889 Paris Exposition.

Winning a competition for a tower 300 meters high, Eiffel's design was chosen over 700 submittals. When it was published in 1886, the design caused a group of 59 artists, including such well-known names as deMaupassant and Dumas, to publish a manifesto, *Protestation Des Artistes* which read in part,... "We...

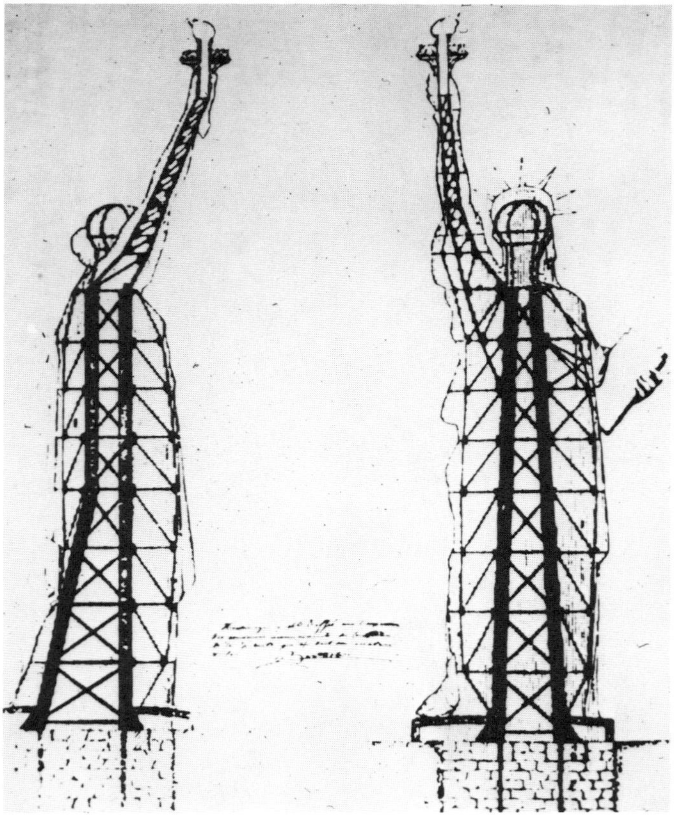

Figure 2

protest with all our strength and wrath . . . against the erection in the heart of our capital of the useless and monstrous Eiffel Tower . . . This arrogant iron mongery . . . hollow candlestick . . . assemblage of ladders . . . disgraceful skeleton . . . even commercial America wouldn't want it."

From the base 125 meters square, the four trussed legs of the tower began to rise in July 1887. On March 30, 1889 the last of 2.5 million rivets was hammered home. Within a year of its opening in May of that year, Eiffel had almost recouped the $1.6 million construction cost from over 2 million visitors.

The tower itself, weighing 8,000 tons, was designed to support the 10,000 visitors weighing 800 tons as well as resisting the forces of the strongest measured winds (100 mph) with a lateral deflection of only 5 inches. It is also interesting to note that a winter freeze can shrink the height of the tower by 6 inches.

Here was Eiffel, the engineer, involved in two works which had as their origin two quite different premises. The first was a sculpture to which the structure brought strength and stability. The second is a structure which has an aesthetic value derived from the cultivated sensibility of its originator. Eiffel said, "The first principle of architectural beauty is that the essential lines of a construction be determined by a perfect appropriateness to its use . . . the curves of the four piers

Figure 3

(of the tower) rising from an enormous base and narrowing toward the top, will give a great impression of strength and beauty".

Sculpture can be defined as the act of representing observed or imagined objects in solid materials and in three dimensions. Starting from an imitation of natural objects, some of which were embodied with music-like abstract beauty, sculpture has changed from classicism to the impressionistic realism of Rodin to become today an art of abstraction. It is this abstract beauty which Eiffel's tower represents. The structure of the Statue of Liberty is interesting from a technical viewpoint. In this work, it is the statue itself which is the art. On the other hand, it is the structure itself that gave life and meaning to the Eiffel Tower (Fig. 3).

In discussing sculpture as structure, we are not concerned with art in the gallery-museum context, but rather with sculpture in the large outdoor environment. It is in this area, that we find the large sculptures of structural significance.

Among the seven wonders of the ancient world, we find two examples of giant sculpture: the statue of Zeus at Olympia and the Colossus of Rhodes (Fig. 4). The first of these, the statue attributed to Phidias, was a forty-foot high marble representation of a seated figure. It undoubtedly presented few structural problems, being essentially a very stable element carved out of stone.

Figure 4

The bronze statue of the Sun God Helios, which is believed to have straddled the harbor of Rhodes, is a more incredible achievement. The sculptor, Chares, a native of Lindus, spent twelve years molding the colossus out of the spoils left by Demetrius Poliorcetes when he raised seige to the city. Standing 100 feet tall with a beacon light in its upraised right hand, the statue stood from 290 B.C., withstanding the forces of the same winds which were powering the ships that passed between its legs—all this, without benefit of the scientific knowledge which permitted Eiffel to design the Statue of Liberty. In the year 225 B.C., the statue was thrown down into the harbor by an earthquake and lay broken for 800 years out of superstitious reverence before being sold by the Saracens in 653 A.D. Even today with a sophisticated knowledge of dynamics, we are sometimes unable to adequately protect ourselves against the destructive forces generated by earthquakes. The sculptor of the Colossus must still be admired for designing a structure that successfully withstood the fickle Aegean winds for almost 65 years.

In Genesis, we read the story of the Tower of Babel . . . "whose top would reach into heaven." Man has always been intrigued with the concept of building towers that reach upward higher and higher. Some of these originated from military considerations such as the towers of Bologna, the highest of which, the Torre Asinelli, reached 320 feet. These were places of refuge from wars and conflict for noble families of the twelfth century. The highest brick tower built in that period is

Figure 5

the 334 foot (102m) Torre del Mangia (1348), built by Minuccio and Francesco di Rinaldo in Siena, so called from the nickname of one of its ravenously hungry bellringers.

Towers responding to religious purposes are for example, the church towers of the Gothic period, the muslim minarets from which the muezzin gives the call to prayer and the eastern pagoda, a multi-storied temple (Fig. 5). The obelisks of ancient Egypt are square, slightly tapered monumental granite pillars, usually ten diameters in height and topped with a 60° pyramidion, which were built to celebrate important personages or deities.

Secular towers were built to satisfy a specific function. Examples are the old clock towers of Europe and the modern towers demonstrating either technical achievement at World's Fairs, such as the Eiffel Tower, the Trylon at the 1939 N.Y. World's Fair or the Needle at the Seattle World's fair, or monuments such as the Washington Monument.

All towers are structural achievements which may or may not have sculptural value. Those that survive over time are usually imbued with artistic-sculptural value. The Washington Monument was originally designed by Robert Mills as a decorated classical obelisk 183m high rising from a circular colonnaded building at the base, planned to be 76m in diameter and 30m high. Ground was broken for the project in 1848 and work was stopped in 1854, being beset by money and political

Figure 6

problems at the time. In 1896, work was restarted by the Corps of Engineers who first decided to reinforce the underdesigned masonry foundation with concrete. This was to be no leaning tower of Washington. The tower was finally topped out at 169 meters, devoid of the originally planned embellishment. The total weight of this marble shaft is 90,000 tons, over ten times greater than the higher Eiffel Tower. Stability of this slender tower to resist wind forces is achieved totally through its dead weight.

The tallest free-standing tower built to date is the CN tower in Toronto. It extends 522m to the upper observation deck and 623m to the top of the antenna. Although obviously a supreme technical achievement, its sculptural value awaits the test of historical perspective.

Moving for a moment from the artist-sculptor to the engineer as sculptor, consider that bridges have always provided the opportunity for creating a high impact, three-dimensional object in the environment. The Roman viaducts such as the Pont du Gard are supreme examples of structural logic with their multi-tiered arches executed with sensitivity toward form and a clear respect for the masonry materials used.

Robert Maillart, the Swiss designer and builder of bridges, best exemplifies a marriage of structure with nature which results in a plastic form that is clearly a man made object yet, "appear(s) to spring out of the shapeless crags with the serene inevitability of Greek Temples"* His bridges all appear to evolve out of slabs: flat, curved and cut like paper dolls, then joined together to produce minimal structural elements. This simplicity results in a structural elegance that is strongly sculptural. This optimization of structure carried out by removing all that is non-functional was also followed by Pier Luigi Nervi in what he called his "form resistant structure." The Gatti Wool Mill (Fig. 6) built in 1951 is an excellent example. Like a delicately woven spider's web, ribs in the slab are developed along isostatic lines. The resultant lace work of ribs presents a minimal expression of structural action and is strongly sculptural. Nervi, by using ferro-cement forms, freed himself from the constraint of standard framework which results in the familiar rectilinear waffle slab.

Another good example of this approach appears in the shaping of columns in the multi-story Banque Lambert conceived by Paul Weidlinger. The sculptural use of the tapered column shape and the hinge derives from purely structural considerations, yet creates its own aesthetic value.

Let us now shift back to the modern artist-sculptor. Two factors make modern large-scale sculpture problematic from the structural viewpoint. First, the sculptor is concerned with forms that appear to defy gravity, as for instance, Isamu Noguchi's "Cube" (Fig. 7) and Barnett Newman's "Broken Obelisk" (Fig. 8). Second, the materials used, such as steel, aluminimum and plastics, are strong and lightweight but very flexible as compared to the traditional stone and bronze. It becomes very difficult to keep in mind these two factors when translating what is often a foot-high model into a 40- or 50-foot high prototype. The linear translation of the scale of one to the other leads generally to difficulties with stability.

The 50-foot high winged sculpture by Picasso for the Chicago Civic Center was conceived from a one-foot high cardboard model. The scale translation introduced not only dimensional problems, but questions of vibration, flutter, buckling and overturning (Fig. 9).

It was necessary to subject a scale model of the final structure to a wind model test prior to the execution of the prototype. These tests caused the artist to slightly modify his concepts.

The group of Four Trees by Dubuffet (Fig. 10) for Chase Manhattan Plaza bears an interesting similarity to the Statue of Liberty. Both are by foreign artists for placement in the United States. Both were structurally designed by eminent engineers of their time; both contain a trussed steel armature which in the case of Dubuffet is of pipes; both were fabricated in France and assembled in New York. The skin material is one difference; the Dubuffet has a fiber glass reinforced polyester skin compared to the copper skin of the Bartholdi.

Going even farther afield is a movement which challenges the ingenuity of the engineer.

Since 1972, SITE, an acronym for Sculpture in the Environment, a group of

*S. Giedion, *Space, Time, & Architecture.*

Figure 7

Figure 8

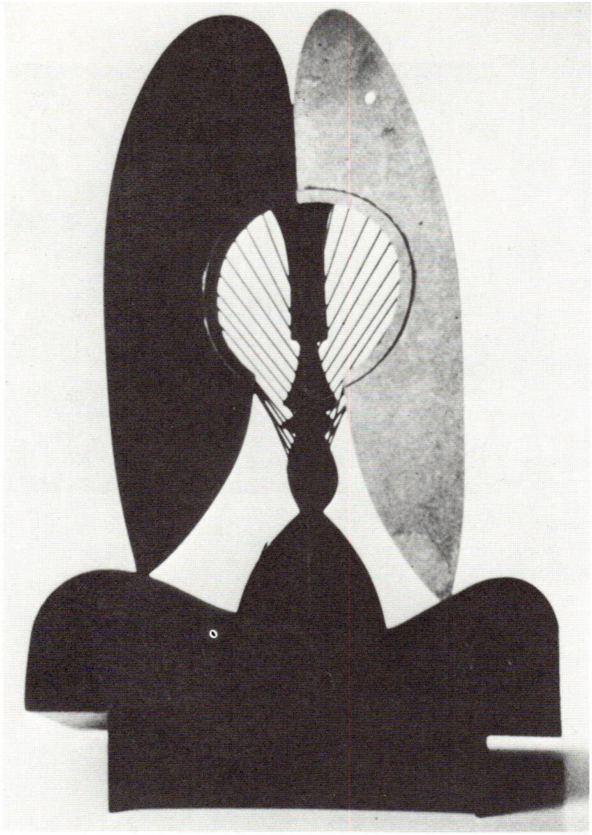

Figure 9

Figure 10

Figure 11

Figure 12

Figure 13

artists, architects and technicians, has been challenging our vision of the conventional shopping center. Starting with an alteration to the Best Products showroom in Richmond, Va., SITE transformed the ordinary facade into a strong hint of decay portrayed in the peeling facade. We are suddenly confronted with disunity and a touch of irony and are drawn into an ordinary showroom expecting further surprises. We may be disappointed, but judging by the sales figures, we do shop there (Fig. 11).

In a similar vein, ruins have always provided instant gratification. Walking through the Acropolis, we project ourselves through time and in our mind's eye complete the picture of what existed based on the remnants we actually see. Why should we wait a thousand years for Alexander to burn Persepolis (325 B.C.) or for a Venetian lieutenant to fire a mortar shell at the Parthenon (1687)?

We can now create the instant ruin with the benefit of modern technology as was done by SITE in Houston, (1974). We no longer can identify the result as representing either "before or after"; i.e., before completion of construction or after start of demolition (Fig. 12). Our sense of order is thrown off balance.

The tilted facade in Towson, Md., is a clear dislocation, a deliberate attempt to wrench the building apart (perhaps to gain entrance?) (Fig. 13).

In all of these seemingly simple-appearing transformations lie structures of some complexity. The structure does not impose—it is not supposed to do so! The structure performs its function, namely to provide support for the sculptural object in the same way Eiffel did for the Statue of Liberty.

The contrast is most clearly illustrated in these examples between structure as sculpture and sculpture as structure. In the first case, structure dominates and provides the message and in the second, structure is the servant contributing an inner strength, but not intruding. In both cases, structure is an integral part of the visual-aesthetic result.

3.4

Two Traditions in Conflict

J. A. KOUWENHOVEN

It was in the various branches of technology that the vernacular tradition first and most uninhibitedly displayed its characteristics, and we have therefore attempted to define it in terms of tools and machines. But even in machine design, as we have seen, it interacted with the tradition of cultivated taste which flowed into our national life from the reservoirs of western European culture. Actually the two traditions mingle from the beginning in all branches of the arts; it is in their interpenetration and in their alternate ascendancy in the work of different men and different periods that the history of American art consists.

To make clear the nature of this interaction, let us turn for a moment to a consideration of the art of building. Throughout the Western world during the nineteenth century there was a disastrous separation between engineering and architecture.[1] Academic architecture, swept along on the flood of classic, Gothic, and Renaissance revivals which culminated in eclecticism, became more and more archaeologically-minded as the century progressed. It was ornament, not construction, that it adopted as its province, and architects seemed increasingly to share the belief of James Ferguson—one of the century's most influential writers on the subject—that "where the engineer leaves off the art of the architect begins." Given the engineer's structural materials, the architect had merely to arrange them "artistically," as the phrase was, and then add ornament.

In the United States it was in such academic architecture that the tradition of cultivated taste found expression. Moreover, it was this tradition that produced some of the most attractive buildings of the century. Richard Upjohn's Trinity Church at the head of Wall Street in New York, Minard Lafever's Holy Trinity in Brooklyn, and James Renwick's St. Patrick's Cathedral on Fifth Avenue are all charmingly successful imitations of medieval Gothic forms, and the national Capitol in Washington, with its vast dome, is an impressive echo of classic styles.

[1] The phrase is Talbot Hamlin's in his magnificent history of *Architecture Through the Ages.*

The very success of such buildings, however, did much to establish in America the dichotomy between architecture and engineering. If Renwick could get away with Gothic nave vaults of papier-mâché painted to look like stone, and if Thomas U. Walter's dome for the Capitol could be so impressive in spite of the structural dishonesty of its iron members, disguised as masonry, why should architecture concern itself with the logic of structural expression?

The development of academic styles necessarily figures largely in the history of the practice of architecture in the United States, and we can learn much about our cultural limitations and aspirations from the way in which we tried to adapt a variety of imported forms and styles to our needs. But, taken by themselves, without reference to the enormous quantity of non-academic and non-professional building which our people produced, these borrowed modes mislead rather than inform us about American civilization, however charming they may be. More significant, from our point of view, are the constructive techniques of the vernacular tradition, which can be seen in their purest form in the technological features of construction—in engineering itself.

National unity, in a land so vast and geographically diverse as ours, could not have been achieved without roads, bridges, canals, railroads, and other means of intercommunication, and farsighted Americans early in our history were well aware that the consolidation and expansion of the nation as a political, social, and economic unit would depend upon technological developments. As early as 1785 Washington was preoccupied with the necessity for developing inland navigation in order to bring the Western settlements in close connection with the Atlantic states. "Without this," he argued, "I can easily conceive they will have different views, separate interests, and other connections." And Thomas Pope, shipbuilder and bridge designer, explicitly stated in his *Treatise on Bridge Architecture* (1811) that no real physical union of the country could take place without "the building of bridges, the digging of canals, and the making of sound turnpike roads."

The speed with which canals and bridges were constructed in America was astonishing to Europeans. With the coming of the canals and railroads it was necessary to span rivers, streams, and chasms cheaply and rapidly, and new techniques were used ingeniously to meet the new needs.[2] David Stevenson, the English civil engineer, traveled widely in this country in the thirties and was struck by the temporary and apparently unfinished state of our canals. "Undressed slopes of cuttings and embankments," he wrote, "roughly built rubble arches, stone parapet-walls coped with timber, and canal locks wholly constructed of that material, everywhere offend the eye accustomed to view European workmanship." But, he added, although the works were wanting in finish, they served their purpose efficiently, and they had the advantage that, as traffic increased, they could be enlarged and improved "without the mortification of destroying expensive and substantial works of masonry." Had some of the Erie barge canal's locks not been originally built of stone, the canal would, he pointed out, have been converted into a ship canal long before. But most construction was of wood, in

[2]It was the "ingenious novelty of invention" displayed in American engineering works to which the Englishman, John Weale, hoped to call attention in the illustrated folio he published in London in 1841. See *The Public Works of the United States of America*, edited by William Strickland, Edward Gill, and Henry R. Campell, London, 1841, with prefatory "Advertisement" by John Weale.

bridges as well as in canal locks and aqueducts, and all kinds of experiments were tried with it. Most important, the techniques developed in wood construction were carried over into later work in more durable materials.

By the seventies, for instance, it had become apparent that the practice of American engineers in iron bridges differed widely from that of their European contemporaries. European bridge builders had always aimed to build structures as strong and safe and durable as possible, cost and speed of construction having been secondary considerations. For hundreds of years before the introduction of iron as a building material, massive stone-arch bridges had been painstakingly erected across European rivers. It had been only natural, therefore, that when in 1793–96 Rowland Burdon built the famous cast-iron bridge at Sunderland, England (based on plans worked out some years before by Thomas Paine, the great propagandist of the American Revolution), he had adapted the methods of stone vaulting to iron construction. The six ribs which formed the 236-foot arch were made up of cast-iron panels which served roughly the same structural function as the wedge-shaped stones in a masonry arch. Even in later bridges European engineers tended to perpetuate in iron the proportions which had been familiar in stonework.

The contrast between European and American practice was made clear in a paper read by Thomas C. Clarke before a meeting of the American Institute of Mining Engineers in 1876. In America the aim of the builders had necessarily been to erect a bridge as rapidly and as cheaply as possible. Lack of capital and the necessity for haste led the early railroad-bridge builder, like the canal builder, to use the most abundant material—wood—and the same influences prompted him to design the bridge so that it could be put together with the utmost rapidity. Hence, as Clarke said, when we began to build our iron bridges, we copied the proportions already established as the most economical in wooden trusses instead of copying the more massive proportions of stone; and rather than rivet the separate parts of the bridge together on the scaffolds (as a mason would cement together the separate stones), we copied the techniques of timber construction—using tenons and sliding joints for the compressive members, and pins and eyebars for those in tension. . . .

The marked feature of the American method, according to Clarke, was the special use of special machine tools, by which the sizes and lengths of all the parts were fitted with the utmost exactness at the place of manufacture. In other words, the parts of the bridge were prefabricated by machinery and could be rapidly assembled on the spot; whereas in the European system of riveted lattice construction assembly was often tediously slow.

The proportions adopted by American designers and their methods of construction both resulted in great economy of material. The result was that when, for instance, English, continental, and American bridge builders submitted competing designs in 1876 for a bridge over the Minamidic River on the Intercolonial Railway of Canada, the American plans called on the average for roughly half the weight of materials required by their competitors. Similarly, when Roebling had completed the Niagara Suspension Bridge (1851–55)—the first large railroad suspension span in the world—he was able to write in his report to the directors of the company: "The work which you did me the honor to entrust to my charge has cost less than

$400,000. The same object accomplished in Europe would have cost four millions without securing a better purpose, or insuring greater safety."

The early builders of wooden-truss bridges, like the early designers of machinery, had worked by rule of thumb. The carpenter-builder had a feel for the strength of timber but no precise knowledge of its capacity to bear loads. Not until 1847—when Squire Whipple, of Utica, New York, published his analysis of bridge building—was there any source of accurate quantitative information on stresses in bridge trusses. The consequence was that some early bridges were uneconomically strong; but the rivalry between the patentees of various wooden trusses in the early days, and the subsequent competition between the various iron-bridge manufacturing companies (like Keystone, Phoenix, and Baltimore) constantly encouraged economy of materials and standardization of design.

In the realm of house building, as in bridge building, the prevalence of wood construction in America as opposed to the stone or brick construction of Europe led to significant differences in architectural fundamentals.

In the twenty-five years immediately preceding the Revolutionary War the tide of Adam's and Chambers' classic revival architecture, which had already swept Gothic, Tudor, and Jacobean ideas out of favor in Britain, reached its height in the American colonies. Almost all the early classic revival buildings in England had been of stone or of brick faced with stucco. But in America the plentifulness of wood had resulted in the development of more skilled carpenters than masons; so the New England builder usually approximated in wood the stonework details he found illustrated in the English books he imported. But modifications of the original designs began to appear almost immediately. Cornices became smaller and more delicate, mantels became daintier, and all the decorative details took on more and more the essential characteristics of wood.

But far more important than these modifications of borrowed forms was the way in which wood construction cooperated with the social system to undermine the cultivated tradition and develop new and more flexible forms. As would be expected, the cultivated classes regarded wood as an inferior material. For one thing it was perishable, and subject to the dangers of storm and fire. But from the earliest days of the Republic there were other and weightier arguments in favor of brick and stone. The article on architecture in *The New and Complete American Encyclopaedia* (New York, 1805) baldly put it thus:

> Considered politically, there is this good attending brick buildings: from durable habitations, in which more money has been spent, and more of the refined tastes gratified, an affection for the soil is increased. . . . But the last and highest consideration which strikes us is that emigration would be less easy, and not so common amongst us, were a finer spirit of building to prevail. The facility with which we *may* move, is a strong incentive to that love of change which it particularly interests us to repress in our citizens.

But change was irresistible under democratic political institutions. Ruskin might appropriately argue in Britain that the right to have a house express one's character and history belonged to its first builder and should be respected by his children. But among the majority of Americans there were few who would have agreed with him that is was "an evil sign" when a people built their houses to last for one generation only. A farmer neighbor of Emerson's, who had been reading the Report of the Agricultural Survey of Massachusetts in 1840, had small respect for the Report's recommendation of stone houses. "They are not so cheap," he said, "not so dry, and not so fit for us. Our roads are always changing their direction, and after a man has built at great cost a stone house, a new road is opened, and he finds himself a mile or two from the highway. Then our people are not stationary, like those of old countries, but always alert to better themselves, and will remove from town as a new market opens or a better farm is to be had, and do not wish to spend too much on their buildings." It was inevitable that in the United States the facility of wood construction would be exploited.

Technologically, the most important contribution in building was the anonymous development of the so-called balloon frame, a revolutionary method of wood construction which appeared early in the nineteenth century in the vernacular realm of utilitarian architecture. No historian of our architecture considered this invention worthy of notice until Siegfried Giedion, in the Norton Lectures at Harvard in 1939, emphasized its importance as marking the point at which industrialization began to penetrate housing. But despite the refusal of academic architecture to concern itself with such a utilitarian development, there are a number of sources for information about the role it played in American life.

For hundreds of years men had made wood frames for houses out of heavy timbers (often a foot or more thick) which were joined together by cutting down the end of one timber to form a tenon which could be fitted into a hole, or mortise, which had been cut out of the other timber. If the joint had to support a pull, rather than a thrust, the two pieces were fastened together with a wooden peg driven through auger holes.

With the invention of nail-making machinery early in the nineteenth century, and the resulting availability of cheap nails, the way was open to the development in the United States of a new construction which soon replaced the old type and which has been used throughout the country ever since. This was the balloon frame (so nick-named, in contempt for its lightness, by traditional carpenters and builders), which the mid-century builders G. E. and F. W. Woodward described as characterized by light sticks, which did not require laborious mortising and tenoning, and (in language very reminiscent of that used in contemporary descriptions of American locomotives) by "a close basket-like manner of construction." The balloon-frame house is nailed together with light studs only two inches by four inches, but is so tied and strengthened that every nail holds to its utmost strength . . . .

Credit for the invention apparently belongs to a carpenter-builder named Augustine Deodat Taylor, who in 1833 built St. Mary's Catholic Church in Chicago—the first balloon-frame building ever built. Giedion and others have credited it to another Chicagoan, but the researches of Walker Field, first published in 1942, seem to establish Taylor as the real originator. Throughout the nineteenth century, however, the new construction developed anonymously.

The first professional architect who recognized its importance, so far as I have discovered, was Gervase Wheeler, an Englishman who came to the States to practice in the 1840s. His book, *Homes for the People* (published in 1855), was the first architectural publication to quote the description of balloon framing which

Solon Robinson had given at a meeting of the American Institute and which had been reported in the New York *Tribune*, January 18, 1855. Robinson himself, who was apparently the first person to describe the new system of construction, was a farmer who had migrated from Connecticut to Indiana in 1834, and he had seen how these lightly framed buildings, erected on the open wind-swept prairies, stood as firm as any of the old frames of New England with posts and beams sixteen inches square. How long they might endure he did not know, but for all practical purposes they were substantial enough to meet the needs of those who were building towns in the wilderness. When another member of the Institute expressed doubts about the long-term durability of the balloon-frame structures, in spite of their admitted strength, Robinson replied: "Sir, we are Christians, you know, and therefore we take no thought for the morrow."

Yet, in spite of Gervase Wheeler's interest in the new method of construction, it was generally ignored by professional architects. As one of the most successful of them put it in 1879:

> It sometimes happens, in localities remote from large cities or large towns, that persons are obliged to do with make-shifts, to get a home at all. It was such a condition of things that led the well-disposed pioneer of the West to adopt the method called "Balloon framing", which is really no framing at all. . . .

It was the carpenter-builders and the farmers who developed it, and its role in American life was best understood by men like Robinson and one of the anonymous authors of *The Great Industries of the United States* (published in Hartford in 1873), who hailed it as "the most important contribution to our domestic architecture which the spirit of economy, and a scientific adaptation of means to ends, have given the modern world." There was, he thought, hardly a better evidence of the American spirit than the prompt adaptation to new conditions reflected in the introduction of this new method of building. And he demonstrated his understanding of its relationship with other aspects of technological advance when he added that "our methods of construction, like our means of transportation, have passed into the railroad phase of development."

Here, then, were the roots of the vernacular tradition in building. Here were the same characteristics which we have already traced in technological design: simplicity, lightness, flexibility, and wide availability.

These balloon-frame buildings were often designed and constructed without reference to any requirements other than those of utility, and they were often appallingly unattractive. Professional architects usually regarded them with horror. To Calvert Vaux, for instance, who, like Gervase Wheeler, had come to the United States with the best available English training and had devoted his life to the realization in this country of his ideals of his art, such houses seemed to have been constructed without any sense of proportion or the slightest apparent desire to make them agreeable objects in the landscape. These "bare, bald white cubes," as he called them in 1857, struck him as monotonous evidence of a life spent "with little or no cultivation of the higher natural perceptions." And like many of his cultivated contemporaries, he set about doing his best to educate the American people in sifting, testing, and improving all suitable architectural forms and modes of the past. He recommends, for instance, Moorish arcades and verandas and

Chinese balconies and trellises added to what he calls the "irregular Italian" or to the "later modifications of the Gothic."

But Vaux, again like a number of his contemporaries, was at least vaguely aware of the basic problem represented by the conflict between the cultivated academic tradition and the square boxes which were springing up in every direction. Republicanism, he argued, "tacitly, but none the less practically, demanded of art to thrive in the open air, in all weathers, for the benefit of all, if it was worth anything, and if not, to perish as a troublesome encumbrance." The balloon-frame houses might trouble him because of their builders' apparent lack of capacity for enjoying what is really desirable in life, but he nevertheless recognized that they were rooted in the lives of the people, "simply and unceremoniously" reflecting both the migratory, independent spirit which pervaded Americans and the economic opportunity which made it possible for almost every storekeeper and mechanic to build his own home.

Vaux was typical, in a sense, of a whole group of professionals to whom it was obvious that the forms and modes of the past could not be slavishly copied in this country. The past, they proclaimed, should be regarded as a servant, not as a master. Yet no matter how devoutly they professed, as William M. Woollett did, that "architectural effect should be obtained by the natural combinations and workings of the constructive portions of the structure, and not by adding or planting on of these features," they nevertheless were capable of the kind of alteration of old houses illustrated in . . . Woollett's book . . . [*Old Homes Made New*, 1878 ed.]. Calvert Vaux was willing enough to quote Emerson's exhortation to the American architect to "study with hope and love the precise thing to be done by him, considering the climate, the soil, the length of the day, the wants of the people, the habit and form of government." Indeed, many professionals felt in a vague sort of way that American architecture would develop some distinctive characteristics, but they generally agreed with Vaux that whatever these characteristics might ultimately turn out to be, they could "hardly be expected to depend much on the employment of really new forms."

Clearly it could not be from men so disposed and so bound to tradition that fundamental innovations would come. Those would continue to develop anonymously in haphazard response to social necessity, as did balloon-framing. But as the anonymous and revolutionary contributions of the vernacular tradition became established in building construction, they inevitably influenced and modified, and were modified by, the ideas and practice of the professional architects. There is no more charming evidence of the interaction of the two traditions than the carpenter-Gothic houses (like the "Lace House" in Blackhawk, Colorado . . .), built in all parts of the country during the middle decades of the century. The fanciful and sometimes highly elaborate scrollwork overlies a balloon-frame structure in much the same way that the carpenter's scrollwork overlay the bold, simple mechanism of many American woodworking machines.

The essential point, however, is that through all the various revivals which were borrowed from Europe by the professional architects in the United States—whether it were Asher Benjamin's Greek style, Alexander Jackson Davis' and Upjohn's Gothic, or Vaux's Italian—certain characteristics recur which distinguish the American examples from their European contemporaries; and these characteristics clearly reflect the recurring influence of the vernacular tradition.

One of these characteristics is the plane surface—the flat wall of wood, or brick, or stone. The simple clapboard wall, for instance, which was evolved in response to the lack of lime in the colonies, has dominated American wood construction for three centuries. There have been, of course, numerous instances in the United States of cluttered and elaborate wall surfaces, in wood as well as in other materials. But in general, even in the worst moments of nineteenth-century eclecticism, American buildings differed from those of England, Germany, or France in being less given to surface richness. Similarly, early in the century, when classic styles had been revived, in Europe it was chiefly the magnificence of Roman forms which had been imitated, while in the United States (except momentarily in a few urban centers in the East) it was the serenity and severity of Greek forms which had appeared in churches, houses, and courthouses throughout the country.

Newspapers, farm journals, family magazines, and many books of the period offer ample evidence of the attitude which is expressed in the buildings themselves. There is a whole literature, for example, on the general subject of rural architecture, some of it by trained architects but much of it simply by farmers or "friends of agriculture" who had enough ingenuity and good sense to plan houses which met practical needs. Almost all of this literature expresses pragmatic contempt for "Gothic castles with piecrust battlements," "fantastical and puerile 'bird cages' with gewgaw carvings," and other follies which, as D. J. Browne told his fellow farmers at an American Institute meeting in the forties, were at variance with the simplicity of our manners, with our climate, and with reason and sound taste.

Nowhere did the interaction of the vernacular and the cultivated tradition find clearer expression than in the popular books on architecture which appeared in the middle of the century. One of the most amusing and enlightening of these was Lewis F. Allen's *Rural Architecture*, originally published in New York in 1852 and reissued several times in the sixties and seventies, a forgotten volume which did more to shape the course of ordinary house building than many a more pretentious and less salty book. Allen was a farmer in western New York State (and an uncle, incidentally, of Grover Cleveland) who had no formal training as an architect, and whose contempt for the professionals sprang from a conviction that they showed no understanding of the purposes to which a rural home should be adapted. He therefore took it upon himself to instruct his neighbors in the fundamentals of house building, so that they would have their own notions about it, "and not be subject to the caprice and government of such as profess to exclusive knowledge." He was concerned, he said, only with the shape, arrangement, and accommodation of the building, not with modes and styles of exterior finish. The latter, so long as they suited those who adopted them, were of little consequence, he felt, and could therefore safely be left to the architects.

The fundamental principles which Allen keeps reiterating throughout his book—whether in connection with farmhouses themselves, or barns, or poultry houses, or rabbit hutches—are all summed up in his definition of good taste. It is a definition which in our day may sound commonplace enough, but which most academic architects of his day overlooked. Good taste, Allen believed, demanded both a fitness to the purpose for which a thing was intended and a harmony between the various parts. Any product of good taste would be both "pleasing to the eye, as addressed to the sense, and satisfactory to the mind, as appropriate to the object for which it is required." No style of architecture or finish could be really *bad*, he insisted, if utility were duly consulted and complied with. Provided there was a harmony amongst them even the meanest buildings on a farm derived a dignity from "the character of utility or necessity which they maintain."

Planting himself on these convictions, he mercilessly attacked the meaningless current styles which the architects of the cultivated tradition were exploiting. At the slightest excuse he laid into "the ambitious cottage, with its covert expression of humility" such as those which men like Vaux and Downing were building. "What," he asks, "are the benefits of a parcel of needless gables and peaked windows, running up like owls ears above the eaves of a house, except to create expense, and invite leakage and decay?" He detested all the "gewgawgery" of the haberdasher-built houses of his time and maintained stoutly that all buildings should show for themselves what they were built of, rather than masquerade as something else.

Being altogether untutored in draftsmanship, Allen had to employ a Buffalo architect to draw up the elevations and plans of his houses for illustrations in his book, and—as might be imagined—the results were not satisfactory. Throughout the text Allen snipes at the architect's renderings. For instance, he objects to the diamond-paned windows with which the draftsman dressed up his design for a poultry house; "but," he sourly remarks, "as he had, no doubt, an eye to the 'picturesque,' we let it pass, only remarking that if we were building the house on our own account, there should be no such nonsense about it . . . ."

But the significance of his book is not confined to its attacks upon frippery and pretense. Its positive contribution is that all the plans for buildings which it offers have their origin in the life to be lived, or the jobs to be done, within them. The layout of buildings, arrangement of rooms, provision for light and ventilation—all are managed with an eye to the comfort, convenience, and pleasure of actual farm living rather than to stylistic design.

Another book which should be better known to students of American architecture is Orson S. Fowler's *A Home for All, or, The Gravel Wall and Octagon Mode of Building*, originally published in New York in 1849. Fowler was, in a way, typical of the reformer-enthusiasts who flourished in the forties and fifties. For a generation he and his brother Lorenzo were the most active boosters of phrenology—the "science" based on the belief that mental faculties and character traits are revealed by the conformation of the skull—and he traveled throughout the country lecturing and selling copies of the books and magazines which he and his brother wrote and published.

It was on one of his Western trips that Fowler came across a method of building which, combined with certain phrenological conceptions of his own, produced a system of building examples of which can still be seen in many Hudson Valley towns and elsewhere. Fowler was convinced that, from all points of view, an octagonal form was the ideal one for a home. For one thing it provided considerably more enclosed space than a square or rectangle of the same circumference, and hence was economically superior. Furthermore, it permitted floor plans which, he argued persuasively, were phrenologically sound and which made housekeeping less of a burden than it had to be in houses of other shapes.

The trouble was, simply, that its obtuse angles were harder to frame than the right angles of conventional houses.

The solution to this problem he found, however, when he saw houses near Janesville, Wisconsin, built of lime, gravel, and sand. But let him tell it in his own words:

> I visited Milton [Wisconsin], to examine the house put up by Mr. Goodrich, the original discoverer of this mode of building, and found his walls as hard as stone itself, and harder than brick walls. I pounded them with the hammer, and examined them thoroughly, till fully satisfied as to their solidity and strength. Mr. Goodrich offered to allow me to strike with a sledge, as hard as I pleased, upon the inside of his parlor walls for six cents per blow, which he said would repair all damages. He said in making his discovery he reasoned thus: Has nature not provided some other building material on these prairies but wood, which is scarce? . . . Let me see what we have. Lime abounds on them everywhere. So does coarse gravel. Will they not do? I will try. He first built an academy not larger than a school house. . . . It stood; it hardened with age. He erected a blacksmith's shop, and finally a block of stores and dwellings; and his plan was copied extensively. And he deserves to be immortalized, for the superiority of his plan must revolutionize building, and especially enable poor men to build their own houses.

Here was a method of construction (in reality an empirical rediscovery of Roman concrete, and one of the earliest uses of this material for domestic architecture in modern times) which was easily adapted to octagonal houses, and Fowler at once set out to sell the idea to his countrymen. His own house at Fishkill, New York, was a sightseers' objective for many years, until seepage from a cesspool made it a typhoid breeder and broke the old man's heart. In the meantime, however, hundreds of houses had been built according to his proposals, and his book did a great deal to promote a method of construction which, quite apart from his octagonal plan, had a share in the development of what later became ferro-concrete architecture.

Like Lewis F. Allen, Fowler had no use for fancy trimmings, and one of the things he liked about his "gravel walls" was the plain surfaces they presented. Nature, he reminded his readers, "never puts on anything *exclusively* for ornament *as such*. She appends only what is useful, and even absolutely *necessary*," and that should be the law of design.

This brings us to another recurring characteristic which—like the plane surface—distinguishes American domestic architecture from its European counterpart:flexibility of ground plan. This flexibility may be traced from its origins in the early eighteenth-century practice of adding ells and lean-tos to the central-chimney houses on up through the nineteenth-century development of sliding doors which came into American house design with the Greek revival in the late 1820s and were extensively used thereafter. Yet to Wilhelm Bode, the German critic, these sliding doors seemed a novelty at the end of the century, when he visited the Chicago World's Fair in 1893.

Openness and flexibility have been characteristic of American house plans for three hundred years, and have to a great extent resulted from the influence of the vernacular tradition. An 1872 book of plans, each of which is designed "to afford an easy opportunity of adding to it in the future without interfering with the construction already in use," was the work of an architect named C. P. Dwyer who in his preface claimed "years of residence in the Great West" and "a knowledge of the various modes of cheap construction" in Europe as well as in America, "where necessity is the superintending architect." It is possible to trace a part of the development in the records of academic architecture, especially in summer cottages and country homes where architect and client felt relatively free to develop independent solutions. But for many of the most daring innovations one must turn to such unlikely and neglected sources as, for example, a volume produced in the sixties by Catherine Beecher and her more eminent sister, Harriet Beecher Stowe. Here, in *The American Woman's Home* (New York, 1869), these worthy and earnest ladies laid down the specifications for what they called "a Christian house; that is, a house contrived [notice the verb] for the express purpose of enabling every member of a family to labor with the hands for the common good, and by modes at once healthful, economical, and tasteful." The house they contrived is a remarkable job. They planned it for mass use, and deliberately acknowledged the industrial environment by arguing that, owing to the railroads, men working in cities could build such a house and rear families in the country. The over-all plan, and every detail, aims—in Miss Beecher's words—at "economizing time, labor, and expense by the close packing of conveniences. . . . "

As in Frank Lloyd Wright's Suntop Homes at Ardmore seventy years later, all the heating, plumbing, and storage facilities are concentrated in a central unit—at the darkest part of the house—leaving the well-lighted outer parts free for family life. The kitchen and stove room, the arrangement of which the authors liken to a cook's galley on a steamship (from which, indeed, they probably got the idea) are an extraordinarily advanced piece of compact, functional planning. The glazed sliding doors between stove room and kitchen serve to shut out heat and smells and to let in light. The whole house is artificially ventilated by a system of flues connected with the stove; the stove warms the air in the flues, thus setting up a current in them, which both draws off the heat and smells of the kitchen (and of the water-closet room on the second floor) and draws fresh air from outdoors to supply all the rooms of the house.

Particularly interesting, since it seems to be the earliest recorded use of a movable partition in an American house, is the screen on rollers by means of which the large room to the left of the entrance could serve as a big, airy sleeping room at night; then in the morning, when the screen is rolled to the middle of the room, as a sitting room on one side of the screen and a breakfast room on the other; and finally, through the day, a large parlor on the front side and a sewing room behind. By means of this movable partition (which, incidentally, was also a "storage wall," with shelves and compartments built in to economize space), and the compact kitchen and stove room, the useless spaces usually devoted to kitchen, entries, hall, back stairs, pantry, etc., would all be eliminated. It is, on the whole, a radical interior plan, broadly prophetic of features of the best twentieth-century design. But the ladies got nowhere with the exterior; that remained strictly conventional. Working from the inside out, they—like Allen and other nameless exponents of the vernacular tradition—did not have the requisite experience and training to evolve a suitable exterior shell for their creation. The chances are, indeed, that they were not much concerned about relating the exterior to the interior. Even among professional designers it was assumed that although the ground plans of a house should be made to conform to the necessities and requirements of those who were to occupy it, those plans did not, as the architect S.B. Reed said, "decide, or even indicate, the style, character, or expense of the outside dress that may be put upon them."

But it was the gradual pressure of flexible and open plans, co-operating with the freedom and facility of the vernacular balloon-frame construction, which encouraged a few architects to abandon the stagnation of academic traditions and to evolve new solutions to the problems of house building.

A writer in *Putnam's Magazine* in 1854 declared that splendors of architecture were not to be looked for in America (except in the shape of bridges and aqueducts) until such time as we learned that twenty or thirty families might live in a palace by pooling their wealth and building one capacious dwelling, while if each built separately they would be compelled to live in inconvenient and unattractive houses. The problem which the writer was facing was one which is still much agitated: how can we provide adequate housing for the masses of our people? As a matter of fact his suggested solution (an idea which he caught from the vast hotels of his pre-apartment-house period) is basically the same as the one which Lewis Mumford and others have advocated in our time: dividing up the expense of major installations, such as heating units and water mains, among a number of families to reduce the per capita investment. But the other approach to the problem—mass-produced portable and prefabricated housing—about which we hear even more today, goes back at least as far.

Balloon-frame construction, as we have seen, early in the last century made substantial houses much more widely available than they had been before. But maximum availability could be achieved only when the building industry adopted mass-production techniques similar to those developed in the manufacture of firearms, machinery, and watches. In recent years, particularly as a result of the war-stimulated demand for emergency housing of workers, we have learned a great deal about prefabrication of houses and about portable and demountable buildings. One gathers from present-day writers on architecture and construction that these are almost exclusively twentieth-century developments. (The John B. Pierce Foundation, for example, published in 1943 a *History of Prefabrication*, by Alfred Bruce and Harold Sandbank, which mentions no case of precut or sectional buildings before 1892). Certainly many important features of modern technique are of recent origin, but the theory of prefabrication itself, and its pioneer practice, go back well into the nineteenth century. The frames of hewn-timber houses and buildings were shipped ready-cut to all parts of the world in the late seventeen hundreds, as we know from early records. The old Seamen's Chapel, or Mission House, in Honolulu, was sent, disassembled, from Boston around Cape Horn in a whaling ship in 1820.

Even some of the most revolutionary current methods were proposed more than a hundred years ago by a weird Pittsburgh genius named John Adolphus Etzler. He described them at considerable length in a long memorial to Congress and in a book called *The Paradise within the Reach of All Men, without Labor, by Powers of Nature and Machinery*, first published in this country in 1833, reissued a number of times in England, and rather scornfully reviewed by Henry Thoreau in the leading article of the *Democratic Review* for November 1843. Etzler described, for instance, how wood, if "cut and ground to dust and then cemented with a liquor," could be molded into any shape and dried so as to become a solid, consistent substance which could be dyed and polished. Thus, he announced, "we may mould and bake any form of any size, entire walls, floors, ceilings, roofs, . . . furnitures, . . . kitchen utensils, pieces of machineries." He proposed the construction of huge, air-conditioned apartment buildings, to be made of "large solid masses, baked or cast in one piece . . . so as to join and hook into each other firmly."

Apparently Etzler's schemes never came to anything much, and the man himself was so completely lost sight of that a popular American writer of the seventies—confused by the numerous English editions of the book—referred to him as "Mr. Etzler, of England." But prefabrication on a less ambitious scale managed to make considerable headway during the next few years. "Barnard's Portable Patent Houses" were exhibited and sold in St. Louis, Missouri in 1847, and were described in the St. Louis *Daily New Era* (June 26, 1847) as "made in single or double sections, . . . suitable for dwellings, stores, offices, shops, &c., . . . of various sizes with one or more rooms, and can at any time be taken down and removed to any part of the town or country at small expense." Thomas P. Kettell observed in 1861 that "the settler on the new lands of the West is now not always required to plunge into the wilderness and rear his first shelter from logs, but may have his house sent from Chicago or other cities by railroad, and put up to await his coming." By 1873 an anonymous writer was able to say that with the opening of the West, with the new methods of transportation, and with the application of machinery to lessening the expenditure of labor, domestic architecture had "partaken fully" of the spirit of the age; the Western prairies, he said, were dotted over with houses which had been "shipped there all made, and the various pieces numbered, so that they could be put up complete, by anyone."

One of the most interesting of the United States exhibits at the Paris Universal Exposition of 1867 was a balloon-frame farmhouse which had been shipped in sections by its manufacturer, Colonel Lyman Bridges, of Chicago, and assembled at the exposition. Bridges had, for a number of years, been in the business of supplying settlers in the West with portable, prefabricated buildings of this type. According to his own testimony at the time, the majority of his customers ordered houses of from two to four rooms, which cost from two hundred to six hundred dollars, but about one out of seven bought more pretentious houses which sold at prices up to a thousand dollars. Schoolhouses, also, were supplied to pioneer villages (usually of a standardized $24' \times 36'$ floor plan—costing a thousand dollars), and stores of various styles and sizes could be had at prices ranging from five hundred to two thousand dollars. It was the colonel's firm, no doubt, which James Parton described (in the chapter on Chicago in his *Triumphs of Enterprise*, 1872) as "happy to furnish cottages, villas, schoolhouses, stores, taverns, churches, courthouses, or towns, wholesale and retail, and to forward them, securely packed, to any part of the country." Similarly, D. N. Skillings and D. B. Flint manufactured and sold in Boston and New York in the early sixties farmhouses, barns, hospitals and barracks for the Union armies, depots for the Adams Express Company, all made in sections, any one of which could be applied to any building of their make (on the general principle of interchangeable parts).

It is difficult to trace the development of prefabrication. Source material for such history (such as catalogues and advertisements) is ephemeral, and no one thus far seems to have bothered to collect the available scraps. But by the seventies the system was well established. Derrom's Ready-Made Houses were widely advertised in 1876. . . . The Ducker Portable House Company, of New York, published an extensive catalogue of buildings made up of strong, light, wood-framed sections—covered with an indurated, waterproof fiber—which locked together without the use of nails, screws, or any other appliance. William H. Wahl, writing in the eighties, said that the manufacture of portable houses had become an important industry in the United States. According to his account, they were extensively used by builders of such public works as railroads and canals; by the Army (as barracks, hospitals, etc.); and by miners, sportsmen, photographers, and others. Railway stations, storehouses, bathing houses, pavilions, fruit stands, summer kitchens, and outbuildings of every description were available, and one could buy substantial summer cottages "of many styles and as elaborately finished outside and inside as may be wished." By 1897, when the Klondike gold rush began, a New York company immediately shipped a large number of houses overland, and began loading a vessel to carry more of them to Seattle via Cape Horn, all of them ready-made in sections so that they could be carried easily in boats up the Yukon or packed on sleds.

Unfortunately these early experiments with prefabricated houses did not lead to a thoroughgoing modernization of the building industry. For one thing carpenters and builders early organized to resist the adoption of a house-manufacturing method which threatened to deprive many of them of jobs. Perhaps equally important, the architects must have quietly done all they could to discourage people from buying the mass-produced products which threatened not only to deprive their profession of work, but also to standardize architectural forms and characteristics which seemed to them incompatible with good taste. We have seen how the vernacular tradition in purely technological design ran afoul of the cultivated tradition during the early days of machine design when inventors had to turn to cabinetmakers and architects for patternmaking. But that was simply the result of the inevitable fumbling for methods and techniques which characterizes early procedure in any new field. House design was quite a different matter; it was not a new field, and it had long been the province of the architects—who had the weight and prestige of the cultivated tradition behind them.

Here, then, and in the development of balloon-framing, were the makings for a showdown between the two traditions. The inevitable clash was marked by all the features which we have come to recognize as misfortunes in the social and artistic history of the United States. On the one hand, the emergent vernacular tradition—neglected as it was by the architects—was free to evolve solutions for problems inherent in the new civilization's environment; but its freedom was bought at the cost of losing all contact with the humane tradition of western European art. On the other hand, the cultivated tradition, refusing to adapt itself to the new environment, turned in upon itself—marrying itself to its own past; and the more it did so, the more anemic and impotent it became. The results of this conflict, at the worst, were on the one hand the bare, unimaginative, depressing houses which stalked both sides of Main Street in Western manufacturing and mining towns, and on the other hand the pointlessly mendacious pseudo-classical and pseudo-Renaissance public buildings which were pompously erected in the proudest cities of the land.

Yet, however much weight the cultivated tradition had, it was a dead weight. In the long run the vernacular, with its creative vitality, was certain to exert increasing influence wherever it appeared—in the United States or abroad. At first, of course, this influence was felt almost exclusively in those areas where the new civilization was least subject to the restraining influence of the older culture. In geographical terms this meant the new industrial centers, particularly in the Midwest and West; in social terms it meant the fields of inexpensive housing and of commercial and industrial structures. It was precisely in these subartistic areas that balloon-frame construction and prefabrication of buildings were developed (both of them, apparently, first in Chicago, and both of them for the satisfaction of plain people's needs); and it was largely in such utilitarian structures as factories, grain elevators, and warehouses that a tradition was developed of direct, unembarrassed simplicity and common-sense adaptation of form to function.

In these areas also there emerged yet another major achievement of the vernacular: the expressive use of the iron or steel skeleton. In order to understand the relationship of this architectural advance to the vernacular tradition as a whole, we must turn for a moment to an even earlier development: the use of iron as a building material.

The first building in which iron columns replaced the masonry of the outer walls as the support of the various floors was a five-story factory erected by James Bogardus in New York in 1848–49. Boulton and Watt in England had used cast-iron beams and columns in interior construction forty-seven years earlier, but Bogardus was the first to employ exterior iron columns as supports in place of brick or stone columns or walls. From 1850 on he designed and built a great number of buildings of this type, using prefabrication techniques much like those which we have already mentioned in the discussion of the construction of American iron bridges. . . .

The discovery of gold in California gave the necessary impetus for the development of Bogardus' invention. Previously he had been unable to persuade American or English capitalists to invest in his project, but when the rush to California set in there was suddenly a huge market. Wrought-iron parts for buildings which were shipped to San Francisco by his English competitors required a month to assemble, but the buildings Bogardus sent out by the shipload could be put up in a day.

Throughout the next thirty years iron was widely used in the United States; but in spite of the fact that Bogardus had hoped from the start to introduce it in domestic architecture, its use (other than in hidden or disguised structural members—as, for instance, in the dome of the Capitol at Washington) was largely confined to warehouses, office buildings, and other structures which were often regarded as beneath the notice of official architecture. Iron, as iron, was suspect.[3]

[3]The professional architect's attitude toward iron as a structural material is plainly revealed in Minard Lafever's *The Architectural Instructor*, New York, 1856, pp. 405–07. Iron, Lafever wrote, was suitable for "warehouses, crystal palaces, banks, and printing establishments" and also "for the interior structure and decoration" of other buildings where fireproof construction was desirable and where floors and galleries required strong support with a minimum of material. For such decorative details as window pediments and cornices, he pointed out, iron when painted and sanded "has the ornate and massive appearance almost of carved stone."

To the custodians of the cultivated tradition the new material offered nothing. Ruskin, after all, had made it clear that "true architecture does not admit iron as a constructive material" because the sense of proportion and the laws of structure had always been based on the use of clay, wood, and stone; a metallic framework would inevitably create a flimsy appearance. Indeed, such works as the cast-iron central spire of Rouen Cathedral and the roofs and pillars of iron railroad stations were, he flatly asserted, "not architecture at all." In 1869 the editors of *Appleton's Journal*, one of the most influential of all the magazines which were trying to improve America's taste in the arts, announced that the use of iron in architecture was "utterly destructive to its dignity." Seven years later Richard Morris Hunt, architect of a number of Fifth Avenue châteaux and Newport villas, and a staunch advocate of "art education for the masses," echoed Ruskin in his official report on the architecture of the Centennial Exhibition when he said that because the main building was built of glass and iron there was "a total absence of anything like monumental grandeur or even apparent substantiality" about it. The author of *Gems of the Centennial Exhibition* (1877) went even further, asserting that there was an "utter lack of beauty and picturesqueness in all iron-frame buildings," and that iron columns and girders, filled in with plate glass, could only produce an inartistic effect.

But to those who, like the authors of *The Great Industries of the United States*, had living contact with the everyday world of machines and trade about them, the use of iron seemed "one of the chief improvements of modern times." It was clear enough to them, to be sure, that the architects of the sixties and seventies had not yet evolved a satisfactory treatment of the new material; but they knew that it had potentialities which should not be neglected. They saw that London's Crystal Palace of 1851 had pointed the way toward exciting new architectural forms and methods of treatment. Here, they wrote, was a material whose tensile strength permitted its use in slender pillars and thin sheets, allowing "an unprecedented proportion of space for windows"; a material which was cheap, handy, strong, safe, and easily movable. No sentimental, Ruskinish objections could, in their view, prevent the increased use of a material with such advantages, and the time would surely come when increased knowledge would lead to its scientific use.

Meanwhile, however, the chief recommendation of iron as an architectural material was the ease with which it could "embody any architectural design." From the architectural rather than the engineering point of view, it was the "iron front" rather than the iron frame that was most interesting, and huge factories were built in Baltimore, New York, Chicago, and other cities to make iron fronts for buildings throughout the country.... Little attention has been paid to this aspect of our architectural history, and the available information has never been collected. Casual research reveals, however, that twenty iron-front buildings, all designed by the architects Van Osdel and Bauman, were built in Chicago in the year 1856, the iron fronts for all but two of which were made (if not designed) byD. D. Badger & Co. of New York. The other two, one of which was a bank, were made by Stone, Boomer, and Bouton of Chicago.

A few fine, intelligently handled examples of iron buildings were built anonymously during the third quarter of the century, such as the warehouses which were torn down to make room for the Jefferson Memorial Park on the old St. Louis waterfront. But most architects who used iron at all continued to treat it as nearly as possible as if it were stone, casting it in Corinthian columns and Romanesque or Gothic arches.... In cities like Chicago and Baltimore there were by 1870 many buildings with iron fronts which were modeled as nearly as possible in the forms and proportions of stone, with only such modifications as the nature of iron made necessary. Such imitation of old forms in new materials is apparently an inevitable stage of development. When sheets of galvanized steel were used to cover the fronts of wooden-frame buildings in the eighties, they were stamped in patterns resembling stone and brick and iron. . . . Worse still, when Chicago was rebuilt after the fire of 1871, and stone was reintroduced in place of the iron which had melted and buckled so quickly in the fire, the forms which had themselves been modified metal imitations of stone construction were now in turn imitated in masonry which was painted to look exactly like cast iron.

Nevertheless, the essential scheme of a metal frame was becoming firmly established, and the way was being prepared for the development of the steel-skeleton architecture which—in skyscrapers and great factories—later appeared as one of the most distinctive achievements of American art. With the increasing use of steel during the seventies and eighties, new techniques of construction had been evolved. In 1874 James B. Eads completed the famous steel-arch bridge across the Mississippi at St. Louis—the first bridge in the world in which steel was extensively used, and a structure which served as an important stimulus to the development of the Chicago School of Architecture in the eighties and nineties. The engineering problems presented by Eads' design and solved by him in carrying it out, and the honesty and majesty of the finished structure, marked it at once as a monument in the history of bridge building. But it was more than that. From this bridge Louis Sullivan, the greatest of the Chicago architects of the end of the century, caught his vision of the power of the creative dreamer—"he who possessed the power of vision needed to harness the intellect, to make science do his will, to make the emotions serve him."

One of the first buildings in which the steel skeleton was used without any self-supporting walls was William LeBaron Jenney's second Leiter Building (1889) at the corner of Van Buren and State streets, Chicago. Unlike Jenney's earlier Home Insurance Company Building (1883–85), which was the first "skyscraper" actually erected, the Leiter Building displays practically no reminiscences of academic architectural styles; as in Jenney's proposed building for the Hercules Manufacturing Company, the huge squares of the steel skeleton give shape and form to the exterior. Wide areas of glass screen the interior from the weather, and there is so little ornament that the eye is not distracted from the clean, strong shape of the building. To many critics it was of no account as architecture, but the anonymous author of *Industrial Chicago* (1891) admired its light, airy, yet substantial appearance and astutely observed that it was constructed "with the same science and all the careful inspection" that would be used in the construction of a steel bridge of the first order.

It would be wrong to give the impression that engineering construction in the United States surpassed the achievements of contemporary Europe. As in the case of machine building, the truth seems to be quite the contrary. According to Giedion, French engineers achieved the most audacious and brilliant constructions during the period from 1855 to 1900, Cottancin's and Dutert's Galerie des Machines and the Eiffel Tower at the Paris Exhibition of 1889 marking the climax and conclusion of a long development. But it was in the United States, and specifically in Chicago during the eighties and nineties, that the science of the engineer and the creative genius of the artist combined to produce a new urban architecture: the clear-cut, open skyscrapers of Jenney, of Burnham and Root, and of Adler and Sullivan.

While emphasizing the engineers' role in developing new techniques and forms we should not, however, lose sight of the fact that architecture, rightly conceived, has always concerned itself with more than mere construction, and that the engineers had by no means taken over the large-scale humanizing and planning functions of the architect. The result inevitably was that whenever vernacular forms were in competition with the work of the best cultivated architects people were likely to prefer the latter. So, for example, the success of the vernacular in Chicago's business district during the eighties and early nineties was for a time almost completely obscured by the academic brilliance of the cultivated tradition as displayed at the Chicago World's Fair of 1893. Returning to the classic forms which had dominated our official architecture in the first decades of the century, the architects erected an impressive group of buildings which, according to the glowing tribute of Hubert H. Bancroft, the Western historian, were "a triumph of the aesthetical." It has often been said that the Chicago Fair set back American architecture thirty years; and in a sense it did. Certainly there was more originality of design in Sullivan's Transportation Building than in the scholarly classic revival buildings which dominated the exhibition. But in another sense the fair was a significant achievement. Never before had Americans seen a group of buildings so skillfully harmonized; nowhere else had they been able to wander down one apparently endless vista of beautifully correlated façades, then turn a corner and face another, and never encounter a discordant detail. It was this over-all planning, this total effect, which made the borrowed, academic style so impressive to the thousands who visited the fair. So long as the fine achievements of the vernacular remained isolated and unrelated phenomena, they inevitably failed to capture the public imagination on so vast a scale.

The century ended with academic architecture in the ascendancy, and with McKim, Mead, and White as the leading practitioners of the art. But despite the elegance and echoed charm of such buildings as the Boston Public Library (1887–95) and New York's Pennsylvania Railroad Station (1906–10), the work of men like McKim, who operated in the cultivated tradition, had less relation to the vital contemporary forces of American life, and to its future, than even the crudest, least ingratiating examples of small-town dwellings or the most materialistically functional office buildings. And meanwhile, however thoroughly Louis Sullivan's work was eclipsed by his Eastern contemporaries, he for one had fused the vital impulses of the vernacular into what he liked to describe as *organic* architecture (a word which he loved for its sense of "a ten-fingered grasp of reality"). The vision he had caught from the great Eads bridge at St. Louis was given concrete expression in his buildings. "With me," he wrote to Claude Bragdon, "architecture is not an art, but a religion and that religion but a part of the greater religion of Democracy." Here, close to the vernacular roots, was the first flowering of an architecture indigenous to modern civilization.

3.5

Art in Engineering—The Need For A New Criticism

DAVID P. BILLINGTON

In concluding a 1970 talk, the Dutch historian of modern art, Hans L. C. Jaffé, proposed the following motto for the culture of the future: "ars naturae magistra—the arts are the master of nature" (1). The present furor over man's interference with nature might lead us to doubt the validity of that motto or at least to misinterpret it as the need for artists to design cities and to make regional plans.

In context, however, Jaffé is speaking of art realized through works of engineering, and in particular public works. Thus, to interpret art or esthetics in terms of civil engineering we can start by describing Jaffé's meaning and seeing how it conditions our ideas of engineering structures.

THREE CATEGORIES OF STRUCTURAL ELEGANCE

To begin, let us distinguish three categories of structures with esthetic appeal, excluding the splendid structures of raw nature—trees, sea shells, and natural bridges.

First, there are those works of man such as primitive rope suspension bridges that are undoubtedly attractive although we know them to have been built out of pure necessity. Although we may have the belief that their designers strove after beauty there is no evidence extrinsic to the works themselves.

Putting this first type of structure under the heading of utility unconscious of beauty, we can contrast it with a second group where the stress for beauty has

This paper represents part of a continuing program at Princeton University entitled Humanistic Studies in Engineering supported by the National Endowment for the Humanities and the Ford Foundation.

overwhelmed the design for utility. These self-conscious structures are usually characterized either by costly facades that cover over the structural form or by sculptural forms unrelated to technical needs.

Between the unconscious and the self-conscious is the third category, those works which are consciously shaped for appearance as well as for utility, where the design engineers strove to relate both desires to one structure.

We can illustrate these three categories by three arch bridges: (a) The Canimar River Bridge in Cuba (Fig. 1), where, to my knowledge, only an economy of materials led to a remarkable lightness and which in this case led to a beauty not intentionally gained by design (2); (b) the Westinghouse Bridge in Pittsburgh (Fig. 2), where a considerable effort was made to create a Roman facade out of the arch form (3); and (c) the Salginatobel arch in Switzerland (Fig. 3) by Robert Maillart, whose clear intention was to create a form of both utility and beauty (4).

But who is to say that the Westinghouse Bridge is less a work of art in engineering than the Salginatobel Bridge? Are there some fixed rules by which anyone can decide and upon which all reasonable men may agree, or is it a matter of individual taste?

ELEMENTS OF CRITICISM

These questions raise the most difficult problem of all and in my view the most important, i.e., can engineers find a way to describe their art convincingly to the general public? In short, is there such a thing as an engineering critic of art in engineering? Or should such criticism be left to the established critics either of art or of technology [i.e., Max Bill on Maillart (5) or Lewis Mumford on Roebling (6)]?

To answer these questions, we can begin by summarizing the intent of criticism in the light of the long critical tradition in art (in the general sense of literature, painting, music, etc.) which puts three burdens on the critic (7).

First he must be able to respond personally and with *feeling* to work of art whether already canonized by past critics or not. That response is individual, it is esthetic, and highly emotional. In short, he feels the merit of the work. But this response would be vague, incommunicable, and eccentric without a second critical requirement, that of *analysis*. The critic needs to interpret his own personal

Figure 1. Canimar River Bridge, Cuba.

Figure 2. Westinghouse Bridge, Covell, Pittsburgh, 1931.

Figure 3. Salginatobel Bridge, Maillart, Switzerland, 1930.

experience to a wide audience, a significant segment of the general public. He must, therefore, take the work apart (in Greek, criticism means to dissect) and systematically make clear to the reader the reasons why a work has artistic value. Thus, he must think through the art work that he has felt to be of merit. But undergirding this analysis must exist a third component of criticism: the *judgment* of works based upon a standard. The critic needs a theory and a theory demands belief or a set of ideals. This theory is explicitly not subject to rigorous proof like the theory of elasticity, rather it makes coherent the individual judgments of merit and thereby checks a whimsical approach to art. In a sense the theory mediates between the subjective feeling and the objective analysis. Strikingly new works may modify a theory, but once it is modified, old works previously praised need not be condemned.

A theory must take into account the conditions existing at the place and at the time a work was realized; therefore the basis for judgment can change both with location and with time. To illustrate the dynamics of judgments, take the new Swiss bridge by Mirko Row (Fig. 4) and compare it to the Salginatobel (Fig. 3). If we praise the Maillart bridge for expressing the arch from hinge to hinge, what shall we say about one where the arch disappears into the roadway? Yet upon accidently coming across this bridge near the upper Rhine, we found it positively exhilarating—not an easy response to analyze—and any theory must somehow account for that exhilaration.

The critic, therefore, has ideals that enliven a theory, he has a methodology for convincing analysis, and finally he has a temperament that exalts in the works he must criticize. Clearly there is both a personal and an impersonal side to criticism. Furthermore, while reaching his wider audience the critic may evoke wide consensus for certain works while stirring up high controversy on others.

Figure 4. Tamins Bridge, Rŏs, Switzerland, 1966.

I do not see how there can be an understanding of art in engineering until there are critics willing to make a special study of this art. The most qualified candidates as critics are structural engineers. Just presenting a collection of novel forms (as many art critics or architectural historians have done) will never convince the public of the existence of any standard. More centrally, without striving for an articulated standard, we engineers cannot hope to make the general public aware of our ideals or of our beliefs about the profession of civil engineering.

Returning to Jaffé's motto: "The arts are the master of nature," we ought, if we follow this motto, to demand of our critics that they make clear what our art is and how it masters nature. This motto is closely related to the ideals of the Dutch movement in modern art known as *de Stijl* (the style) and I shall try to indicate how such ideals developed as well as how they might be used as a basis for analysis and judgment on works of engineering art. To make these ideals clear requires some brief description of previous and competing theories.

THEORIES OF BUILDING

First, there was the belief, basic to Renaissance architecture, that building is an abstract science and that each part has to be interrelated following some system of mathematical ratios (8). These ratios are geometrical, not structural; that is, they follow an esthetic premise whose closest analogy is music. Thus the Renaissance buildings of the 15th and 16th centuries and many designs since have been founded upon a *geometric theory* which stresses harmony, numerical ratios, and simple geometric shapes.

If the first theory related building to mathematics, the second one evolved in the 17th and 18th centuries when natural science was gaining a cultural dominance. Building then was assumed to be beautiful if it expressed physical laws, everything from statics for the ribbed structure to fluids for the gargoyled rainspouts. The high point in the *rationalist theory* occurred with the publication of Viollet-le-Duc's lectures in the 1870's (9). While the Renaissance builders looked to the classical periods of Greece and Rome, the rationalist builders took medieval Gothic as their ideal. Form, being structurally correct, would thus be visually beautiful.

Both the geometrical and rationalist theories were based on analogies to rigorous subjects whose laws severely restricted the permissible structural forms, a restriction which fit reasonably well to the limited construction materials (wood, masonry, and forged iron) available to pre-19th century builders. But with the introduction of industrialized iron, reinforced concrete, and structural steel, the technical possibilities changed dramatically and designers felt suddenly freed from the constrictions of rigorous standards derived from mathematics or physical science. Many 19th and 20th century designers began to think of building as based first upon some vague analogy to the life sciences and second upon an equally vague relationship to 20th century abstract art. Technology provided them with the means to create an anti-technological architecture which was closer to sculpture than to structure. This *sculptural theory*, a kind of romantic protest against the dry dictates of mathematical science, is based, in part, upon the belief that each new structure should be a new monument. At its best this belief displays spectacular imagination (for example, Maillart's Cement Hall); at its worst it leads to almost unbuildable structure [e.g., the Sidney Opera House (10)].

Having in mind the geometrical, rational, and sculptural theories, we are prepared to discuss a fourth theory that has some affinity to each of the others and yet presents a new synthesis of harmony, reason, and emotion with structure. This theory, although implicit in many countries during the past 100 years, was made most explicit in Holland beginning about 1917 when, as neutrals in World War I, Dutch painters, sculptors, architects, and city planners coalesced around a new journal they called *de Stijl* (The Style) which lasted from 1917 to 1931 (11).

MODERN DUTCH IDEALS FOR ART AND BUILDING

This movement extolled the beauty of abstract art, like many modern art groups, but it went further by proclaiming such art to be a model for all modern life and in particular for new construction. Artists would cease to exist as such; instead they would all become "builders of the new life" (11, p. 191). By trying to think through the meaning of abstract art, intellectual artists, like the painter Piet Mondrian, came to see their own work as both art and manifesto. But the manifesto, rather than accenting the political, was literally constructive because, with architects like Oud and city planners like van Eesteren in the group, it was possible for this ideal to become immediately a buildable reality. By 1918 Oud had become city architect of Rotterdam and by 1919 van Eesteren became chief civil engineer-town planner to the City of Amsterdam.

The new element in the *de Stijl* movement was a realization that esthetics and engineering could come together in the static structures for public use. As Mondrian stated it: "Use and beauty *purify each other* in architecture" and again "Today structural and aesthetic purity are merged in a new way" (11, pp. 169–170). For the *de Stijl* group, the union of technology and art did not mean gimmicks, machines, electronics, or movement; rather it meant explicitly buildings and bridges and even implicitly, in Holland, dykes and polders. In short, these artists, stripped of the need to represent nature in painting or sculpture, could suddenly see the works of a new static technology (which also had no need to copy natural forms or decors) as a central part of the world to be turned into art. As van Eesteren said in the retrospective 1927 issue of *de Stijl* (11, pp. 227–228):

> The artists could see that our age is formless, i.e., that our age has found no synthesis of thought and work. But they could sense the beginnings of a synthesis in the products of technique, which have all arisen from the same mode of thought, from technology. The engineer and the modern artist, although unknown to each other, already had much in common.
>
> Artists began to approach matter differently. In their compositions of material they began to denaturalize the subject-matter (Cubists, Dadaists).
>
> The engineer did the same in his construction. Iron, a material that was formerly forged and kneaded into more or less playful forms, became a new material for him. He invented steel, experimented with it for spans, and began, after he had made the necessary safety calculations, to construct with it. He designed constructions in which forces of tension and compression cancelled each other out—he constructed an

> equilibrium of tensions. In place of the medieval smith with his sense of play came the engineer with intuition controlled and guided by the intellect. The engineer also denaturalized material. Intelligence constructs. Material is the means of achieving the end with a minimum of material and a maximum of result. The development of technique gave man a hitherto unknown degree of freedom in relation to material.

Out of this sometimes vague polemic (often confusing to engineers) there arose gradually the belief that the engineers' construction could be art as long as they avoided fancy (the overly sculptural) and monotony (the overly rational). In short, purified abstract art made purified concrete structure an esthetic ideal. Both were based upon technical thought informed by the search for use and beauty.

The modern Dutch artists sensed that modern engineering structure had its own imperatives arising from modern materials, steel and concrete, and modern construction, industrialization and high labor costs. Thus these artists saw that new building forms should conform to those new imperatives. Still these new forms would not automatically be art; they had to be designed; i.e., someone had to think out how to combine good engineering practice with an esthetic sensitivity. In *de Stijl* art the new forms consisted of abstract, denatured designs using the primary colors and straight-line forms characteristic of man-made constructions rather than natural colors (greens and brown) with the curving forms of the natural landscape (trees and streams).

Mondrian painted abstractions of these man-made ideas (Fig. 5), Vantongerloo made comparable sculptures (Fig. 6), van Doesberg and van Eesteren designed related building forms (Fig. 7) and one architect, Rietveld, designed a house whose exterior and interior form (Fig. 8) carried modern art into modern design.

Figure 5. *Composition with Red, Yellow, and Blue*, Mondrian, Holland, 1922.

Figure 6. Composition, Vantongerloo, 1925.

Figure 7. *De Stijl* architectural project, van Doesburg and van Eesteren, Holland, 1923.

Figure 8. Schröder house interior, Rietveld, Utrecht, Holland, 1923–1924.

AN EXAMPLE OF STRUCTURAL ART

Perhaps nowhere have the ideas evolved for modern art been better put into practice than in the Berenplaat Water Treatment Plant for the city of Rotterdam where a thoroughly modern public works complex has been consciously built to be a work of art (12). Unlike other movements in abstract art, *de Stijl* arises in part from the efficiency of the Dutch landscape and led to the attempt to synthesize beauty and efficiency in Dutch public works such as the water treatment plant (Fig. 9 and 10).

The entire plant is both a modern automated water utility and a modern composition of structural art. This combination was achieved consciously but

Figure 9. Water treatment plant, Berenplaat, dosage building, Holland, 1967.

without facade. The structure is the sculpture; and even though the structural cost was increased about 5% by avoiding the cheapest solution, the result, a kind of technological park to be visited, is close to a maximum for the minimum of materials.

The dosage building in the Berenplaat (Fig. 9) shows a strong rectilinear form; the filter building with its *de Stijl* colored interior uses the repetitive thin shells made of straight lines, i.e., hyperbolic paraboloids (Fig. 10, 11, and 12); and the separate iron-sulfate storage silo (Fig. 13), although clearly a sculptural structure with its flat-formed hyperbolic paraboloids, is designed to symbolize how at this plant "use and beauty purify each other" in structure. This silo, the smallest major structure of the plant, doubles as sculpture to proclaim a utility designed to purify by cleaning the water and beautifying the environment. In Holland this double purity is so natural that their language has one word for both, *schoon*, meaning clean and beautiful.

Returning again to Jaffé's motto, "The arts are the master of nature," this silo symbolizes the ideal that our public structures have the potential to become works of art so that they they may master nature in the dual sense of satisfying a public need while at the same time and with the same structure enhancing the public environment. Too often our tendency is to build an unimaginative structure and then add an expensive piece of scuplture in front of it rather than think of the two as one. The Berenplaat is significant because it strives for beauty under the discipline of structural engineering. The plant expresses the ideal of engineers controlling the design to make works of art.

THE IDEAL OF STRUCTURAL ART

I take this ideal of double purpose to be the basis for a modern theory or standard against which to judge contemporary public works. This ideal, of the inseparable

Figure 10. Water treatment plant, Berenplaat, filter building exterior, Holland, 1967.

Figure 11. Water treatment plant, Berenplaat, filter building interior, Holland, 1967.

Figure 12. Water treatment plant, Berenplaat, filter building, interior column, Holland, 1967.

Figure 13. Water treatment plant, Berenplaat, iron-sulfate storage silo, Holland, 1967.

Figure 14. Schwandbach Bridge, Maillart, Switzerland, 1933 (Drawing by W. F. Shellman, Jr.).

essences of use and beauty, is in turn based upon the belief that there is no such thing as an optimum design for a public works project. Beauty will not unconsciously arise out of a search for economy alone. Rather there are personal choices for the engineer, and he is judged on them. Furthermore, the design ought to exhibit harmony—as the Renaissance designers emphasized—but that harmony need not be strictly mathematical; rather the structure should improve the outer environment and show as well some inner integration of its elements (see Fig. 3).

Design must not violate rational structure. Although it need not be the cheapest, it should be competitive. Maillart's mountain bridges were usually chosen because they were cheaper than all other proposals (Fig. 14, the Schwandbach Bridge); being also more beautiful, they are fine examples for a modern theory of structure (4). Finally, in combining use and beauty, the designing engineer need not avoid some sculpting of the form, but that departure from strict structure needs always to be controlled by construction expense. It is relatively easy to propose novel forms where cost is no object. In a democratic society, where cost for public works must be justified to the public, the search for a unity of use and beauty, while difficult and demanding dedication, ought continually to be pursued.

When easy sculptural fancy pushes up costs, it needs to be publicly criticized and where thoughtless monotony dulls the environment, it needs to be publicly criticized; conversely, where fine design embellishes without waste, it needs to be praised. But most important, for the usual design, neither really good nor willfully bad, criticism is badly needed to point a way: to give the designer higher goals and the general public greater expectations.

So much of past criticism was based upon building for church, villa, and the private patron. Now, as the Dutch have for so long recognized, attention must be centered on works for the common good, commissioned and paid for by the public. For these structures to serve fully, there is a need for a new criticism, the main burden for which falls inevitably upon those civil engineers who care about the profession and the public it serves.

CONCLUSIONS

I have tried to identify the elements of criticism necessary to the beginning of a continuing evaluation of public works structures. Above all, the need is for critics to come from the profession of civil engineering, to develop case studies and at the same time to search for a theory of structural art appropriate to our time and our society. Throughout I have intimated that ideas about the structures of civil engineering have a long and rich heritage upon which the modern critic can draw if he will make it a serious study (13).

Notes

1. Jaffé, H. L. C., "Civil Works, Modern Art, and the Life in the Netherlands," *Selected Papers Civil Engineering: History, Heritage, and the Humanities*, Vol. 1, Department of Civil and Geological Engineering, Princeton University, Princeton, N.J., August 1971.
2. On this point I was wrong and the discussions of this paper revealed the designer at Canimar, Luis Saenz, to have been deeply concerned about esthetics: see Luis P. Saenz, "Discussion,"

Engineering Issues, ASCE, Vol. 100, No. E13, July 1974, pp. 261–264. For a brief discussion of Saenz's work see Ignacio Martin, "Good Esthetics a 'Happy Accident,'" *Civil Engineering*, December 1974, pp. 71–73. This article was stimulated also by my paper and a shortened version of it which appeared in *Civil Engineering*, October 1973, pp. 36–40.

3. The photo shows the bridge at an angle which emphasizes its mass. However, seen in pure profile, as the Canimar bridge was shown, the Westinghouse Bridge appears much lighter even though the heavy and unnecessary pillars at the arch springings still are evident. This example emphasizes the importance of judging bridge esthetics on the basis of a full view of the work rather than just on one or several carefully composed photographs.
4. Stettler, E., "Reflections on Maillart," *Maillart Papers*, Second National Conference on Civil Engineering: History, Heritage, and the Humanities, Princeton, N.J., 1973.
5. Bill, M., *Robert Maillart*, Praeger Publishers, New York, 1969 (first published in Switzerland, 1949).
6. Mumford, L., *Brown Decades*, Part II, Dover Publications, New York, 1955 (originally published 1931), pp. 57–106.
7. Stauffer, D. A. (editor), *The Intent of the Critic*, Bantam Books, New York, 1966 (originally published in 1941), comprising essays by Edmund Wilson, Norman Foerster, John Crowe Ransom, and W. H. Auden along with the editor's introductory essay from which the present discussion of criticism is largely derived.
8. Wittkower, R., *Architectural Principles in the Age of Humanism*, Random House, New York, 1962, p. 101.
9. Schuyler, M., *American Architecture and Other Writings*, W. H. Jordy and R. Coe, (editors), 2 Vols., Harvard University Press, Cambridge, Mass.; Belknap Press, 1961; see especially the introduction which described the influence of Viollet-le-Duc on the criticism of architecture and structural engineering.
10. Giedion, S., *Space, Time and Architecture*, 5th ed., Harvard University Press, Cambridge, Mass., 1967. The late author added to this last edition a remarkable defense of the Sidney Opera House whose form is perhaps the most extreme example in our century of the almost unbuildable structure.
11. Jaffé, H. L. C., *De Stijl*, Thames and Hudson, London, England, 1970; originally published in Germany, in 1967, and drawing heavily on the author's definitive treatise *De Stijl 1917–1931: The Dutch Contribution to Modern Art*, Meulenhoff, Amsterdam, Netherlands, 1956.
12. Aarnoudse, J., "Het Drinkwaterproduktiebedrijf Berenplaat: (The Water Treatment Plant on the Berenplaat Island near Rotterdam)," *Cement*, Vol. 17 No. 2, 1965, pp. 75–82.
13. Since this article first appeared, I have pursued the subject of esthetics farther by means of a detailed study of Robert Maillart's works. See D. P. Billington, *Robert Maillart's Bridges: The Art of Engineering*, Princeton University Press, Princeton, N.J., 1979. My major goal there was to show that Maillart achieved remarkable beauty by esthetic choices that never violated efficiency or utility. The works of two contemporary Swiss engineers, in the same tradition, Christian Menn and Heinz Isler, show how this ideal of design has produced structures of great beauty in the late 20th century. See *The Bridges of Christian Menn*, Princeton University Art Museum, Princeton, N.J., 1978, and *Heinz Isler as Structural Artist*, Princeton University Art Museum, Princeton, N.J., 1980. Both writings include catalogs for art museum exhibitions prepared by the Art Museum and the School of Engineering and Applied Science, Princeton University.

3.6

The Message of Structure

MARIO G. SALVADORI

There can be structure without architecture, but no architecture without structure. There can be aesthetics without architecture, but no architecture without aesthetics. But is there an influence of structure on aesthetics? And hence: should the structural engineer be interested in aesthetics?

To clarify this issue, one may ignore the various definitions of the beautiful given by aestheticians through the centuries, much as these definitions are a reflection of the culture that accepted them. One may simply notice that aesthetic tenets have changed through the ages, that the same piece of architecture was considered a masterpiece in a certain era and demoted to a second rate achievement later. We thus recognize aesthetic tenets as a set of dynamic laws with nothing absolute about them. And yet, humanity has always been keen on achieving aesthetic results, even in the most humble examples of construction, because the satisfaction of aesthetic needs is one of the basic human aspirations. On this basis alone, a structural engineer cannot ignore those aspects of the profession that are influenced by and influence, in turn, aesthetic results in construction.

It is easy to prove that aesthetically satisfying buildings can be designed even if structural laws are partially neglected due to ignorance or disregard. For example, the translation into stone of a wood temple led the Greeks to the creation of one of the masterpieces of architecture, the Parthenon, which—if judged from a purely structural viewpoint—is anything but "correct." On the other hand, some great engineers, among them Pier Luigi Nervi, have preached that structuralists do not have to bother about aesthetics because if a building is "correctly designed from a structural perspective, beauty will fallout from the correctness of the structure." The innumerable examples of entirely correct structures that are essentially ugly unfortunately prove this theory incorrect. (That engineers like Nervi or Maillart

---

This paper appeared in a slightly different form as Preprint No. 2823, ASCE Convention, Dallas, Texas, April 25–29, 1977.

have reached aesthetically wonderful results is due to their innate feeling for beauty rather than as a consequence of their structural genius.)

Finally, in considering the aesthetics of a building, one must carefully distinguish between those buildings in which the structure is relatively unimportant and hence not uniquely determined, and those in which the structure is essential in determining the appearance of the building. There are buildings in which, in the limit, the appearance is almost exclusively determined by structural considerations. A one-family house can be built with a wood, steel or a concrete structure, while a suspension bridge is uniquely determined by its structural requirements. All buildings between these two extremes have a structural component that is bound to influence, in varying degrees, their appearances and thus influence the aesthetic responses of their users. By the same token, the constraints due to aesthetics often influence the structural solutions adopted by the engineer at the request of the designing architects. It is through this interplay of two persons of different background that the final compromised solution is achieved, as all practitioners know only too well.

During the last few years, in the discussion of aesthetics, a new point of view has become widely accepted by aestheticians, architects, engineers and other creators involved in structured buildings: that of the so-called semiotic message of the finished building. Semiotics, a branch of philosophy developed over the last 60 or 70 years, considers any and all products of the human mind from the viewpoint of communication, particularly non-verbal communication. Communication takes place by means of messages. Verbal messages are usually directly concerned with the meaning of the message, using the elements of a language to achieve communication. Thus, a verbal message is totally dependent on the culture in which the language is understood and in which it continuously evolves. Non-verbal communication can be either directly related to a specific message that could as well be expressed verbally, such as the communication of a road sign, which may be internationally understood, but stands for a specific message ("do not park"). (Fig. 1) Or it may be essentially the by-product of an artifact, whose main purpose was *not* to send a message, but to satisfy a given function. Clothing, for example, is made to cover the human body, but it also signifies status (Figs. 2 and 3).

In architecture we often find both kinds of non-verbal messages. The push-buttons in an elevator or the windows in a building perform specific functions, while at the same time they send out messages related either to their own functions or to some other purpose. The message of the elevator buttons has to do exclusively with their function of moving the elevator to a specific level. The window, on the other hand, through its shape and dimensions may indicate something beyond its intrinsic purpose of transmitting light. The barred window of a jail says very clearly: "This building is a jail," while the ornamented windows of a Renaissance palace state unequivocally the status of the person occupying the palace.

It is obvious from these simple examples that the semiotic message is always deeply embedded in a culture. A savage entering into the lobby of a modern high-rise building in the United States would not understand the meaning of the elevator buttons; neither would a citizen of the jungles of Central Africa understand the social status expressed by the windows of the Renaissance palace.

Figure 1

Finally, like any verbal message, the non-verbal message varies in meaning with time: the message of the pyramids was basically religious in nature at the time of the pharaohs. It was magic and scientific, and later became military and glorious for Napoleon and his troops. Now it is a mixture of artistic, sociological and structural messages for the modern visitor.

The use of structure in architecture leads to a two-fold way of introducing into it a semiotic message. When structure is (as it is almost always, at present) an essential but *not visible* part of the building, it constitutes the foundation of the architectural message, but leaves it often without an obvious structural content. A building may depend essentially on its structure to achieve purely aesthetic results, but this may not be apparent to the layperson or even to the casual onlooker with a structural background. On the other hand, particularly in those buildings in which the structure is almost uniquely determined by the requirements of statics, the structure itself transmits a message *strictly related to structural action*, which, while not independent of the architectural message, has a meaning of its *own*.

Not much has been written on the semiotic message of structure, so that the following remarks must be considered as tentative; a simple beginning in a field in rapid evolution.

It is my belief that purely structural messages originate in our intuitive understanding of structural behavior and that this understanding stems both from our physical experience with structural actions in our daily lives and from our

Figure 2

perception of structural forms in nature. To exemplify the first kind of experience, one may mention the lifting of weights by means of ropes—say, the lifting of water from a well. This is an action that goes back to the beginning of civilization and gives us an immediate understanding of the tensile properties of natural or artificial ropes. Similarly, the primordial breaking of tree branches by bending, to obtain wood for a fire, or the resistance of our muscles in going either up or down a mountain, make it clear that muscular reactions as well as physical observations have given the race an immediate feeling for structural action much before a quantitative grasp of it was obtained by some kind of scientific approach.

The visual perception of natural structures has been basic in extending our intuitive understanding of man-made structures. The branches of a tree, acted upon by their own weight and snow or wind, initially suggest the shape and behavior of cantilevers, with their larger dimensions at the root and their slender dimensions at the tip. The trunk of a tree confirms our vague comprehension of the need for larger cross-sections at the root of a cantilever. At the same time, it introduces us to the needs created by the axial vertical loads acting on tall columns, which accumulate from top to bottom. I believe it is due to these primordial experiences that while we feel an instinctive puzzlement at the sight of columns larger at the top than at the bottom, we accept naturally the shape of a classical Greek column, even though its actual dimensions do not correspond to structural

Figure 3

requirements. Similarly, I know that we would all consider a cantilever beam with a larger section at its tip than at its root to be "ugly," since it is an insult to a structural behavior of which we might not consciously be aware, but that nature has taught us from the beginning of the race (Figs. 4 and 5).

A similar reaction of shock occurs whenever we are confronted with, say, a large mass in the shape of an inverted pyramid (Fig. 6). Mountains, due to the action of gravity, are all shaped as right-side-up pyramids. The Egyptian pyramids have a shape, geometrically idealized, but basically identical to that of all the high mountains we have ever seen. A modern sculpture in the shape of an inverted pyramid does not "say" to the layperson how and why it stands up: it tells him or her that some "magic trick" has been used to achieve such an unnatural result. It is this unnaturalness that arouses in us a sense of uneasy surprise, and not the inner

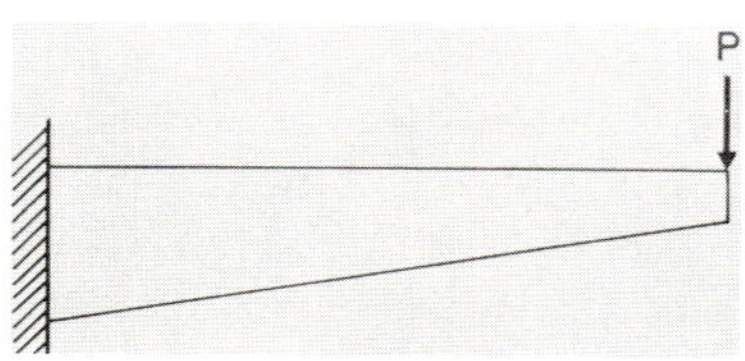

Figure 4

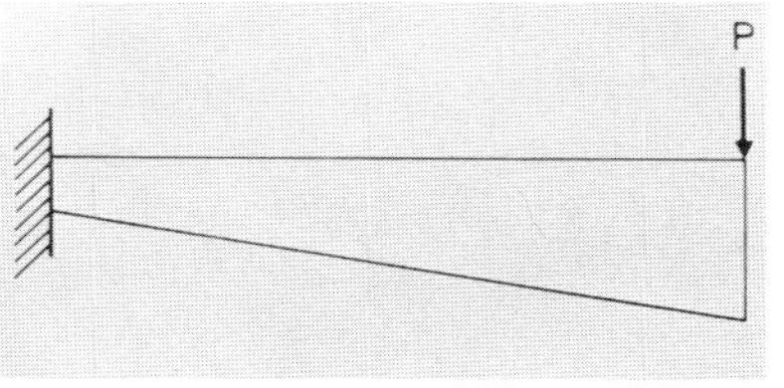

Figure 5

feeling of balance related to daily experiences with "honest" static behavior.

Natural arches have taught us that when stone is used to span a gap, a downward curvature is ideal to achieve this goal with a material so strong in compression that it can support an entire mountain, but so weak in tension that it can be cut by primitive methods using tensile stress. This approach can be extended to two-dimensional structures such as domes: we need only refer to the times we lived in natural caves with curved inner surfaces that gave us a feeling of arch action in space, although they did not make clear to us the amount of material needed to create a curved roof.

On the other hand, the vine catenaries linking tree to tree showed us the need for an upward curvature in a tension structure, a requirement that even modern engineering techniques cannot eliminate.

It would seem, therefore, that to those of us who are not experts in the technical field of structures, the semiotic message of structure itself can come only from a series of atavistic intuitions and that the accumulation of these intuitions results in a set of reactions having an aesthetic content. This is why, I believe, the lay

Figure 6

individual considers a correctly-designed cantilever beam as "lovely" and a cantilever with incorrect structural dimensions as "ugly." By the same token, what beam engineers call "simply-supported" may be correctly designed with a bigger depth at mid-span, because greater bending strength is needed at this point. But it is hard to "understand" for this layperson, who has seldom, if ever, seen "simply-supported beams" in nature and considers such a beam "ugly." As a counterproof of this statement, one must notice that most natural beams are strongly-connected to their supports and so a beam with haunches looks "right" to the inexperienced eye. Moreover, the haunched beam has the visual profile of a low arch and seems ideally-suited to span distances. It is something the expert would agree with once he or she considers the principal stress lines of any beam and realizes that, indeed, arch action is bound to take over when the beam becomes overloaded. It would seem that, in this case, the human eye is conditioned to see the "ultimate" shape of the beam, following the basic principle that a structure will not collapse unless it really *has to*.

Shifting our attention to some structures which have appeared only recently on the scene, such as steel frames hinged at their foot or two or three-hinged arches, we notice that their message is mostly equivocal to the lay reader because their structural behavior is not easily understood. Consequently, it appears that such structures do not have much aesthetic content for the lay observer, and neither do they speak beauty to the experts. While the curved shape of an arch, whatever its boundary conditions, has a strong aesthetic charge, a frame appears mostly as a purely utilitarian structure. One would be hard put to mention a building with aesthetics that have been enhanced by such a structural element.

A similar reaction is elicited by those *correct* structural shapes that have their justifications in subtle physical phenomena. For example, a compression member whose shape is dictated by *buckling* is hardly viewed as beautiful or ugly: most of the time, it is looked upon as a component of a mechanical set of elements. In fact, machines have for a long time been considered ugly, because while their shapes are totally logical from a structural and from a functional point of view, they cannot be grasped as part of the universe of aesthetics by the untrained eye. It is true that some great painters of the beginning of this century, like Picabia, did introduce machine elements into their world (Fig. 7). But they were careful to dissociate the industrial image from the world of mechanical reality, so that machines became mere symbols of a new order rather than objects having a visual content related to the part they play in the real world.

The same type of consideration can be made in the three-dimensional world of structures. Domes and their natural counterparts, the caves, have already been mentioned. The reference in nature to elements with gaussian curvature of constant sign is quite common. Sea-shells (Fig. 8) are a symbol of protection, but in addition have a strong aesthetic content whenever they are ribbed in any of the great variety of ways in which nature has shown us how to stiffen a curved three-dimensional element (Fig. 9).

It is interesting to notice in this context that *scale* does not seem to impair the semiotic message of a man-made structure of large dimensions. The reference of the message to the common experiences of the race is unaffected by changes in size. A typical example of a common message independent of scale is given by the

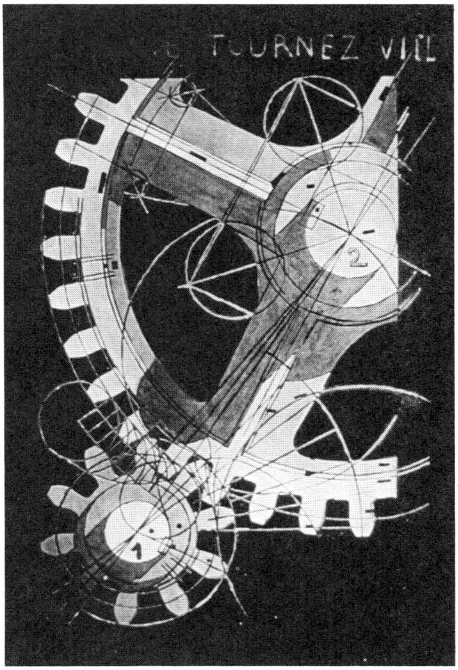

Figure 7

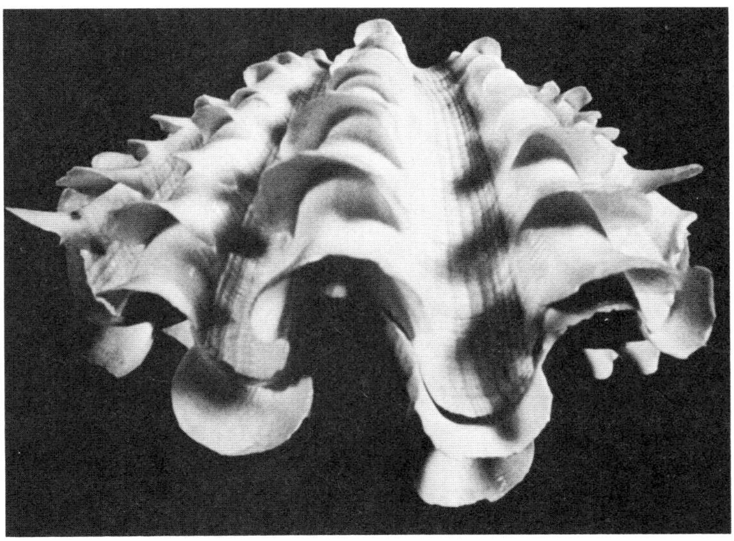

Figure 8

Figure 9

comparison of almost any tensile structure to a spider web. Because of the efficiency of tension stress fields, tensile structures are always exceptionally light in appearance and, whatever their shape and size, are considered "elegant" by the lay observer. Besides the obvious reference to wild vines and spider webs, one must add that tensile structures very often have funicular shapes; that is, shapes dictated by the loads they carry. This implies to the reader of their message a naturalness and a correctness that at times may not be evident, or may not exist, in a structure that uses bending or compression to perform its function—so much so that if one, following Gaudi, uses an upside down (anti-funicular) shape in compression, dictated by the corresponding tensile funicular shape, one may reach shapes technically correct, but whose message may or may not be one of beauty. An exception to this reaction is to be found in the particularly beautiful shape of the shallow cable of a suspension bridge, which becomes by inversion the shape of a shallow arch, a shape we often encounter in nature.

It is interesting to notice, at this point, that no domed structure is lighter than a pneumatic roof (Fig. 10). But its natural shape, which should look "elegant" to the inexperienced eye because of its lightness and translucency, is not comprehended aesthetically as such and is more a source of puzzlement than of appreciation. It goes without saying that, from the outside, most pneumatic structures spell heaviness and that this is an obstacle to their general acceptance. The traditional

Figure 10

message of the stone dome in compression does not allow us, as yet, to understand the recent pneumatic structural message, which therefore confuses us both structurally and aesthetically.

It is fascinating to notice that this confusion does not arise in tents. These "pure" tensile structures have their counterparts in nature. They have also been seen by the race for thousands of years. Hence, both real tents and similar tensile roofs (like inverted, cable-suspended domes) elicit the kind of aesthetic appreciation showered on most curved structures modeled after natural examples.

One should not consider the understanding of structural behavior a necessary condition for the aesthetic appreciation of a structure. The Parthenon has been mentioned in this light. But a more striking example of the unimportance of structural understanding is given by the shape of a roof universally admired while seldom understood structurally. The hyperbolic paraboloid, one of the most efficient structural roof forms (at least, when correctly supported) is blessed by a feature often appearing in nature: a negative gaussian curvature, that is, a geometrical twist that allows the shape to be generated either by the translation of two parabolas or by the translation of straight lines. Modern sculpture exhibits such a feature in a large percentage of pieces and at all scales. The geometrical twist may appear in a small detail or may be the basis for the general shape of the sculpture. Similarly, hyperbolic paraboloids can be used correctly in a more or less horizontal position to create roofs (Fig. 11) or in a structurally unjustifiable vertical setting as walls, but their aesthetic message is always one of beauty. The reaction of a twelve-year-old, who had never seen such shape in a model of a roof before, may indicate some of the associations dictated by the semiotic message of the hyperbolic paraboloid. He first indicated the similarity of the model to a horse saddle, but then, on second thought, he added, "it also looks like a bird in flight." This confirms the need for natural references in justifying the aesthetic reaction to special forms, whether they have a correct structural content or not.

Figure 11

In some cases even an unconscious understanding of the correctness of a structure may well enhance its aesthetic content. The lines of principal stress, often expressed in a structure, help the intuitive understanding of its behavior. Even in a case as complicated as that of a ribbed slab with ribs along the lines of the principal bending moments (which facilitate the comprehension of plate behavior without the help of a torsional mechanism), the pattern of the ribs becomes a source of aesthetic satisfaction to those who do not even dream that torsion exists in a plate. Similarly, the lines of principal stress in a beam make its bending behavior clearer by expressing it in terms of much "easier" states of stress, tension and compression. They also explain the visual reaction of lay observers to complex structural elements whenever the principal stress lines are made more or less evident, as in some domes.

One may wonder whether the puzzling message of certain structures may ever lead to an aesthetic acceptance of their behavior. The shape of prestressed concrete elements, governed by the tension in their invisible tendons, may contradict what one would expect to be their "logical" shape. And one may ask what the reaction of future generations will be, from an aesthetic point of view, to masses floated in space by electromagnetic fields of force (Fig. 12). These totally invisible fields have no reference in nature and produce results amazing to the viewer in that they seem to defy some basic laws of nature. This defiance is both puzzling and shocking, since there can be no doubt about the complete satisfaction obtained from a semiotic message, when it appears *both* aesthetically and structurally acceptable.

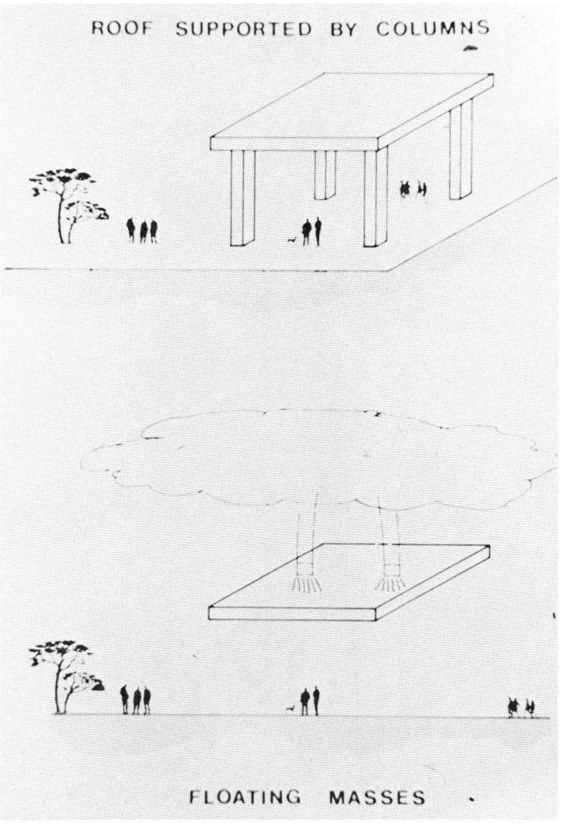

Figure 12

It is this completeness, this harmony between the visual needs of beauty and the respect for the basic laws of nature that has dictated in the past, and dictates even more in the present, the tendency to exhibit the structure of a building. The same completeness is the source of our admiration for the Roman vaults, on one hand, and for the Hancock Insurance Building in Chicago on the other (Fig. 13).

As we have shown, the semiotic message of a structure is strictly related to our personal experience and, culturally, to the experience of the race. That such experience can, within a short number of years, change our aesthetic appreciation of a structure is proven by the classical example of the Tour Eiffel in Paris. This extraordinary steel structure, erected by an engineer of genius on the occasion of an exposition in 1889, had a purely utilitarian purpose and was to be dismantled at the closing of the exhibition. The campaign against its erection involved some of the most respected representatives of French culture of the time, including famous writers, poets, painters and politicians. But, as is often the case with supposedly demountable structures, the Eiffel Tower was not dismounted and—only 23 years after its erection—became the theme of a famous series of paintings

Figure 13

by Delaunay (Fig. 14), showing its acceptance from a purely aesthetic viewpoint. It did not take much longer for the tower to become not just *one* of the symbols of that center of world culture called Paris, but *its very* symbol. And a few years later the tower, all on its own structural steam, became the semiotic symbol of France. In this extraordinary case, the total semiotic message stems directly and uniquely from a structural message: the Eiffel Tower is a structural masterpiece in which nothing was conceded to decoration, nothing was used to hide its necessary sinews. Its acceptance indicates an amazing reversal of public opinion and shows the possibility of a pure aesthetic message on the part of structure. A similar interaction between structural and aesthetic needs led to the nakedness of the towers of the George Washington Bridge, against the opposition of a large part of the New York intelligentsia and of the designing engineer himself: it is another indication of the rapid change in cultural meaning of the semiotic message of structure (Fig. 15).

The Beaubourg Museum (Fig. 16) has just been inaugurated in Paris. The dismay

Figure 14

Figure 15

Figure 16.

of the art world at the erection of an art shelter whose aesthetic message is based not only on its structure, but on its mechanical systems, must be accepted in the light of our recent historical past. One cannot forecast that the Beaubourg will become the new symbol of Paris or of modern art, but one should not be surprised if the incorporation of mechanical elements into its aesthetic message were to lead to a widening of the vocabulary of architecture and become accepted as a matter of course in a few generations or even a few years. After all, if the last operas of Mozart were criticized as "noise, not music" by the Vienna critics of the time, if the French painters of the beginning of our century were called beasts, if *Finnegan's Wake* is still mostly unread, why should the message of the structural and mechanical systems enjoy a different destiny?

Selected Additional Readings

"Painters of Industry," *Fortune*, December 1949, pp. 131–142.

R. L. Bailey, *Disciplined Creativity for Engineers*, Ann Arbor Science Publishers, Ann Arbor, Mich., 1978.

D. P. Billington, *Robert Maillart's Bridges—The Art of Engineering*, Princeton University Press, Princeton, N.J., 1979.

J. Bronowski, "The Creative Process," *Scientific American*, September 1958, pp. 59–65.

C. Condit, "Sullivan's Skyscrapers as the Expression of 19th Century Technology," *Technology and Culture*, winter 1959, pp. 78–93.

C. H. Ellis, *Railway Art*, New York Graphic Society, New York, 1977.

J. B. Eklund, "Art Opens the Way for Science," *Chemical & Engineering News*, June 5, 1978, pp. 25–32.

E. S. Ferguson, "The Mind's Eye: Non-Verbal Thought in Technology," *Science*, August 26, 1977, pp. 827–835.

C. T. Grimm, "Rationalized Esthetics in Civil Engineering Design," *Journal of the Structural Division*, American Society of Civil Engineers, September 1975, pp. 1813–1822.

G. Hambraeus, "Phases of Creativity in Science and Technology," *Science, Technology and the Human Prospect*, C. Starr and P. C. Ritterbush (editors), Pergamon Press, New York, 1979.

H. A. Klein, "Pieter Bruegel the Elder as a Guide to 16th Century Technology," *Scientific American*, March 1978, pp. 134–140.

E. B. Mock, *The Architecture of Bridges*, The Museum of Fine Art, New York, 1949.

L. Mumford, *Art and Technics*, Columbia University Press, New York, 1952.

J. Robinson, *Highways and Our Environment*, McGraw-Hill Book Company, New York, 1971.

C. S. Smith, "Art, Technology and Science: Notes on Their Historical Interaction," *Technology and Culture*, October 1970, pp. 493–549.

C. S. Smith, "Forward," *Art and Technology, A Symposium on Classical Bronzes*, S. Doeringer, D. Mitten, and A. Steinberg (editors), The M.I.T. Press, Cambridge, Mass., 1970, pp. v–vii.

J. Wosk, "Artists on Technology," *Technology Review*, December/January 1980, pp. 66–74.

PART 4

Philosophy and Engineering

To comprehend what is, this is the task of philosophy, because what is, is reason.

G. W. F. Hegel

In Breughel's Icarus, for instance: how everything turns away
Quite leisurely from the disaster; the plowman may
Have heard the splash, the forsaken cry,
But for him it was not an important failure;...

W. H. Auden, "Musée des Beaux Arts"

"Every science begins as philosophy and ends as art" (1). Philosophy in its most general terms can be said to be the asking of the questions: Why? and How? Science is the seeking of answers to these questions, while art is the application of the answers. From this perspective, engineering, with its fundamentals rooted in science and its applications as the practice of an art, has many close connections to philosophy. But what is the relation between philosophy and engineering? Literally, the word *philosophy* means "love of wisdom," and in our cultural past the pursuit of wisdom or knowledge, the how and why, was philosophy regardless of the area of activity. Every discipline has its philosophical aspects, and philosophy, in this sense. This is reflected in the terminal academic degree of Ph.D. (Doctor of Philosophy). The Ph.D. refers not only to the practice of a discipline but also to reflection upon the methods and goals of that practice as it relates to other activities and to human existence in general.

Philosophy becomes relevant to engineering when it is recognized that engineering is practiced, not in isolation from other activities, but rather as a human function performed in society. The engineer interacts with others in the society using a body of specialized knowledge in a variety of ways that have consequences for all. Engineering may well be a paradigm for philosophical reflection, for it is in engineering that both scientific and craft knowledge are applied to human needs, wants, goals, and desires. Furthermore, engineering is guided in its activity by these same human concerns. In the technologically based society of our modern Western world, we might argue that no individual should be more concerned with philosophical reflection than the engineer.

Only relatively recently has philosophy emerged as a separate discipline. Prior to this, all learned men were philosophers through their recognition as seekers of wisdom and truth. Currently, philosophy as a discipline includes five recognized divisions: (a) logic—the study of valid reasoning; (b) aesthetics—the study of the phenomena of beauty; (c) social and ethical theory—concern with human conduct and the social organization; (d) epistemology—the theory of knowledge; and (e) metaphysics which, though difficult to define precisely, may be considered as the study of first principles or ultimate reality of mind and matter.

Our concern with philosophy in this book is with the role technology plays or should play in our complex world of today and with the ethical concerns that confront modern technology and its practitioners. Both are the basis for many multifaceted questions that can be introduced at best in the limited selections included in this part.

There is little question about the impact of technology on humanity and the world in which we live. With the growth of human dependence on machines to produce the food, materials, and products that form the basis for Western civilization, technology and its results have had profound effects on society. The very growth of human dependence on technological contributions has altered the concept of our relation to nature. Though nature has healed many of the wounds we have inflicted and has been forgiving of our attacks, the ever-increasing force of technological impact has caused changes.

What will be the long-range results of these effects? Can technology continue on its present path without concern for the future of the Earth? These questions become more critical when we recognize now that we know the significance of the technological shock to our world. We no longer can excuse many of our most harmful actions by "we didn't know." And what of the force of technology on humanity? Our skills hold out possibilities to change people by controlling human development, reproduction, and behavior. Can or should technology continue unrestricted in developments that will influence the world as a whole, today and in the future, thus possibly effecting future generations? Possession of the tools of knowledge carry an obligation to use them properly. Being human, we must always be aware we can use our tools to make unredeemable mistakes. These and many other similar questions and concerns make up the larger question of what is the role technology should play in our complex world. They are reflected in the rapidly expanding literature that addresses the part technology should or ought to play in shaping public policy on the major issues of our time that, in many instances, have technological overtones.

Former chief justice of the Supreme Court Earl Warren (2) spoke of ethics as a law beyond the law created by, and indispensable to, civilization. Ethics, as we normally think of it, is concerned with those human actions that deal with the general question: What should I do? But ethics is not a code of laws nor a series of hard and fast rules for general application. Nor can ethics be equated to moral theology, though certainly many of the concepts of theology are reflected in our view of what is good and what we should do to achieve that end.

Ethics has been a concern of the professional since the days of ancient Greece when Hippocrates is credited for formulating the Hippocratic oath for the medical practitioner. Today we define a profession, in part, by the acceptance, by the professional group, of a specific set of ethical standards. But these standards, or codes of ethics, are simply duties or obligations that a member of a profession accepts as binding when voluntarily joining a professional organization that has a code of ethics for conduct. These duties and obligations may include ethical considerations but more generally are guidelines for the professional relations between a professional and society, his or her clients, employer, and peers. Many of these obligations fall under a more general heading of social responsibility. This broad obligation includes the responsibility of the individual engineer and the profession, and extends to government through control over technology in the service of society. The questions of social responsibility are complex and are inextricably linked to ethics through the dilemma posed by the question: Should we build it simply because we can? or by the statement: It's not what I can do, but what I ought to do. This is illustrated further by a quotation from the presidential address of A. Z. Hill to the British Association for the Advancement of Science, as quoted by Trevor Williams (3):

> The conquest of disease has led to a vast increase in the world's population. The result may be starvation, unrest, and even the end of civilization. If ethical principles deny our right to do evil that good may happen, are we justified in doing good when the foreseeable consequence is evil? The forces of good and evil depend not on the scientist [engineer] but on the moral judgement of the whole community.

The questions of social responsibility and professional ethics are receiving more and more attention from the engineering profession, its practitioners, and various government agencies. The reasons for this renewel of interest is probably the direct result of the public disclosure of unethical behavior by engineers and other professionals in the recent past. This reason has been reinforced by questions of the legality of portions of several of the professional codes of ethics that had been posed by the U.S. Department of Justice and by a renewed study of their codes by a number of professional societies. Beyond this, and perhaps more important to the purpose of this part, broad questions of social responsibilty and the role of technology/engineering have been raised in public debate. Such topics include aircraft safety (DC-10 incidents), nuclear power (Brown's Ferry and Three Mile Island), bridge collapse (Tampa Bay Sunshine Bridge), firesafety in high-rise construction (MGM Grand Hotel, Las Vegas), super tankers, the Alaska pipeline, and too many others.

The educated engineer today and in the future must be sensitive to concerns and questions like those previously reviewed. The study of philosophy, to a greater extent than available in this book, is both a desirable and a justifiable activity for engineers. "Philosophy's results concern us all vitally" and "Philosophy's queerest arguments tickle our sense of subtlety and ingenuity" (4).

"Technology and Responsibility: Reflections on the New Task of Ethics" was written by Hans Jonas, a distinguished philosopher and contributor of many articles to the literature of ethics. In this article, he presents very well the changing role of technology and the accompanying need to modify consideration of the ethics of the use of modern technology. Jonas traces the concept of ethics as governing an individual's relation to another within a limited time and space frame, to a recognition that a person's actions today may have influences far beyond the immediate vicinity in both time and distance. In addition, many modern activities are not always those of individuals but often of a group. The existence of the changing scope and concern of ethical action finds an excellent case in point with current and prospective technology. Jonas defines the problem of the role of technology within an ethical context through consideration of the broader question of what should be the role of technology in the world of today and tomorrow.

Norman Balabanian, a well-known and active electrical engineer and computer scientist, has extended the concern of technology's role in a world that is more and more dependent on it. The title of his essay "Presumed Neutrality of Technology" refers to a frequently expressed concept that technology is itself neutral and the results of its use (good or bad) are a function of humanity's use of it. Balabanian sees technology as the tool of political and economic interests that control its use (or misuse) for the benefit of a few to the detriment of the many. His article calls for "an altering [of] the very nature of technology" and the adoption of "a new style of living which is in keeping with the natural world." Critics of this article have pointed out that "in the pre-technical age, life, even though in harmony with nature, was short and brutal" (5).

In his essay "Making, Doing, and Thinking," Karl Pavlovic, a modern philosopher, is concerned with the narrow definition of technology. He expands on a concept introduced by Ortega y Gasset that the advantages, threats, and limitations of technology should be considered in a broader context. He proposes a redirection of the discussion of technology to include its relation to all aspects of the social and material life to achieve the most desirable uses of technology for the future.

"Living on a Lifeboat" by Garrett Hardin presents a serious analysis of the extent of humanity's capability to continue to grow in a world of limited resources. Hardin, a biologist, is concerned primarily with the question of population and the distribution of limited resources within a world of population explosion. Technology is often called on to provide the capability to extend the ability to support a growing population and, simultaneously, is taken to task for despoiling nature and the environment in the effort to do so. Hardin's commentary provides a view from a different perspective of the dilemma facing engineers (and all humanity) on how we can apply our technology and other skills to maintain life without doing evil in attaining that end. The ethics of the lifeboat are brought into focus in a modern setting.

Many engineering societies have codes of ethics or codes of conduct. We have chosen to present the version of the Engineers' Council for Professional Development (ECPD) as representative. This code is also the most widely adopted one within the engineering profession. Note, however, that a recent combination of engineering societies under the name of the American Association of Engineering Societies is continuing a study of engineering codes of ethics with the goal of establishing a common code for all engineers.

Engineering codes of ethics suffer from many shortcomings. These include the frequent inclusion of articles that are more business etiquette than ethics, and there is an element of self-service that lingers in the background of the codes. Unquestionably, ethical concerns vary greatly for the engineer in practice and the engineer in private employment. For example, the former may be more frequently concerned with the ethical propriety of political contributions, while the latter is likely to face an ethical dilemma when the obligations to an employer are in conflict with other obligations to the public welfare. In the past decade, the U.S. Department of Justice has raised a number of questions about the legality of professional codes of ethics, including the former prohibition against competitive bidding and restrictions on the permissible form of advertising of professional services. Engineering codes have changed many times since the first one: the Institution of Civil Engineers, formulated in 1910. Each change has reflected the changing concept of the role of the profession. Currently, all codes (ECPD version, for example) feature the obligation of professionals to use their talents for the benefit of the public. This is a worthy aim and objective, but it raises numerous questions: Who is the engineer's public? What determines when a result is a benefit? Engineering practice is generally a team effort: What is the responsibility of the individual engineer within the team? The ECPD Code of Ethics should be studied to establish what is being asked of those who accept it, with questions similar to those posed continually in your own mind.

Albert Flores has provided the final article in this part. He attempts to establish the philosophical basis for codes of ethics. The basic question considered is the justification for engineering codes of ethics rather than the motives that have led to the adoption of such codes or the causes which have influenced engineers to accept them. Flores explores in more detail many of the questions raised in the preceding paragraphs. He brings to his analysis a wide experience gained with the program in engineering ethics education that has been so successful at Rensselaer Polytechnic Institute.

In the final analysis, ethics is founded in human values which, though generally similar in any given culture, will vary among individuals. Ethical decisions are difficult, and "uncertainty and imprecision beset hard moral choices" (6). Unfortunately, engineering experience gained in technical development provides little help, for the very nature of engineering brings the engineer increased problems of moral responsibility as he or she advances in the profession.

References

1. Durant, W., *The Story of Philosophy*, Simon & Schuster, New York, 1961, p. 2.
2. Address at Louis Marshall Award Dinner of the Jewish Theological Seminary of America, New York, November 12, 1962.
3. Williams, Trevor (editor), *History of Technology, Vol. VII—The Twentieth Century, c. 1900 to c. 1950*, Part I, Clarendon Press, Oxford, England, 1978, p. 1473.

4. William James as quoted by Florman, S. C., *Engineering and the Liberal Arts*, McGraw-Hill Book Company, New York, 1968, p. 168.
5. Weinberg, A. M., "Technological Optimism," *Society*, March/April 1980, p. 18.
6. Bok, S., *Lying: Moral Choice in Public and Private Life*, Pantheon Books, New York, 1978, p. 54.

4.1

Technology and Responsibility: Reflections on the New Tasks of Ethics

HANS JONAS

All previous ethics—whether in the form of issuing direct enjoinders to do and not to do certain things, or in the form of defining principles for such enjoinders, or in the form of establishing the ground of obligation for obeying such principles—had these interconnected tacit premises in common: that the human condition, determined by the nature of man and the nature of things, was given once for all; that the human good on that basis was readily determinable; and that the range of human action and therefore responsibility was narrowly circumscribed. It will be the burden of my argument to show that these premises no longer hold, and to reflect on the meaning of this fact for our moral condition. More specifically, it will be my contention that with certain developments of our powers the *nature of human action* has changed, and since ethics is concerned with action, it should follow that the changed nature of human action calls for a change in ethics as well: this not merely in the sense that new objects of action have added to the case material on which received rules of conduct are to be applied, but in the more radical sense that the qualitatively novel nature of certain of our actions has opened up a whole new dimension of ethical relevance for which there is no precedent in the standards and canons of traditional ethics.

The novel powers I have in mind are, of course, those of modern *technology*. My first point, accordingly, is to ask how this technology affects the nature of our acting, in what ways it makes acting under its dominion *different* from what it has been through the ages. Since throughout those ages man was never without technology, the question involves the human difference of *modern* from previous technology. Let us start with an ancient voice on man's powers and deeds which in an archetypal sense itself strikes, as it were, a technological note—the famous Chorus from Sophocles' *Antigone.*

Many the wonders but nothing more wondrous than man.
This thing crosses the sea in the winter's storm,
making his path through the roaring waves.
And she, the greatest of gods, the Earth—
deathless she is, and unwearied—he wears her away
as the ploughs go up and down from year to year
and his mules turn up the soil.

The tribes of the lighthearted birds he ensnares, and the races
of all the wild beasts and the salty brood of the sea,
with the twisted mesh of his nets, he leads captive, this clever man.
He controls with craft the beasts of the open air,
who roam the hills. The horse with his shaggy mane
he holds and harnesses, yoked about the neck,
and the strong bull of the mountain.

Speech and thought like the wind
and the feelings that make the town,
he has taught himself, and shelter against the cold,
refuge from rain. Ever resourceful is he.
He faces no future helpless. Only against death
shall he call for aid in vain. But from baffling maladies
has he contrived escape.

Clever beyond all dreams
the inventive craft that he has
which may drive him one time or another to well or ill.
When he honors the laws of the land and the gods' sworn right
high indeed is his city; but stateless the man
who dares to do what is shameful.

(lines 335–370)

This awestruck homage to man's powers tells of his violent and violating irruption into the cosmic order, the self-assertive invasion of nature's various domains by his restless cleverness; but also of his building—through the self-taught powers of speech and thought and social sentiment—the home for his very humanity, the artifact of the city. The raping of nature and the civilizing of himself go hand in hand. Both are in defiance of the elements, the one by venturing into them and overpowering their creatures, the other by securing an enclave against them in the shelter of the city and its laws. Man is the maker of his life *qua* human, bending circumstances to his will and needs, and except against death he is never helpless.

Yet there is a subdued and even anxious quality about this appraisal of the marvel that is man, and nobody can mistake it for immodest bragging. With all his boundless resourcefulness, man is still small by the measure of the elements: precisely this makes his sallies into them so daring and allows those elements to tolerate his forwardness. Making free with the denizens of land and sea and air, he yet leaves the encompassing nature of those elements unchanged, and their generative powers undiminished. Them he cannot harm by carving out his little dominion from theirs. They last, while his schemes have their shortlived way. Much as he harries Earth, the greatest of gods, year after year with his plough—she is ageless and unwearied; her enduring patience he must and can trust, and to her cycle he must conform. And just as ageless is the sea. With all his netting of the salty brood, the spawning ocean is inexhaustible. Nor is it hurt by the plying of ships, nor sullied by what is jettisoned into its deeps. And no matter how many illnesses he contrives to cure, mortality does not bow to his cunning.

All this holds because man's inroads into nature, as seen by himself, were essentially superficial, and powerless to upset its appointed balance. Nor is there a hint, in the *Antigone* chorus or anywhere else, that this is only a beginning and that greater things of artifice and power are yet to come—that man is embarked on an endless course of conquest. He had gone thus far in reducing necessity, had learned by his wits to wrest that much from it for the humanity of his life, and there he could stop. The room he had thus made was filled by the city of men—meant to enclose, and not to expand—and thereby a new balance was struck within the larger balance of the whole. All the well or ill to which man's inventive craft may drive him one time or another is inside the human enclave and does not touch the nature of things.

The immunity of the whole, untroubled in its depth by the importunities of man, that is, the essential immutability of Nature as the cosmic order, was indeed the backdrop to all of mortal man's enterprises, including his intrusions into that order itself. Man's life was played out between the abiding and the changing: the abiding was Nature, the changing his own works. The greatest of these works was the city, and on it he could confer some measure of abidingness by the laws he made for it and undertook to honor. But no long-range certainty pertained to this contrived abidingness. As a precarious artifact, it can lapse or go astray. Not even within its artificial space, with all the freedom it gives to man's determination of self, can the arbitrary ever supersede the basic terms of his being. The very inconstancy of human fortunes assures the constancy of the human condition. Chance and luck and folly, the great equalizers in human affairs, act like an entropy of sorts and make all definite designs in the long run revert to the perennial norm. Cities rise and fall, rules come and go, families prosper and decline; no change is there to stay, and in the end, with all the temporary deflections balancing each other out, the state of man is as it always was. So here too, in his very own artifact, man's control is small and his abiding nature prevails.

Still, in this citadel of his own making, clearly set off from the rest of things and entrusted to him, was the whole and sole domain of man's responsible action. Nature was not an object of human responsibility—she taking care of herself and, with some coaxing and worrying, also of man: not ethics, only cleverness applied to her. But in the city, where men deal with men, cleverness must be wedded to morality, for this is the soul of its being. In this intra-human frame dwells all traditional ethics and matches the nature of action delimited by this frame.

Let us extract from the preceding those characteristics of human action which are relevant for a comparison with the state of things today.

1. All dealing with the non-human world, i.e., the whole realm of *techne* (with the exception of medicine), was ethically neutral—in respect both of the object and the subject of such action: in respect of the object, because it impinged but little on the self-sustaining nature of things and thus raised no question of permanent injury to the integrity of its object, the natural order as a whole; and in respect of the agent subject it was ethically neutral because *techne* as an activity conceived itself as a determinate tribute to necessity and not as an indefinite, self-validating advance to mankind's major goal, claiming in its pursuit man's ultimate effort and concern. The real vocation of man lay elsewhere. In brief, action on non-human things did not constitute a sphere of authentic ethical significance.
2. Ethical significance belonged to the direct dealing of man with man, including the dealing with himself: all traditional ethics is *anthropocentric.*
3. For action in this domain, the entity "man" and his basic condition was considered constant in essence and not itself an object of reshaping *techne.*
4. The good and evil about which action had to care lay close to the act, either in the praxis itself or in its immediate reach, and were not a matter for remote planning. This proximity of ends pertained to time as well as space. The effective range of action was small, the time-span of foresight, goal-setting and accountability was short, control of circumstances limited. Proper conduct had its immediate criteria and almost immediate consummation. The long run of consequences beyond was left to chance, fate or providence. Ethics accordingly was of the here and now, of occasions as they arise between men, of the recurrent, typical situations of private and public life. The good man was he who met these contingencies with virtue and wisdom, cultivating these powers in himself, and for the rest resigning himself to the unknown.

All enjoinders and maxims of traditional ethics, materially different as they may be, show this confinement to the immediate setting of the action. "Love thy neighbor as thyself"; "Do unto others as you would wish them to do unto you"; "Instruct your child in the way of truth"; "Strive for excellence by developing and actualizing the best potentialities of your being *qua* man"; "Subordinate your individual good to the common good"; "Never treat your fellow man as a means only but always *also* as an end in himself"—and so on. Note that in all these maxims the agent and the "other" of his action are sharers of a common present. It is those alive now and in some commerce with me that have a claim on my conduct as it affects them by deed or omission. The ethical universe is composed of contemporaries, and its horizon to the future is confined by the foreseeable span of their lives. Similarly confined is its horizon of place, within which the agent and the other meet as neighbor, friend or foe, as superior and subordinate, weaker and stronger, and in all the other roles in which humans interact with one another. To this proximate range of action all morality was geared.

It follows that the *knowledge* that is required—besides the moral will—to assure the morality of action, fitted these limited terms: it was not the knowledge of the scientist or the expert, but knowledge of a kind readily available to all men of good will. Kant went so far as to say that "human reason can, in matters of morality, be easily brought to a high degree of accuracy and completeness even in the most ordinary intelligence";[1] that "there is no need of science or philosophy for knowing what man has to do in order to be honest and good, and indeed to be wise and virtuous. . . . [Ordinary intelligence] can have as good hope of hitting the mark as any philosopher can promise himself";[2] and again: "I need no elaborate acuteness to find out what I have to do so that my willing be morally good. Inexperienced regarding the course of the world, unable to anticipate all the contingencies that happen in it," I can yet know how to act in accordance with the moral law.[3]

Not every thinker in ethics, it is true, went so far in discounting the cognitive side of moral action. But even when it received much greater emphasis, as in Aristotle, where the discernment of the situation and what is fitting for it makes considerable demands on experience and judgment, such knowledge has nothing to do with the science of things. It implies, of course, a general conception of the human good as such, a conception predicted on the presumed invariables of man's nature and condition, which may or may not find expression in a theory of its own. But its translation into practice requires a knowledge of the here and now, and this is entirely non-theoretical. This "knowledge" proper to virtue (of the "where, when, to whom, and how") stays with the immediate issue, in whose defined context the action *as the agent's own* takes its course and within which it terminates. The good or bad of the action is wholly decided within that short-term context. Its moral quality shines forth from it, visible to its witnesses. No one was held responsible for the unintended later effects of his well-intentioned, well-considered, and well-performed act. The short arm of human power did not call for a long arm of predictive knowledge; the shortness of the one is as little culpable as that of the other. Precisely because the human good, known in its generality, is the same for all time, its realization or violation takes place at each time, and its complete locus is always the present.

All this has decisively changed. Modern technology has introduced actions of such novel scale, objects, and consequences that the framework of former ethics can no longer contain them. The *Antigone* chorus on the *deinotes*, the wondrous power, of man would have to read differently now; and its admonition to the individual to honor the laws of the land would no longer be enough. To be sure, the old prescriptions of the "neighbor" ethics—of justice, charity, honesty, and so on—still hold in their intimate immediacy for the nearest, day by day sphere of human interaction. But this sphere is overshadowed by a growing realm of collective action where doer, deed, and effect are no longer the same as they were in the proximate sphere, and which by the enormity of its powers forces upon ethics a new dimension of responsibility never dreamt of before.

Take, for instance, as the first major change in the inherited picture, the critical *vulnerability* of nature to man's technological intervention—unsuspected before it

[1]Immanuel Kant, *Groundwork of the Metaphysic of Morals,* preface.
[2]*Op. cit.,* chapter 1.
[3]*Ibid.* (I have followed H. J. Paton's translation with some changes.)

began to show itself in damage already done. This discovery, whose shock led to the concept and nascent science of ecology, alters the very concept of ourselves as a causal agency in the larger scheme of things. It brings to light, through the effects, that the nature of human action has *de facto* changed, and that an object of an entirely new order—no less than the whole biosphere of the planet—has been added to what we must be responsible for because of our power over it. And of what surpassing importance an object, dwarfing all previous objects of active man! Nature as a human responsibility is surely a *novum* to be pondered in ethical theory. What kind of obligation is operative in it? Is it more than a utilitarian concern? Is it just prudence that bids us not to kill the goose that lays the golden eggs, or saw off the branch on which we sit? But the "we" that here sits and may fall into the abyss is all future mankind, and the survival of the species is more than a prudential duty of its present members. Insofar as it is the fate of *man*, as affected by the condition of nature, which makes us care about the preservation of nature, such care admittedly still retains the anthropocentric focus of all classical ethics. Even so, the difference is great. The containment of nearness and contemporaneity is gone, swept away by the spatial spread and time-span of the cause-effect trains which technological practice sets afoot, even when undertaken for proximate ends. Their irreversibility conjoined to their aggregate magnitude injects another novel factor into the moral equation. To this take their cumulative character: their effects add themselves to one another, and the situation for later acting and being becomes increasingly difficult from what it was for the initial agent. The cumulative self-propagation of the technological change of the world thus constantly overtakes the conditions of its contributing acts and moves through none but unprecedented situations, for which the lessons of experience are powerless. And not even content with changing its beginning to the point of unrecognizability, the cumulation as such may consume the basis of the whole series, the very condition of itself. All this would have to be co-intended in the will of the single action if this is to be a morally responsible one. Ignorance no longer provides it with an alibi.

Knowledge, under these circumstances, becomes a prime duty beyond anything claimed for it heretofore, and the knowledge must be commensurate with the causal scale of our action. The fact that it cannot really be thus commensurate, i.e., that the predictive knowledge falls behind the technical knowledge which nourishes our power to act, itself assumes ethical importance. Recognition of ignorance becomes the obverse of the duty to know and thus part of the ethics which must govern the ever more necessary self-policing of our out-sized might. No previous ethics had to consider the global condition of human life and the far-off future, even existence, of the race. Their now being an issue demands, in brief, a new conception of duties and rights, for which previous ethics and metaphysics provide not even the principles, let alone a ready doctrine.

And what if the new kind of human action would mean that more than the interest of man alone is to be considered—that our duty extends farther and the anthropocentric confinement of former ethics no longer holds? It is at least not senseless anymore to ask whether the condition of extra-human nature, the biosphere as a whole and in its parts, now subject to our power, has become a human trust and has something of a moral claim on us not only for our ulterior sake but for its own and in its own right. If this were the case it would require quite some rethinking in basic principles of ethics. It would mean to seek not only the human good, but also the good of things extra-human, that is, to extend the recognition of "ends in themselves" beyond the sphere of man and make the human good include the care for them. For such a role of stewardship no previous ethics has prepared us—and the dominant, scientific view of *Nature* even less. Indeed, the latter emphatically denies us all conceptual means to think of Nature as something to be honored, having reduced it to the indifference of necessity and accident, and divested it of any dignity of ends. But still, a silent plea for sparing its integrity seems to issue from the threatened plenitude of the living world. Should we heed this plea, should we grant its claim as sanctioned by the nature of things, or dismiss it as a mere sentiment on our part, which we may indulge as far as we wish and can afford to do? If the former, it would (if taken seriously in its theoretical implications) push the necessary rethinking beyond the doctrine of action, i.e., ethics, into the doctrine of being, i.e., metaphysics, in which all ethics must ultimately be grounded. On this speculative subject I will here say no more than that we should keep ourselves open to the thought that natural science may not tell the whole story about Nature.

Returning to strictly intra-human considerations, there is another ethical aspect to the growth of *techne* as a pursuit beyond the pragmatically limited terms of former times. Then, so we found, *techne* was a measured tribute to necessity, not the road to mankind's chosen goal—a means with a finite measure of adequacy to well-defined proximate ends. Now, *techne* in the form of modern technology has turned into an infinite forward-thrust of the race, its most significant enterprise, in whose permanent, self-transcending advance to ever greater things the vocation of man tends to be seen, and whose success of maximal control over things and himself appears as the consummation of his destiny. Thus the triumph of *homo faber* over his external object means also his triumph in the internal constitution of *homo sapiens*, of whom he used to be a subsidiary part. In other words, technology, apart from its objective works, assumes ethical significance by the central place it now occupies in human purpose. Its cumulative creation, the expanding artificial environment, continuously reinforces the particular powers in man that created it, by compelling their unceasing inventive employment in its management and further advance, and by rewarding them with additional success—which only adds to the relentless claim. This positive feedback of functional necessity and reward—in whose dynamics pride of achievement must not be forgotten—assures the growing ascendancy of one side of man's nature over all the others, and inevitably at their expense. If nothing succeeds like success, nothing also entraps like success. Outshining in prestige and starving in resources whatever else belongs to the fullness of man, the expansion of his power is accompanied by a contraction of his self-conception and being. In the image he entertains of himself—the potent self-formula which determines his actual being as much as it reflects it—man now is evermore the maker of what he has made and the doer of what he can do, and most of all the preparer of what he will be able to do next. But not you or I: it is the aggregate, not the individual doer or deed that matters here; and the indefinite future, rather than the contemporary context of the action, constitutes the relevant horizon of responsibility. This requires imperatives of a new sort. If the realm of making has invaded the space of essential action, then morality must invade the realm of making, from which it had formerly stayed aloof, and must do so in the form of public policy. With issues of such inclusiveness and such lengths of anticipation public policy has never had to deal before. In fact, the changed nature of human action changes the very nature of politics.

For the boundary between "city" and "nature" has been obliterated: the city of men, once an enclave in the non-human world, spreads over the whole of terrestrial nature and usurps its place. The difference between the artificial and the natural has vanished, the natural is swallowed up in the sphere of the artificial, and at the same time the total artifact, the works of man working on and through himself, generates a "nature" of its own, i.e., a necessity with which human freedom has to cope in an entirely new sense. Once it could be said *Fiat justitia, pereat mundus*, "Let justice be done, and may the world perish"—where "world,' of course, meant the renewable enclave in the imperishable whole. Not even rhetorically can the like be said anymore when the perishing of the whole through the doings of man—be they just or unjust—has become a real possibility. Issues never legislated on come into the purview of the laws which the total city must give itself so that there will be a world for the generations of man to come.

That there *ought* to be through all future time such a world fit for human habitation, and that it ought in all future time to be inhabited by a mankind worthy of the human name, will be readily affirmed as a general axiom or a persuasive desirability of speculative imagination (as persuasive and as undemonstrable as the proposition that there being a world at all is "better" than there being none): but as a *moral* proposition, namely, a practical *obligation* toward the posterity of a distant future, and a principle of decision in present action, it is quite different from the imperatives of the previous ethics of contemporaneity; and it has entered the moral scene only with our novel powers and range of prescience.

The *presence of man in the world* had been a first and unquestionable given, from which all idea of obligation in human conduct started out. Now it has itself become an *object* of obligation—the obligation namely to ensure the very premise of all obligation, i.e., the *foothold* for a moral universe in the physical world—the existence of mere *candidates* for a moral order. The difference this makes for ethics may be illustrated in one example.

Kant's categorical imperative said: "Act so that you *can* will that the maxim of your action be made the principle of a universal law." The "can" here invoked is that of reason and its consistency with itself: *Given* the existence of a community of human agents (acting rational beings), the action must be such that it can without self-contradiction be imagined as a general practice of that community. Mark that the basic reflection of morals here is not itself a moral but a logical one: The "I *can* will" or "I *cannot* will" expresses logical compatibility or incompatibility, not moral approbation or revulsion. But there is no self-contradiction in the thought that humanity would once come to an end, therefore also none in the thought that the happiness of present and proximate generations would be bought with the unhappiness or even non-existence of later ones—as little as, after all, in the inverse thought that the existence or happiness of later generations would be bought with the unhappiness or even partial extinction of present ones. The sacrifice of the future for the present is *logically* no more open to attack than the sacrifice of the present for the future. The difference is only that in the one case the series goes on, and in the other it does not. But that it *ought to go on*, regardless of the distribution of happiness or unhappiness, even with a persistent preponderance of unhappiness over happiness, nay, even of immorality over morality[4]—this cannot be derived from the rule of self-consistency *within* the series, long or short as is happens to be: it is a commandment of a very different kind, lying outside and "prior" to the series as a whole and its ultimate grounding can only be metaphysical.

An imperative responding to the new type of human action and addressed to the new type of agency that operates it might run thus: "Act so that the effects of your action are compatible with the permanence of genuine human life"; or expressed negatively: "Act so that the effects of your action are not destructive of the future possibility of such life"; or simply "Do not compromise the conditions for an indefinite continuation of humanity on earth"; or most generally: "In your present choices, include the future wholeness of Man among the objects of your will."

It is immediately obvious that no rational contradiction is involved in the violation of this kind of imperative. I *can* will the present good with sacrifice of the future good. It is also evident that the new imperative addresses itself to public policy rather than private conduct, which is not in the causal dimension to which that imperative applies. Kant's categorical imperative was addressed to the individual, and its criterion was instantaneous. It enjoined each of us to consider what would happen *if* the *maxim* of my present action were made, or at this moment already were, the principle of a universal legislation; the self-consistency or inconsistency of such a *hypothetical* universalization is made the test for my *private* choice. But it was no part of the reasoning that there is any probability of my private choice *in fact* becoming universal law, or that it might contribute to its becoming that. The universalization is a thought-experiment by the private agent to test the immanent morality of his action. Indeed, real consequences are not considered at all, and the principle is one not of objective responsibility but of the subjective quality of my self-determination. The new imperative invokes a different consistency: not that of the act with itself, but that of its eventual *effects* with the continuance of human agency in times to come. And the "universalization" it contemplates is by no means hypothetical—i.e., a purely logical transference from the individual "me" to an imaginary, causally unrelated "all" ("*if* everybody acted like that"); on the contrary, the actions subject to the new imperative—actions of the collective whole—have their universal reference in their actual scope of efficacy: they "totalize" themselves in the progress of their momentum and thus are bound to terminate in shaping the universal dispensation of things. This adds a *time* horizon to the moral calculus which is entirely absent from the instantaneous logical operation of the Kantian imperative: whereas the latter extrapolates into an ever-present order of abstract compatability, our imperative extrapolates into a predictable real *future* as the open-ended dimension of our responsibility.

Similar comparisons could be made with all the other historical forms of the ethics of contemporaneity and immediacy. The new order of human action requires a commensurate ethics of foresight and responsibility, which is as new as are the issues with which it has to deal. We have seen that these are the issues posed by the works of *homo faber* in the age of technology. But among those novel works we haven't mentioned yet the potentially most ominous class. We have considered *techne* only as applied to the non-human realm. But man himself has

[4]On this last point, the biblical God changed his mind to an all-encompassing "yes" after the Flood.

been added to the objects of technology. *Homo faber* is turning upon himself and gets ready to make over the maker of all the rest. This consummation of his power, which may well portend the overpowering of man, this final imposition of art on nature, calls upon the utter resources of ethical thought, which never before has been faced with elective alternatives to what were considered the definite terms of the human condition.

a. Take, for instance, the most basic of these "givens," man's mortality. Who ever before had to make up his mind on its desirable and *eligible* measure? There was nothing to choose about the upper limit, the "threescore years and ten, or by reason of strength fourscore." Its inexorable rule was the subject of lament, submission, or vain (not to say foolish) wish-dreams about possible exceptions—strangely enough, almost never of affirmation. The intellectual imagination of a George Bernard Shaw and a Jonathan Swift speculated on the privilege of not having to die, or the curse of not being able to die. (Swift with the latter was the more perspicacious of the two.) Myth and legend toyed with such themes against the acknowledged backgound of the unalterable, which made the earnest man rather pray "teach us to number our days that we may get a heart of wisdom" (Psalm 90). Nothing of this was in the realm of doing and effective decision. The question was only how to relate to the stubborn fact.

But lately, the dark cloud of inevitability seems to lift. A practical hope is held out by certain advances in cell biology to prolong, perhaps indefinitely extend the span of life by counteracting biochemical processes of aging. Death no longer appears as a necessity belonging to the nature of life, but as an avoidable, at least in principle tractable and long-delayable, organic malfunction. A perennial yearning of mortal man seems to come nearer fulfillment. And for the first time we have in earnest to ask the question "How desirable is this? How desirable for the individual, and how for the species?" These questions involve the very meaning of our finitude, the attitude toward death, and the general biological significance of the balance of death and procreation. Even prior to such ultimate questions are the more pragmatic ones of who should be eligible for the boon: persons of particular quality and merit? of social eminence? those that can pay for it? everybody? The last would seem the only just course. But it would have to be paid for at the opposite end, at the source. For clearly, on a population-wide scale, the price of extended age must be a proportional slowing of replacement, i.e., a diminished access of new life. The result would be a decreasing proportion of youth in an increasingly aged population. How good or bad would that be for the general condition of man? Would the species gain or lose? And how *right* would it be to preempt the place of youth? Having to die is bound up with having been born: mortality is but the other side of the perennial spring of "natality" (to use Hannah Arendt's term). This had always been ordained; now its meaning has to be pondered in the sphere of decision.

To take the extreme (not that it will ever be obtained): if we abolish death, we must abolish procreation as well, for the latter is life's answer to the former, and so we would have a world of old age with no youth, and of known individuals with no surprises of such that had never been before. But this perhaps is precisely the wisdom in the harsh dispensation of our mortality: that it grants us the eternally renewed promise of the freshness, immediacy and eagerness of youth, together with the supply of otherness as such. There is no substitute for this in the greater accumulation of prolonged experience: it can never recapture the unique privilege of seeing the world for the first time and with new eyes, never relive the wonder which, according to Plato, is the beginning of philosophy, never the curiosity of the child, which rarely enough lives on as thirst for knowledge in the adult, until it wanes there too. This ever renewed beginning, which is only to be had at the price of ever repeated ending, may well be mankind's hope, its safeguard against lapsing into boredom and routine, its chance of retaining the spontaneity of life. Also, the role of the *memento mori* in the individual's life must be considered, and what its attenuation to indefiniteness may do to it. Perhaps a non-negotiable limit to our expected time is necessary for each of us as the incentive to number our days and make them count.

So it could be that what by intent is a philanthropic gift of science to man, the partial granting of his oldest wish—to escape the curse of mortality—turns out to be to the detriment of man. I am not indulging in prediction and, in spite of my noticeable bias, not even in valuation. My point is that already the promised gift raises questions that had never to be asked before in terms of practical choice, and that no principle of former ethics, which took the human constants for granted, is competent to deal with them. And yet they must be dealt with ethically and by principle and not merely by the pressure of interests.

b. It is similar with all the other, quasi-utopian powers about to be made available by the advances of biomedical science as they are translated into technology. Of these, *behavior control* is much nearer to practical readiness than the still hypothetical prospect I have just been discussing, and the ethical questions it raises are less profound but have a more direct bearing on the moral conception of man. Here again, the new kind of intervention exceeds the old ethical categories. They have not equipped us to rule, for example, on mental control by chemical means or by direct electrical action on the brain via implanted electrodes—undertaken, let us assume, for defensible and even laudable ends. The mixture of beneficial and dangerous potentials is obvious, but the lines are not easy to draw. Relief of mental patients from distressing and disabling symptoms seems unequivocally beneficial. But from the relief of the *patient*, a goal entirely in the tradition of the medical art, there is an easy passage to the relief of *society* from the inconvenience of difficult individual behavior among its members: that is, the passage from medical to social application; and this opens up an indefinite field with grave potentials. The troublesome problems of rule and unruliness in modern mass society make the extension of such control methods to non-medical categories extremely tempting for social management. Numerous questions of human rights and dignity arise. The difficult question of preempting versus enabling care insists on concrete answers. Shall we induce learning attiudes in school children by the mass administration of drugs, circumventing the appeal to autonomous motivation? Shall we overcome aggression by electronic pacification of brain areas? Shall we generate sensations of happiness or pleasure or at least contentment through independent stimulation (or tranquilizing) of the appropriate centers—independent, that is, of the objects of happiness, pleasure, or content and their attainment in personal living and achieving? Candidacies could be multiplied. Business firms might become interested in some of these techniques for performance-increase among their employees.

Regardless of the question of compulsion or consent, and regardless also of the question of undesirable side-effects, each time we thus bypass the human way of dealing with human problems, short-circuiting it by an impersonal mechanism, we have taken away something from the dignity of personal selfhood and advanced a further step on the road from responsible subjects to programmed behavior systems. Social functionalism, important as it is, is only one side of the question. Decisive is the question of what kind of individuals the society is composed of—to make its existence valuable as a whole. Somewhere along the line of increasing social manageability at the price of individual autonomy, the question of the worthwhileness of the whole human enterprise must pose itself. Answering it involves the image of man we entertain. We must think it anew in light of the things we can do to it now and could never do before.

c. This holds even more with respect to the last object of a technology applied on man himself—the genetic control of future men. This is too wide a subject for cursory treatment. Here I merely point to this most ambitious dream of *homo faber*, summed up in the phrase that man will take his own evolution in hand, with the aim of not just preserving the integrity of the species but of modifying it by improvements of his own design. Whether we have the right to do it, whether we are qualified for that creative role, is the most serious question that can be posed to man finding himself suddenly in possession of such fateful powers. Who will be the image-makers, by what standards, and on the basis of what knowledge? Also, the question of the moral right to experiment on future human beings must be asked. These and similar questions, which demand an answer before we embark on a journey into the unknown, show most vividly how far our powers to act are pushing us beyond the terms of all former ethics.

The ethically relevant common feature in all the examples adduced is what I like to call the inherently "utopian" drift of our actions under the conditions of modern technology, whether it works on non-human or on human nature, and whether the "utopia" at the end of the road be planned or unplanned. By the kind and size of its snowballing effects, technological power propels us into goals of a type that was formerly the preserve of Utopias. To put it differently, technological power has turned what used and ought to be tentative, perhaps enlightening plays of speculative reason into competing blueprints for projects, and in choosing between them we have to choose between extremes of remote effects. The one thing we can really know of them is their extremism as such—that they concern the total condition of nature on our globe and the very kind of creatures that shall, or shall not, populate it. In consequence of the inevitably "utopian" scale of modern technology, the salutary gap between everyday and ultimate issues, between occasions for common prudence and occasions for illuminated wisdom, is steadily closing. Living now constantly in the shadow of unwanted, built-in, automatic utopianism, we are constantly confronted with issues whose positive choice requires supreme wisdom—an impossible situation for man in general, because he does not possess that wisdom, and in particular for contemporary man, who denies the very existence of its object: viz., objective value and truth. We need wisdom most when we believe in it least.

If the new nature of our acting then calls for a new ethics of long-range responsibility, coextensive with the range of our power, it calls in the name of that very responsibility also for a new kind of humility—a humility not like former humility, i.e., owing to the littleness, but owing to the excessive magnitude of our power, which is the excess of our power to act over our power to foresee and our power to evaluate and to judge. In the face of the quasi-eschatological potentials of our technological processes, ignorance of the ultimate implications becomes itself a reason for responsible restraint—as the second best to the possession of wisdom itself.

One other aspect of the required new ethics of responsibility for and to a distant future is worth mentioning: the insufficiency of representative government to meet the new demands on its normal principles and by its normal mechanics. For according to these, only *present* interests make themselves heard and felt and enforce their consideration. It is to them that public agencies are accountable, and this is the way in which concretely the respecting of rights comes about (as distinct from their abstract acknowledgment). But the *future* is not represented, it is not a force that can throw its weight into the scales. The non-existent has no lobby, and the unborn are powerless. Thus accountability to them has no political reality behind it yet in present decision-making, and when they can make their complaint, then we, the culprits, will no longer be there.

This raises to an ultimate pitch the old question of the power of the wise, or the force of ideas not allied to self-interest, in the body politic. What *force* shall represent the future in the present? However, before *this* question can become earnest in practical terms, the new ethics must find its theory, on which do's and don'ts can be based. That is: before the question of what *force*, comes the question of what *insight* or value-knowledge shall represent the future in the present.

And here is where I get stuck, and where we all get stuck. For the very same movement which put us in possession of the powers that have now to be regulated by norms—the movement of modern knowledge called science—has by a necessary complementarity eroded the foundations from which norms could be derived; it has destroyed the very idea of norm as such. Not, fortunately, the feeling for norm and even for particular norms. But this feeling becomes uncertain of itself when contradicted by alleged knowledge or at least denied all sanction by it. Anyway and always does it have a difficult enough time against the loud clamors of greed and fear. Now it must in addition blush before the frown of superior knowledge, as unfounded and incapable of foundation. First, Nature had been "neutralized" with respect to value, then man himself. Now we shiver in the nakedness of a nihilism in which near-omnipotence is paired with near-emptiness, greatest capacity with knowing least what for. With the apocalyptic pregnancy of our actions, that very knowledge which we lack has become more urgently needed than at any other stage in the adventure of mankind. Alas,urgency is no promise of success. On the contrary, it must be avowed that to seek for wisdom today requires a good measure of unwisdom. The very nature of the age which cries out for an ethical theory makes it suspiciously look like a fool's errand. Yet we have no choice in the matter but to try.

It is a question whether without restoring the category of the sacred, the category most thoroughly destroyed by the scientific enlightenment, we can have an ethics able to cope with the extreme powers which we possess today and constantly increase and are almost compelled to use. Regarding those consequences imminent enough still to hit ourselves, fear can do the job—so often the best substitute for genuine virtue or wisdom. But this means fails us towards the more distant prospects, which here matter the most, especially as the beginnings seem mostly innocent in their smallness. Only awe of the sacred with its unqualified veto is independent of the computations of mundane fear and the solace of uncertainty about distant consequences. But religion as a soul-determining force is no longer there to be summoned to the aid of ethics. The latter must stand on its worldly feet—that is, on reason and its fitness for philosophy. And while of faith it can be said that it either is there or is not, of ethics it holds that it must be there.

It must be there because men act, and ethics is for the ordering of actions and for regulating the power to act. It must be there all the more, then, the greater the powers of acting that are to be regulated; and with their size, the ordering principle must also fit their kind. Thus, novel powers to act require novel ethical rules and perhaps even a new ethics.

"Thou shalt not kill" was enunciated because man has the power to kill and often the occasion and even inclination for it—in short, because killing is actually done. It is only under the *pressure* of real habits of action, and generally of the fact that always action already takes place, without *this* having to be commanded first, that ethics as the ruling of such acting under the standard of the good or the permitted enters the stage. Such a *pressure* emanates from the novel technological powers of man, whose exercise is given with their existence. *If* they really are as novel in kind as here contended, and if by the kind of their potential consequences they really have abolished the moral neutrality which the technical commerce with matter hitherto enjoyed—then their pressure bids to seek for new prescriptions in ethics which are competent to assume their guidance, but which first of all can hold their own theoretically against that very pressure. To the demonstration of those premises this paper was devoted. If they are accepted, then we who make thinking our business have a task to last us for our time. We must do it in time, for since we act anyway we shall have some ethic or other in any case, and without a supreme effort to determine the right one, we may be left with a wrong one by default.

4.2

Presumed Neutrality of Technology

NORMAN BALABANIAN

Contemporary technological society is experiencing a profound multidimensional crisis. It was not always so. From at least the time of the Industrial Revolution in England, a general feeling of optimism pervaded Western society. It was commonly believed that the growth of scientific knowlege knew no limits, and that scientific knowledge could always be applied to the problems of society. Since science and technology were so successful in producing marvelous inventions, they could eventually solve any human problem.

This attitude was exemplified at the Chicago *Century of Progress* World's Fair in 1933. The motto of the Fair was: *Science Finds—Industry Applies—Man Conforms.* The guidebook to the Fair amplified further: "Science discovers, genius invents, industry applies, and man adapts himself, or is molded by, new things. . . . Individuals, groups, entire races of men fall into step with science and technology."

Such sentiments were not expressed in sorrow, but in obvious satisfaction. The irony that human beings should bend willingly to the dictates of a technological imperative, when for ages they have struggled to be free of human tyrants, seemed to escape the "happy technologists" of that day. A similar outlook survives among technological optimists of today, typified by Simon Ramo in *Century of Mismatch:*

> We must now plan on sharing the earth with machines. . . . But much more important is that we share a way of life with them . . . We become partners. The machines require for their optimum performance, certain patterns of society. We too have preferred arrangements. But we want what the machines can furnish, and so we must compromise. We must alter the rules of society so that we and they can be compatible.

There is no "compromise" here; it is not that the machine will be constructed to be compatible with human processes, but that humanity must conform to the machine and take on the machine's way of life.

For some time now, and at an increasingly rapid pace, many people have begun to realize that the benefits flowing from science and technology to contemporary society have been purchased at a very high price. The social problems associated with science and technology, which point up the nature of the crisis, are well-known. A brief, nonexhaustive taxonomy of crisis-level problems includes:

- *Environmental:* pollution, resource depletion, excess population;
- *Medical/Health:* dangers to health and safety from industrial production processes and from industrial products;
- *Psychological/Emotional:* substitution of machine values for human values; transformation of the nature of work from craftsmanship to meaninglessness, leading to worker alienation; a feeling of citizen powerlessness in the face of a complexity said to be understandable only by an expert elite, and thus, alienation from politics; social malaise exhibited in symptoms of increasing crime, senseless vandalism, anxiety, disharmony and tension, apathy and loss of a feeling of community;
- *Military:* potential nuclear annihilation through MAD (Mutual Assured Deterrence);
- *Social:* increasing centralization, bureaucratization, authoritarianism; diminishing real returns on immense capital requirements.

What do commentators mean when they use the term *technology*? The first image conjured by the term is a machine, a physical object. This is an inadequate conception; like the term *society, technology* is an abstract concept. Society is not simply a collection of people but subsumes the interrelationships among them. In the same way, technology is not simply a collection of machines but encompasses relationships among them and their uses. Tools, looms, x-ray machines, nuclear reactors, refrigerators, and automobiles are all elements of technology. But if such objects are all that are comprehended by the term *technology*, the imagination is impoverished indeed. Just as a single word, or collection of words, cannot adequately represent the rich texture of language, so also a single machine, or collection of machines, cannot adequately represent contemporary technology.

There are two serious omissions from the simple-minded notion of technology. One of these is the concept of *know-how*. Accumulated knowledge is as much a part of technology as a machine. In the view of Harvey Brooks, technology is nothing but certain kinds of know-how; technology "is not hardware but knowledge, including the knowledge not only of how to fabricate hardware to predetermined specifications and functions, but also of how to design administrative processes and organizations to carry out specified functions, and to influence human behavior toward specified ends."

A more important omission is the concept of organization, of *system:* the organized structures; the mechanisms of management and control; the processes of production; the specific designs of the overall organizations and systems within which the machines are embedded; the linkages that tie together the physical objects—the "hardware"—with the social institutions—the "software". Technology is not just the computer, for example, but large-scale computer networks linked through telecommunications systems; operating and managing systems; data banks, the knowledge and programs to manipulate them, and the power implicit in their control. Any analysis of contemporary technological society which fails to understand this is deeply flawed.

MODELS AND PREMISES

In going about their everyday lives, people carry in their heads models or paradigms of what the world is like, what society is like. Walter Lippmann noted that individuals create for themselves a *pseudoenvironment*, an internal representation of the world, built up over a lifetime. People's perceptions of events are determined by the images, the preconceptions, the premises that underlie this pseudoenvironment. Most people tend not to question these preconceptions, even when the consequences of acting in accordance with their images of reality lead to anomalies that the paradigm cannot reconcile.

In science, when such anomalies develop, creative thinkers question the old paradigm and its premises and develop a new image of the world based on a new way of looking at things. For example, Newtonian mechanics gave way to relativistic mechanics; the economic system of mercantilism of the fifteenth and sixteenth centuries gave way to the classical capitalist economic system of the eighteenth and nineteenth centuries, which in turn has given way to the neoclassical and Keynesian economic systems. The Keynesian analysis evolved because the world depression of the 1930s was a traumatic anomaly that the previous paradigm could not explain.

In everyday life, on another level, many people are aware of the anomalies between the rhetorically-cultivated presuppositions of the social system and the adverse effects of the realities they experience on the receiving end. Such people seriously question the premises on which the social system is founded; and sometimes this questioning leads to a violent "paradigm shift"—through revolution. But, more often than not, their questioning is "contained." Historically, whenever a paradigm shift has been necessary (in science or social structure) it has been resisted by those with an interest in the old paradigm (intellectual, emotional, or financial). It is difficult for people who have invested a lifetime in the service of a particular world-view to switch to a different outlook. Even if they intellectually understand the need for it, their previous life leaves them unprepared to carry on comfortably in the new paradigm.

Contemporary technological society is now facing anomalies which the social and economic preconceptions cannot explain. These preconceptions are built into the social structure and culture. But reality cries out for a paradigm shift. It is essential to identify and critically examine the premises—sometimes explicit, but often hidden—that undergird the inadequate contemporary model. Some of the major premises are:

> *Self-Seeking Values:* The preferred behavior mode for human beings is pursuit of one's own self-interest. The foundation of the capitalist ethos is that such self-seeking behavior will lead to social good through the operation of an "invisible hand."

- *Elastic Wants:* Human wants are infinitely elastic; they expand without limit. It is necessary to have continual economic growth in order to satisfy them.
- *Man's Domination of Nature:* The physical world is there for man to subdue, conquer, dominate, subjugate, and exploit. By scientific knowledge, said Descartes, "We may be able to make ourselves masters and possessors of nature."
- *Neutrality of Technology:* Technology is morally and politically neutral; it is a mere tool which can be used for good or evil. If it is not used properly, man is to blame.
- *Freedom of Choice:* Individuals in our free market system have free choice. The root cause of such ills as wasteful consumption, urban congestion, pollution, and the design of inappropriate products lies in the free choice exercised by autonomous individuals.

Each of these undergirding pillars fails to withstand critical examination. They must all be rejected if transformation to a humane society is a goal. The first three refer to the presumed nature of human beings. Although I will devote some brief thoughts to them, the bulk of my analysis will be reserved for the last two.

The premise or assertion that self-seeking behavior is the preferred mode for humans tends to encourage an aggressive, contentious, noncooperative spirit; it cultivates self-aggrandizement, greed, and envy—looking out for Number One at the expense of the community. Its proponents say this is human nature and cannot be changed. The social structure fostered by this outlook is hierarchical, with individuals engaged in a scramble for status on the ladder of success, elbowing their lonely way to the top.

HUMAN NATURE

The capitalist culture and social system require individuals to act in a self-centered, contentious manner. The proper operation of the system *demands* it. How convenient, then, to ascribe this attitude to basic human nature! Writing in 1930, John Maynard Keynes said: "For at least another hundred years we must pretend to ourselves and to everyone that fair is foul and foul is fair; for foul is useful and fair is not. Avarice and usury and precaution must be our gods." Keynes thought that self-seeking should be encouraged because it was useful to the operation of the capitalist system. But he at least recognized that it was not laudable, and that you had to work at it to make avarice and greed appear as gods.

Is the greedy, status-seeking pursuit of self-interest a consequence of unalterable human nature? Demonstrating that it is not merely requires finding counterexamples. Have humans ever acted selflessly? If we could find any such cases (and we need only look to ordinary people in common situations to find many counterexamples) we would have to conclude that the proposition that self-seeking is unalterable human nature is, at the least, not proved. With a bit of further thought, we would have to say that some humans act selflessly, others selfishly; that any human sometimes acts one way, sometimes the other; that individuals often have conflicts within themselves as to whether they should respond to a given situation in a self-seeking manner or in a cooperative, communitarian manner. For each example of greedy, self-centered human behavior one can find an example of altruistic, other-directed, cooperative behavior. Without the cooperative and symbiotic working together of its millions of cells, the human body itself could not function. Far from being by nature selfish, humans can just as validly be assumed to be cooperative by nature. How they actually behave in given situations depends on their socialization, on the reinforcements they obtain for their behavior. If the culture, the social order, continually reinforces them for self-seeking behavior, for personal aggrandizement, the chances are they will act this way more often than not. An ouside observer would then notice that most people, most of the time, behave in noncooperative, self-seeking, status-enhancing ways. The observer who concludes that such behavior is intrinsic human nature would be a naive observer, indeed.

Sometimes the happy technologist will unknowingly concede this. Samuel Florman writes: "Man, for all his angelic qualities, is self-seeking and competitive." There it is: a concession that humans have noble impulses as well as base ones. The real question becomes: which qualities should be cultivated and reinforced? It is not a question of changing human nature to something it is not, but arranging conditions so that people can more often exhibit, and behave in accordance with, their "angelic qualities" rather than their base ones.

Economists postulate that human wants are infinite and insatiable—no sooner has one want been served than another is stimulated; humans are not capable of saying "enough." Florman states that "contemporary man is not content because he *wants* more than he can ever have." This being the case, goes the argument, it is essential to maintain economic growth and increasing levels of consumption. The serious flaw in these assertions is the failure to distinguish between those wants that are basic and absolute needs—such as food, clothing, shelter, sex—which humans will experience independent of the condition of other human beings around them, and those which are social and relative. It is clearly untrue that basic human needs cannot be satisfied. The sight of food is not tempting to a person who has just finished dinner. It is also clear that some basic human needs are not at present satisfied for a substantial fraction of the world's population. To this extent, growth is still needed; not generalized growth, but increase in those areas of production intended to satisfy basic needs of the poor.

The second category of wants may well be insatiable; but what is their nature? These wants are experienced only in a relative sense, only if a feeling of superiority to others is achieved, only if vanity and status are enhanced. These wants require continual comparison with others and feverish activity in pursuit of inequality. They are not ennobling but base. In a sane society, their pursuit would be discouraged. But in a society dominated by the capitalist ethos, they are encouraged and cultivated through high-powered promotion and persuasion. Herbert Marcuse refers to such wants as "false needs," not to deny that they exist but as a judgment of their worthiness. The flourishing of false needs is a reflection not on human nature but on the values consciously cultivated by the social system. It makes no sense actively to promote ego trips, feelings of vanity and prestige, desires for superiority and status, and then to demand economic growth in order to satisfy these desires—at the expense of the crises now being faced. Furthermore, the effort of satisfying these wants is doomed to failure. If people's satisfactions depend almost entirely on status, ego gratification, and feelings of superiority, then increasing levels of consumption cannot yield increasing satisfaction to society as a whole. The superiority and enhanced status of some implies the inferiority and reduced status of others.

The idea that human beings should have dominion over the earth and all that it contains (now extended to the entire universe) was an early tenet of Western civilization; it even carried Biblical sanction. And mankind has not been reticent in carrying out this injunction. But there are at least two fallacies in this outlook.

The first relates to depletion. Clearly, intrusive human activities carried out over a period of time can have, and have had, devastating effects on nature and on human society. In the past, the wholesale destruction of forests, improper cultivation leading to soil erosion, and other similar activities have had major effects on the present conditions of many parts of the earth. They have even partly caused flourishing civilizations to vanish. But the scale, intensity, and quality of current human interventions in nature dwarf those of the past. The disappearance of much of the earth's wealth seems to be imminent. Whatever utility the concept of domination, subjugation, conquest, exploitation, and control of nature by humans may have had in the past, it is now counterproductive. Rather than subjugating nature, human beings must learn to live in harmony with it, to cooperate with its processes. That does not mean that people must lie down supinely before nature, but that their activities should take nature's processes into account. Humans should avoid building their structures over faults in the earth's plate, for example. They should avoid building their towns in flood plains. Cooperating with nature also means understanding its limits and minimizing those activities that put pressure on them or use up its irreplaceable parts.

The second fallacy is the mistaken assumption that nature is one thing and human beings something else. That is false, of course; human beings are as much a part of nature as mountains and birds. If the earth were destroyed, human beings would cease to exist also. Human domination and exploitation of nature implies human domination and exploitation of other humans. Although this proposition was accepted in the past, slavery, imperialism, and other institutions of exploitation and domination should no longer be tolerated. Their rejection also implies the rejection of the parent concept: the subjugation of nature by humans.

Although champions of "advanced" technology—happy technologists—may approach their subjects from different perspectives, there is a common refrain to their individual verses that amounts to a litany of technology: technology is just a passive tool whose consequences depend on the use to which it is put; if technology is used harmfully, man is to blame; there are no values embodied in technology; technology plays an entirely passive role with respect to issues of power and control; the prime reasons for introducing innovations in production processes are increased efficiency and productivity; the prime reason for introducing innovations in products is to satisfy human needs. This litany of the happy technologist constitutes an ideology that is a collection of errors, illusions, and mystification presenting an inverted, truncated, distorted reflection of reality. It is also a set of values characteristic of a group; the integrated assertions, theories, and aims that constitute a sociopolitical program. The ideology of technology has purposes quite remote from explaining reality to members of society. It fails to take political power and economic interests into account and thus masks their predominant role. It promotes a model which ascribes to technology an objectivity, a value-neutrality, which technology does not in fact possess. A useful clue to the ideological nature of a statement purporting to be explanatory is the ascribing of action to vague collective nouns and pronouns, as in: "*Man*" is to blame; "*Mind* determines the shape and direction of technology"; "If technology is sometimes used for bad ends, *all* bear responsibility . . . "; " . . . make technology work as *our* servant"; "Thus, *we* manufacture millions of products to enhance *our* physical comfort and convenience . . . But in doing this, *we* overlook the need to plan ahead".

Are we *all* equal in responsibility, or are some more equal than others? Who are the "we" who do the manufacturing? Is that the same "we" who forgot to plan? Doesn't somebody's profit enter the picture at all? Surely some specific minds shape technology, not an abstract "mind". The preceding manner of speaking conceals the existence of specific, powerful corporations whose activities in pursuit of their interests are major factors in the problems of our contemporary society.

NEUTRALITY OF TECHNOLOGY

Some insight into the role played by technological ideology can be obtained by analogy, through an examination of the role of economics in our society. Like any other science, says John Kenneth Galbraith, the purpose of economics is understanding; in this case, understanding the economic system—how it works, the nature of money, etc. But economics also has an instrumental function—to serve the goals of those with power. It creates images in the minds of people—thus contributing to their model of society—which are not at all consonant with the reality of the economic system, at least with the more than fifty percent of the economy represented by what Galbraith calls the "planning system," the large corporations and their activities. The instrumental function is to induce people to behave as if the image were the reality. It is to conceal from people the true nature of most of the economy as a planned system (not free enterprise), with the planning done by a handful of large corporations in their own interest. Like economics, technology also has two goals. Quite apart from their purpose "to enhance our physical comfort and convenience," technology and technical expertise serve an ideological, instrumental function. This function is, again, image-making and concealment, the covering of political and economic power in a cloak of technical objectivity. The image is created that decisions and actions serving the interests of those in power are simply the consequences of objective facts, carried out for such objective reasons as efficiency.

The ideological assertion is that technological innovations and their introduction serve the objective goals of efficiency, increase in productivity, and human needs satisfaction. To those holding the images of the dominant paradigm, this seems reasonable. But it does not stand up under close scrutiny. In his study of the development of the textile industry during the Industrial Revolution in Britain, David Dickson shows that the rise of the factory system of production, the organization of work in factories, was largely a managerial necessity, rather than a technological one. It was done for the purpose of "curbing the insolence and dishonesty of men." The rising capitalist class made no bones about the introduction of specific machines having as its major purpose the subduing and disciplining of workers. Speaking of one invention in the textile industry, Andrew Ure, an early champion of the factory system, says, "This invention confirms the great doctrine already propounded, that when capital enlists science in her service, the refractory hand of labour will always be taught docility." Technological innovation was not so much determined by concern for production efficiency as by management's desire to maintain fragmentation of workers, authoritarian forms of discipline, hierarchical structure, and regimentation. These same concerns are evident to the present day. After analyzing the demands that contemporary corporations impose on employees, Richard C. Edwards concludes in *Contested Terrain* that the complex hierarchy of the modern corporation grew not from the demands of technology but from the desire for greater control of workers.

To a large extent, technological innovations in consumer products arise as a consequence of research and development (R&D) activities. Almost all R&D activities directed toward product development are carried out in the laboratories of large, technology-intensive corporations. The goals of these corporations, simply stated, are survival, growth in sales, and growth in profits. All the activities of corporations—production, marketing, sales, *and* R&D—are carried out to reach these goals. Product innovation, no less than production or marketing, serves corporate purposes and would be carried out independent of social need.

Lack of need would not suffice to thwart the corporation's desire to increase sales and profit; if the market does not exist, it is developed—created, cultivated, and nourished. Once the decision has been made to introduce an innovation in furtherance of corporate objectives, "developing the market" is thrown into high gear. The entire arsenal of persuasion is unleashed to convince potential consumers that the new innovation will not only perform the specific function for which the product is designed, but it will also enhance status, satisfy vanity, increase sexual appeal, etc. The result is to intensify these base impulses, and then to prey upon people's expectations that such impulses can be satisfied by the product. People are thus sold the idea that their self-worth is measured by the goods they possess, in general, and by this specific product, in particular.

Another dimension related to technological innovations appears when one seeks to determine the factors that guide the introduction of an innovation. The paramount consideration, of course, is profit; but why introduce one thing rather than another? On a superficial level, it seems that trial-and-error has something to do with it. In 1978, for example, the processed food industry introduced to the market some ten thousand new products. Of these, about eighty percent failed. Was there a societal need for these products? Was "lower cost to the user" a criterion for developing, producing, and marketing them? In fact, pricing of all the corporation's products must reflect losses from those that failed. Hence, processed food prices generally must rise as a result of these innovations, quite contrary to the ideologically stated reason for product innovation.

The preceding analysis evaporates the claims of technological objectivity and value-neutrality. The nature of a society's technology is intimately related to issues of power and control, and reflects the dominant paradigm in terms of which reality is interpreted. A society in which economic growth is highly valued necessitates a particular kind of technology—one with a high level of innovation, quite independent of social need. Policies leading to economic expansion have to be reflected in the particular form of technology through which this expansion is achieved. Hierarchical forms of social control become reflected in the technology. The presumed neutrality of technology then lends legitimacy to social policies, however repressive.

Like other happy technologists, Ramo over and over again explicitly claims the value-neutrality of technology. But, without realizing it, he makes an amazing concession contradicting this position and acknowledging an instrumental function for technology. In reviewing the space program and justifying having spent large resources on this program, he concludes: "The pattern that we are developing—to be far-sighted, to be bold, to want to pioneer, to be willing to take some risks, to carry with us as a part of our way of life the exploration of the unknown—it is these habits that we cultivated when we carried out the space program." This is a remarkable concession that a specific technological development had an agenda quite unrelated to the primary goal; that the program served the ideological purpose of cultivating certain habits. Can it be denied that less glorious-sounding attitudes than the ones admitted by Ramo—greed, status-seeking, self-aggrandizement, contentiousness—are on the agenda? The general truth cannot be escaped: technology serves an instrumental function.

INDIVIDUAL AUTONOMY

Not only is technology passive and neutral, according to the happy technologist, but whatever evil consequences are associated with the deployment of technology result from autonomous individuals exercising their free choice. Not only that, but people perversely go on using technology even though this fact lends to environmental degradation, depletion of resources, and other unpleasant things they themselves experience. A sampling of their assertions follows:

John Gardner: Everyone lampoons modern technology but no one is prepared to give up his refrigerator.

Samuel Florman: However much we deplore our automobile culture, clearly it has been created by people making choices, not by a runaway technology.

Melvin Kranzberg: That kind of spatial freedom (of the past in rural areas) vanished from the onrush of urbanism; but people apparently want to live together in large agglomerations.

Alvin Weinberg: A social problem exists because many people behave, individually, in a socially unacceptable way . . . Too many people drive cars in Los Angeles with its curious meteorology, and so Los Angeles suffocates from smog.

Emmanuel Mesthene: The negative effects of technology . . . are traceable less to some mystical autonomy presumed to lie in technology and much more to the autonomy that our political and economic institutions grant to individual decision making.

Simon Ramo: National control is really the only answer, and that can come only if a majority of Americans are willing to accept interference in their freedom to choose automobiles.

These assertions constitute a second litany. To be generous, one would ascribe the origins of this assertion litany not to a deliberate agenda, ideologically promoted, but to a fundamental misunderstanding of the nature of technology. As noted earlier, the systems within which the separate components of technology (the machines) are incorporated are its essential features. Failure to understand that the term *technology* subsumes organized and integrated systems implies a profound misunderstanding of the nature of contemporary technology.

The term *lifestyle* designates the manner in which individuals go about their daily activities at home, at work, at play, etc. In American society there is a general perception that lifestyle is a matter of individual choice, at least for a vast majority. Disregarding economic means for a moment, people think that one can choose to lead a "bohemian" lifestyle or a "straight" one, to wear flashy clothes or sober ones, etc. But is one free to choose to have a refrigerator or not? Is it a simple matter of lifestyle choice or do other institutional arrangements of society impinge with demands of their own? Is it the kind of choice referred to by Justice Holmes when he said: "In its evenhanded majesty, the law forbids rich and poor alike to sleep under a bridge"?

A refrigerator (including freezer) performs several functions. It stores perishable food (a necessity) and cools beer or produces ice for cooling drinks (a comfort or luxury). The latter category is not an essential function. The desirability of cold beer, for example, is culturally or socially induced; other cultures (the British) find warm beer more desirable, so people in those societies do not need a refrigerator to perform this particular function. Consider another society—even a developed European society like France a decade or two ago—in which it is possible for people to purchase their perishable foods on a daily basis in markets or small shops, easily accessible and within walking distance of their homes, even in the largest cities. This option is not available to most contemporary Americans. The supermarket as a social institution, not within walking distance of most people, has its own imperatives. One buys for a week of eating, not for a day, so storage in a refrigerator becomes essential to living. It is a necessity induced by a lifestyle over which individuals have little control—an institutionalized lifestyle. To chide individuals for recalcitrance or perversity for their unwillingness to give up a refrigerator is to profoundly misjudge the nature of contemporary technology and its induced social change. No value judgement about the merits of different lifestyles is implied in this scenario. It is irrelevant to the argument whether or not a supermarket/refrigerator society has advantages over the other. The only question is: do individuals have autonomy to freely choose one or the other?

A similar and even stronger case can be made concerning the automobile. The question, again, is one of choice. Those who claim the existence of free choice gaze out at society as it exists at a given period of time, with the social structure and state of technology as they exist. Within this framework, they argue, individuals can choose. They can choose to buy this model car or that, this color or that, this upholstery material or that, this option or that. The one fact that belies all of this apparent freedom is that the majority of individuals cannot choose *not* to buy a car—if they also want to participate in the normal life of the society, such as going to work, buying food and clothing, etc.

Many communities in the U.S. had electric railway systems that served admirably in the first third of this century. It was not the autonomous choice of individuals that killed these existing urban and interurban mass public transportation systems and prevented their expansion and improvement. In more than one case they were purchased by automotive corporations (the major culprit being General Motors) and converted to motor buses, and then allowed to die in order to promote the use of the private automobile. A case in point is Los Angeles. It had an extensive electric streetcar and interurban rail system in the early part of this century. During the 1920s the Pacific Electric Railway operated 1200 miles of interurban rail service. When the population of the area was only one million in 1924, the system carried a volume of one-hundred and nine million passengers. (By comparison, 45 years later when the population was eight to nine times greater, public transit buses carried only one hundred ninety million passengers annually.) The reason we have smog in Los Angeles, says Weinberg, is because too many people drive cars. A much more accurate reason is because General Motors bought the Pacific Electric Railway system and destroyed it. The single most important cause for not only the smog, but the fact that over half the land area of Los Angeles—including freeways, streets, driveways, parking lots, gas stations, sales rooms, etc.—is dedicated to the automobile, is the power of large corporations.

Simon Ramo decries federal "interference" in setting safety regulations, emission standards, etc., because it would control people's "freedom to choose automobiles." But many government laws and regulations have had, and continue to have, a major impact in instituting, cultivating and maintaining the U.S. automobile culture. Among these are the investment tax credit that encouraged auto manufacturers to produce cars and the oil depletion allowance encouraged oil companies to produce fuel for the cars. But of far greater significance was the establishment and funding method of the federal interstate highway system in the late fifties. This was one of the most-far-reaching acts of interference by government. Yet ideologically committed technologists like Ramo do not cry "control" when the actions of government favor the corporate interests.

People were not asked to debate the merits of different transportation systems and then choose what they favored. Specific corporate interests, in order to promote their own welfare, caused the current state of the U.S. transportation system. The design of the transportation system is not a consequence of two hundred million citizens exercising their free choice, but a handful of powerful groups pursuing their own interest. It is not descriptive of reality to say now that people have free choice of their mode of transportation. To set up the social system so that individuals are compelled to buy cars just in order to be active members of society, and then to sneer at them because they are unwilling to give up their cars, is to add insult to injury. It is like blaming the victim for the crime. For most people, most of the time, driving a car is not discretionary; it is mandatory.

Even within the context of a market and the regulation of technological developments by the market, is it possible accurately to describe the current status of specific technologies (e.g., the transportation system) as the consequence of untrammelled individual choice guiding the invisible hand? Market prices can be kept artificially low by transferring some of the costs associated with production or use from the manufacturer and/or user to third parties. This can be done in at least two ways—by subsidies from the government or by failure to account for "external costs" or negative "externalities." Both of these processes have operated extensively to distort price structures. Vast sums have been transferred to corporations in subsidies by the federal government, either directly (through grants and low-interest loans) or indirectly (through the taxing mechanism or having the government assume responsibility for certain components of technological systems, like highways and airports). For example, subsidies have been running at $10 billion annually just to the traditional energy industries; subsidies to the oil corporations alone have totalled more than $75 billion.

Economic activity involves producers and consumers, sellers and buyers. External costs are those undesirable effects upon other people resulting from these activities. Such costs can be tangible—such as cleaning bills and medical bills due to the effects of economic activity on the environment—or intangible, yet real. Inestimably huge external costs, both privately and socially borne, have been transferred to others. Purchasing decisions are influenced by prices that are artificially depressed in such ways. If this circumstance permits a large-scale technological development to take place, which then induces major changes in the way people live, would it be meaningful to assert that the detailed forms of the resulting society are consequences of individual free choice?

A cogent and compelling analysis of these issues is provided by James Carroll:

> ... technology often embodies and expresses political value choices that, in their operations and effects, are binding on individuals and groups, whether such choices have been made in political forums or elsewhere. ... technological processes in contemporary society have become the equivalent of a form of law—that is, an authoritative or binding expression of social norms and values from which the individual or group may have no immediate recourse.

Furthermore, says Carroll, there are often no appropriate political processes for "identifying and debating the value choices implicit in what appear to be technical alternatives." Technological processes are technically complex and occur in administrative organizations (either government agencies or corporations) to which citizens have no access. Ordinary people on the outside have neither the opportunity nor the means to identify the value questions, far less to resolve them in a public forum. *De facto*, the locus of the political value choices becomes the technological processes themselves; there is no public debate and issues are posed and resolved in technical terms. Individuals have no autonomous choice in the matter.

Is Kranzberg serious when he offers pure individual choice as the explanation for the specific way that Americans are distributed around the country? Did people leave the Dust Bowl in the thirties because of free choice? Were individuals consulted on the mechanization of the farms which drove farm workers away from the rural areas? The locations of industries will greatly influence where people live; were individuals consulted as to where specific industries should be located? Did the interests of huckstering land developers have nothing to do with luring people to southern California? Was it individual choice or the manipulation of specific corporate interests that caused the specific lay-out of Los Angeles where it is necessary to drive long distances just to carry on everyday activities?

The answers to all these questions are obvious. The design of cities, the locations of services, places of employment, shopping centers, etc., are all based on the private motor car as the dominant mode of transportation. Individuals in most places have little choice; they cannot decide to buy either at the supermarket or the corner grocery—they must buy at the supermarket because there are no longer many small stores. They cannot choose either to take mass transportation or to drive a car—they are compelled to drive a car. They cannot choose to buy unpackaged, unwrapped, unboxed products because such products do not exist in most places. They cannot choose to recycle much of the waste materials (paper, glass, metal) they are compelled to bring into their homes because no recycling facilities are available in most places.

NECESSARY HARMONY

Technology is not a neutral, passive tool devoid of values; it takes the shape of and, in turn, helps to shape, its embedding social system. The ideologically-promoted neutral-tool/use-abuse model of technology conceals the issues of economic and political power relationships among different groups in society, and thus serves the instrumental function of legitimating the dominant ideology. In addition, contemporary technology, far from increasing freedom, limits individual autonomy and imposes a style of living about which individuals have little choice.

The needs to which technology is said to respond are induced by the social system. In a social order with different values and goals, the needs would be different and so the nature of technology would be correspondingly different. Thus, the crisis of contemporary society cannot be resolved if the contemporary form of technology remains dominant; it must be replaced by a technology of a different nature. Since technology is intimately tied to matters of political power and social control, changing the technology implies a profound change in the social order.

The crisis is not a crisis *within* technological society which can be overcome by patching up the system, but a crisis of the technological system *itself*. The major question is not *who* is to control the means of production, but *what* the means of production shall be; and *what* shall be produced. It is not *where* to locate the nuclear power plants, but *whether* to have nuclear power at all. It is not merely a question of possibly limiting growth but of radically altering the very nature of technology. Contemporary technology is based on a narrowly conceived economic efficiency, on social control, and on profit. I would characterize the alternate technology needed as *harmonious* and based on different criteria. Harmonious technology would respect ecological values and be in symbiosis with nature. This does not mean that there would be no human intervention in nature, just that such interventions would not be destructive and exploitative, but in harmony with ecological values; consequently, harmonious technology would rely mainly on renewable energy and be minimally consumptive of nonrenewable resources. Harmonious technology would be responsive to direct social needs and would not require a hierarchical, exploitative, and alienating relationship among human beings. It would not oppress people nor treat them as appendages to machines, but would be satisfying to work with. Harmonious technology would value durability and quality of products, decentralization of production, diversity—rather than monoculture—in agriculture, and pluralism in lifestyle and culture.

Accomplishing this transformation requires a new consciousness that sees the interrelations among the physical, biological, and social spheres; collectively they constitute a system of which humanity is a part. A new style of living which is in harmony with the natural world is also needed. A harmonious technology in a harmonious society is the goal. An appropriate motto might be: *Science Discovers—Humanity Decides—Technology Conforms.*

4.3

Making, Doing, and Thinking

KARL RICHARD PAVLOVIC

A subject to be ardently discussed in the coming years is that of the advantages, the threat, and the limitations of technology. (1)

These words were written approximately 45 years ago by the Spanish philosopher José Ortega y Gasset. Technology is presently a "live" issue. Legal battles are fought, political initiatives are launched, politicians make speeches and dispense money, in the media panegyrics of technology and polemics against it appear, lectures are given, books are written, courses are taught—the focus of all this activity is technology. If we can call all this activity a discussion, then the discussion that Ortega foresaw is now fully upon Western society.

> I have always considered it the task of the writer to foresee in leisurely anticipation the problems which will be confronting his readers in days to come, and to provide them in time, i.e., before the discussion opens, with clear ideas about the question, so as to enable them to enter the fray with the serene mind of people who, in principle, see the solution. (2)

We can further say that in this discussion serene minds are not much in evidence.

Either writers such as Ortega failed in their duty, or no one listened, or this business of rational discourse's replacing the talk that is merely the accompanying score to the clash of force is all in vain. There is not much point in discussing the third possibility. The irony is that people who have experienced the efficacy of rational discusssion, i.e., precisely those people already persuaded, would be the only beneficiaries, for those people who are not so acquainted, i.e., those whom I would hope to persuade, would in all likelihood remain unpersuaded.

As for the first possibility, since Ortega wrote these words, there has been no lack of books on technology. It seems more likely that no one listened. It is told that a man publicly challenged the Buddha to reveal to him the meaning of existence. The Buddha serenely lifted one finger. At that the man jumped up, exclaiming that at last he had seen, and ran off. The disciple Ananda then turned to the Buddha and asked, "What do you suppose he saw?" The Buddha replied, "He is like a good horse—one has but to show him the whip." More representative of human nature is the story of the man who bought a mule that purportedly understood English. After failing to find any evidence of this ability, he returned to the seller loudly proclaiming the mule to be the stupidest he had ever encountered. The seller apologized, saying he had forgotten to tell the buyer one thing of importance. He then picked up a club, struck the mule squarely between the eyes, and commanded it to run three times round a circle—which it did. "He's not stupid," said the seller, "you just have to get his attention first." People like Ortega responded to the mere sight of the whip; the rest of us needed the whip upon our backs, or the club between the eyes, if you prefer.

The whip is now upon our backs, and by that I do mean that we are being flogged by technology—although it may well be that we are. I mean, rather, that as a group we are suffering from uncertainty about technology, and this uncertainty is taking up a great deal of time and energy. When the shoe pinches, one looks for a cobbler. But until then one's attention is taken up with, e.g., the water coming through the leak in the roof. Now is perhaps a more propitious time to take up some of the ideas of cobblers like Ortega.

It is almost too obvious to bear saying even one more time—we live today in a technological society. At the time that Ortega was writing, the technological society had just begun to appear on the horizon of the future—or so it seemed. Distance also makes for serenity of mind. Today technology is an all-pervasive fact of almost everyone's lives, but not always in the same way. Some people have a heavily vested interest in technology, while others not so. Some people have suffered and some have benefited from technology. Others see their future heavily bound up with technology. All these different people bring different perspectives to the discussion. Because technology is not just a game in our lives, it is not likely that it will suddenly tomorrow be decided to the satisfaction of all whether technology is a problem or, if so, what the solution will be. It is not likely then that a few words on paper, even if read and agreed to by everyone, will alter this. Interests besides our interest in the truth are bound up with the question.

On the other hand, while a few words are unlikely to resolve our uncertainty about technology and bring a close to the discussion, words may facilitate the discussion. The latter is the intent of these words and those to follow. One of Ortega's points was that we are not altogether clear in our own minds about what technology really is. Moreover, even when we believe that we know, he thought we do not.

We do not clearly see and understand technology, and there is a reason why this is so. It is not because our eyesight is poor or our minds feeble. It is because we are in the habit of not looking in a certain direction. There is no longer any particular reason for this habit—it is just a habit. I hope to explain and display this habit to you and, thus, to bring you to the point where you can decide whether you will continue with it.

We are inclined to think of *things* as *material, tangible* things, and we are inclined to distinguish things as either *natural* (made by nature) or *artificial* (made by man). We are also inclined to think of *making* as a specific activity distinct from other kinds of activities. Finally, we are inclined to think of *technology* as a process whereby *things* are *made*. We have these inclinations of thought from the Greeks of the classical period (the end of the 1st millenium B.C.). All of this seems, perhaps, too obvious to deserve comment. Nonetheless, there is something wrong with it. Moreover, we all "know" some or all of the parts of it that are wrong. Before I talk about that, however, I want to say something more about how these ideas came to be put together in this way.

Greek society of the classical period was ethnically stratified. There was an overclass of citizens, members of a group of closely related tribes that came from somewhere in the north earlier and conquered the territory of the Greek peninsula. There was an underclass of slaves, composed of members of the indigenous tribes of the area. Originally it appears that the Greeks were organized into what we would call a warrior aristocracy, but by the classical period they had evolved a variety of forms of political organization that ran from monarchy to the participatory democracy of Athens. Corresponding to this social division was a division of activity. Citizens *did*, whether it was overseeing agriculture, trade, craft industry, the running of the state, or making war. Slaves *made*, worked with their hands, whether it was growing crops, keeping accounts, craft production, etc. Of course, the division was not hard and fast, and small citizen farmers might well work at times in their own fields and a master craftsman might well be a citizen. Nonetheless, this is the way they themselves saw the matter. "Making" was seen by them as the making of material things with the hands, predominantly things that did not exist naturally. The ability, the capability to make things was called an "art" (Greek: *techne*, from which we have the word "technology"), and thus technology was the process of making material, artificial things. The social division added another aspect to this understanding. Technology, the art and act of making, was a matter for slaves, an inferior activity. "Making" and "doing," two distinct activities done by different people and having distinctly different results, one activity superior to the other.

The classical period added a third to these two activities: thinking. A number of scholars have tried to figure out why it should be reasonable to say that the Greeks *started* thinking in the classical period (3). Before this time the evidence suggests that for the Greeks there were only the two activities, making and doing. Good, effective, successful *making* was a matter of proper training, the proper habits in production. Good, effective, successful *doing* was a matter of proper breeding in the old sense of heredity. One *did* on instinct—instincts that one received from one's ancestors. There is evidence to suggest that during the classical period matters began to go wrong, particularly in the area of doing, to such an extent that people started asking How? and Why? they did certain things that before they had simply done without further ado (4). This then is what we mean when we say they started thinking. In addition, it was not everyone who engaged in this activity. As the Greeks saw it, there were three different activities: making, doing, and thinking. These activities were performed by different kinds of people: slaves, citizens, and sophists or philosophers. These activities had three different results: material things, actions, and explanations.

This all occurred over 2000 years ago, and yet it persists to a remarkable degree. We divide work into "blue-collar" work and "white-collar" work. Blue-collar workers work with their hands and predominantly *make material* things. White-collar workers *do*. White-collar workers are further subdivided into those who *manage* in private business and government and those who *think*, academicians. Blue-collar work is thought to be inferior to white-collar work. A person in nonacademic areas whose job consists primarily of thinking is frequently called a "professor." Engineering's rise from a craft to a profession has been paralleled by increasing numbers' of engineers ending their careers in management positions. In our society today the lines drawn by the Greeks persist.

As for the evidence against all this that I mentioned earlier, everyone knows that we think in the way I have just described and yet everyone also "knows" that this is not the way things are. Everyone knows that making, doing, and thinking are inextricably bound up together. Everyone knows that in order to *make* something, you must *do* something and that more often than not in making something you come to points where you must *think* about what you are *making* and about what to *do* next. Further, everyone knows that there are things other than the kind of thing you can hold in your hand or see with your eye. And since you know that, you also know that in doing something, you also make some thing, if only the action itself.

What does all this have to do with technology? Let us go back and look at what we and the Greeks commonly call technology, the making of material things. We desire to have something, we perceive a lack, we manipulate things that do exist, and the result is the material object we desire. It may end there, or that material object may in turn become something to manipulate to produce some other material object that did not heretofore exist. But we find this process at work with other sorts of things besides just material things (5). It seems that the only thing that distingushes what we commonly call technology from other instances of this process is that technology deals with material objects. When we create social institutions, we perceive a need, we manipulate things into a new arrangement and produce something that did not heretofore exist. Similarly with our theories about how and why we do things and make things: the social, political, ethical, and scientific conceptual frameworks with which we order our thoughts and actions and judge the things that we produce. While it does seem odd that with so many similarities and only one difference we do not call these other processes also technology, it would be unnecessarily pedantic to insist that we extend the use of the term. What is important, however, is that when we speak of technology, we do not lose sight of its similarities with these other human activities and theirs with it.

How often in discussions of technology do we talk and think about technology as something essentially foreign (either good or bad) added on to our lives? How often do we see it thus, rather than as one instance, one expression, of a pervasive human activity that consists of bringing into being what did not heretofore exist, whether it be a social institution, a pattern of behavior, a conceptual framework such as a theory, a material object, or the sentences we utter? And how often are our actions guided by the former conception rather than by the latter?

How often in discussions of technology and in the actions of technologists (principally engineers) do we talk and act as if technology were just a making, rather than also a doing and a thinking (6)? To make something is to do something that affects what is made and that also affects those people and things that come in contact with it. To make something is to solve a problem, and to solve a problem correctly or successfully often requires that one think about what problem is to be solved. How often, because we think of technology as something apart from other human activities, do we think only of the technological problems to be solved?

Finally, how often in discussions of technology do we fail to think of our political and social institutions and of our conceptual frameworks as products of essentially the same process that gives us the products of technology?

The point that Ortega and writers like him were trying to make is that what we call technology is one instance of a process that seems to almost define human activity in general. To that extent it is very similar to other human activities and is a subsystem within a great total system of human activity. This realization will not resolve and bring to an end the current discussion of technology. It will not immediately solve any of the specific problems or issues being discussed. These are, however, the general facts of technology that everyone "knows." It is only that we are in the habit of talking and acting at variance with these facts. If the facts are ignored, they will take their revenge. Taking facts into account does not in itself guarantee a solution—it only increases the probability that a satisfactory solution will be found. If technology is inextricably making, doing, and thinking, then how successful will we be if we talk and act as if it were only making? If technology is intimately, structurally tied up with other human activities, then how successful will we be if we talk and act as if it were merely something added on to life, either for good or for bad?

Notes

1. "Man the Technician," in *History as a System* by José Ortega y Gasset, (New York, Norton, 1961), p. 87.
2. *Ibid.*
3. It is necessary to have a concept of oneself before one can recognize thinking as an activity. Bruno Snell argues in *The Discovery of the Mind* that it was only at this time that the Greeks began to construct a concept of self.
4. Arthur W. H. Adkins describes in *Merit and Responsibility* the details of what went wrong and how the Greeks responded.
5. This is the insight that Ortega tried to bring out in "Man the Technician." It is an essay that should be read by everyone. In essence he recapitulates human history as the working of this process. A philosophically very sophisticated analysis of the process can be found in *Being and Time*, secs. 14–18, by Martin Heidegger.
6. A very provocative description of what can happen when technology is understood only as making can be found in *Labor and Monopoly Capital* by Harry Braverman.

4.4

Living on a Lifeboat

GARRETT HARDIN

Susanne Langer (1942) has shown that it is probably impossible to approach an unsolved problem save through the door of metaphor. Later, attempting to meet the demands of rigor, we may achieve some success in cleansing theory of metaphor, though our success is limited if we are unable to avoid using common language, which is shot through and through with fossil metaphors. (I count no less than five in the preceding two sentences.)

Since metaphorical thinking is inescapable it is pointless merely to weep about our human limitations. We must learn to live with them, to understand them, and to control them. "All of us," said George Eliot in *Middlemarch*, "get our thoughts entangled in metaphors, and act fatally on the strength of them." To avoid unconscious suicide we are well advised to pit one metaphor against another. From the interplay of competitive metaphors, thoroughly developed, we may come closer to metaphor-free solutions to our problems.

No generation has viewed the problem of the survival of the human species as seriously as we have. Inevitably, we have entered this world of concern through the door of metaphor. Environmentalists have emphasized the image of the earth as a spaceship—Spaceship Earth. Kenneth Boulding (1966) is the principal architect of this metaphor. It is time, he says, that we replace the wasteful "cowboy economy" of the past with the frugal "spaceship economy" required for continued survival in the limited world we now see ours to be. The metaphor is notably useful in justifying pollution control measures.

Unfortunately, the image of a spaceship is also used to promote measures that are suicidal. One of these is a generous immigration policy, which is only a particular instance of a class of policies that is in error because they lead to the tragedy of the commons (Hardin 1968). These suicidal policies are attractive because they mesh with what we unthinkingly take to be the ideals of "the best people." What is missing in the idealistic view is an insistence that rights and responsibilities must go together. The "generous" attitude of all too many people results in asserting inalienable rights while ignoring or denying matching responsibilities.

For the metaphor of a spaceship to be correct the aggregate of people on board would have to be under unitary sovereign control (Ophuls 1974). A true ship always has a captain. It is conceivable that a ship could be run by a committee. But it could not possibly survive if its course were determined by bickering tribes that claimed rights without responsibilities.

What about Spaceship Earth? It certainly has no captain, and no executive committee. The United Nations is a toothless tiger, because the signatories of its charter wanted it that way. The spaceship metaphor is used only to justify spaceship demands on common resources without acknowledging corresponding spaceship responsibilities.

An understandable fear of decisive action leads people to embrace "incrementalism"—moving toward reform by tiny stages. As we shall see, this strategy is counterproductive in the area discussed here if it means accepting rights before responsibilities. Where human survival is at stake, the acceptance of responsibilities is a precondition to the acceptance of rights, if the two cannot be introduced simultaneously.

LIFEBOAT ETHICS

Before taking up certain substantive issues let us look at an alternative metaphor, that of a lifeboat. In developing some relevant examples the following numerical values are assumed. Approximately two-thirds of the world is desperately poor, and only one-third is comparatively rich. The people in poor countries have an average per capita GNP (Gross National Product) of about $200 per year; the rich, of about $3,000. (For the United States it is nearly $5,000 per year.) Metaphorically, each rich nation amounts to a lifeboat full of comparatively rich people. The poor of the world are in other, much more crowded lifeboats. Continuously, so to speak, the poor fall out of their lifeboats and swim for a while in the water outside, hoping to be admitted to a rich lifeboat, or in some other way to benefit from the "goodies" on board. What should the passengers on a rich lifeboat do? This is the central problem of "the ethics of a lifeboat."

First we must acknowledge that each lifeboat is effectively limited in capacity. The land of every nation has a limited carrying capacity. The exact limit is a matter for argument, but the energy crunch is convincing more people every day that we have already exceeded the carrying capacity of the land. We have been living on "capital"—stored petroleum and coal—and soon we must live on income alone.

Let us look at only one lifeboat—ours. The ethical problem is the same for all, and is as follows. Here we sit, say 50 people in a lifeboat. To be generous, let us assume our boat has a capacity of 10 more, making 60. (This, however, is to violate the engineering principle of the "safety factor." A new plant disease or a bad change in the weather may decimate our population if we don't preserve some excess capacity as a safety factor.)

The 50 of us in the lifeboat see 100 others swimming in the water outside, asking for admission to the boat, or for handouts. How shall we respond to their calls? There are several possibilities.

One. We may be tempted to try to live by the Christian ideal of being "our brother's keeper," or by the Marxian ideal (Marx 1875) of "from each according to his abilities, to each according to his needs." Since the needs of all are the same, we take all the needy into our boat, making a total of 150 in a boat with a capacity of 60. The boat is swamped, and everyone drowns. Complete justice, complete catastrophe.

Two. Since the boat has an unused excess capacity of 10, we admit just 10 more to it. This has the disadvantage of getting rid of the safety factor, for which action we will sooner or later pay dearly. Moreover, *which* 10 do we let in? "First come, first served?" The best 10? The neediest 10? How do we *discriminate*? And what do we say to the 90 who are excluded?

Three. Admit no more to the boat and preserve the small safety factor. Survival of the people in the lifeboat is then possible (though we shall have to be on our guard against boarding parties).

The last solution is abhorrent to many people. It is unjust, they say. Let us grant that it is.

"I feel guilty about my good luck," say some. The reply to this is simple: *Get out and yield your place to others.* Such a selfless action might satisfy the conscience of those who are addicted to guilt but it would not change the ethics of the lifeboat. The needy person to whom a guilt-addict yields his place will not himself feel guilty about his sudden good luck. (If he did he would not climb aboard.) The net result of conscience-stricken people relinquishing their unjustly held positions is the elimination of their kind of conscience from the lifeboat. The lifeboat, as it were, purifies itself of guilt. The ethics of the lifeboat persist, unchanged by such momentary aberrations.

This then is the basic metaphor within which we must work out our solutions. Let us enrich the image step by step with substantive additions from the real world.

REPRODUCTION

The harsh characteristics of lifeboat ethics are heightened by reproduction, particularly by reproductive differences. The people inside the lifeboats of the wealthy nations are doubling in numbers every 87 years, those outside are doubling every 35 years, on the average. And the relative difference in prosperity is becoming greater.

Let us, for a while, think primarily of the U.S. lifeboat. As of 1973 the United States had a population of 210 million people, who were increasing by 0.8% per year, that is, doubling in number every 87 years.

Although the citizens of rich nations are outnumbered two to one by the poor, let us imagine an equal number of poor people outside our lifeboat—a mere 210 million poor people reproducing at a quite different rate. If we imagine these to be the combined populations of Colombia, Venezuela, Ecuador, Morocco, Thailand, Pakistan, and the Philippines, the average rate of increase of the people "outside" is 3.3% per year. The doubling time of this population is 21 years.

Suppose that all these countries, and the United States, agreed to live by the Marxian ideal, "to each according to his needs," the ideal of most Christians as well. Needs, of course, are determined by population size, which is affected by reproduction. Every nation regards its rate of reproduction as a sovereign right. If our lifeboat were big enough in the beginning it might be possible to live *for a while* by the Christian-Marxian ideals. *Might.*

Initially, in the model given, the ratio of non-Americans to Americans would be one to one. But consider what the ratio would be 87 years later. By this time Americans would have doubled to a population of 420 million. The other group (doubling every 21 years) would now have swollen to 3,540 million. Each American would have more than eight people to share with. How could the lifeboat possibly keep afloat?

All this involves extrapolation of current trends into the future, and is consequently suspect. Trends may change. Granted: but the change will not necessarily be favorable. If—as seems likely—the rate of population increase falls faster in the ethnic group presently inside the lifeboat than it does among those now outside, the future will turn out to be even worse than mathematics predicts, and sharing will be even more suicidal.

RUIN IN THE COMMONS

The fundamental error of the sharing ethics is that it leads to the tragedy of the commons. Under a system of private property the man (or group of men) who own property recognize their responsibility to care for it, for if they don't they will eventually suffer. A farmer, for instance, if he is intelligent, will allow no more cattle in a pasture than its carrying capacity justifies. If he overloads the pasture, weeds take over, erosion sets in, and the owner loses in the long run.

But if a pasture is run as a commons open to all, the right of each to use it is not matched by an operational responsibility to take care of it. It is no use asking independent herdsmen in a commons to act responsibly, for they dare not. The considerate herdsman who refrains from overloading the commons suffers more than a selfish one who says his needs are greater. (As Leo Durocher says, "Nice guys finish last.") Christian-Marxian idealism is counterproductive. That it *sounds* nice is no excuse. With distribution systems, as with individual morality, good intentions are no substitute for good performance.

A social system is stable only if it is insensitive to errors. To the Christian-Marxian idealist a selfish person is a sort of "error." Prosperity in the system of the commons cannot survive errors. If *everyone* would only restrain himself, all would be well; but it takes *only one less than everyone* to ruin a system of voluntary restraint. In a crowded world of less than perfect human beings—and we will never know any other—mutual ruin is inevitable in the commons. This is the core of the tragedy of the commons.

One of the major tasks of education today is to create such an awareness of the dangers of the commons that people will be able to recognize its many varieties, however disguised. There is pollution of the air and water because these media are treated as commons. Further growth of population and growth in the per capita conversion of natural resources into pollutants require that the system of the commons be modified or abandoned in the disposal of "externalities."

The fish populations of the oceans are exploited as commons, and ruin lies ahead. No technological invention can prevent this fate: in fact, all improvements in the art of fishing merely hasten the day of complete ruin. Only the replacement of the system of the commons with a responsible system can save oceanic fisheries.

The management of western rangelands, though nominally rational, is in fact (under the steady pressure of cattle ranchers) often merely a government-sanctioned system of the commons, drifting toward ultimate ruin for both the rangelands and the residual enterprisers.

WORLD FOOD BANKS

In the international arena we have recently heard a proposal to create a new commons, namely an international depository of food reserves to which nations will contribute according to their abilities, and from which nations may draw according to their needs. Nobel laureate Norman Borlaug has lent the prestige of his name to this proposal.

A world food bank appeals powerfully to our humanitarian impulses. We remember John Donne's celebrated line, "Any man's death diminishes me," But before we rush out to see for whom the bell tolls let us recognize where the greatest political push for international granaries comes from, lest we be disillusioned later. Our experience with Public Law 480 clearly reveals the answer. This was the law that moved billions of dollars worth of U.S. grain to food-short, population-long countries during the past two decades. When P.L. 480 first came into being, a headline in the business magazine *Forbes* (Paddock and Paddock 1970) revealed the power behind it: "Feeding the World's Hungry Millions: How it will mean billions for U.S. business."

And indeed it did. In the years 1960 to 1970 a total of $7.9 billion was spent on the "Food for Peace" program, as P.L. 480 was called. During the years 1948 to 1970 an additional $49.9 billion was extracted from American taxpayers to pay for other economic aid programs, some of which went for food and food-producing machinery. (This figure does *not* include military aid.) That P.L. 480 was a give-away program was concealed. Recipient countries went through the motions of paying for P.L. 480 food—with IOUs. In December 1973 the charade was brought to an end as far as India was concerned when the United States "forgave" India's $3.2 billion debt (Anonymous 1974). Public announcement of the cancellation of the debt was delayed for two months: one wonders why.

"Famine—1974!" (Paddock and Paddock 1970) is one of the few publications that points out the commercial roots of this humanitarian attempt. Though all U.S. taxpayers lost by P.L. 480, special interest groups gained handsomely. Farmers benefited because they were not asked to contribute the grain—it was bought from them by the taxpayers. Besides the direct benefit there was the indirect effect of increasing demand and thus raising prices of farm products generally. The manufacturers of farm machinery, fertilizers, and pesticides benefited by the farmers' extra efforts to grow more food. Grain elevators profited from storing the grain for varying lengths of time. Railroads made money hauling it to port, and shipping lines by carrying it overseas. Moreover, once the machinery for P.L. 480

was established an immense bureaucracy had a vested interest in its continuance regardless of its merits.

Very little was ever heard of these selfish interests when P.O. 480 was defended in public. The emphasis was always on its humanitarian effects. The combination of multiple and relatively silent selfish interests with highly vocal humanitarian apologists constitutes a powerful lobby for extracting money from taxpayers. Foreign aid has become a habit that can apparently survive in the absence of any known justification. A news commentator in a weekly magazine (Lansner 1974), after exhaustively going over all the conventional arguments for foreign aid—self-interest, social justice, political advantage, and charity—and concluding that none of the known arguments really held water, concluded: "So the search continues for some logically compelling reasons for giving aid . . . " In other words, *act now, justify later*—if ever. (Apparently a quarter of a century is too short a time to find the justification for expending several billion dollars yearly.)

The search for a rational justification can be short-circuited by interjecting the word "emergency." Borlaug uses this word. We need to look sharply at it. What is an "emergency"? It is surely something like an accident, which is correctly defined as *an event that is certain to happen, though with a low frequency* (Harden 1972a). A well-run organization prepares for everything that is certain, including accidents and emergencies. It budgets for them. It saves for them. It expects them—and mature decision-makers do not waste time complaining about accidents when they occur.

What happens if some organizations budget for emergencies and others do not? If each organization is solely responsible for its own well-being, poorly managed ones will suffer. But they should be able to learn from experience. They have a chance to mend their ways and learn to budget for infrequent but certain emergencies. The weather, for instance, always varies and periodic crop failures are certain. A wise and competent government saves out of the production of the good years in anticipation of bad years that are sure to come. This is not a new idea. The Bible tells us that Joseph taught this policy to Pharaoh in Egypt more than 2,000 years ago. Yet it is literally true that the vast majority of the governments of the world today have no such policy. They lack either the wisdom or the competence, or both. Far more difficult than the transfer of wealth from one country to another is the transfer of wisdom between sovereign powers or between generations.

"But it isn't their fault! How can we blame the poor people who are caught in an emergency? Why must we punish them?" The concepts of blame and punishment are irrelevant. The question is, What are the operational consequences of establishing a world food bank? If it is open to every country every time a need develops, slovenly rulers will not be motivated to take Joseph's advice. Why should they? Others will bail them out whenever they are in trouble.

Some countries will make deposits in the world food bank and others will withdraw from it: there will be almost no overlap. Calling such a depository-transfer unit a "bank" is stretching the metaphor of *bank* beyond its elastic limits. The proposers, of course, never call attention to the metaphorical nature of the word they use.

THE RATCHET EFFECT

An "international food bank" is really, then, not a true bank but a disguised one-way transfer device for moving wealth from rich countries to poor. In the absence of such a bank, in a world inhabited by individually responsible sovereign nations, the population of each nation would repeatedly go through a cycle of the sort shown in Fig. 1. P_2 is greater than P_1, either in absolute numbers or because a deterioration of the food supply has removed the safety factor and produced a dangerously low ratio of resources to population. P_2 may be said to represent a state of overpopulation, which becomes obvious upon the appearance of an "accident," e.g., a crop failure. If the "emergency" is not met by outside help, the population drops back to the "normal" level—the "carrying capacity" of the environment—or even below. In the absence of population control by a sovereign, sooner or later the population grows to P_2 again and the cycle repeats. The long-term population curve (Hardin 1966) is an irregularly fluctuating one, equilibrating more or less about the carrying capacity.

A demographic cycle of this sort obviously involves great suffering in the restrictive phase, but such a cycle is normal to any independent country with inadequate population control. The third century theologian Tertullian (Hardin 1969a) expressed what must have been the recognition of many wise men when he wrote: "The scourges of pestilence, famine, wars, and earthquakes have come to be regarded as a blessing to overcrowded nations, since they serve to prune away the luxuriant growth of the human race."

Only under a strong and farsighted sovereign which theoretically could be the people themselves, democratically organized—can a population equilibrate at some set point below the carrying capacity, thus avoiding the pains normally caused by periodic and unavoidable disasters. For this happy state to be achieved it is necessary that those in power be able to contemplate with equanimity the "waste" of surplus food in times of bountiful harvests. It is essential that those in power resist the temptation to convert extra food into extra babies. On the public relations level it is necessary that the phrase "surplus food" be replaced by "safety factor."

But wise sovereigns seem not to exist in the poor world today. The most anguishing problems are created by poor countries that are governed by rulers insufficiently wise and powerful. If such countries can draw on a world food bank in times of "emergency," the population *cycle* of Fig. 1 will be replaced by the population *escalator* of Fig. 2. The input of food from a food bank acts as the pawl of a ratchet, preventing the population from retracing its steps to a lower level. Reproduction pushes the population upward, inputs from the world bank prevent its moving downward. Population size escalates, as does the absolute magnitude of "accidents" and "emergencies." The process is brought to an end only by the total collapse of the whole system, producing a catastrophe of scarcely imaginable proportions.

Such are the implications of the well-meant sharing of food in a world of irresponsible reproduction.

I think we need a new word for systems like this. The adjective "melioristic" is applied to systems that produce continual improvement; the English word is derived from the Latin *meliorare*, to become or make better. Parallel with this it

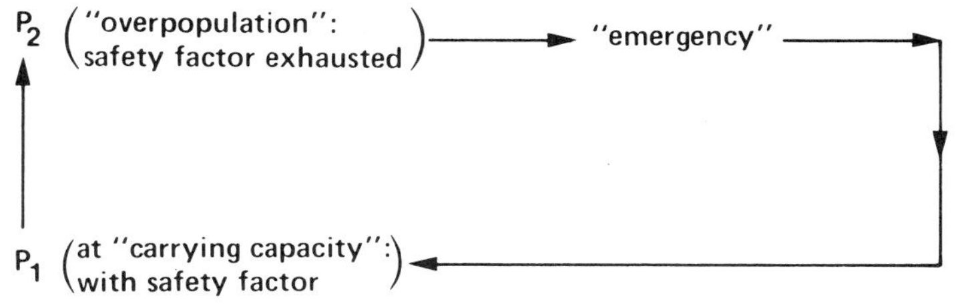

Figure 1. The population cycle of a nation that has no effective, conscious population control, and which receives no aid from the outside. P_2 is greater than P_1.

would be useful to bring in the word *pejoristic* (from the Latin *pejorare*, to become or make worse). This word can be applied to those systems which, by their very nature, can be relied upon to make matters worse. A world food bank coupled with soveriegn state irresponsibility in reproduction is an example of a pejoristic system.

This pejoristic system creates an unacknowledged commons. People have more motivation to draw from than to add to the common store. The license to make such withdrawals diminishes whatever motivation poor countries might otherwise have to control their populations. Under the guidance of this ratchet, wealth can be steadily moved in one direction only, from the slowly breeding rich to the rapidly breeding poor, the process finally coming to a halt only when all countries are equally and miserably poor.

All this is terribly obvious once we are acutely aware of the pervasiveness and danger of the commons. But many people still lack this awareness and the euphoria of the "benign demographic transition" (Hardin 1973) interferes with

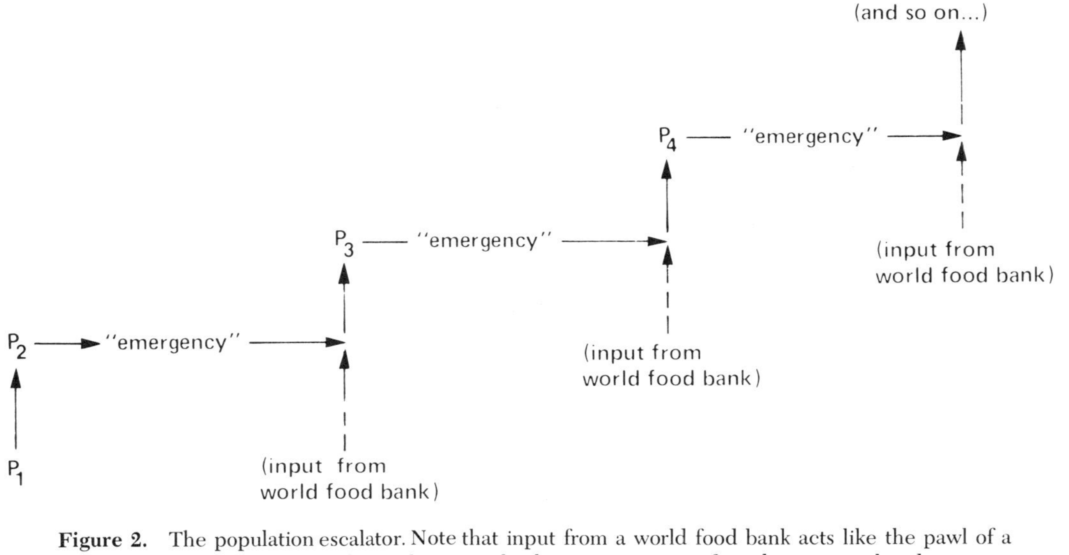

Figure 2. The population escalator. Note that input from a world food bank acts like the pawl of a ratchet, preventing the normal population cycle shown in Figure 1 from being completed. P_{n+1} is greater than P_n, and the absolute magnitude of the "emergencies" escalates. Ultimately the entire system crashes. The crash is not shown, and few can imagine it.

the realistic appraisal of pejoristic mechanisms. As concerns public policy, the deductions drawn from the benign demographic transition are these:

1. If the per capita GNP rises the birthrate will fall; hence, the rate of population increase will fall, ultimately producing ZPG (Zero Population Growth).
2. The long-term trend all over the world (including the poor countries) is of a rising per capita GNP (for which no limit is seen).
3. Therefore, all political interference in population matters is unneccessary; all we need to do is foster economic "development"—*note the metaphor*—and population problems will solve themselves.

Those who believe in the benign demographic transition dismiss the pejoristic mechanism of Fig. 2 in the belief that each input of food from the world outside fosters development within a poor country, thus resulting in a drop in the rate of population increase. Foreign aid has proceeded on this assumption for more than two decades. Unfortunately it has produced no indubitable instance of the asserted effect. It has, however, produced a library of excuses. The air is filled with plaintive calls for more massive foreign aid appropriations so that the hypothetical melioristic process can get started.

The doctrine of demographic laissez faire implicit in the hypothesis of the benign demographic transition is immensely attractive. Unfortunately there is more evidence against the melioristic system than there is for it (Davis 1963). On the historical side there are many counter-examples. The rise in per capita GNP in France and Ireland during the past century has been accompanied by a rise in population growth. In the 20 years following the Second World War the same positive correlation was noted almost everywhere in the world. Never in world history before 1950 did the worldwide population growth reach 1% per annum. Now the average population growth is over 2% and shows no signs of slackening.

On the theoretical side, the denial of the pejoristic scheme of Fig. 2 probably springs from the hidden acceptance of the "cowboy economy" that Boulding castigated. Those who recognize the limitations of a spaceship, if they are unable to achieve population control at a safe and comfortable level, accept the necessity of the corrective feedback of the population cycle shown in Figure 1. No one who knew in his bones that he was living on a true spaceship would countenance political support of the population escalator shown in Fig. 2.

ECO-DESTRUCTION VIA THE GREEN REVOLUTION

The demoralizing effect of charity on the recipient has long been known. "Give a man a fish and he will eat for a day; teach him how to fish and he will eat for the rest of his days." So runs an ancient Chinese proverb. Acting on this advice the Rockefeller and Ford Foundations have financed a multipronged program for improving agriculture in the hungry nations. The result, known as the "Green Revolution," has been quite remarkable. "Miracle wheat" and "miracle rice" are splendid technological achievements in the realm of plant genetics.

Whether or not the Green Revolution can increase food production is doubtful

(Harris 1972, Paddock 1970, Wilkes 1972), but in any event not particularly important. What is missing in this great and well-meaning humanitarian effort is a firm grasp of fundamentals. Considering the importance of the Rockefeller Foundation in this effort it is ironic that the late Alan Gregg, a much-respected vice-president of the Foundation, strongly expressed his doubts of the wisdom of all attempts to increase food production some two decades ago. (This was before Borlaug's work—supported by Rockefeller—had resulted in the development of "miracle wheat.") Gregg (1955) likened the growth and spreading of humanity over the surface of the earth to the metastasis of cancer in the human body, wryly remarking that "cancerous growths demand food; but, as far as I know, they have never been cured by getting it."

"Man does not live by bread alone"—the scriptural statement has a rich meaning even in the material realm. Every human being born constitutes a draft on all aspects of the environment—food, air, water, unspoiled scenery, occasional and optional solitude, beaches, contact with wild animals, fishing, hunting—the list is long and incompletely known. Food can, perhaps, be significantly increased: but what about clean beaches, unspoiled forests, and solitude? If we satisfy the need for food in a growing population we necessarily decrease the supply of other goods, and thereby increase the difficulty of equitably allocating scarce goods (Hardin 1969b, 1972b).

The present population of India is 600 million, and it is increasing by 15 million per year. The environmental load of this population is already great. The forests of India are only a small fraction of what they were three centuries ago. Soil erosion, floods, and the psychological costs of crowding are serious. Every one of the net 15 million lives added each year stresses the Indian environment more severely. *Every life saved this year in a poor country diminishes the quality of life for subsequent generations.*

Observant critics have shown how much harm we wealthy nations have already done to poor nations through our well-intentioned but misguided attempts to help them (Paddock and Paddock 1973). Particularly reprehensible is our failure to carry out post-audits of these attempts (Farvar and Milton 1972). Thus have we shielded our tender consciences from knowledge of the harm we have done. Must we Americans continue to fail to monitor the consequences of our external "do-gooding"? If, for instance, we thoughtlessly make it possible for the present 600 million Indians to swell to 1,200 millions by the year 2001—as their present growth rate promises—will posterity in India thank *us* for facilitating an even greater destruction of *their* environment? Are good intentions ever a sufficient excuse for bad consequences?

IMMIGRATION CREATES A COMMONS

I come now to the final example of a commons in action, one for which the public is least prepared for rational discussion. The topic is at present enveloped by a great silence which reminds me of a comment made by Sherlock Holmes in A. Conan Doyle's story "Silver Blaze." Inspector Gregory had asked, "Is there any point to which you would wish to draw my attention?" To this Holmes responded:

"To the curious incident of the dog in the night-time."
"The dog did nothing in the night-time," said the Inspector.
"That was the curious incident," remarked Sherlock Holmes.

By asking himself what would repress the normal barking instinct of a watch dog Holmes realized that it must be the dog's recognition of his master as the criminal trespasser. In a similar way we should ask ourselves what repression keeps us from discussing something as important as immigration?

It cannot be that immigration is numerically of no consequence. Our government acknowledges a *net* inflow of 400,000 a year. Hard data are understandably lacking on the extent of illegal entries, but a not implausible figure is 600,000 per year (Buchanan 1973). The natural increase of the resident population is now about 1.7 million per year. This means that the yearly gain from immigration is at least 19%, and may be 37%, of the total increase. It is quite conceivable that educational campaigns like that of Zero Population Growth, Inc., coupled with adverse social and economic factors—inflation, housing shortage, depression, and loss of confidence in national leaders—may lower the fertility of American women to a point at which all of the yearly increase in population would be accounted for by immigration. Should we not at least ask if that is what we want? How curious it is that we so seldom discuss immigration these days!

Curious, but understandable—as one finds out the moment he publicly questions the wisdom of the status quo in immigration. He who does so is promptly charged with *isolationism, bigotry, prejudice, ethnocentrism, chauvinism,* and *selfishness.* These are hard accusations to bear. It is pleasanter to talk about other matters, leaving immigration policy to wallow in the cross-currents of special interests that take no account of the good of the whole—*or of the interests of posterity.*

We Americans have a bad conscience because of things we said in the past about immigrants. Two generations ago the popular press was rife with references to *Dagos, Wops, Pollacks, Japs, Chinks,* and *Krauts*—all pejorative terms which failed to acknowledge our indebtedness to Goya, Leonardo, Copernicus, Hiroshige, Confucius, and Bach. Because the implied inferiority of foreigners was *then* the justification for keeping them out, it is *now* thoughtlessly assumed that restrictive policies can only be based on the assumption of immigrant inferiority. *This is not so.*

Existing immigration laws exclude idiots and known criminals; future laws will almost certainly continue this policy. But should we also consider the quality of the average immigrant, as compared with the quality of the average resident? Perhaps we should, perhaps we shouldn't. (What is "quality" anyway?) But the quality issue is not our concern here.

From this point on, *it will be assumed that immigrants and native-born citizens are of exactly equal quality,* however quality may be defined. The focus is only on quantity. The conclusions reached depend on nothing else, so all charges of ethnocentrism are irrelevant.

World food banks move food to the people, thus facilitating the exhaustion of the environment of the poor. By contrast, unrestricted immigration moves people to the food, thus speeding up the destruction of the environment in rich countries. Why poor people should want to make this transfer is no mystery: but why should rich hosts encourage it? This transfer, like the reverse one, is supported by both selfish interests and humanitarian impulses.

The principal selfish interest in unimpeded immigration is easy to identify: it is the interest of the employers of cheap labor, particularly that needed for degrading jobs. We have been deceived about the forces of history by the lines of Emma Lazarus inscribed on the Statue of Liberty:

> Give me your tired, your poor
> Your huddled masses yearning to breathe free,
> The wretched refuse of your teeming shore,
> Send these, the homeless, tempest-tossed, to me:
> I lift my lamp beside the golden door.

The image is one of an infinitely generous earth-mother, passively opening her arms to hordes of immigrants who come here on their own initiative. Such an image may have been adequate for the early days of colonization, but by the time these lines were written (1886) the force for immigration was largely manufactured inside our own borders by factory and mine owners who sought cheap labor not to be found among laborers already here. One group of foreigners after another was thus enticed into the United States to work at wretched jobs for wretched wages.

At present, it is largely the Mexicans who are being so exploited. It is particularly to the advantage of certain employers that there be many illegal immigrants. Illegal immigrant workers dare not complain about their working conditions for fear of being repatriated. Their presence reduces the bargaining power of all Mexican-American laborers. Cesar Chavez has repeatedly pleaded with congressional committees to close the doors to more Mexicans so that those here can negotiate effectively for higher wages and decent working conditions. Chavez understands the ethics of a lifeboat.

The interests of the employers of cheap labor are well served by the silence of the intelligentsia of the country. WASPS—White Anglo-Saxon Protestants—are particularly reluctant to call for a closing of the doors to immigration for fear of being called ethnocentric bigots. It was, therefore, an occasion of pure delight for this particular WASP to be present at a meeting when the points he would like to have made were made better by a non-WASP speaking to other non-WASPS. It was in Hawaii, and most of the people in the room were second-level Hawaiian officials of Japanese ancestry. All Hawaiians are keenly aware of the limits of their environment, and the speaker had asked how it might be practically and constitutionally possible to close the doors to more immigrants to the islands. (To Hawaiians, immigrants from the other 49 states are as much of a threat as those from other nations. There is only so much room in the islands, and the islanders know it. Sophistical arguments that imply otherwise do not impress them.)

Yet the Japanese-Americans of Hawaii have active ties with the land of their origin. This point was raised by a Japanese-American member of the audience who asked the Japanese-American speaker: "But how can we shut the doors now? We have many friends and relations in Japan that we'd like to bring to Hawaii some day so that they can enjoy this beautiful land."

The speaker smiled sympathetically and responded slowly: "Yes, but we have children now and someday we'll have grandchildren. We can bring more people here from Japan only by giving away some of the land that we hope to pass on to our grandchildren some day. What right do we have to do that?"

To be generous with one's own possessions is one thing; to be generous with posterity's is quite another. This, I think, is the point that must be gotten across to those who would, from a commendable love of distributive justice, institute a ruinous system of the commons, either in the form of a world food bank or that of unrestricted immigration. Since every speaker is a member of some ethnic group it is always possible to charge him with ethnocentrism. But even after purging an argument of ethnocentrism the rejection of the commons is still valid and necessary if we are to save at least some parts of the world from environmental ruin. Is it not desirable that at least some of the grandchildren of people now living should have a decent place in which to live?

THE ASYMMETRY OF DOOR-SHUTTING

We must now answer this telling point: "How can you justify slamming the door once you're inside? You say that immigrants should be kept out. But aren't we all immigrants, or the descendants of immigrants? Since we refuse to leave, must we not, as a matter of justice and symmetry, admit all others?"

It is literally true that we Americans of non-Indian ancestry are the descendants of thieves. Should we not, then, "give back" the land to the Indians; that is, give it to the now-living Americans of Indian ancestry? As an exercise in pure logic I see no way to reject this proposal. Yet I am unwilling to live by it; and I know no one who is. Our reluctance to embrace pure justice may spring from pure selfishness. On the other hand, it may arise from an unspoken recognition of consequences that have not yet been clearly spelled out.

Suppose, becoming intoxicated with pure justice, we "Anglos" should decide to turn our land over to the Indians. Since all our other wealth has also been derived from the land, we would have to give that to the Indians, too. Then what would we non-Indians do? Where would we go? There is no open land in the world on which men without capital can make their living (and not much unoccupied land on which men with capital can either). Where would 209 million putatively justice-loving, non-Indian, Americans go? Most of them—in the persons of their ancestors—came from Europe, but they wouldn't be welcomed back there. Anyway, Europeans have no better title to their land than we to ours. They also would have to give up their homes. (But to whom? And where would *they* go?)

Clearly, the concept of pure justice produces an infinite regress. The law long ago invented statutes of limitations to justify the rejection of pure justice, in the interest of preventing massive disorder. The law zealously defends property rights—but only *recent* property rights. It is as though the physical principle of exponential decay applies to property rights. Drawing a line in time may be unjust, but any other action is practically worse.

We are all the descendants of thieves, and the world's resources are inequitably distributed, but we must begin the journey to tomorrow from the point where we are today. We cannot remake the past. We cannot, without violent disorder and suffering, give land and resources back to the "original" owners—who are dead anyway.

We cannot safely divide the wealth equitably among all present peoples, so long as people reproduce at different rates, because to do so would guarantee that our grandchildren—everyone's grandchildren—would have only a ruined world to inhabit.

MUST EXCLUSION BE ABSOLUTE?

To show the logical structure of the immigration problem I have ignored many factors that would enter into real decisions made in a real world. No matter how convincing the logic may be it is probable that we would want, from time to time, to admit a few people from the outside to our lifeboat. Political refugees in particular are likely to cause us to make exceptions: We remember the Jewish refugees from Germany after 1933, and the Hungarian refugees after 1956. Moreover, the interests of national defense, broadly conceived, could justify admitting many men and women of unusual talents, whether refugees or not. (This raises the quality issue, which is not the subject of this essay.)

Such exceptions threaten to create runaway population growth inside the lifeboat, i.e., the receiving country. However, the threat can be neutralized by a population policy that includes immigration. An effective policy is one of flexible control.

Suppose, for example, that the nation has achieved a stable condition of ZPG, which (say) permits 1.5 million births yearly. We must suppose that an acceptable system of allocating birthrights to potential parents is in effect. Now suppose that an inhumane regime in some other part of the world creates a horde of refugees, and that there is a widespread desire to admit some to our country. At the same time, we do not want to sabotage our population control system. Clearly, the rational path to pursue is the following. If we decide to admit 100,000 refugees this year we should compensate for this by reducing the allocation of birth-rights in the following year by a similar amount, that is downward to a total of 1.4 million. In that way we could achieve both humanitarian and population control goals. (And the refugees would have to accept the population controls of the society that admits them. It is not inconceivable that they might be given proportionately fewer rights than the native population.)

In a democracy, the admission of immigrants should properly be voted on. But by whom? It is not obvious. The usual rule of a democracy is votes for all. But it can be questioned whether a universal franchise is the most just one in a case of this sort. Whatever benefits there are in the admission of immigrants presumably accrue to everyone. But the costs would be seen as falling most heavily on potential parents, some of whom would have to postpone or forego having their (next) child because of the influx of immigrants. The double question *Who benefits? Who pays?* suggests that a restriction of the usual democratic franchise would be appropriate and just in this case. Would our particular quasi-democratic form of government be flexible enough to institute such a novelty? If not, the majority might, out of humanitarian motives, impose an unacceptable burden (the foregoing of parenthood) on a minority, thus producing political instability.

Plainly many new problems will arise when we consciously face the immigration question and seek rational answers. No workable answers can be found if we ignore population problems. And—if the argument of this essay is correct—so long as there is no true world government to control reproduction everywhere it is impossible to survive in dignity if we are to be guided by Spaceship ethics. Without a world government that is sovereign in reproductive matters mankind lives, in fact, on a number of sovereign lifeboats. For the foreseeable future survival demands that we govern our actions by the ethics of a lifeboat. Posterity will be ill served if we do not.

References

Anonymous. 1974. *Wall Street Journal* 19 Feb.

Borlaug, N. 1973. Civilization's future: a call for international granaries. *Bull. At. Sci.* **29**: 7–15.

Boulding, K. 1966. The economics of the coming Spaceship earth. **In** H. Jarrett, ed. *Environmental* Quality in a Growing Economy. Johns Hopkins Press, Baltimore, Md.

Buchanan, W. 1973. Immigration statistics *Equilibrium* **1**(3): 16–19.

Davis, K. 1963. Population. *Sci. Amer.* **209**(3): 62–71.

Farvar, M. T., and J. P. Milton. 1972. *The Careless Technology*, Natural History Press, Garden City, N.Y.

Gregg, A. 1955. A medical aspect of the population problem. *Science* **121**: 681–682.

Hardin, G. 1966. Chap. 9 in *Biology: Its Principles and Implications*, 2nd ed. Freeman, San Francisco.

______. 1968. The tragedy of the commons, *Science* **162**: 1243–1248.

______. 1969a Page 18 in *Population, Evolution, and Birth Control*, 2nd ed. Freeman, San Francisco.

______. 1969b. The economics of wilderness. *Nat. Hist.* **78**(6): 20–27.

______. 1972a. Pages 81–82 in *Exploring New Ethics for Survival: The Voyage of the Spaceship* Beagle, Viking, New York.

______. 1972b. Preserving quality on Spaceship Earth. In J. B. Trefethen, ed. *Transactions of the Thirty-Seventh North American Wildlife and Natural Resources Conference.* Wildlife Management Institute, Washington, D.C.

______. 1973. Chap. 23 in *Stalking the Wild Taboo.* Kaufmann, Los Altos, Cal.

Harris, M. 1972. How green the revolution. *Nat. Hist.* **81**(3): 28–30.

Langer, S. K. 1942. *Philosophy in a New Key.* Harvard University Press, Cambridge, Mass.

Lansner, K. 1974. Should foreign aid begin at home? *Newsweek*, 11 Feb., p. 32.

Marx, K. 1875. Critique of the Gotha program. Page 388 in R. C. Tucker, ed., *The Marx-Engels Reader*, Norton, New York, 1972.

Ophuls, W. 1974. The scarcity society. *Harpers* **248**(1487): 47–52.

Paddock, W. C. 1970. How green is the green revolution? *BioScience* **20**: 897–902.

Paddock, W., and E. Paddock. 1973. *We Don't Know How.* Iowa State University Press, Ames, Iowa.

Paddock, W., and P. Paddock. 1967. *Famine—1975!* Little, Brown, Boston.

Wilkes, H. G. 1972. The green revolution. *Environment* **14**(8): 32–39.

4.5

Code of Ethics of Engineers

ENGINEERS' COUNCIL FOR
PROFESSIONAL DEVELOPMENT

THE FUNDAMENTAL PRINCIPLES

Engineers uphold and advance the integrity, honor and dignity of the engineering profession by:

I. Using their knowledge and skill for the enhancement of human welfare;

II. Being honest and impartial, and serving with fidelity the public, their employers and clients;

III. Striving to increase the competence and prestige of the engineering profession; and

IV. Supporting the professional and technical societies of their disciplines.

THE FUNDAMENTAL CANONS

1. Engineers shall hold paramount the safety, health and welfare of the public in the performance of their professional duties.
2. Engineers shall perform services only in the areas of their competence.
3. Engineers shall issue public statements only in an objective and truthful manner.
4. Engineers shall act in professional matters for each employer or client as faithful agents or trustees, and shall avoid conflicts of interest.
5. Engineers shall build their professional reputation on the merit of their services and shall not compete unfairly with others.
6. Engineers shall act in such a manner as to uphold and enhance the honor, integrity and dignity of the profession.

Approved by the Board of Directors, October 5, 1977.

7. Engineers shall continue their professional development throughout their careers and shall provide opportunities for the professional development of those engineers under their supervision.

SUGGESTED GUIDELINES FOR USE WITH THE FUNDAMENTAL CANONS OF ETHICS

1. Engineers shall hold paramount the safety, health and welfare of the public in the performance of their professional duties.
   a. Engineers shall recognize that the lives, safety, health and welfare of the general public are dependent upon engineering judgments, decisions and practices incorporated into structures, machines, products, processes and devices.
   b. Engineers shall not approve nor seal plans and/or specifications that are not of a design safe to the public health and welfare and in conformity with accepted engineering standards.
   c. Should the Engineers' professional judgment be overruled under circumstances where the safety, health, and welfare of the public are endangered, the Engineers shall inform their clients or employers of the possible consequences and notify other proper authority of the situation, as may be appropriate.
      (c.1) Engineers shall do whatever possible to provide published standards, test codes and quality control procedures that will enable the public to understand the degree of safety or life expectancy associated with the use of the design, products and systems for which they are responsible.
      (c.2) Engineers will conduct reviews of the safety and reliability of the design, products or systems for which they are responsible before giving their approval to the plans for the design.
      (c.3) Should Engineers observe conditions which they believe will endanger public safety or health, they shall inform the proper authority of the situation.
   d. Should Engineers have knowledge or reason to believe that another person or firm may be in violation of any of the provisions of these Guidelines, they shall present such information to the proper authority in writing and shall cooperate with the proper authority in furnishing such further information or assistance as may be required.
      (d.1) They shall advise proper authority if an adequate review of the safety and reliability of the products or systems has not been made or when the design imposes hazards to the public through its use.
      (d.2) They shall withhold approval of products or systems when changes or modifications are made which would affect adversely its performance insofar as safety and reliability are concerned.
   e. Engineers should seek opportunities to be of constructive service in civic

affairs and work for the advancement of the safety, health and well-being of their communities.

f. Engineers should be committed to improving the environment to enhance the quality of life.

2. Engineers shall perform sevices only in areas of their competence.
   a. Engineers shall undertake to perform engineering assignments only when qualified by education or experience in the specific technical field of engineering involved.
   b. Engineers may accept an assignment requiring education or experience outside of their own fields of competence, but only to the extent that their services are restricted to those phases of the project in which they are qualified. All other phases of such project shall be performed by qualified associates, consultants, or employees.
   c. Engineers shall not affix their signatures and/or seals to any engineering plan or document dealing with subject matter in which they lack competence by virtue of education or experience, nor to any such plan or document not prepared under their direct supervisory control.

3. Engineers shall issue public statements only in an objective and truthful manner.
   a. Engineers shall endeavor to extend public knowledge, and to prevent misunderstandings of the achievements of engineering.
   b. Engineers shall be completely objective and truthful in all professional reports, statements, or testimony. They shall include all relevant and pertinent information in such reports, statements, or testimony.
   c. Engineers, when serving as expert or technical witnesses before any court, commission, or other tribunal, shall express an engineering opinion only when it is founded upon adequate knowledge of the facts in issue, upon a background of technical competence in the subject matter, and upon honest conviction of the accuracy and propriety of their testimony.
   d. Engineers shall issue no statements, criticisms, nor arguments on engineering matters which are inspired or paid for by an interested party, or parties, unless they have prefaced their comments by explicitly identifying themselves, by disclosing the identities of the party or parties on whose behalf they are speaking, and by revealing the existence of any pecuniary interest they may have in the instant matters.
   e. Engineers shall be dignified and modest in explaining their work and merit, and will avoid any act tending to promote their own interests at the expense of the integrity, honor and dignity of the profession.

4. Engineers shall act in professional matters for each employer or client as faithful agents or trustees, and shall avoid conflicts of interest.
   a. Engineers shall avoid all known conflicts of interest with their employers or clients and shall promptly inform their employers or clients of any

business association, interests, or circumstances which could influence their judgment or the quality of their services.

b. Engineers shall not knowingly undertake any assignments which would knowingly create a potential conflict of interest between themselves and their clients or their employers.

c. Engineers shall not accept compensation, financial or otherwise, from more than one party for services on the same project, nor for services pertaining to the same project, unless the circumstances are fully disclosed to, and agreed to, by all interested parties.

d. Engineers shall not solicit nor accept financial or other valuable considerations, including free engineering designs, from material or equipment suppliers for specifying their products.

e. Engineers shall not solicit nor accept gratuities, directly or indirectly, from contractors, their agents, or other parties dealing with their clients or employers in connection with work for which they are responsible.

f. When in public service as members, advisors, or employees of a governmental body or department, Engineers shall not participate in considerations or actions with respect to services provided by them or their organization in private or product engineering practice.

g. Engineers shall not solicit nor accept an engineering contract from a governmental body on which a principal, officer or employee of their organization serves as a member.

h. When, as a result of their studies, Engineers believe a project will not be successful, they shall so advise their employer or client.

i. Engineers shall treat information coming to them in the course of their assignments as confidential, and shall not use such information as a means of making personal profit if such action is adverse to the interests of their clients, their employers, or the public.

(i.1) They will not disclose confidential information concerning the business affairs or technical processes of any present or former employer or client or bidder under evaluation, without his consent.

(i.2) They shall not reveal confidential information nor findings of any commission or board of which they are members.

(i.3) When they use designs supplied to them by clients, these designs shall not be duplicated by the Engineers for others without express permission.

(i.4) While in the employ of others, Engineers will not enter promotional efforts or negotiations for work or make arrangements for other employment as principals or to practice in connection with specific projects for which they have gained particular and specialized knowledge without the consent of all interested parties.

j. The Engineer shall act with fairness and justice to all parties when administering a construction (or other) contract.

k. Before undertaking work for others in which Engineers may make improvements, plans, designs, inventions, or other records which may justify copyrights or patents, they shall enter into a positive agreement regarding ownership.

l. Engineers shall admit and accept their own errors when proven wrong and refrain from distorting or altering the facts to justify their decisions.

m. Engineers shall not accept professional employment outside of their regular work or interest without the knowledge of their employers.

n. Engineers shall not attempt to attract an employee from another employer by false or misleading representations.

o. Engineers shall not review the work of other Engineers except with the knowledge of such Engineers, or unless the assignments or contractual agreements for the work have been terminated.

(o.1) Engineers in governmental, industrial or educational employment are entitled to review and evaluate the work of other engineers when so required by their duties.

(o.2) Engineers in sales or industrial employment are entitled to make engineering comparisons of their products with products of other suppliers.

(o.3) Engineers in sales employment shall not offer nor give engineering consultation or designs or advice other than specifically applying to equipment, materials or systems being sold or offered for sale by them.

5. Engineers shall build their professional reputation on the merit of their services and shall not compete unfairly with others.

a. Engineers shall not pay nor offer to pay, either directly or indirectly, any commission, political contribution, or a gift, or other consideration in order to secure work, exclusive of securing salaried positions through employment agencies.

b. Engineers should negotiate contracts for professional services fairly and only on the basis of demonstrated competence and qualifications for the type of professional service required.

c. Engineers should negotiate a method and rate of compensation commensurate with the agreed upon scope of services. A meeting of the minds of the parties to the contract is essential to mutual confidence. The public interest requires that the cost of engineering services be fair and reasonable, but not the controlling consideration in selection of individuals or firms to provide these services.

(c.1) These principles shall be applied by Engineers in obtaining the services of other professionals.

d. Engineers shall not attempt to supplant other Engineers in a particular employment after becoming aware that definite steps have been taken toward the others' employment or after they have been employed.

(d.1) They shall not solicit employment from clients who already have Engineers under contract for the same work.

(d.2) They shall not accept employment from clients who already have Engineers for the same work not yet completed or not yet paid for unless the performance or payment requirements in the contract are being litigated or the contracted Engineers' services have been terminated in writing by either party.

(d.3) In case of termination of litigation, the prospective Engineers before accepting the assignment shall advise the Engineers being terminated or involved in litigation.

e. Engineers shall not request, propose nor accept professional commissions on a contingent basis under circumstances under which their professional judgments may be compromised, or when a contingency provision is used as a device for promoting or securing a professional commission.

f. Engineers shall not falsify nor permit misrepresentation of their, or their associates' academic or professional qualifications. They shall not misrepresent nor exaggerate their degree of responsibility in or for the subject matter of prior assignments. Brochures or other presentations incident to the solicitation of employment shall not misrepresent pertinent facts concerning employers, employees, associates, joint ventures, or their past accomplishments with the intent and purpose of enhancing their qualifications and work.

g. Engineers may advertise professional services only as a means of identification and limited to the following:

(g.1) Professional cards and listings in recognized and dignified publications, provided they are consistent in size and are in a section of the publication regularly devoted to such professional cards and listings. The information displayed must be restricted to firm name, address, telephone number, appropriate symbol, names of principal participants and the fields of practice in which the firm is qualified.

(g.2) Signs on equipment, offices and at the site of projects for which they render services, limited to firm name, address, telephone number and type of services, as appropriate.

(g.3) Brochures, business cards, letterheads and other factual representations of experience, facilities, personnel and capacity to render service, providing the same are not misleading relative to the extent of participation in the projects cited and are not indiscriminately distributed.

(g.4) Listings in the classified section of telephone directories, limited to name, address, telephone number and specialities in which the firm is qualified without resorting to special or bold type.

h. Engineers may use display advertising in recognized dignified business and professional publications, providing it is factual, and relates only to engineering, is free from ostentation, contains no laudatory expressions or implication, is not misleading with respect to the Engineers' extent of participation in the services or projects described.

i. Engineers may prepare articles for the lay or technical press which are factual, dignified and free from ostentations or laudatory implications.

Such articles shall not imply other than their direct participation in the work described unless credit is given to others for their share of the work.

j. Engineers may extend permission for their names to be used in commercial advertisements, such as may be published by manufacturers, contractors, material suppliers, etc., only by means of a modest dignified notation acknowledging their participation and the scope thereof in the project or product described. Such permission shall not include public endorsement of proprietary products.

k. Engineers may advertise for recruitment of personnel in appropriate publications or by special distribution. The information presented must be displayed in a dignified manner, restricted to firm name, address, telephone number, appropriate symbol, names of principal participants, the fields of practice in which the firm is qualified and factual descriptions of positions available, qualifications required and benefits available.

l. Engineers shall not enter competitions for designs for the purpose of obtaining commissions for specific projects, unless provision is made for reasonable compensation for all designs submitted.

m. Engineers shall not maliciously or falsely, directly or indirectly, injure the professional reputation, prospects, practice or employment of another engineer, nor shall they indiscriminately criticize another's work.

n. Engineers shall not undertake nor agree to perform any engineering service on a free basis, except professional services which are advisory in nature for civic, charitable, religious or non-profit organizations. When serving as members of such organizations, engineers are entitled to utilize their personal engineering knowledge in the service of these organizations.

o. Engineers shall not use equipment, supplies, laboratory nor office facilities of their employers to carry on outside private practice without consent.

p. In case of tax-free or tax-aided facilities, engineers should not use student services at less than rates of other employees of comparable competence, including fringe benefits.

6. Engineers shall act in such a manner as to uphold and enhance the honor, integrity and dignity of the profession.

   a. Engineers shall not knowingly associate with nor permit the use of their names nor firm names in business ventures by any person or firm which they know, or have reason to believe, are engaging in business or professional practices of a fraudulent or dishonest nature.

   b. Engineers shall not use association with non-engineers, corporations, nor partnerships as 'cloaks' for unethical acts.

7. Engineers shall continue their professional development throughout their careers, and shall provide opportunities for the professional development of those engineers under their supervision.

a. Engineers shall encourage their engineering employees to further their education.

b. Engineers should encourage their engineering employees to become registered at the earliest possible date.

c. Engineers should encourage engineering employees to attend and present papers at professional and technical society meetings.

d. Engineers should support the professional and technical societies of their disciplines.

e. Engineers shall give proper credit for engineering work to those to whom credit is due, and recognize the proprietary interests of others. Whenever possible, they shall name the person or persons who may be responsible for designs, inventions, writings or other accomplishments.

f. Engineers shall endeavor to extend the public knowledge of engineering, and shall not participate in the dissemination of untrue, unfair or exaggerated statements regarding engineering.

g. Engineers shall uphold the principle of appropriate and adequate compensation for those engaged in engineering work.

h. Engineers should assign professional engineers duties of a nature which will utilize their full training and experience insofar as possible, and delegate lesser functions to subprofessionals or to technicians.

i. Engineers shall provide prospective engineering employees with complete information on working conditions and their proposed status of employment, and after employment shall keep them informed of any changes.

4.6

The Philosophical Basis of Engineering Codes of Ethics

ALBERT FLORES

What is the philosophical basis of engineering codes of ethics? The importance of this question is obvious, for without an acceptable answer, we can hardly expect that engineers would willingly submit to the requirements that these codes impose on their conduct. The injunctions that are contained in these codes are intended to restrict engineering practice to areas that are morally acceptable, hence they involve a direct limitation of an engineer's freedom. Obviously, our freedom is one of our most cherished possessions, and its restriction is permissible and acceptable only if there exist compelling reasons that justify such a restriction. Indeed, engineering codes could claim little authority over engineering practice unless there are good reasons to accept their precepts and incorporate them into one's routine activities. At issue, then, is the question: Why should engineers adopt their profession's code of conduct?

The question being asked is not a question about why engineers might be *motivated* to adopt professional codes, nor is it aimed at providing an explanation of the *causes* that account for engineers' acceptance of these codes. These are, no doubt, interesting questions, and their answers are clearly relevant to questions concerning why engineers would willingly restrict their professional practice to the limits prescribed in their ethical codes. So, for example, engineers may be motivated to observe these limits because they fear the censure of their colleagues or because claiming allegiance to a code of ethics may enhance their reputation. They may believe that following its injunctions will relieve them of the guilt feelings they have when they fail to obey its prescriptions or that, generally, obedience will bring them rewards, make them happy, or allow them to avoid the responsibilities associated with making their own decisions. These are just some of the explanations that we can construct to account for engineers' behavior, but none can *justify* this behavior. For they all refer to the causes of behavior and not to the reasons why these behaviors are chosen. Questions that ask Why? in reference to how people act, are ambiguous in that they fail to adequately distinguish between reasons and causes (1). Unfortunately, our conduct is as much governed by motives, feelings, and causes as it is by reasons. And it is with the latter that we are here concerned. In brief, the kind of answer we are seeking is one that justifies adopting a code of ethics, and justifications are constructed from reasons and arguments, not by reference to motives and causes.

In moral philosophy the adoption of a moral code of conduct has been justified in either of two general ways: arguments that refer to the beneficial consequences that flow from following such a code illustrate what is known as a *consequentialist* justification. John Stuart Mill's extended argument in *On Liberty* showing why it is socially beneficial to respect the freedom of individuals to pursue their own interests unfettered by society, so long as they cause no harm to others, is a paradigm of a consequentialist justification (2). On the other hand, there are nonconsequentialist, or *deontological*, arguments (from the Greek *deon* meaning "ought"). These arguments appeal to principles of obligation or rules expressing duties in order to justify adopting a particular code. Accordingly, Kant has argued that no set of beneficial consequences can, for example, justify the telling of a lie, because lying violates a basic moral obligation or duty that we have to be truthful in word and deed (3). It is, in short, in the nature of the act of lying itself and not in its possible harmful consequences that we find justification for its prohibition, according to those who adopt this deontological mode of justification. It would, however, be mistaken to conclude that truth-telling is not also justified by an appeal to the good consequences that following such a principle would produce. Indeed, many of our most cherished moral principles are justified in either of these two ways. In attempting to provide an answer to the question of why engineers should adopt their professional codes of conduct, we will examine arguments illustrating both these lines of justification.

However, before we proceed to this task, there is an important preliminary task to which we must now attend. Although engineering codes of ethics purport to define the standards of ethical performance expected of professional engineers, in reality these codes contain numerous elements which have nothing to do with ethics. In addition to stating the moral duties which engineers are expected to honor, these codes contain principles addressing business practices, professional etiquette, and the special interests of the profession that are intended to regulate the behavior of engineers, but not in areas that can be reasonably called areas of *ethical* concern. Questions concerning whether an engineer should or should not advertise or engage in competitive bidding, for example, do not inherently raise moral issues, though they are certainly concerns which the engineering profession has an interest in overseeing. Whether it is wise to include such nonmoral concerns in a code of "ethics" is an engaging, but separate, matter. In any case, since our interest here is in understanding the reasons why engineers should adopt the *moral* principles embodied in these codes, it is important that we distinguish moral constraints from those nonmoral injunctions the codes contain. Thus the justificatory arguments which we shall discuss are aimed only at supporting those sections of the profession's codes that deal with matters of ethics.

This objective is complicated by the fact that in contrast to most major professions, engineering is organized into over one hundred and fifty separate societies and professional associations, and nearly all these organizations have enacted separate codes of ethics for the members. Hence, there is no single code of ethics that is "the profession's" code. But although the existence of a number of distinct codes can be a source of some important difficulties for a profession, these codes are fortunately similar in the respect that they all incorporate a set of basic ethical principles that are easily identifiable (4). They include the following negative and positive duties. Engineers are obliged to use their skills and knowledge in ways that do not injure, harm, or endanger the welfare and safety of the public, their clients, or employers. They are obliged, in addition, to be honest, fair, and loyal in all their dealings. And finally, they are expected to practice their profession with the dignity and honor befitting their special status as professionals, respecting the moral integrity of all those with whom they deal. In short, engineers are morally required to do no harm and to pursue a set of positive moral values in the routine conduct of their professional practice. These summarize the moral duties that engineering codes contain.

According to the deontological approach to ethics, the moral principles that are contained in engineering codes are justified only if they conform with our *duty* as moral agents. Kant, for example, argued that obligation or duty is the basic moral concept, and it is in terms of this concept that the moral rightness or wrongness of an action is defined. An action is "right" provided that it conforms with our moral duty and "wrong" when it is contrary to duty. Implicit in each is the presumption that we should do what is right and avoid doing wrong. And this is just another way of expressing what has come to be known as taking "the moral point of view." Since our concern is with justifying the ethical principles in engineering codes, can it be shown that the principles mentioned in the preceding section *should* be followed because they direct us to do what is morally right? That is, do they express what are our duties as prescribed by a rationally conceived morality? If we can show this, we would then have a sufficient reason to justify adopting these codes.

To begin with, consider the negative duty to do no harm in the application of engineering skills. Although this duty is normally stated in positive terms, as a duty to protect the public's safety, health, and welfare, it clearly intends that engineers should avoid using their skills in ways that may harmfully jeopardize the public's safety, health, or welfare. Obviously, we all believe and feel that it is wrong to cause harm to another, but what reasons ground this belief?

An important feature of Kant's moral theory is that the rightness or wrongness of an act is an intrinsic property of the act itself, and hence its moral quality is independent of its effectiveness to bring about desired ends or the special interests of those who perform it. Moreover, morality, as conceived by Kant, is rational and universal. That is, any action which is right or wrong at all, would be right or wrong for any rational individual, regardless of the circumstances. Hence if it is wrong for engineers to cause harm to another, then it is wrong for anyone to cause harm to another. And the wrongness of such an action rests precisely in the fact that we cannot consistently will that everyone can freely do harm to another and at the same time continue to pursue causing such harm ourselves. In other words, a morally acceptable principle of conduct is one that is characterized by the logical rule that if anyone accepts it as a principle justifying their action, then rational consistency demands that it must be acceptable that *everyone* else make it their principle of conduct, too. This rule, which Kant calls the *categorical imperative*, is put forth as the basic criterion for judging the morality of reasons or principles. Thus it requires that any reason or principle offered to justify an action must be capable of being "universalized"; that is, it must be capable of being willed as a universal law of morality. And if it fails the test, then morality as rationally conceived imposes upon us a duty to avoid such an action, because it is morally wrong. Clearly we cannot will that it is acceptable that everyone is free to cause harm to another, without inconsistency, since no one can rationally accept as a moral principle that others are morally justified in harming them. So, it is everyone's moral duty to do no harm, and by implication this is a duty which engineers are required by ethics to accept, as well.

A similar argument can be constructed for each of the positive duties that engineering codes of ethics require of engineers. Thus, for example, engineers are enjoined to be honest in their relations with clients, colleagues, employers, and with the public at large. Honesty simply means being truthful in word and deed, or avoiding falsehood and deceit when we speak and act. Does the duty to be honest satisfy the categorical imperative; that is, can we will it as a universal principle of morality that everyone be honest, without inconsistency? To see that we can, it is helpful to examine its denial, for if allowing everyone to be *dis*honest violates the demands of rational consistency, then dishonesty is wrong and, conversely, honesty must be morally right, hence our duty. The categorical imperative implies that *if* I am justified in being deceitful in my relations with others, then everyone else is also justified to act on this ground. But such a universal principle would make communication impossible, either by word or deed, for no one could know whether what is said or done is true or not. And if there can be no communication, then we cannot in principle be deceitful, since deception presupposes the possibility of successful communication in order to convey to another what is untrue. But, of course, the possibility of communication is exactly what is denied by implication when dishonesty is universalized. Thus dishonesty is morally wrong, and it is, therefore, our duty to avoid being dishonest. In brief, morality requires each of us to be honest in our relations with others.

And if this is what is required of every rational individual, then it is likewise required of professional engineers. It is, therefore, appropriate that engineering codes impose a moral obligation of honesty on engineers and reasonable that engineers adopt such a code. With some effort we can similarly establish the moral validity of those other positive duties to be loyal and fair which the profession's codes contain.

It should be clear by now that one of the distinctive marks of this deontological, or nonconsequentialist, approach to the justification of codes of conduct is its logical formalism. In other words, an important consideration which this approach emphasizes is that acceptable moral principles must meet the test of rational consistency. However, it is not just consistency with one's self, but of necessity it must encompass the entire community of freely choosing moral agents if it is to be a code which is rationally acceptable and morally valid.

But the injunction that engineers should "uphold and advance the integrity, honor and dignity of the engineering profession" (5) appears to be merely a provincial consideration and not an injunction that we can reasonably expect *every* member of the moral community to observe. Yet it would be strange to deny that engineers have these kinds of moral obligations to their profession. Can we justify these duties in a manner similar to the arguments offered above?

Not strictly, although a nonconsequentialist justification of "institutional objectives" can be constructed in terms of the role-defined duties required of individuals who are members of institutions, such as the engineering profession, provided that the institution itself can be shown to contribute to the well-being of society (6). So, for example, although the value of the engineering profession as an institution can be established by reference to the value of the goods it produces for society, the special duties that engineers have to their profession are justified *not* by their functional utility but arise as a logical consequence of the fact that these individuals are members of a socially important institution that imposes on its members obligations critical to fulfilling its purpose or function. Like most institutions, the engineering profession is constituted by rules that define the activities which its members are required to perform in order to satisfactorily accomplish its social function or purpose. By accepting membership into a profession or any other institution, one voluntarily takes on a set of role-defined duties above and beyond one's general moral obligations, which nonmembers are not expected to satisfy. So, engineers' duties to uphold the honor and dignity of their profession may be viewed as duties that are logically required by the very concept of a "professional engineer." Thus it is in virtue of one's role as a professional that we find a morally acceptable reason for the professional duties engineering ethical codes impose on their members.

Except for the argument just given, the arguments offered above are formally representations of the Golden Rule which requires that "we do unto others as we would have them do unto us." Although this is clearly a persuasive reason for incorporating the profession's ethical codes into one's practice, we can as well appeal to the good consequences that flow from these principles of conduct in order to further substantiate accepting their injunctions.

And what could be more plausible than that our moral duty requires of us to promote the greatest balance of good over evil? That is, the rightness of an action is justified by reference to the good consequences it produces, and an action that on the whole produces bad consequences is, consequently, morally wrong. Whereas according to the deontologist an action is right or wrong insofar as it conforms with our moral duty, according to this consequentialist approach, the rightness or wrongness of actions is a function of the goodness or badness of its consequences. Mill has, for example, defended the following principle as a way of determining the moral rightness of actions: "actions are right in proportion as they tend to promote happiness, wrong as they tend to produce the reverse of happiness" (7). Although Mill uses the term "happiness" to indicate pleasure and the absence of pain, it is, perhaps, more appropriate to follow Aristotle's notion of happiness as the well-being or good of the individual, or society in general. Thus, according to this "principle of utility," the duty of every moral agent is the promotion of the greatest good for the greatest number. At issue, then, is whether the moral injunctions contained in the engineering profession's codes of ethics contribute to this end.

Although it is admittedly difficult to calculate all the various consequences that result from engineering activities, it is reasonable to assert that if these activities are characterized by honesty, fairness, and loyalty, far greater good would result than if they were not. Indeed, not only does good result from being honest, etc., but much evil is avoided, as well. Furthermore, there is no doubt that pursuing the practice of engineering so that one avoids causing injury, harm, or death to others is morally justified by the fact that such harm is an inherent evil, while intentionally avoiding it is inherently productive of much good. To put the matter more generally, engineers should follow the morally relevant injunctions contained in their profession's codes of ethics because the effect of such a conduct is a substantial increase in the general well-being of society and it avoids countless bad consequences that would naturally result if such principled conduct were not followed or consistently rejected. Thus the philosophical basis of engineering codes of ethics is found not only in their consistency with the duties morality requires of us all but also in the goodness of the consequences that result from restricting engineering practice to the limits they set down.

Although the philosophical ground of engineer's codes is rather straightforward, this should not mislead us to conclude that so long as we strictly abide by its provisions our actions will be rationally or morally justified. Nor should we conclude that we will be able to know what we should and should not do in specific cases. Engineers may have moral obligations to protect the public's welfare, to be honest and loyal, etc., but these are only *prima facie* duties. That is, they describe engineers' actual duties if and only if all other things are equal such that there are no special qualifying, morally relevant circumstances that would justify overriding them, in a particular case. To see what specifically is meant here, consider the following example:

> A salaried production engineer discovers in the routine performance of his duties that an inferior grade of material is being used in the manufacture of a product he supervises. Engineering experience tells him that the safety of the product will be seriously compromised unless it is produced with the quality of material identified in its specifications. However, after bringing this to the attention of management, he learns that it is precisely the economic benefit that derives from using this lower-grade material that allows his employers to remain competitive in the marketplace. In short, it is highly unlikely that they will change the grade of the material to meet the product's specifications, because to do so would have very serious negative economic effects on the manufacturer's profitability. Management appears, moreover, willing to accept the risks of product liability suits because it is more cost efficient than using the required materials.

What should the engineer in this case do?

If we look to the profession's codes of ethics, the answer may *seem* evident: the engineer has a duty to protect the public's health and welfare. But, in addition, he is obliged to be loyal to colleagues and employer; to practice his profession with dignity and integrity; and to be competent in the performance of his duties. Clearly, in this case these duties come into direct conflict with each other such that the engineer is faced with a moral dilemma involving conflicting obligations to the public, his employer, and his profession, which the profession's codes only intensify, but do not answer. Since the engineer cannot simultaneously satisfy all the demands that his profession's codes make on him in this situation, the interesting question it poses is how should this dilemma be resolved so that we know what our actual duty is in cases like this.

On this question, however, the codes are silent. And that is as it should be, because insofar as they are written in general terms to provide a common guide for professional practice, professional codes cannot be expected to function as an algorithm that gives specific answers to every practical moral problem an engineer might confront. But the lesson this suggests is that not only are there good reasons for engineers to accept the moral requirements dictated by their profession's codes of ethics, but it is also important that engineers understand the nature of the philosophical ground justifying these codes if they are to know how to deal effectively with the kinds of ethical problems posed by this and similar cases. Without this understanding, engineers are likely to continue to perceive their professional codes as just so many more rules that limit and infringe on their freedom, presenting ethical problems that have, unfortunately, been easier to ignore than to confront.

Notes

1. For a more detailed discussion of this distinction, see G. E. M. Anscombe, *Intention*, 2nd ed. (Ithaca, N.Y.: Cornell University Press, 1963); and A. I. Melden, *Free Action* (London: Humanities Press, 1961); for an interesting critical perspective, see Donald Davidson, "Actions, Reasons, and Causes," *The Journal of Philosophy*, Vol. 60 (1963).
2. By way of an aside, Mill also argues here, that all men have an obligation, once they have been educated, to ask for the *reasons* of an action, and if no reasons can be given to justify it, then they are obliged to forsake the action.
3. Immanuel Kant, *Foundations of the Metaphysics of Morals*, reprinted in *Ethical Theories*, 2nd ed., edited by A. I. Melden (Englewood Cliffs, N.J.: Prentice-Hall, 1955).
4. The problems of multiple codes and the identification of core ethical concepts is discussed by Andrew G. Oldenquist and Edward E. Slowter in "Proposed: A Single Code of Ethics for All Engineers," reprinted in *Ethical Problems in Engineering, 2nd edition, Vol. 1 Readings*, edited by Albert Flores (Troy, N.Y.: Rensselaer Polytechnic Institute, 1980), pp. 44–47.
5. Engineers' Council for Professional Development, *Code of Ethics of Engineers* (1974), fundamental principles (New York: Engineers' Council for Professional Development).
6. For a further helpful discussion of how duties arise out of institutions, see Alan Gerwith, "Obligation: Political, Legal, Moral," in *Political and Legal Obligation*: NOMOS 12, edited by J. Poland Pennock and John W. Chapman (New York: Atherton Press, 1970).
7. John Stuart Mill, *Utilitarianism* (New York: Bobbs-Merrill, 1957). Philosophers have traditionally distinguished between applying this principle to justify particular actions (act utilitarianism) and justifying general moral principles and rules by means of it (rule utilitarianism). However, unlike deontological views that stress the importance of rules in the determination of our duties, the latter view stresses the utility of adopting such rules in promoting the well-being of society.

Selected Additional Readings

R. M. Anderson et al., *Divided Loyalties*, Purdue University, Lafayette, Ind., 1980.

A. Ashkinazy, "Are Engineers Responsible for the Uses and Effects of Technology?" *Professional Engineer*, August 1972, pp. 46–47.

R. J. Baum, "The Limits of Professional Responsibility," *Values and the Public Works Professional*, University of Missouri Press, Columbia, Mo., 1980.

R. J. Baum, "A Philosophical/Historical Perspective on Contemporary Concerns and Trends in the Area of Science and Values," *Science, Technology, & Human Values*, Harvard University, Cambridge, Mass., 1974.

D. L. Bazelon, "Risk and Democracy," *Science*, July 20, 1979, p. 277–280.

L. B. Cebic, "On the Nature of Ethical Codes: An Outline of a Cookbook," a paper presented at the National Conference on Engineering Ethics, Rensselear Polytechnic Institute, Troy, N.Y., June 20–22, 1980.

D. W. Ewing, *Freedom Inside the Organization*, E. P. Dutton, New York, 1977.

F. Ferré, "Beyond Technology," an address delivered to the Departments of Nuclear and Chemical Engineering, Purdue University, Lafayette, Ind., February 25, 1975.

S. C. Florman, *The Existential Pleasures of Engineering*, St. Martin's Press, New York, 1976.

G. Hardin, "The Tragedy of the Commons," *Science*, December 13, 1968, pp. 1243–1248.

H. Jonas, "Toward a Philosophy of Technology," *Hastings Center Report*, The Hastings Center, Hastings-on-Hudson, N.Y., February 1979, pp. 34–43.

J. McCabe, "Some Ethical, Moral and Legal Factors of Professional Conduct," *Ethics, Professionalism and Maintaining Competence*, Speciality Conference of the American Society of Civil Engineers, March 1977, American Society of Civil Engineers, New York, pp. 38–50.

M. I. Mantell, *Ethics and Professionalism in Engineering*, Macmillan Publishing Co., New York, 1964.

C. Mitchum and R. Mackey (editors), *Philosophy and Technology*, The Free Press, A Division of Macmillan Publishing Co., New York, 1972.

C. Nelson and S. Peterson, "Ethical Decisions for Engineers: Systematic Avoidance and the Need for Confrontation," *Civil Engineering Education*, Vol. 1, American Society of Civil Engineers, New York, pp. 628–634.

J. H. Schaub and K. Pavlovic (editors), *Engineering Professionalism and Ethics*, John Wiley & Sons, New York, 1982.

M. W. Thring, "Our Responsibility to Mankind," *The Chartered Mechanical Engineer*, September 1970, pp. 348–351, 359.

PART 5

History and Engineering

What has been is what will be, and what has been done is what will be done; and there is nothing new under the sun.

Ecclesiastes 1:9

History is philosophy teaching by examples.

Dionysius of Halicarnassus, Roman Antiquities

The history of humanity is the history of technology. From the very beginning humans have been distinguished from other animals by the ability to reason and to apply that reasoning capability to produce tools, devices, structures, and so forth, for the unique purpose of advancing human development. This is recognized by the description of humankind as both homo sapiens (man the thinker) and as homo faber (man the tool maker). These two descriptive classifications are indistinguishable from each other and reflect the role of technology and engineering in human development, for they define the essential characteristics of engineering: the use of ingenuity to make or to build.

Though there are intrinsic differences in definitions between engineering and technology, these two activities of humanity are similar, and moderns tend to use these words interchangeably. They will be used in this part in the sense of interchangeability, for engineering is dependent on the technical skills we call technology; and modern technology has drawn much of its usefulness from engineering development. Now perhaps it is desirable to distinguish between science and engineering also, for these two fields are so closely associated. There is no clear-cut dividing line between them, but generally science seeks knowledge of the universe, while technology (engineering) strives to develop and apply that of it which is useful. Theodore von Karman (1) is quoted as distinguishing between the two with a couplet:

The scientist explores what is;
The engineer creates what has not been.

There is no facet of life untouched in some way by the contributions of engineering. Technology has been described as the most powerful factor in the development of civilization (2). Whether one accepts this broad statement in its totality or not, undoubtedly technology is a distinctive and significant part of the development of modern culture. In this context, we use "culture" in the definition advanced by the anthropologist E. B. Tyler (3):

> Culture . . . is the complex whole which includes knowledge, belief, art, morals, law, custom, and any other capabilities and habits acquired by man as a member of society.

You might well ask: Why should an engineer study history, or even the history of technology or engineering? Many authors have tried to answer this question, with precious little agreement. Educated engineers must understand the complex interplay of their art with the numerous social, economic, and political factors that influence all technical projects. This is so they may respond better to society's charge and discharge their own professional responsibilities. A study of history offers us insights to this need, for certainly we should know where we have been to plan where we wish to go. History may not repeat itself as often claimed, but a knowledge of the past will provide guidance for the present and an insight to the future. Beyond this, E. J. Cleary (4) offers four very pragmatic reasons why engineers should study the history of technology: (a) history provides a memory bank; (b) provides an essential ingredient to facilitate problem solving; (c) offers basic clues for the conduct of future planning; and (d) provides the background to nurture an awareness of professional traditions and responsibilities.

This part will introduce a student to the history of engineering through a selection of articles and extracts written by prominent authors, historians, and engineers. The articles selected do not cover the full range of topics that might have been included nor do they relate to the variety of activities included in the broad scope of engineering. Rather our objective is to illustrate how technology has contributed to human development and to suggest the significance of engineering in Western culture. The selected additional readings listed at the end of the part will give some guidance to a few further items of interest in the fascinating story of technology and its impact on the history of humanity.

The opening article in this series is appropriate, both chronologically and to establish the environment for those that follow. Peter Drucker is a prominent author, economist, and management consultant who has worked closely with engineers throughout his career. His article "The First Technological Revolution and Its Lessons" discusses the interaction of technology and the social development of humanity in the irrigation states of the early civilizations of Mesopotamia and the surrounding areas. Irrigation was one of the first, if not *the* first, of humanity's technical achievements in the transition from life as a nomad to a stable life. The magnitude of the irrigation venture established the basis for cooperative action among individuals. It also fostered the resulting social concepts of the individual as a citizen, the specialization of labor, organized government, and organized knowledge. This interpretation of the influence of an engineering accomplishment of a magnitude unsurpassed at that time illustrates that technology and cultural development have been closely related since this early time in recorded history. Though Drucker's thesis is not accepted by all, it is supported by Arnold Toynbee in his last book, *Mankind and Mother Earth* (5):

> The mastering of the jungle-swamp was a social, far more than a technological, achievement. Of course, all of mankind's technological achievements have been social achievements, too.

The growth and accomplishments of the Roman Empire have been studied by historians for many generations. The decline of that empire has been studied with equal intensity though there is far less agreement on the causes of the so-called Fall of the Roman Empire. Gilfillan, a historian of antiquity, has proposed an interesting and novel explanation in his essay "Lead Poisoning and the Fall of Rome." Romans mined, smelted, and cast lead in large quantities for water pipes, cooking urns, and a variety of other uses generally available only to the rich and ruling classes. The physiological effects of lead on the human body are well known today and, even with modern detailed medical knowledge, remain cause of widespread concern. Gilfillan suggests that the extensive use of lead pipe, cooking utensils, and a popular sweetener made only in lead pots literally poisoned the elite and ruling class of the empire. The result was a decline, over several generations, of the class of dedicated and trained administrators that had led Rome to her preeminent position in the known world. Though the Gilfillan hypothesis is attractive in its simplicity and in the extensive documentation of lead poisoning among the elite of the Roman Empire, it is based upon a generally discredited concept that the poor and slave classes could not produce creative and productive leaders from within their ranks. The paper, however, presents an interesting example of an early historical interplay between technology (mining, metallurgy, water supply, food preservation, etc.) and public health with the resulting impact on society as a whole.

To modern society, technology is a coin with two sides; one side offers the benefits that have improved living conditions through better housing, improved clothing, faster transportation and communications, greater medical knowledge and care, and the overall improvement in the quality of life. The reverse side of the coin is the negative impact that technology has had on the environment, the extreme horrors of war in our time, the danger today's achievements may present to tomorrow's generations, and many other harmful or at least questionable impacts on the continued improvement of the human condition in the world in which we live. Tools to attempt to predict the results of new technology are being sought in many directions. The question is continually posed—Just because we *can* build it, should we? One approach to this problem is the use of technological assessment.

Lynn White, Jr., one of the most respected historians of our time, recognizes the need and potential value of technological assessment but asks in his article "Technological Assessment from the Stance of a Medieval Historian" if "we know how to assess the impact of a proposed technological change, whether it be a new invention or a new canal across Central America?" To illustrate his concern, White has chosen to consider in retrospect the assessment of several technological changes of the medieval era. The result is a fascinating insight into the technology of Medieval Europe and the effects (those intended and those unexpected) that resulted from a number of technical advances. In each of the several cases discussed, technology and society are obviously interdependent and inseparable.

Railroads and bridges made a major contribution to the Age of Industry that had its beginnings about 1750. The rapid growth of technological changes, engineering developments, and the beginnings of a working relationship between science and engineering led to widespread expansion of the populated areas of the Western world and the resulting need for better communications. The railroad was the first of a series of major transportation developments that contributed to the need and helped to satisfy it. With the growth of the railroad system, bridge engineering became more essential to meet the demands of heavier vehicles moving faster and creating previously unknown stresses and vibrations. Fortunately, new materials and new techniques were developed parallel with the railroad and the need for the bridges it generated. Bridges for uses other than railroads were equally in demand to span longer and longer distances. The development of railroads and bridges provided one of the most thrilling segments in the history of engineering. Three articles in this part are devoted to a portion of that story.

T. C. Cochran's essay "The Social Impact of the Railroad" was drawn from *The Railroad and the Space Program* which resulted from an American Academy of Arts and Science study of the impact of the space program on American society. The study was based on a historical analogy approach using the impact of the railroad in the nineteenth century as analogous to what might be expected from the technological changes introduced by the space program in the mid-twentieth century. The social effect of the railroad was far reaching on the course of American history. A study of this impact illustrates once again that engineering achievements are social achievements and that these accomplishments have both positive and negative aspects. The lessons presented by Cochran can be extended readily to other engineering developments: past, present, and future.

James B. Eads and John A. Roebling were two outstanding engineers of the last century. Both achieved what their peers called the impossible. They led the way in establishing new engineering concepts for design, analysis, and construction. Both Eads and Roebling had the creative vision to foresee that the problems for which they were seeking solutions were not only engineering problems but contained elements that would influence the total social, political, and economic growth of the nation. Their works reflect this awareness.

As these engineers were outstanding in the nineteenth century, so were their major works recognized among the leading engineering accomplishments of that period. The Great St. Louis Bridge was built in 1874 by Eads. The Brooklyn Bridge, opened in 1882, was conceived and designed by John Roebling but completed by his son, George Washington Roebling, with great contributions from George's wife, Emily, following her husband's crippling accident. Both bridges served far more than merely spanning a stream. The Eads Bridge was the first successful bridge across the Mississippi River. It provided both railroad and carriage connections. Literally it served to tie the nation together and demonstrate that previously unconquerable natural obstacles could be overcome. The Brooklyn Bridge, the longest suspension bridge of its time, provided the vehicular connection between the first (New York) and the third (Brooklyn) largest cities in this nation at the time of construction. The impact of this juncture played an incalculable role in the growth of what is now Greater New York City.

Both bridges offered many examples of innovation and led to new concepts in engineering analysis, design, and construction. Both have performed their functions well, and continue to do so today. Both are world recognized as engineering achievements of significance. The story of the concept, construction, and impact of each reads like a novel of triumph over adversity. "James B. Eads and the Great St. Louis Bridge" was prepared for the Eads Bridge Centennial in 1974 by The Engineers' Club of St. Louis under the authorship of Joseph E. Vollmar, Jr. It provides an excellent summary of the engineering accomplishment called "impractical and impossible" by many and its significance. It also provides an insight into the amazing character of James B. Eads. "The Plan" is a chapter from David McCullough's widely acclaimed book *The Great Bridge*. This selection describes the many obstacles Roebling had to overcome before his idea for the world's longest suspension bridge was accepted. It illustrates very well the factors (technological and other) that influence the creation of a "great bridge."

Irrigation was perhaps humanity's first technological accomplishment (see the Drucker article, section 5.1). It has remained a major concern as population pressures caused farmers to turn more and more frequently to the more difficult areas of the world for farmland. In the United States, reclamation and irrigation have a long history, predating the arrival of the European explorers. The magnitude of needed irrigation projects (as immigration and settlement brought water problems to the fore in the latter part of the last century in the Southwest) necessitated government participation and control. The selection "Reclamation Origins" is extracted from *Water for the West—The Bureau of Reclamation, 1902–1977*, by the historian Michael C. Robinson, published by the Public Works Historical Society. As the title suggests, this article summarizes the reclamation activities that preceded the establishment of the Reclamation Service in 1902. Once again, the interplay among forces of nature; the needs of people; and the social, political, and economic pressures of an era are demonstrated to be an integral part of the history of engineering and the history of humanity.

Engineering has contributed in many ways to the benefit of society as a whole. The examples we provided in the first section of this part are concerned with some of these contributions, but engineering has made as many significant contributions to the lives of individuals. The last two selections illustrate these effects. "Power by Persuasion," by Kingery, Berg, and Schillinger, describes the extension of electrical power to the rural areas of this country. This was spearheaded by the efforts of Emil Lehmann, an agricultural engineer at the University of Illinois during the 1920s. Lehmann was convinced that farm life and agricultural production would improve with the farmer's access to electricity as an energy source. His efforts in convincing the farm community and the power companies to make electricity available on the farm led the way to a modern agricultural revolution based upon improvements in farm machinery, crop production, farm buildings, and home life. The work he accomplished in Illinois started a movement toward rural electrification that culminated in the Rural Electrification Adminstration and the extension of the many benefits of electrical power to the nonurban areas in America.

The growth of home-directed technology was of equal value. Much of it was based on electrical applicances. Ruth Schwartz Cowan, a modern social historian, traces this story as part of her article "Women and Technology in American Life." This is the final selection of this part. Cowan's article goes much further, though. It provides an excellent analysis of the role of technology in the changing life of women as workers, citizens, and homemakers.

All selections offered in this part can provide, at best, a mere introduction to the relationship of technology-engineering to cultural development throughout recorded history. There are many other aspects of this interaction that we could discuss. For example: Consider the impact of the telephone or the automobile upon the modern way of life or the role sanitation and water supply have played in urban development. Certainly, the maturation of metallurgy has influenced the path of history in as many ways as the steam engine, the machine-tool industry, and chemical processing, among others. This part then is a sampling of the powerful factor technology-engineering has been (for better or worse) in the civilization and culture we have inherited, and are continuing to evolve.

References

1. von Karman, Theodore, *The Function of the Engineer & the Scientist*, National Society of Professional Engineers, Washington, D.C., November 1962, cover.
2. Hamilton, S. B., "Why Engineers Should Study History," *Transactions*, Newcomen Society, Vol. 25, 1945–1946, 1946–1947, pp. 1–10.
3. E. B. Tyler, quoted by Woodruff, W., in "The Impact of Technology on Western Culture," a lecture given at the Science Museum of Victoria, Melbourne, Australia, December 11, 1975.
4. Cleary, E. J., "Remarks at the Third Annual Business Meeting," Public Works Historical Society, September 11, 1977, Chicago, Illinois.
5. Toynbee, A., *Mankind and Mother Earth*, Oxford University Press, New York, 1976, p. 51.

5.1

The First Technological Revolution and Its Lessons

PETER F. DRUCKER

Aware that we are living in the midst of a technological revolution, we are becoming increasingly concerned with its meaning for the individual and its impact on freedom, on society, and on our political institutions. Side by side with messianic promises of utopia to be ushered in by technology, there are the most dire warnings of man's enslavement by technology, his alienation from himself and from society, and the destruction of all human and political values.

Tremendous though today's technological explosion is, it is hardly greater than the first great revolution technology wrought in human life seven thousand years ago when the first great civilization of man, the irrigation civilization, established itself. First in Mesopotamia, and then in Egypt and in the Indus Valley, and finally in China there appeared a new society and a new polity: the irrigation city, which then rapidly became the irrigation empire. No other change in man's way of life and in his making a living, not even the changes under way today, so completely revolutionalized human society and community. In fact, the irrigation civilizations were the beginning of history, if only because they brought writing.

The age of the irrigation civilization was pre-eminently an age of technological innovation. Not until a historical yesterday, the eighteenth century, did technological innovations emerge which were comparable in their scope and impact to those early changes in technology, tools, and processes. Indeed, the technology of man remained essentially unchanged until the eighteenth century insofar as its impact on human life and human society is concerned.

But the irrigation civilizations were not only one of the great ages of technology. They represent also mankind's greatest and most productive age of social and

---

This article was Dr. Drucker's presidential address to the Society for the History of Technology, presented on December 29, 1965, in San Francisco.

political innovation. The historian of ideas is prone to go back to ancient Greece, to the Old Testament prophets, or to the China of the early dynasties for the sources of the beliefs that still move men to action. But our fundamental social and political institutions antedate political philosophy by several thousand years. They all were conceived and established in the early dawn of the irrigation civilizations. Any one interested in social and governmental institutions and in social and political processes will increasingly have to go back to those early irrigation cities. And, thanks to the work of archeologists and linguists during the last fifty years, we increasingly have the information, we increasingly know what the irrigation civilizations looked like, we increasingly can go back to them for our understanding both of antiquity and of modern society. For essentially our present-day social and political institutions, practically without exception, were then created and established. Here are a few examples.

1. The irrigation city first established government as a distinct and permanent institution. It established an impersonal government with a clear hierarchical structure in which very soon there arose a genuine bureaucracy—which is of course what enabled the irrigation cities to become irrigation empires.

Even more basic: the irrigation city first conceived of man as a citizen. It had to go beyond the narrow bounds of tribe and clan and had to weld people of very different origins and blood into one community. This required the first super-tribal deity, the god of the city. It also required the first clear distinction between custom and law and the development of an impersonal, abstract, codified legal system. Indeed, practically all legal concepts, whether of criminal or of civil law, go back to the irrigation city. The first great code of law, that of Hammurabi, almost four thousand years ago, would still be applicable to a good deal of legal business in today's highly developed, industrial society.

The irrigation city also first developed a standing army—it had to. For the farmer was defenseless and vulnerable and, above all, immobile. The irrigation city which, thanks to its technology, produced a surplus, for the first time in human affairs, was a most attractive target for the barbarian outside the gates, the tribal nomads of steppe and desert. And with the army came specific fighting technology and fighting equipment: the war horse and the chariot, the lance and the shield, armor and the catapult.

2. It was in the irrigation city that social classes first developed. It needed people permanently engaged in producing the farm products on which all the city lived; it needed farmers. It needed soldiers to defend them. And it needed a governing class with knowledge, that is, originally a priestly class. Down to the end of the nineteenth century these three "estates" were still considered basic in society.[1]

But at the same time the irrigation city went in for specialization of labor resulting in the emergence of artisans and craftsmen: potters, weavers, metal workers, and so on; and of professional people: scribes, lawyers, judges, physicians.

---

[1]See the brilliant though one-sided book by Karl A. Wittvogel, *Oriental Despotism: A Comparative Study of Total Power*, (New Haven, Conn., 1957).

And because it produced a surplus it first engaged in organized trade which brought with it not only the merchant but money, credit, and a law that extended beyond the city to give protection, predictability, and justice to the stranger, the trader from far away. This, by the way, also made necessary international relations and international law. In fact, there is not very much difference between a nineteenth-century trade treaty and the trade treaties of the irrigation empires of antiquity.

3. The irrigation city first had knowledge, organized it, and institutionalized it. Both because it required considerable knowledge to construct and maintain the complex engineering works that regulated the vital water supply and because it had to manage complex economic transactions stretching over many years and over hundreds of miles, the irrigation city needed records, and this of course meant writing. It needed astronomical data, as it depended on a calendar. It needed means of navigating across sea or desert. It therefore had to organize both the supply of the needed information and its processing into learnable and teachable knowledge. As a result, the irrigation city developed the first schools and the first teachers. It developed the first systematic observation of natural phenomena, indeed, the first approach to nature as something outside of and different from man and governed by its own rational and independent laws.

4. Finally, the irrigation city created the individual. Outside the city, as we can still see from those tribal communities that have survived to our days, only the tribe had existence. The individual as such was neither seen nor paid attention to. In the irrigation city of antiquity, however, the individual became, of necessity, the focal point. And with this emerged not only compassion and the concept of justice; with it emerged the arts as we know them, the poets, and eventually the world religions and the philosophers.

This is, of course, not even the barest sketch. All I wanted to suggest is the scope and magnitude of social and political innovation that underlay the rise of the irrigation civilizations. All I wanted to stress is that the irrigation city was essentially "modern," as we have understood the term, and that, until today, history largely consisted in building on the foundations laid five thousand or more years ago. In fact, one can argue that human history, in the last five thousand years, has largely been an extension of the social and political institutions of the irrigation city to larger and larger areas, that is, to all areas on the globe where water supply is adequate for the systematic tilling of the soil. In its beginnings, the irrigation city was the oasis in a tribal, nomadic world. By 1900 it was the tribal, nomadic world that had become the exception.

The irrigation civilization was based squarely upon a technological revolution. It can with justice be called a "technological polity." All its institutions were responses to opportunities and challenges that new technology offered. All its institutions were essentially aimed at making the new technology most productive.

I hope you will allow me one diversion.

The history of the irrigation civilizations has yet to be written. There is a tremendous amount of material available now, where fifty years ago we had, at best, fragments. There are splendid discussions available of this or that irrigation civilization, for instance of Sumer. But the very big job of recreating this great achievement of man and of telling the story of his first great civilization is yet ahead of us.

This should be pre-eminently a job for historians of technology such as we profess to be. At the very least the job calls for a historian with high interest in, and genuine understanding of, technology. The essential theme around which this history will have to be written must be the impacts and capacities of the new technology and the opportunities and challenges which this, the first great technological revolution, presented. The social, political, cultural institutions, familiar though they are to us today—for they are in large measure the institutions we have been living with for five thousand years—were all brand-new then, and were all the outgrowth of new technology and of attempts to solve the problems the new technology posed.

It is our contention in the Society for the History of Technology that the history of technology is a major, distinct strand in the web of human history. We believe that the history of mankind cannot be properly understood without relating to it the history of man's work and man's tools, that is, the history of technology. Some of our colleagues and friends—let me mention only such familiar names as Lewis Mumford, Fairfield Osborne, Joseph Needham, R. J. Forbes, Cyril Stanley Smith, and Lynn White—have in their own words brilliantly demonstrated the profound impact of technology on political, social, economic, and cultural history. But while technological change has always had impact on the way men live and work, surely at no other time has technology so literally influenced civilization and culture as during the first technological revolution, that is, during the rise of the irrigation civilizations of antiquity.

Only now, however, is it possible to tell the story. No longer can its neglect be justified. For the facts are available, as I stated before. And we now, because we live in a technological revolution ourselves, are capable of understanding what happened then—at the very dawn of history. There is a big job to be done: to show that the traditional approach to our history—the approach taught in our schools—in which "relevant" history really begins with the Greeks (or with the Chinese dynasties), is short-sighted and distorts the real "ancient civilization."

I have, however, strayed off my topic: the question I posed at the beginning, what we can learn from the first technological revolution regarding the impacts likely to result on man, his society, and his government from the new industrial revolution, the one we are living in. Does the story of the irrigation civilization show man to be determined by his technical achievements, in thrall to them, coerced by them? Or does it show him capable of using technology to his own, to human ends, and of being the master of the tools of his own devising?

The answer which the irrigation civilizations give us to this question is threefold.

1. Without a shadow of doubt, major technological change creates the need for social and political innovation. It does make obsolete existing institutional arrangements. It does require new and very different institutions of community, society, and government. To this extent there can be no doubt: technological change of a revolutionary character coerces; it *demands innovation.*

2. The second answer also implies a strong necessity. There is little doubt, one would conclude from looking at the irrigation civilizations, that specific technological changes demand equally specific social and political innovations. That the basic institutions of the irrigation cities of the Old World, despite great cultural difference, all exhibited striking similarity may not prove much. After all, there probably was a great deal of cultural diffusion (though I refuse to get into the quicksand of debating whether Mesopotamia or China was the original innovator). But the fact that the irrigation civilizations of the New World around the Lake of Mexico and in Maya Yucatan, though culturally completely independent, millennia later evolved institutions which, in fundamentals, closely resemble those of the Old World (e.g., an organized government with social classes and a permanent military, and writing) would argue strongly that the solutions to specific conditions created by new technology have to be specific and are therefore limited in number and scope.

In other words, one lesson to be learned from the first technological revolution is that new technology creates what a philosopher of history might call "objective reality." And objective reality has to be dealt with on *its* terms. Such a reality would, for instance, be the conversion, in the course of the first technological revolution, of human space from "habitat" into "settlement," that is, into a permanent territorial unit always to be found in the same place—unlike the migrating herds of pastoral people or the hunting grounds of primitive tribes. This alone makes obsolete the tribe and demands a permanent, impersonal, and rather powerful government.

3. But the irrigation civilizations can teach us also that the new objective reality determines only the gross parameters of the solutions. It determines where, and in respect to what, new institutions are needed. It does not make anything "inevitable." It leaves wide open *how* the new problems are to be tackled, what the purposes and values of the new institutions are to be.

In the irrigation civilizations of the New World the individual, for instance, failed to make his appearance. Never, as far as we know, did these civilizations get around to separating law from custom nor, despite a highly developed trade, did they invent money, and so on.

Even within the Old World, where one irrigation civilization could learn from the others, there were very great differences. They were far from homogeneous even though all had similar tasks to accomplish and developed similar institutions for these tasks. The different specific answers expressed above all different views regarding man, his position in the universe, and his society—different purposes and greatly differing values.

Impersonal bureaucratic government had to arise in all these civilizations; without it they could not have functioned. But in the Near East it was seen at a very early stage that such a government could serve equally to exploit and hold down the common man and to establish justice for all and protection for the weak. From the beginning the Near East saw an ethical decision as crucial to government. In Egypt, however, this decision was never seen. The question of the purpose of government was never asked. And the central quest of government in China was not justice but harmony.

It was in Egypt that the individual first emerged, as witness the many statues, portraits, and writings of professional men, such as scribes and administrators, that have come down to us—most of them superbly aware of the uniqueness of the individual and clearly asserting his primacy. It is Early Egypt, for instance, which records the names of architects who built the great pyramids. We have no names for the equally great architects of the castles and palaces of Assur or Babylon, let alone for the early architects of China. But Egypt suppressed the individual after a fairly short period during which he flowered (perhaps as part of the reaction against the dangerous heresies of Ikhnaton). There is no individual left in the records of the Middle and New Kingdoms, which perhaps explains their relative sterility.

In the other areas two entirely different basic approaches emerged. One, that of Mesopotamia and of the Taoists, we might call "personalism," the approach that found its greatest expression later in the Hebrew prophets and in the Greek dramatists. Here the stress is on developing to the fullest the capacities of the person. In the other approach—we might call it "rationalism," taught and exemplified above all by Confucius—the aim is the molding and shaping of the individual according to pre-established ideals of rightness and perfection. I need not tell you that both these approaches still permeate our thinking about education.

Or take the military. Organized defense was a necessity for the irrigation civilization. But three different approaches emerged: a separate military class supported through tribute by the producing class, the farmers; the citizen-army drafted from the peasantry itself; and mercenaries. There is very little doubt that from the beginning it was clearly understood that each of these three approaches had very real political consequences. It is hardly coincidence, I believe, that Egypt, originally unified by overthrowing local, petty chieftains, never developed afterwards a professional, permanent military class.

Even the class structure, though it characterizes all irrigation civilizations, showed great differences from culture to culture and within the same culture at different times. It was being used to create permanent castes and complete social immobility, but it was also used with great skill to create a very high degree of social mobility and a substantial measure of opportunities for the gifted and ambitious.

Or take science. We now know that no early civilization excelled over China in the quality and quantity of scientific observations. And yet we also know that early Chinese culture did not point toward anything we would call "science." Perhaps because of their rationalism the Chinese refrained from generalization. And though fanciful and speculative, it is the generalizations of the Near East or the mathematics of Egypt which point the way toward systematic science. The Chinese, with their superb gift for accurate observation, could obtain an enormous amount of information about nature. But their view of the universe remained totally unaffected thereby—in sharp contrast to what we know about the Middle Eastern developments out of which Europe arose.

In brief, the history of man's first technological revolution indicates the following:

1. Technological revolutions create an objective need for social and political

innovations. They create a need also for identifying the areas in which new institutions are needed and old ones are becoming obsolete.

2. The new institutions have to be appropriate to specific new needs. There are right social and political responses to technology and wrong social and political responses. To the extent that only a right institutional response will do, society and government are largely circumscribed by new technology.
3. But the values these institutions attempt to realize, the human and social purposes to which they are applied, and, perhaps most important, the emphasis and stress laid on one purpose as against another, are largely within human control. The bony structure, the hard stuff of a society, is prescribed by the tasks it has to accomplish. But the ethos of the society is in man's hands and is largely a matter of the "how" rather than of the "what."

For the first time in thousands of years, we face again a situation that can be compared with what our remote ancestors faced at the time of the irrigation civilization. It is not only the speed of technological change that creates a "revolution," it is its scope as well. Above all, today, as seven thousand years ago, technological developments from a great many areas are growing together to create a new human environment. This has not been true of any period between the first technological revolution and the technological revolution that got under way two hundred years ago and has still clearly not run its course.

We therefore face a big task of identifying the areas in which social and political innovations are needed. We face a big task in developing the institutions for the new tasks, institutions adequate to the new needs and to the new capacities which technological change is casting up. And, finally, we face the biggest task of them all, the task of insuring that the new institutions embody the values we believe in, aspire to the purposes we consider right, and serve human freedom, human dignity, and human ends.

If an educated man of those days of the first technological revolution—an educated Sumerian perhaps or an educated ancient Chinese—looked at us today, he would certainly be totally stumped by our technology. But he would, I am sure, find our existing social and political institutions reasonably familiar—they are after all, by and large, not fundamentally different from the institutions he and his contemporaries first fashioned. And, I am quite certain, he would have nothing but a wry smile for both those among us who predict a technological heaven and those who predict a technological hell of "alienation," of "technological unemployment," and so on. He might well mutter to himself, "This is where I came in." But to us he might well say, "A time such as was mine and such as is yours, a time of true technological revolution, is not a time for exultation. It is not a time for despair either. It is a time for work and for responsibility."

5.2

Lead Poisoning and the Fall of Rome

S. C. GILFILLAN

Our novel theory concerns what mattered most about the fall of Rome. This was not the empire's military defeat and break-up. Those events might even have been of benefit to mankind. What concerns us here is the decay of Rome, the decline of every aspect of her civilization, except technology; the debasement of her arts; the cessation of novelty and progress; the complete disappearance of science after the second century; the retreat from almost all that had spelled "the glory that was Greece and the grandeur that was Rome." There are many indications that Greece had been smitten in the same way a few centuries earlier, when her genius suddenly declined and her population shrank, and that it was the importation of Greek winemaking and cookery into the Roman mansions that chiefly ruined the Roman aristocracy; but the proof as to Greece is left for better Greek scholars.

We present here a completely new explanation for the Roman decay. It is derived not from the realm of philosophy of history, but from such firm sciences as toxicology, vital statistics, archeology, and evidence such as old bones and old Roman recipes and lead-lined pots for brewing poisons esteemed as delicious by the ancient well-to-do. We are well aware that the fall of Rome has been a favorite theme not only for professional philosophers of history, but for every curbstone amateur, and that each tends to ascribe that downfall to whatever evil he most opposes today. And we know that the fall of Rome was really a net result of many causes, such as the rising civilization of the barbarians. It included salutary as well as destructive forces, so that a true answer must include them all, and not just lead poisoning.

NATURE AND ORIGINS OF LEAD POISONING

Lead poisoning, although still a potent industrial disease, is unfamiliar to most. Continued daily intake of lead or any of its compounds by mouth or lungs in an amount greater than about 1 mg per day may be dangerous--and 1 mg is only 1/28,300 of an avoirdupois ounce.

The varying symptoms of chronic plumbism include at first, most conspicuously, an obstinate and painful constipation and colic that can be relieved by pressure around the naval and by opium. These were symptoms the Romans knew and recorded. Anemia, pallor, emaciation, a metallic taste in the mouth, dark stools, loss of appetite, and pains in the joints are other signs. Further stages include paralysis of the extremities, starting with a peculiar wrist drop, headache, insomnia, blindness and mental disturbances extending to insanity. "Crazy as a painter" was a common phrase not long ago when those craftsmen usually had lead poisoning. A wide range of further baleful effects occur in extreme cases, including death. Those most interesting to us are various interferences with the fertility of men and especially of women,[1] whom sufficient lead brings to sterility, miscarriage, stillbirth, or premature labor; children born to such women are apt to die shortly after birth.

Children also seem to be especially susceptible to plumbism, and the results are usually or perhaps always some degree of permanent physical or mental damage. Premature birth also often produces mental retardation. American city hospitals treat a succession of lead-poisoned children, usually from the slums, where they developed pica, eating flecks of lead paint from old walls. (Lead is banned from modern interior paint.) Also, infants crawl on the floor, then put their dusty hands in their mouths, whence it is likely the high death rate of children of lead workers originates. The well-to-do Romans had lead paints, and the favorite color of their rich walls was the famous Pompeian red, produced by minium, a salt of lead or of mercury—also poisonous.[1]

THE ARGUMENT

Our reasoning involves the chain of facts listed below. Some have already been proved by familiar and conclusive evidence and will be but briefly treated. Some of the points are not well known, but have been proved by Hofmann,[8, 9] Kobert,[12, 13] and other specialist authorities and have been set forth with adequate documentation and argued with sufficient cogency.[5] The major ideas that have been personally added, and that give the argument its import are shown below in italics and are developed as needed.

(1) Roman culture declined and lost progressiveness, except in technology. (2) Rome's cultural progress, except in technology, depended upon the upper class. (3) The Roman upper class, beginning about the second or first century B.C., died out with extreme rapidity; each generation being perhaps a fourth of the previous one, largely from rearing few children. (All possible causes for this extermination are examined.) (4) Lead poisoning of the wives and mistresses of upper-class Romans occurred chiefly through their diet, after the introduction of Greek cookery around 150 B.C. and the relaxation of the old rule against wives drinking wine. The most significant sources of the lead were wine, grape syrup, and preserved fruit. (5) Lead in women produces sterility, miscarriage, stillbirth, heavy child-mortality; it brings permanent mental impairment in children. (6) *This lead poisoning was sufficient to produce the observed decimation of the upper class;*

nothing else was, venereal disease being then unimportant. (7) *The Roman free poor and the slaves had much less lead in their diet.* (8) They did have other checks to their reproduction, especially poverty, and positive interferences with slave reproduction. (9) Despite these they must have maintained their number without much decline, because the Roman population as a whole did. (10) *It follows (from items 3–6) that whatever qualities enabled Roman individuals to make money, or to marry or mate with money, were rigorously bred out of the race and culture by lead and lesser forces.* (11) *Technical invention (aside from architecture, civil engineering, and the art of war) was in ancient times confined to the artisan class, so invention continued, albeit retarded.*

Belief in eugenics is not necessary for acceptance of our thesis, for whatever personal qualities come not from heredity, must come from education and other environment, and the upper-class environment was extinguished along with the aristocratic scions. The son of a slave or peasant might succeed to a vacant chiefship, but he could not have had the education, family traditions, nor prestige of the scion. In short, Roman culture lost its source of progress, when and because its elite, endowed by heredity and/or upbringing with the powers to make or mate with money, was exterminated by various factors, above all by lead in the diet of the upper-class women.

That the Romans must often have suffered lead poisoning was perceived as long ago as 1824[7] from examination of their recipes and their pots and water pipes. In 1883 the Austrian toxicologist K. B. Hofmann, and fellow German Rudolf Kobert (1887–1909), abundantly proved and explained this thesis by hundreds of citations from classical authors on ancient culinary dietary customs and illnesses and by archeological evidence in the form of leaden containers used for food and drink. Kobert's student, Rosenblatt,[16] analyzed ancient bones for lead and found the saturnine metal in bones from the classical period, and not from before and after. In 1919 the Canadian physician L. G. Stevenson wrote an unpublished doctoral dissertation on the history of lead poisoning,[19] adding further medical explanations. We shall rely especially on the able, exhaustive work of Hofmann, Kobert and Stevenson.[8–10, 12, 13, 19]

All this has attracted scarcely a word of attention in the 81 years since Hofmann proved ancient plumbism because students have totally missed the class element, which alone gives to ancient plumbism its interest and social significance. If the Romans and Greeks were poisoned by lead, they did not die out and we have not heard of their being sickly, but perceiving that all the rich Romans and their tablemates were made almost childless, leaving the inheritance of the world to the less capable, illuminates the famous problem of the decline of Roman culture.

Why were the ancient rich so reckless as to drink lead compounds daily? They had scarcely an idea of their toxicity. They knew plumbism (since Nicander in the second century B.C.) only as a disease of lead workers, the first or second industrial disease to be discovered. Also their physicians knew about acute poisoning by lead, and were cautious about lead in medicines. But as to lead poisoning from water or food we can find hardly more than a reference or two, chiefly that of Vitruvius in his first-century A.D. treatise on water supply, echoed by Galen in the second century. It was certainly not a well-known danger.

But were not the poor dying out at the same time, for the same reason? In Greece, to some extent, they were, but hardly in Rome. In both areas the poor were constantly reinforced in some degree by war-captured slaves, although the reproduction of these—especially that of the Northern captives and other gang laborers in the Roman world—was much interfered with in various ways. The diet of the poor was not so badly poisoned with lead under the Romans. It is safe to say that the poor had not much wine, olive oil, grape syrup, other drinks made from fruit or honey, or preserved fruit—all the chief distinctive dietary sources of lead for the best people. Nor did they have the luxuries of cosmetics, or lead paint. They drank the same water, which would be leaded if it came through lead pipes and was acid from decaying vegetation or from volcanic, basaltic soil, or was neutral, like a rainwater. Hard, alkaline water takes up no lead, but lead pipes were presumably less used in the country, where more of the poor and the unintellectual always live. The diet of the poor was mostly porridge, made from barley and other grains and perhaps acorns, but it also included bread, legumes, and other vegetables, greens, vinegar, and a little meat, fish, milk, fresh or dried fruit, and wine. Less cooking might be done than for the rich, since wood fuel was expensive, the nearby forests being denuded, and transportation bad from lack of horseshoes and suitable harness. For cooking and liquid food storage the poor would be likely to use cheap earthenware, everywhere locally made, rather than expensive lead-lined bronze carried from afar.

Cooking involving heat greatly heightens the take-up of lead especially if the pot be scraped from time to time to prevent burning as is needed with thick liquids. The ancient writer Columella[4] said this scraping had to be done in making grape syrup (used for sweetening and preserving), which the well-to-do prepared by boiling down grape juice to as little as one third the original volume. Sugar, imported from India, hardly existed, leaving for sweetening only costly honey and the various fruits.

ANCIENT USES OF LEAD

To understand further how lead could be ingested in so many ways by the ancient aristocracy, we should consider its position among the metals and other substances that might have been substitutes for it, especially as containers of acid food.

Lead was a metal the ancients produced in superabundance, and hence used wholesale. It is the easiest of all metals (save sometimes iron) to mine, smelt and fashion, and needs only moderate heat. It was used in Anau I, (4200 B.C.). Later, especially with the invention of the cupellation process in the third millenium B.C., lead became usually a by-product of silver mining. Today the reverse is true; silver usually is a by-product of mining lead and other metals. In the great silver mines of Laurion, the fountain of Athens' prosperity, 300 oz. of lead were produced for every 1 oz. of silver. Sometimes they threw the lead away. It has been estimated[9] that 2,100,000 tons of it were produced there in 3 centuries. A Greek wine-ship of about 150 B.C., recently salvaged, had tacked on it a rather useless 200 tons of cast lead sheathing. Again, 1200 tons were used to lay an aqueduct under one valley. Most of this dangerous metal was consumed in a great variety of domestic ways, often in contact with food or water, since there was so little idea of its poisonous nature. They had easily learned about copper poisoning from bronze, because this is detectable at once by color, taste, and sickness. Lead poisoning is far less evident and immediate. So the bronze vessel was replaced by a leaden or lead-lined one, save when by rare luck they used expensive pure tin. Most often plating was done with a mixture of the two metals. Lead was used for water pipes and containers, cups, toys, statuettes, tokens, coins, lids, sieves, household repairs, solder, paint, cosmetics, external medicines, boxes, markers, writing tablets, coffins, and roofing, and many alloyed it with copper (lead bronze) and with tin (pewter) and used it in many other ways.

Romans put lead into their wine in as many as 14 different ways, as we shall describe below. The final, saturnine touch of perfection was to warm the wine before serving it in cold weather. Since the houses were scarcely heated we can imagine they reckoned many cold days. The warming was naturally done in a lead-lined bronze apparatus with a charcoal brazier. In early Republican times, up to the Gracchi, wine had been scarce and precious; but after the conquest of Greece it became abundant for the wealthy, along with other luxuries.

The aristocracy's most significant source of lead was the grape syrup described above, which they called, irregularly, *sapa* or *defrutum*, according to the degree of boiling down of the unfermented grape juice (must), or sometimes by its Greek names, *car(o)enum, hepsema,* or *sireion (siraeum).* Such syrup is still used for flavoring and sweetening in some Mediterranean countries and in American food manufacturing, so that we have all consumed it as "grape concentrate." Pliny[7] gives directions for making it: "Leaden and not-bronze pots should be used." And the like said Cato, [3] Palladius,[15] and Columella.[4] We have cited the latter's injunction for stirring. Pliny[7] evidently envisioned a solid leaden vessel, for he warns against leaving a pebble in it, lest this melt a hole through the pot.

This slow poison, this delicious syrup, was frequently put into wine to give it bouquet, strength, and keeping quality. The Romans often boiled the must before fermenting it to wine, since with their insanitary methods the wine often would not last more than a year, if that. The lead had two virtues, only one of which they understood correctly—that it gave a slightly sweetish taste. More importantly, it poisoned the microorganisms that cause fermentation and souring. Not having the modern art of sterilization by canning, they knew not that they were also sterilizing themselves. Grape syrup was further used for preserving olives, plums, cherries, apples, pears, peaches, and quinces—all cooked in lead. The syrup figures in a large number of the luxury dishes in Apicius' cookbook. Grapes were preserved for a year simply by placing them in rainwater boiled down to one-third.[17] Why such a virtue in boiled down rainwater, which should be only pure water, we should think? Evidently because it took up lead in long boiling, not being hard water—and perhaps it had also been collected from a lead roof and stored in a leaden cistern.

A weak drink for invalids, *adynamon,* was made by boiling down grape syrup with seawater. It was a favorite of courtesans. With myrrh the syrup made a special drink for women. Wine boiled with wormwood might be used for a hangover. Grape syrup with onion made an abortifacient.

Democritus said that if the must is too sour its acidity must be blunted by minium, lead oxide. Some minor amounts of lead in wine were from storage or drinking in leaden vessels, from the water always mixed with it, from the grape syrup used to cement the amphora's stopper, from mending an amphora with leaden clamps inside and out (but there was a coating of pitch), and perhaps leaden parts in the wine-press. Similar additions of lead, deliberate or ignorant, have continued occasionally almost until present times although under legal attack as early as the 1100's.

Some recognition of the toxicity of the wine, though not of the lead in it, perhaps appears in the rule of the earlier Republic,[7] that wives must not drink wine. People may have observed that wine-bibbing wives had fewer children. The rule ceased to be observed in later Republican times, when wine had become no longer a rare luxury but an everyday drink of the well-to-do. There was also an ancient tradition that fat women were more fertile than thin ones. This would be the reverse of the attitude today, but would be true in ancient times, since emaciation is one of the symptoms of lead poisoning. A fat woman would be one who had somehow avoided lead. There were also ancient statements[1] that certain Greek wines produced sterility, miscarriage, constipation, headache, or insomnia—all symptoms of plumbism.

That man has a sweet tooth is evidenced by the American consumption of sugar—nearly 100 lb. per capita per annum. So the ancients who could afford grape syrup or the limited supply of honey, used them freely. Pliny and Celsus said that raw honey was a good laxative, but that cooked, it was antidiarrheic. This could only happen through the action of lead.[13] Dioscurides (v. 16) said that hydromel, *aqua mulsa*, was honey mixed with twice its volume of spring water, then boiled down to a third to prevent fermentation. A sour honey preparation called oxymel, Pliny says,[13] was made by heating honey with seawater 10 times. At any rate, two of its uses were as a contraceptive and an abortifacient.

Another drink, whose name we might translate as wine-mead, was composed of must, water, and honey, fermented. Celsus and Pliny say again that drunk cold it was laxative, but when hot, stopped diarrhea.[13] Such opposite effects after cooking an acid food show that lead was involved: the possible escapes through cooking in tin-plate or pottery were not utilized.

Kobert gathered the bones of 22 people from the ancient world. His student, Rosenblatt,[16] found lead in 2 from Roman Carthage, and 2 from the same city before the Roman conquest, including a child in a stone coffin, hence rich. The remaining 18 came from outside the Roman Empire in time or place and showed no lead, although 5 early ones showed tin, and 1, copper. The chemistry of 1906 hardly permitted determinations in bone of more than the presence or absence of a metal. Later the dithizone and other methods enabled much more sensitive and accurate analyses for lead. So, in 1959, necromancers Specht and Fischer[18] called up Pope Clement II, who had died 9 centuries before, and was buried in a stone coffin in Bamberg. The amount of lead they found in a rib was so great that he must have died of plumbism. So I have set out to gather bones from the Roman Empire and Greece before, which can be identified from the conditions of burial as definitely rich or poor, and with further data desirable. My first samples are being analyzed by the Kettering Laboratory in the University of Cincinnati, which has been carrying out accurate studies of the occurrence of lead in bones today. Distinction will be made according to which bone is analyzed, since they store lead unequally, and also according to the subjects' age and sex, and whether circulating ground waters would have deposited any lead found, and how much. If it shall appear that, on the average, the rich of the Imperial period were markedly poisoned by lead, and the poor not too much for reproduction, clinical necromancy will have added a final seal of proof to evidence logically sufficient without it. Returns from the tests so far received on the bones of 40 persons are consistent with the theory here presented. The complete returns will lead to a subsequent report perhaps devoted to that single item.

BIRTH AND DEATH RATES

Sociologists always want census statistics. We shall try to oblige, with what little can be found from the ancient world,[11] viz., for an aristocratic quarter of Roman Troy (Ilion) and for Miletus—both Greek. In the Trojan quarter, aside from the widowed and the new citizens there were 101 men over age 19, of whom 66 were unmarried. Of the 35 married, 18 were childless, 10 had 1 child, and 7 had 2 or 3—a total of 27 children from 101 men counted. They were doubtless younger than those making up our adult population, and their families not yet complete in some cases, but the deceased, departed and illegitimate children would not be counted. However we refigure it, there was evidently a low rate of surviving children, and an extraordinary amount of celibacy. The Milesian data are from the third to first century B.C. and show that the naturalized men, an old group probably, had 1.36 children apiece.

From the top of Roman society, emperors and such, we have abundant data on the few legitimate children. Few of the emperors had a son to succeed him. So the Antonines established the excellent system of adopting a fit man for heir. But at last Marcus Aurelius had a son, and by folly appointed him, the fool Commodus. If he had been mentally damaged by lead in childhood, it would not be surprising. So then the adoptive system was wrecked, and thereafter the "system" was a strong invitation to civil war every time an emperor died or someone chose to assassinate him. Here lies some explanation of Rome's political decay, and a contrast with the medieval and modern routine of thrones descending almost always automatically to the king's eldest son or another close relative.

The Romans recognized what was happening. The bachelor Horace persuaded Augustus, who had only a daughter, to pass laws penalizing aristocratic bachelors, and granting special privileges, the *Jus trium liberorum*, to fathers of 3 or more children. But these laws were little enforced, despite long continuance on the books and efforts of various emperors to revive them. There were also payments to poor men with children. In time the *Jus trium liberorum* came to be awarded even to childless men and bachelors. Ovid, the "singer of tender loves," as he rightly called himself, provides a case typical of the Augustan age. Of equestrian birth, with one brother who died young, he had four marriages and one surviving daughter.

For a group to survive, its birthrate must equal or exceed its death rate. So we seek the death rate of the aristocratic Romans and find some statistics based on tombstone inscriptions—their presence evidence of a class able to afford that modest luxury, though they were often freedmen. Indicated is an expectation of life at birth of 22–25 years[21] comparing with 72 for white Americans today. Expectation was longest about 5 years after birth. Such a high death rate suggests two things: First, that a very high birthrate would have been requisite to preserve such a race, as opposed to the very few births indicated by our literary and census data. (Perhaps the aristocratic birthrate was one-fourth of what would have been necessary to maintain their number. Multiply one-fourth by itself 18 times—for the 18 generations of our history—and one obtains the result that of the original aristocracy there would remain but a trace, and about as little from most of the successive leadership cadres promoted as replacements from the most capable of the plebs.) Second, the high death rate as well as the low birthrate strongly suggests lead poisoning. Many must have died of it.

Three factors with which we must also reckon were deliberate restriction of offspring by birth control and abortions; homosexuality; and above all the extraordinary frequency of celibacy. First as to contraception—the science of the process was so little understood, and the worthless and the uncertain methods so mixed with the effective, that I do not think voluntary restriction was the major factor in the infertility. A method might work 9 times out of 10, and still plenty of children be conceived. Abortion was not rare and, like contraception, sometimes involved lead. Homosexuality was certainly frequent, though not so much as in Greece. But so long as people had heterosexual relations even occasionally—e.g., being married, there would still have been many children, just as with uncertain birth control. Celibacy, including intervals between the frequent divorces, was certainly an important fact. But why did so many men prefer celibacy? We may find lead a cause among others, thus: The chief purpose of marriage was to rear children; otherwise the ancient man would prefer full freedom, as Metellus Macedonicus said in 131 B.C. in an often quoted speech.[11] Children were wanted both for patriotic and religious motives—to continue the family. As with Hindus and Chinese, this religious motive declined. In addition, as they looked about them they would have perceived that none of their prosperous fellows had many children, that most who wanted to beget a child could not do so by the wife, and that a child born would like as not never grow up. So when the desire for children decreased, and also the possibility of having them, men largely gave up marrying: Why try to have a son and heir? Why marry?

The Roman moralists often declaimed against selfishness as the cause of celibacy and the rearing of few offspring. But consider that a moralist must blame people for immoral choices, or he is no moralist. He does not deplore the facts of nature, or what seem to be such—e.g., the infertility of old women, or the high death rate of children, or the fact that the wealthy eat this and that which is tasty and seems harmless.

The continued decimation of the Roman aristocracy is a historic fact, variously attested. But decimation is not the right word, since it properly means the killing of one-tenth, whereas it appears about three-quarters were eliminated in each generation on the higher hills of Roman society. One line of evidence is the disappearance of the old family or "gentile" names of Republican days, like Julius and Cornelius. With the families all gone about as early as the Republic, the Romans changed the basis of their naming system from strict inheritance to personal choice or parental bestowal, so that the name test cannot be used later. By Nero's time, Tacitus said that nearly all the knights and most of the senators betrayed a servile origin.

Still other possible causes, indeed all possibilities anent this aristothanasia, or death of the best, must be inquired into. Political proscriptions were certainly one factor, especially those of Sulla, Marius, and Octavius, who slaughtered their prominent political opponents by thousands. The later emperors were less wholesalers in massacre of leaders. But it was always the prominent, the leaders by brains, devotion, energy, and money, who were most often politically proscribed. Still, such leaders are usually men of middle age or older, whose consorts were probably so hampered by age and accumulated lead that they would have had no more children anyway. And the Roman proscriptors did not slaughter the wife and children with the father, as savage tyrants have done.

BINDING UP THE ARGUMENT

By this time the reader should be thoroughly confused, as to what happened and why, among the many reasons advanced for the aristothanasia that occurred. But though the items be somewhat conjectural, their total sum should be clear—viz., that the Roman upper class came to have a birthrate far below its death rate, resulting in a six-centuries-continued aristothanasia, killing off the brightest and winsomest from the poorer class, who in each generation were promoted to wealth and extinction, like the prize animals garlanded with flowers as they were led to slaughter at an altar. Whatever the relative strength of the causes, the result, aristothanasia, stands established.

And a second fact should be clear—that among all those causes one new one stands proved as certainly important: that the well-to-do were poisoned by lead, and thereby their fertility must have been much diminished, their death often hastened, and their surviving children often damaged mentally. This fact, in its class aspect,* was not known before.

Thirdly, the aristothanasia entailed that the ultimately surviving race, which in time had to fill all the posts from lowest to highest, was the screened-out remnant, composed of those who had least capacity to get money, whether by merit or rapacity, or by virtue of the beauty and charm that would enable a poor girl to marry money.

The beatitude said, "Blessed are the meek, for they shall inherit the earth." They did, and the result was the dysgenic and cultural disaster of the Dark Ages and Byzantine stagnation.

NEGATIVE ASPECTS

The reader has likely been thinking he has heard of still other causes for the fall of Rome—malaria, soil exhaustion, the plagues, the rising civilization and strength of the northern and eastern barbarians, the pushers behind them from the desiccation of central Asia by climatic change, the view of Gibbon against Christianity, or of political conservatives that it was the result of an overgrown bureaucracy. There may be truth in all these claims. But the fall of Rome is not what we are talking about, nor her military defeat and break-up, nor her declines in population and wealth. We are here concerned only with the decay of Roman culture, the

*An additional possibility, which we should not call a fact, is that some degree of hereditary immunity to lead was finally built up in the less afflicted lower class.

debasement of her arts, the ending of hopefulness and progress, and the connection of these declines of outward, evidenced intellectuality, with a loss of innate, inherited brains. Or to satisfy our environmentalists, we are concerned with the debasement of education, cultural traditions, and prestige of a ruling class, which under ancient social conditions could be transmitted only from wealthy parents to their own children.

It may be objected that culture, whether dependent on education or inborn ability, is also inseparably connected with economic conditions, wealth and political power. For culture, ideas, brains, create wealth and population and spread empires; and vice versa, if wealth and population decline, or Saxon pirates overrun a Roman Britain, culture is bound to suffer. We answer: Half true.

There is a connection, but not an unvarying one, between culture and those more material and commonplace sides of life; a connection but not a close and overriding one. A few instances will make this clear. Athens at her great century had culture supremely; she surpassed Rome at her greatest. And yet look at the material comparison: Athens was a city of about 120,000, and Rome 10 times that. The size of their empires compared similarly: the Athenians' homes were of adobe, the Romans' of masonry—from palaces to 10-story tenements, Rome's general technic was a little better, and her civil engineering works astonishing, while Athens could show nothing more than her walls. And yet with all that material superiority, Rome did not equal Athens in culture. Or compare poor and tiny Iceland with similarly disadvantaged Newfoundland, and with fertile and pleasant Portugal having 48 times Iceland's population. The intellectual contributions of Newfoundland and Portugal have for centuries been near zero, but Iceland has maintained a vigorous intellectual life and a stream of notables, as Ellsworth Huntington has shown.[11] The only plausible explanation is that the selection of Iceland's Norse settlers in the ninth century peopled that desolate island with one of the best races on earth. Again, among the individuals known to us personally, we recognize that culture, character, and brains are in large measure independent of their parental educational opportunities and their own wealth and power.

We conclude, therefore, that Rome's material losses—in population, empire, wealth, and health—damaged her culture somewhat, but are far from explaining the depth of its ruin. And vice versa, the decline of culture, and of the brains which any eugenist believes lie behind culture, helped to explain the civil wars, inroads of barbarians, and all other events in the material sphere. But fading culture and brains are by no means the whole explanation of the material losses.

In Renaissance and modern centuries, other selective factors as to who got born and survived have come along.[20] Smallpox, syphilis, and gonorrhea have appeared, as well as doctors and food for all, and increasingly birth control, selective on the basis of intelligence and to some extent according to religion, urbanism, and wealth. The general result is that now some kinds of people have seven or a dozen children, other kinds three, two, or one, and others none. So the imperial Roman breeding system has returned to America and all the advanced countries with only the difference that our well-to-do are not dying out absolutely (they have about enough children merely to continue their number), while the poor, and likewise the backward nations, are about doubling in population in each generation. The capable are not being killed off, they are only being swamped.

What has been presented here are but fragments of the accumulated mass of evidence connecting lead poisoning with the fall of the Roman power. The total accumulation warrants a full-size text, and such a book is in preparation.

A few fragments here presented should be sufficient to provide some conviction that lead poisoning was destructive in the Roman Empire. For a long while lead poisoning has been widely recognized as a race poison. Now, it is to be recognized as once a class poison. Many influences contributed to the fall of Rome. In the total number, lead poisoning is to be reckoned the major influence in the ruin of the Roman culture, progressiveness, and genius.

References

1. Athenaeus of Naucratis: *The Deipnosophists* (2nd or 3rd century A.D.) Gulick trans., p. 139, and Pliny 14:19, 22, 28.
2. Cantarow, A., and Trumper, M. *Lead Poisoning*, Williams & Wilkins, Baltimore, 1944, pp. 84, 142.
3. Cato, M. Porcius, the Elder: *De Re Rustica*, Cap. 105, 107, 122.
4. Columella, L. J. M. *De Re Rustica xii:* 19, 20.
5. Gilfillan, S. C., "The inventive lag in classical Mediterranean society." *Technol. & Culture* 3:85, 1962.
6. Hart, H., "Technology and Social Change." Metropolitan Life Insurance Co. *Statistical Bulletin*, Oct. 1917, pp. 1–3.
7. Henderson, A. *History of Ancient and Modern Wines.* 1824, p. 339, quoting Pliny the Elder in *Natural History* 14:14, 27, 23:2, 34:48, 49.
8. Hofmann, K. B., "Die Getränke der Griechen und Römer vom hygienischen Standpunkte." *Arch. Gesch. Medizin 6:26*, 269, 1883.
9. Hofmann, K. B., *Das Blei bei den Völkern des Altertums.* 1885. (Also in *Sammlung Gemeinverständlicher Wissenschaftlichen.* R. Virchow-Holzendorff, Ed.).
10. Hofmann, K. B., *Vorträge* 20:565, 1885.
11. Huntington, E., *The Character of Races.* Scribner, New York, 1923, pp. 264–300.
12. Kobert, R., *Ueber den Zustand der Arzneikunde vor 18 Jahrhunderten*, 1886.
13. Kobert, R., "Chronische Bleivergiftung im klassischen Altertum." In *Beiträge aus der Geschichte der Chemie.* P. Diergart, Ed., 1906–9, pp. 103–19.
14. Landry, A., "Qq. Aperçus Concernant la dépopulation dans l'Antiquité Gréco-Romaine." In *Revue Historique* 177-1-33, 1936, pp. 2, 3, citing on Troy Ch. Michel: *Recueil d'inscriptions grecques* No. 667; on Miletus Wiegand: Milet, Ergebnisse, Bd. 1, Heft 3, pp. 55–105; on Metellus, p. 11.
15. Palladius (4th Century): Bk. 11, Oct. 18.
16. Rosenblatt, M., *Beiträge zur Kenntnis der chronischen Bleivergiftung.* 1906, 107 pp.
17. Soyer, A., *The Pantropheon*, or *History of Food and Its Preparation.*
18. Specht, W., and Fischer, K., "Vergiftungsnachweis an den Resten einer 900 Jahre alten Leichen," *Arch. Kriminol.* 124:61, 159.
19. Stevenson, L. G., *A History of Lead Poisoning.* Doctoral dissertation, Johns Hopkins University, 1949, pp. 1–12, 404–16.
20. Weyl, N., and Possony, S., *The Geography of Intellect*, Regnery, 1963.
21. Wilcox, W. F., "Length of life in the early Roman empire," in *Proceedings of Congrès Internationale de la Population*, Paris, II (1937), pp. 14–22.

5.3

Technology Assessment from the Stance of a Medieval Historian

LYNN WHITE, JR.

On October 13, 1972, the American federal government established in Washington an Office of Technology Assessment to advise Congress on legislative problems related to new technology and its probable impact. This act reflects an ambivalence toward engineering innovation that has been rare during the last thousand years in the Occidental culture of which we are part. Both pagan and Christian antiquity, of course, had been dubious about technology. St. Augustine, the most penetrating mind of a groping age, expressed amazement at the ingenuity and variety of the arts, yet feared that the good coming from them may be counterbalanced by the evil of "so many poisons, weapons and military machines" in addition to superfluities and vanities.[1] The Latin Middle Ages, by contrast, developed an almost entirely affirmative view of technological improvement. This new attitude is clearly detectable in the early ninth century, and by 1450 engineering advance had become explicitly connected with the virtues: it was integral to the ethos of the West.[2]

People are organized into cultures by the basic presuppositions—often unverbalized—that they share: their axioms. They put their intelligence, energy, and money into what they corporately consider good. The results are as varied as the majestic pyramids of pharaonic Egypt, the sadistic games of the arenas of the Roman West, and the family-centering, but globally focused, television sets of the contemporary industrialized world. Medieval Europe came to believe that technological progress was part of God's will for man. The result was an increasing thrust

---

This was the presidential address delivered by Mr. White at the annual meeting of the American Historical Association, San Francisco, December 28, 1973.

of invention that has been extrapolated, without interruption or down-curve, into our present society.

Never was the sense of the virtuousness of technology more vivid than in nineteenth-century America. In 1853 an English mission exploring the sources of industrial success on this side of the Atlantic concluded, in awed tones, that

> the real secret of American productivity is that American society is imbued through and through with the desirability, the rightness, the morality of production. Men serve God in America, in all seriousness and sincerity, through striving for economic efficiency.[3]

Clearly this Victorian investigative team did not know that they were observing an attitude that had been held by the common medieval and puritan ancestors both of themselves and of the renegade colonists: in England it was already in decay. In New England, however, it survived until toward the end of the century, when its demise was signaled in Henry Adam's dichotomy of Dynamo and Virgin, a tragic vision of reality most curiously presented against the backdrop of a totally misunderstood Middle Ages.[4]

Today the medieval axiom of the rightness of technological progress has been challenged in the entire Western world, and not merely by mystics and eccentrics of the Blake and Thoreau ilk. To the secular among us it seems quaint; to the religious it is blasphemy. The most extreme repudiation of it is that by Jacques Ellul, a French Calvinist professor at Bordeaux, to whom "Technology" appears to be the new name for Antichrist, a demonic force that is completely out of hand.[5] Few will go so far; yet among us, few likewise still share the old confidence that all problems produced by changing engineering will be solved automatically by remedial forms of technology, quite without the intrusion of public policy based on ethical and esthetic sensibility. Hence the establishment by Congress of the Office of Technology Assessment.

We must have assessment of technology: our national crises of energy, exhaustion of natural resources, and pollution of air, water, and soil interlock with global crises of armaments, population, and food. The real question is: Do we know how to assess a proposed technological change, whether it be a new invention or a new canal across Central America?

Technology assessment today is a discipline largely concerned with weapons systems, industrial production, power networks, traffic patterns, marketing problems, and large engineering projects. The systems analysis that is its method is based almost entirely on costs-benefits calculations that are narrowly construed because they are designed to answer military and business questions formulated within a rather limited range set by those commissioning the studies. Their failure to ask about wider social and other costs and benefits has led to unfortunate effects that impinge increasingly on the lives of millions, with resulting popular backlash against engineers and scientists, soldiers and business leaders.

Some of the most perceptive systems analysts are pondering today how to incorporate into their procedures for decision the so-called fragile or nonquantifiable values to supplement and rectify their traditional quantifications. Unhappy clashes with aroused groups of ecologists have proved that when a dam is being proposed, kingfishers may have as much political clout as kilowatts. How do you apply cost-benefit analysis to kingfishers? Systems analysts are caught in Descartes's dualism between the measurable *res extensa* and the incommensurable *res cogitans*, but they lack his pineal gland to connect what he thought were two sorts of reality. In the long run the entire Cartesian assumption must be abandoned for recognition that quantity is only one of the qualities and that all decisions, including the quantitative, are inherently qualitative. That such a statement to some ears has an ominously Aristotelian ring does not automatically refute it.

There is a second present defect in the art of technology assessment: the lack of a sense of depth in time; this may be called the Hudson Institute syndrome. It is understandable not only because most systems analysts are trained either in engineering or in the social sciences that normally take a flat contemporary view of phenomena, but also because the concrete problems set before systems analysts for solution look toward future action and discourage probing the genesis of things. Since history deals with nonrepetitive events, historians cannot help in specific ways to answer questions concerning assessment of technology in our time. I believe, however, that contemporary technology assessment will become sophisticated and more successful only if those who practice it are made aware of the complexity and ramifications of the effects of technological changes in the past. History can offer no solutions, but it may help to guide an acute mind toward kinds of questions that in the present state of systems analysis tend to be overlooked. Above all it may illuminate the limitations as well as the possibilities of assessing technology.

To show what I mean, let me present a rapid and necessarily superficial review of a few Western medieval innovations and their impacts. To what extent could a prescient medieval futurologist have foreseen what was going to happen?[6]

Sometime between 1150 and 1167 alcohol, or, more exactly, brandy, was first distilled from wine as a pharmaceutical at Salerno, the site of Europe's most famous medical school.[7] Drunkeness, of course, had produced moral and social problems ever since Noah took shore leave (Gen. 9:21–27), and it is not likely that anyone in the later twelfth century could have thought that the new medicine would amplify a traditional vice. The considerable medieval literature on *aqua ardens* or *aqua vitae* stresses the beneficial effect of alcohol for facial tic, chronic headache, stomach trouble, falling or graying hair, worms, epilepsy, cancer, sterility, sciatica, arthritis, and bad breath. In general, it was good for people who, in terms of the humoral physiology of that age, were considered to have a "cold" temperament. Then in the midst of such medical advice, one treatise lets down its guard: "Brandy when drunk makes a man happy and sociable":[8] in other words, this is the "fun" medicine.

Especially in Northern Europe the winters were chilly enough to make anyone believe that he was of a "cold" complexion and that such a medicament was essential to his health. Apothecaries were making it in ever more considerable quantities, and during the fifteenth century, when it began to be produced from a beer mash as well as from wine, it became quite cheap. Drunkenness and consequent public disorder increased alarmingly: for example, at Frankfurt am Main decrees trying to cope with the problem—but obviously in vain—were issued in 1361, 1391, 1433, 1456, and 1487.[9]

India and Sunnite (as distinct from parts of Shiite) Islam arrived at a negative assessment of intoxicants long before alcohol was distilled. In 1919 the United

States reached the same conclusion, but so many socially evil side effects of that decision emerged that by 1933 the consensus was reversed. Alcohol remains fun for many, and a disaster for many. As the annual meetings of this Association demonstrate, it adds considerably to conviviality and perhaps even to the flow of ideas. Statistics, however, indicating that half the traffic deaths in this country are connected with either drunken drivers or drunken pedestrians make one wonder about the beneficence of the Salernitan gift to mankind. A study group eight centuries ago, equipped with entire foresight, would have failed at an assessment of alcohol as today we fail.

In our own generation some of the most successful technology assessments have dealt with weapons. How would similar efforts have fared in the Middle Ages, or how did they?

In the later eleventh century the West developed a new and more powerful form of crossbow, presumably made possible by a firmer trigger.[10] Anna Comnena tells us that at the time of the First Crusade the Byzantines regarded it as a Frankish novelty;[11] eventually it spread as far as South India, where it was known as the *paranqi*, or "Frankish" bow.[12] The wounds from its bolts were terrible, and in 1139, at the Second Lateran Council, Innocent II banned it on moral grounds, except for use against infidels.[13] The prohibition was unenforceable: in the heat of warfare, every foe seemed at least a crypto-infidel. The nonquantifiable value of compassion was indeed fragile in the face of the crossbow's measurable ability to shoot further and hit harder than any other portable missile thrower before the English longbow appeared in the later thirteenth century. As between clear negative and affirmative assessments of the crossbow, the latter won out.

Edward I of England learned to respect the longbow in the hands of the Southern Welsh, both his foes and his allies. He and his staff appraised it correctly: using it in Wales and Scotland, he worked out new tactics and combinations of forces that made the English army almost invincible until toward the end of the Hundred Years' War.[14] In the hands of a skilled archer, cloth-yard shafts of great striking force could be launched several times more rapidly than an equally powerful crossbow could be reloaded and shot. The supply of archers remained sufficient for nearly a century before it began to dry up because of a change in the recreational patterns of English commoners. In 1365 Edward III—a ruler of great military acumen—commanded all sheriffs to suppress bowling, quoits, handball, football, club ball, hockey, cockfighting, "and other vain games of no value," and to see to it that on Sundays and holidays Englishmen of the lower orders should practice with bows and arrows. In 1388 tennis and dice were added to the list of banned sports, and similar measures were enacted into the sixteenth century.[15] Nevertheless, the long decline of English archery continued: in 1549 Bishop Hugh Latimer thundered that "we have taken up whoring in towns instead of shooting in the fields."[16] When in 1595 the longbow was officially discarded by Elizabeth's army in favor of the musket, it was still technically the superior weapon.[17] The musket, however, was reasonably effective in the hands of less well-trained soldiers, and that fact was decisive.

Could Edward I have foreseen that Englishmen might not always shoot at the butts, and could he have taken successful measures to prevent the decay of archery before it proved irreversible? To tell the truth, we still do not know with any certainty the reasons for that decay. Archery doubtless prospered best in hamlets so small that they lacked a tavern or enough inhabitants for many group games. A relevant factor, long at work by Edward I's time, was the gradual shift, especially in the more fertile areas, from the ox to the swifter horse as the ordinary beast for plowing. This reduced by one-half the time spent going and coming between stable and plowland and encouraged peasants to abandon hamlets, while working the same fields, and to agglomerate into larger villages where life was more lively.[18] Moreover, the shrinkage of population during the second half of the fourteenth century led to abandonment to pasture of much cultivated land of marginal productivity and small settlements: the now scarcer peasant labor was concentrated on the more profitable soils, and thus in larger villages.[19] It is doubtful that any number of royal decrees could have perpetuated archery indefinitely in the face of the new tempo and variety of popular life in late medieval England. In any event, three centuries after Edward, Elizabeth was forced by the dearth of skilled personnel to resort to the technically inferior musket.

What about medieval artillery? The torsion artillery of Hellenistic and Roman times was often unsatisfactory because the skeins of hair or sinews powering the machines stretched in wet weather and lost their resilience. During the later Carolingian period the West received from Islam—or else by way of Islam—a new sort of stone thrower consisting of a horizontally pivoted beam with a sling at the shooting end and ropes at the other that were pulled simultaneously by a gang of men.[20] This was an artillery for all seasons, and it probably replaced the classical types speedily. The pattern remained unchanged until toward 1200, when some engineer, a European it would seem,[21] realized that labor could be saved, size increased, and accuracy of aim improved if a great counterweight were substituted for the gang of men who pulled the ropes. Such machines, called trebuchets, developed amazing capacities in the early thirteenth century. The counterweight and firing arm being uniform, properly calibrated stone balls of equal weight would hit the same spot on a fortification at each shot.[22] A great trebuchet designed by Bishop Durand of Albi in 1244 to besiege the Cather stronghold of Montségur threw a forty-kilogram missile at its walls at twenty-minute intervals, day and night, for weeks and proved to be the key, opening defenses that had seemed impregnable.[23]

Had an Office of Technology Assessment been asked to present a report on gunpowder artillery when it first appeared at Florence in 1326,[24] the measuring rod of effectiveness would have been the trebuchet. The earliest cannon were crude, cumbersome, and inefficient. They were costly to make and costly to supply with their chemical fuel. They could not be aimed with any great exactness; they were slow to load and to fire; they could rarely hit the same spot on a fortification twice because of irregular composition and combustion of the powder. The shaking of gunpowder during transport made the lighter charcoal particles rise to the top, with the result that early gunpowder had to be carefully remixed just before it was used: a perilous process during battle. We have no evidence that corned gunpowder—designed to prevent this difficulty and also to assure even combustion by introducing small air spaces—was known before 1429.[25] Any rational technology assessment of the cannon in 1326 or for a hundred years later would have concluded: "Stick to the trebuchet."

Yet Europe did not stick to the trebuchet: by the end of the century trebuchets had been practically discarded.[26] Why? Probably because of the cannon's splendid roar and flash, and also because the extravagance of it made it a status symbol. The greater cost-benefit efficiency of the trebuchet as a weapon succumbed to the nonquantifiable values of vanity and political visibility. It was decades after the eclipse of the trebuchet that the prolonged labors of guncasters, industrial chemists, and gunners finally transformed cannon into an artillery as effective as the kind it had replaced. One doubts that any futurologist of 1326, contemplating those cannon at Florence, could have anticipated so complex a course for the new weapon.

There is at least one instance of the social impact of a medieval invention that might easily have been assessed intelligently, only to have the prognosis overturned by later events. As Edward Rosen of the City University of New York has shown, eyeglasses were invented by a man who was living in the Lucca-Pisa area in the 1280s.[27] They were a boon to the presbyopic, and their use spread rapidly. Any contemporary, if he had thought about it, could have seen that this would enable aging men in teaching, the book trade, law, bureaucracy, banking, commerce—indeed, in any occupation that demanded frequent reading—to keep actively at work longer than otherwise would have been possible. Eyeglasses thus would slow the promotion of younger men, lead to discontent among them, and produce a generation gap. Indeed, *a priori*, the common use of spectacles certainly tended to produce such a slowing of youthful careers. However, the first half of the fourteenth century was a period of such turmoil that life expectancies seem to have decreased even before the catastrophe of the Black Death.[28] While young and old perished alike, the higher general incidence of death led to increased velocity in promotions and thus masked the reverse effects of the introduction of eyeglasses.

There are some technological developments that seem initially so modest, and that grow so quietly, that even in our own day no one would wonder about their impact until long after the effects were irreversible. Leroy J. Dresbeck of Western Washington State College has recently clarified such an instance in his study of the chimney flue and mantled fireplace.[29]

For heating, the Romans used braziers and hypocausts (radiant heating). Braziers burned charcoal and were thus costly to operate; moreover, in a room tightly closed against the weather they were dangerous: the Emperor Julian, wintering in Paris, was once nearly killed by carbon monoxide from a brazier.[30] Hypocausts were inflexible and wasteful of fuel because they heated the entire mass of the masonry of floors or walls. In the variable climate of Northern Europe, and especially in winter, they would not do. During the early Middle Ages, whether in hovel or royal hall, people centered their lives around a fireplace in the middle of a room with a high, louvered roof to carry out the smoke. Unfortunately it carried out much of the heat as well.

By the ninth century the central fireplace was occasionally moved to a corner of the room and was covered by a hood or mantle to catch the smoke and take it out through a hole, or even up a chimney. Experimentation with the design of chimney stacks led to the discovery that a draft of air can be brought down the flue to reverse itself and draw off the smoke while leaving much of the heat to be radiated into the room. Moreover, replacement of louvers by chimneys meant that fireplaces could be located on any level of a multistoried building and not merely under a roof directly beneath the sky. In the eleventh century chimneys and mantled fireplaces became common in the dwellings of the great. By the end of the twelfth, even the poor were enjoying them.

Much more than simple comfort was involved. In the days of the old central fireplace, to keep warm in Northern Europe everyone from lord and lady to humblest servant lived and ate together in the great hall, and slept there, too, normally in curtained compartments. Society was hierarchical, but the strata knew each other intimately. With the new flexibility of heating made possible by chimney and mantled fireplace, privacy could be implemented. Lord and lady increasingly ate, lived, and slept in withdrawing rooms. As affluence increased, noble residences were redesigned so that rank after rank of the social structure could enjoy the new sense of individuation in its life style. To Dresbeck's remark that the chimney may have affected the art of love more than the troubadors did,[31] one may add that it may likewise have fostered the individualism of the later Middle Ages more than all the humanists. Yet a high social price was paid for the new ideal of the idiosyncratic person. As communication between classes decreased, class consciousness and snobbery grew. By the 1370s William Langland was assessing the chimney bitterly:

> Woe is in the hall each day in the week.
> There the lord and lady like not to sit.
> Now every rich man eats by himself
> In a private parlor to be rid of poor men,
> Or in a chamber with a chimney,
> And leaves the great hall.[32]

The chimney is as important as any other single factor in the shift from medieval to modern Occidental attitudes, and not all of this process was good. I doubt, however, whether anyone much earlier than Langland could have assessed properly its less desirable effects, and by that time the process could not be turned back.

Indeed, technology assessment becomes an enterprise of almost terrifying immediacy when we realize that our most intimate psychic structure may at times be influenced by seemingly minor external innovations. The modern American family is often a paidocracy, child centered to a degree unknown elsewhere; yet our pattern is the completion of one long developing in Europe. Philippe Ariès's *Centuries of Childhood*[33] is a blundering and often perverse book that nevertheless establishes that in the Middle Ages no one paid much attention to children before about the age of seven, when they were admitted to the world of grown-ups. He believes that it was not until the later sixteenth or the seventeenth century that our culture discovered the "sweetness, simplicity and drollery" of small children, began to enjoy and coddle them, to accept them on their own terms, and to dress them in special clothes that were not simply small versions of adult attire. I myself would push the beginnings of the change somewhat earlier, but must recognize that in the early fifteenth century a family like the Pastons paid amazingly little attention to their younger offspring.[34]

It is assumed that this indifference was an effort to defend adult emotions against the grim fact of infant mortality. No one could affort to invest great emotional capital in a child whose chances of survival were 50 per cent or less. Not until five to seven years had proved a certain durability could one risk great affection or interest. I believe that this hypothesis makes sense, but it leaves unexplained the reasons for the presumed improvement in child survival that at last led parents to venture lavishing affection upon the very young.

I see no clues in sanitation, diet, or medicine.[35] Explanation must be found elsewhere. Partly because their bodies are so small, little people are peculiarly vulnerable to cold and resulting pulmonary infections. We have already noted that in the later Middle Ages houses of all classes were better heated than before, and the increased glazing of windows helped to retain the heat. But the snugness of clothing was also much improved. The first functional buttons appeared in central Germany in the 1230s,[36] and by the fourteenth century they were revolutionizing costume design. In the dress of modern American and European children knit textiles are basic. The first evidence of a knit body garment (as distinct from a few specimens of socks, gloves, and so forth)[37] is on an altarpiece from Buxtehude near Hamburg painted in the last decade of the fourteenth century. It depicts the Virgin Mary knitting a shirt on four needles for the Christ child[38] . . . It is a safe surmise that the development of knitting, along with functional buttons and heating devices, helped to keep more little children alive, and thus played a large part in fostering modern attitudes toward them. Late medieval mothers and grandmothers with clacking needles undoubtedly assessed knitting correctly as regards infant comfort and health, but that in the long run a new notion of relationships within the family would thereby be encouraged could scarcely have been foreseen.

How far into the future can even the sharpest eye look? Let me sketch another sequence that starts with textiles but ramifies curiously.

With the old toplike spindle the production of thread was slow and laborious. Especially after the horizontal heddle-treadle loom displaced the vertical loom in the eleventh to twelth centuries,[39] in the production of simple unpatterned cloth one weaver could use the product of many spinners. The first known spinning wheel appears in a Chinese painting of about 1035.[40] During the thirteenth century[41] this invention reached Europe. We may be sure that the people making and selling textiles recognized at once—although none seems to have recorded his views—that the new instrument, by speeding yarn production, considerably reduced the labor component of the final cost of plain cloth. In the very competitive pan-European textile market of that age, this meant a lowering of prices and consequent increase of consumption. Linen was particularly affected, both because it was normally unpatterned and because it was seldom dyed, only bleached by exposure to sunlight. A good merchandiser of the late thirteenth century might have foreseen what in fact happened in the fourteenth century; an immense increase in the use of linen shirts, underwear, bed linen, towels, and even vast coifs of starched and folded linen decking the heads of fashionable ladies.

Contemporary technology assessment of the spinning wheel might reasonably have probed even further. Increased use of linen meant more linen rags, and probably cheaper. Linen rags were the best material for making paper. This meant that the burgeoning new paper industry could expand production, lower prices, and increase consumption.

I greatly doubt that even the keenest systems analyst, looking at the potential of the new spinning wheel in the thirteenth century—or perhaps for three generations thereafter—could have gone further than this. Yet with the wisdom of hindsight we can perceive that the second impact of the spinning wheel was not on the textile industry but on the book business.

To produce a large Bible took the skins of between two and three hundred sheep or calves. The preparation of parchment and vellum was arduous and the finished product expensive. By a happy chance we know that in 1280 at Bologna paper was already six times cheaper than parchment.[42] Although I have no proof, it is probably that the relative cost of paper continued to decline. Except for deluxe volumes, paper was increasingly used in making manuscripts. This meant that the wages of the scribe became by far the greatest cost in manufacturing a book. There was every incentive to experiment with mechanical means of writing and, when a method was found, to make the considerable capital investment needed to operate it. This was Gutenberg's accomplishment. Its presupposition was the spinning wheel.

Elizabeth Eisenstein of the American University has recently revivified in so salutary a way discussion of printing's impact upon our culture[43] that nothing need be said here. While conclusions may differ about specifics, there is agreement that the Western world was deeply shaken by the printed book. It is curious, but consonant with the axiom of the goodness of technological change, that among contemporary Europeans—save for a very few snobbish bibliophiles—enthusiasm for printing was complete.

Not so in Islam: here the technology assessment of printing was so negative that it lasted for centuries. The techniques were easily available: Maronite, Greek, Armenian, and Jewish subjects of Muslim rulers were operating presses for their own purposes long before there was any printing in Turkish, Persian, or Arabic.[44] Many parts of Islam were sophisticated and creative, as the monuments of sixteenth-century Istanbul or seventeenth-century Isfahan amply demonstrate. There was no general allergy to borrowings from Europe, as the adoption of gunpowder artillery shows.[45] The avoidance of the printing press appears to have been deliberate and selective. So far as I can discover, no one has yet explored what Muslims said to each other about printing and their opposition to it. I personally suspect that the leaders of Islamic society felt—perhaps subliminally—that cheap books would eventually destroy the elitist world that they valued. If so, they were correct, as the later history of the West proves.

This little cluster of case studies that I have offered was selected not at random, but rather to show different levels of complexity, or different kinds of relationships, that may be found in the assessment of a new technology. My thesis is that technology assessment, if it is not to be dangerously misleading, must be based as much, if not more, on careful discussion of the imponderables in a total situation as upon the measurable elements. Systems analysis must become cultural analysis, and in this historians may be helpful.

Notes

1. *De civitate Dei* 22.24, in *Corpus christianorum, series latina*, 48 (Turnhold, 1955): 848–49.
2. See Lynn White, jr., "The Iconography of *Temperantia* and the Virtuousness of Technology," in T. K. Rabb and J. E. Seigel, eds., *Action and Conviction in Early Modern Europe: Essays in Memory of*

E. H. Harbison (Princeton, 1969), 197–219; and White, "Cultural Climates and Technological Advance in the Middle Ages," *Viator*, 2 (1971): 171–201.

3. Quoted by Charled L. Sanford, "The Intellectual Origins and New-Worldliness of American Industry," *Journal of Economic History*, 18 (1958): 16; see also page 14 for Patrick Tracy Jackson's dictum that "the village steeple is an unfailing companion of the waterwheel."
4. Lynn White, jr., "Dynamo and Virgin Reconsidered," *American Scholar*, 27 (1958): 183–94; reprinted in White, *Machina ex Deo: Essays in the Dynamism of Western Culture* (Cambridge, Mass., 1968), 57–73.
5. Jacques Ellul, *The Technological Society*, tr. J. Wilkinson (New York, 1964).
6. There appears to be no study of the development of the more modern social-psychological kinds of insight, as contrasted with the legal-political Greco-Roman way of analyzing situations and trends. That seeds of the new method were sprouting in the later thirteenth century is shown by Brian Tierney, *Medieval Poor Law: A Sketch of Canonical Theory and Its Application in England* (Berkeley, 1959), 44–67. Canonists at that time began shifting the focus of discussions of charity from its effect upon the spiritual well-being of the donor to its impact upon the recipient.
7. E. O. von Lippmann, "Zur Geschichte des Alkohols," *Chemiker-Zeitung*, 44 (1920): 625. Despite the efforts of Mohammed Yahia Haschmi to show the contrary, there is no firm evidence that alcohol was discovered in Islam earlier than in Europe: *aqua vitae* was given its pseudo-Arabic name by Paracelsus. "Sur l'histoire de l'alcool," *Actes du XIIe Congrès international de l'histoire des sciences, 1968*, 3a (1971): 69–72.
8. "Item gebrant win gedruncken machet den menschen frolich und wohl gemüt." Gundolf Keil, "Der deutsche Branntweintraktat des Mittelalters," *Centaurus*, 7 (1960):84.
9. Robert J. Forbes, *Short History of the Art of Distillation* (Leiden, 1948), 90–91.
10. The earliest picture of it that I have found—Christian Spanish, dated 1086—stresses the trigger; see Pedro de Palol and Max Hirmer, *Early Medieval Art in Spain* (New York [1967]), 58 and pl. XIV.
11. Anna Comnena, *Alexiad* 10.8.6, ed. B. Leib (Paris, 1945), 2: 217–18.
12. J. Hornell, "South Indian Blow-Guns, Boomerangs and Crossbows," *Journal of the Royal Anthropological Institute of Great Britain and Ireland*, 54 (1924): 344–45.
13. "Artem autem illam mortiferam et Deo odibilem ballistariorum et sagittariorum adversus Christianos et Catholicos exerceri de cetero sub anathemate prohibemus." J. D. Mansi, *Sacrorum conciliorum nova et amplissima collectio* (Venice, 1776), 21: 533.
14. John E. Morris, *The Welsh Wars of Edward I* (Oxford, 1901), 99–104.
15. Austin Lane Poole, "Recreations," in A. L. Poole, ed., *Medieval England* (new ed.; Oxford, 1958), 625.
16. Hugh Latimer, *Sermons*, ed. G. E. Corrie (Cambridge, 1844), 197.
17. See Thomas Esper, "The Replacement of the Longbow by Firearms in the English Army," *Technology and Culture*, 6 (1965): 393.
18. See Lynn White, jr., *Medieval Technology and Social Change* (Oxford, 1962), 67–68.
19. Maurice W. Beresford finds little evidence of immediate and catastrophic abandonment of settlements following the epidemics. *The Lost Villages of England* (London, 1954), especially 158–66.
20. Donald R. Hill, "Trebuchets," *Viator*, 4 (1973): 99–114.
21. Muslims often called the largest trebuchets "Frankish machines" (manjaniq faranji).
22. Such trebuchet balls were being produced according to engineers' specifications in England by 1244; see John Harvey, *English Mediaeval Architects* (London, 1954), 111.
23. Fernand Niel, *Montségur: le site, son histoire* (Grenoble, 1962), 222.
24. Carlo M. Cipolla reproduces the Florentine document of February 11, 1326, referring to the purchase of "pillas seu palloctas ferreas et canones de mettalo." *Guns and Sails in the Early Phase of European Expansion, 1400–1700* (London, 1965), 32 pl. An English manuscript of 1327 shows a picture of a cannon; thereafter the evidence is massive.
25. See A. O. von Essenwein, *Quellen zur Geschichte der Feuerwaffen* (Leipzig, 1872), 25.
26. Philippe Contamine concludes that the expansion of French royal artillery was particularly rapid

in 1395–1400; he cites Christine de Pisan's list of material needed for a great siege as including 8 trebuchets of 2 models, but also 128 cannon supplied with 30,000 pounds of powder; for the projected siege of Calais at just that time (1406) the French army procured at least 20,000 pounds of powder. *Guerre, état et société à la fin du moyen âge: Études sur les armées des rois de France, 1337–1494* (Paris, 1972), 229 n. 115, 665–66.

27. Edward Rosen, "The Invention of Eyeglasses," *Journal of the History of Medicine and Allied Sciences*, 11 (1956): 13–46, 182–218.
28. For a learned summary, see Josiah C. Russell, "Effects of Pestilence and Plague, 1315–1385," *Comparative Studies in Society and History*, 8 (1966): 464–73. There is some evidence that mature men were especially vulnerable to the plague.
29. L. J. Dresbeck, "The Chimney and Fireplace: A Study in Technological Development Primarily in England During the Middle Ages" (Ph.D. dissertation, University of California, Los Angeles, 1971); summary in "The Chimney and Social Change in Medieval England," *Albion*, 3 (1971): 21–32.
30. *The Works of the Emperor Julian*, tr. W. C. Wright (London, 1913), 2: 430–33.
31. Dresbeck, "The Chimney and Fireplace," 207.
32. This passage is found only in the B text: William Langland, *The Vision of Piers the Plowman: The "Crowley" Text; or Text B*, ed. Walter W. Skeat, Early English Text Society, 38 (London, 1869), passus X, lines 93–98.
33. Philippe Ariès, *Centuries of Childhood: A Social History of Family Life*, tr. R. Baldick (New York, 1962). The French original has a more exact title: *L'enfant et la vie familiale sous l'ancien regime* (Paris, 1960). For an acute analysis of Ariès's not-always-coherent thesis, see David Hunt, *Parents and Children in History: The Psychology of Family Life in Early Modern France* (New York, 1970), 27–51. Urban T. Holmes, "Medieval Children," *Journal of Social History*, 2 (1968): 164–72, warns that Ariès's picture of the plight of the medieval child is too grim; however, Richard C. Trexler, "Infanticide in Florence: New Sources and First Results," *History of Childhood Quarterly*, 1 (1973); 98–116, indicates a remarkable amount of deliberate infanticide in fifteenth-century Florence, particularly directed against girl infants.
34. See Ann S. Haskell, "The Paston Women on Marriage in Fifteenth-Century England," *Viator*, 4 (1973): 469.
35. That the first two European treatises exclusively devoted to distinguishing the medical problem of children from those of adults—Paolo Bagellardi, *Libellus de infantium aegritudinibus* (Padua, 1472) and Bartholomaeus Metlinger, *Ein Regiment der jungen Kinder* (Augsberg, 1473)—appeared only a year apart both north and south of the Alps, assuredly reflects the rather sudden development of a new attitude toward small children, but the practical medical effect of such treatises, as distinct from the effect of the new attitude, on infant mortality is in doubt.
36. Some buttons were used in antiquity for ornament, but apparently not for warmth. The first functional buttons are shown ca. 1235 on the "Adamspforte" of Bamberg Cathedral and in 1239 on a relief at Bassenheim; see Erwin Panofsky, *Deutsche Plastik des 11. bis 13, Jahrhundert* (Munich, 1924), plate 24; Herman Schnitzler, "Ein unbekanntes Reiterrelief aus dem Kreise des Naumburger Meisters," *Zeitschrift des Deutschen Vereins für Kunstwissenschaft*, 2 (1935): 413, fig. 13.
37. A. Latour is dubious about claims that certain finds in Coptic graves of the fifth to seventh centuries are in fact knitted; he holds that the first firm evidence of knitting is late medieval, notably a pair of knit gloves that belonged to Pope Clement V (d. 1314). "The Stocking," *CIBA Review*, no. 106 (1954): 3800.
38. Hamburg, Kunsthalle. On the date and context see Wilhelm Worringer, *Die Anfänge der Tafelmalerei* (Leipzig, 1924), 193–94, fig. 59, and Alfred Strange, *Deutsche Malerei der Gotik* (Nedeln, 1934), 2: 145–50, figs. 178, 181. The knitted shirt is the seamless garment that Christ wore at his crucifixion; legend says that his mother made it and that it grew as he did. The precocious infant turns from reading scriptural prophecy to meditate upon the vision of two angels holding instruments of his passion. A whipping top lies abandoned at his side. We lack an adequate history of toys: such a study would tell us much about adult attitudes toward children.
39. The earliest European evidence of the horizontal loom appears in the Talmudic commentary of Rabbi Rashi (d. 1105), written in Northern France; see Eleanora Carus-Wilson, "Haberget: A

Medieval Textile Conundrum," *Medieval Archaeology*, 13 (1969[1971]): 165. Rashi indicates that it is used by professional male weavers, in contrast to the old vertical loom used by women weavers.

40. Joseph Needham, *Science and Civilization in China*, vol. 4, pt. 2 (Cambridge, Eng. 1965): 758–59.
41. For the recent literature, see Walter Endrei, "Changements dans la productivité lainière au moyen âge," *Annales: économies, sociétés, civilisations*, 26 (1971): 1291–99.
42. C. M. Briquet, *Les filigranes* (2d ed.; Leipzig, 1923), 2: 317.
43. Among her several articles, see especially Elizabeth L. Eisenstein, "The Advent of Printing and the Problem of the Renaissance," *Past and Present*, no. 45 (1969): 19–89, and no. 52 (1971): 135–44; "The Advent of Printing in Current Historical Literature," AHR, 75 (1969–70): 727–43; and "L'avènement de l'imprimerie et la Réforme," *Annales: économies, sociétés, civilisations*, 26 (1971): 1355–82.
44. For the beginnings of Hebrew and Syriac printing under Islamic rule, see Jean Muller and Ernst Roth, *Aussereuropäische Druckereien im 16. Jahrhundert: Bibliographie der Drucke* (Baden-Baden, 1969), 53, 57–60, 63, 65–67. Armenian and Greek printing in the Near East began in the seventeenth century.
45. See the remarkable study by David Ayalon, *Gunpowder and Firearms in the Mameluke Kingdom: A Challenge to a Medieval Society* (London, 1956).

5.4

The Social Impact of the Railroad

THOMAS C. COCHRAN

In this chapter the term social will be narrowed arbitrarily to exclude economic and business effects of the railroad, because other chapters cover these areas, but the scope will be extended sufficiently to cover certain psychological effects. This use of "social" conforms to that usually employed by historians in defining the field of "social history." For purposes of analysis it is convenient to divide the social effects of the railroad into three categories: demographic, institutional, and social-psychological.

DEMOGRAPHIC EFFECTS

The major demographic effects of the railroad were increased spatial mobility and redistribution of population.

Spatial mobility, or migration, has been one of the major influences in shaping American culture. As Everett Lee has demonstrated, nearly all of the characteristics attributed by Frederick Jackson Turner to the influence of the frontier are also typical of the behavior required for successfully playing the role of in-migrant, whether to rural or urban communities.[1] Hence, any factor affecting the nature of migration is an important influence on social change.

Each improvement in inland transportation presumably increased the average range of migration. Even in the late colonial period newcomers who were not bound to labor usually found land within a hundred miles of the port of debarkation; and freed servants, farmers' sons, or previous settlers selling and

[1]Everett S. Lee, "The Turner Thesis Reexamined," *American Quarterly, XIII* (Spring, 1961), 77–83.

moving, probably went less than a hundred miles. Moves from the Lancaster, Pennsylvania, area to a "western" farm across the Susquehanna, or from eastern to western Massachusetts, may tentatively be assumed to be modal distances. There were, of course, longer journeys by certain groups, such as the Scotch-Irish immigrants of the 1740's, who in some instances moved from western Massachusetts to western Pennsylvania, but it seems safe to regard these as exceptional.

The river steamboat, and the opening of western lands previously controlled by Indians, extended the range of migration along the relatively small areas near waterways. Migrants who could afford to cross the Appalachians, an exhausting and expensive journey, might go considerable distances down the Ohio River. The same was true of those reaching Buffalo and sailing westward on the Great Lakes. But again, this long-distance migration was probably a small part of the total. The hard-surfaced road, after the mid-1790's, facilitated local movement in the populous East, and on a few roads penetrating into the trans-Appalachian West. But the median, as distinct from the average distance of movement, may not have been much greater than in the colonial period.

For all periods, it is impossible from published records to ascertain exact figures on the average length of travel of the population moving into a given area, such as a county, before and after the coming of the railroad. Place of birth is recorded in the censuses, but not place of last residence. Study of the "birthplace by county" data for Iowa in the censuses of 1870 and 1880 has revealed four propositions about the effect of railroad construction on population distribution:[2]

1. The railroads completely overshadowed other routes of transportation as determinants of migration.
2. They stimulated greatly the growth of areas connected with centers of population; this stimulus was most pronounced in the case of the foreign born, intermediate in the case of persons born in other states, and least in the case of persons born in the state.
3. The growth of the foreign-born population was due, in the case of the Germans, to new migrants to the area, and in the case of the Irish, to "drop-offs" from the construction crew. (Other nationalities were not sufficiently represented in the population of Iowa at the time to permit such analysis.)
4. In the first few years following the construction of a railroad, the population of a county received a relatively larger share of migrants from more distant states than either before or after that time.

[2]Based on research by G. Putnam Barber, University of Pennsylvania. The four propositions can be clearly shown in the data for Iowa by use of what Barber calls the index of migration impact. This index is constructed by taking the ratio of percentage increase in the population from a certain birthplace in a given county to the increase for the state as a whole. The index fluctuates around 100; values greater than 100 indicate that the county gained population from a given birthplace faster than the state as a whole. Strict interpretation of the index would require that one assume that the force of mortality, and the propensity to remigrate, were equal in their impact on persons of all birthplaces in all counties. This strict assumption can be relaxed, however, if all that is necessary is a general impression of the effects of differing patterns of migration.

As a result of Barber's research and Curti's study of Trempealeau County, Wisconsin, it appears that the extension of both the median and average length of each migration as a result of the railroad cannot be documented.[3] The fact that the secular trend of foreign immigration was rising at the same time (1845–1890) that the railroad was penetrating the interior strongly supports the deduction of an over-all increase in distance of migration but does not touch on the problem of the length of each specific move by the previously resident population.

Another effect of the railroad operated to increase the usual distance of migration. From 1850, the major direction of migration was not from south to north or east to west but from country to city. The migrant leaving a farm to go to a city must, in general, have selected his destination on the basis of imagined opportunity.[4] This means that he would not necessarily have gone to the nearest city, for instance from a farm in central Vermont to Rutland, but rather to a city that was expanding rapidly and offered many job opportunities, such as Buffalo, some three hundred miles away.

A minor result of the coming of the railroad was the creation in the trans-Mississippi West of large pockets of immigrants from the same country where the native language and customs would be preserved for at least a generation. Some groups, such as the Pennsylvania Germans of the colonial period, and the Cincinnati, St. Louis, and Milwaukee German concentrations established by water transport, had preceded the railroad. But the latter spread such pockets into the rural areas of the Mississippi and Missouri valleys.

This conclusion leads directly to another major aspect of the railroad impact: the location of population in areas that had not been agriculturally profitable with previous forms of transportation. The railroad with its ability to spread in most directions, particularly over the level plains, could open up more of the country than could water transportation.

It was, of course, greatly inferior in this respect to motorized transport over hard-surfaced roads. But even without the competition of the railroad, motor vehicles would probably not have been economically successful before the late 1860's or 1870's. Such transportation required light steam or gasoline engines, and while experiments with steam vehicles were under way in England in the early 1860's, and the development of the internal combustion engine in Germany from 1865 to 1885 would no doubt have been greatly stimulated, commercial adaptation would have taken time. Macadam roads, on the other hand, were technically adequate and could have been built rapidly in most parts of the United States.[5]

The demographic effect of the railroad was, therefore, a more uniform deployment of population over land resources than would have been possible previously, but a distribution that still did not give farmers the best conceivable combination of the factors of fertility and nearness to the consumer. Long fingers of population

[3]Merle E. Curti and Others, *The Making of an American Community* (Stanford: Stanford University Press, 1959), pp. 62–64.

[4]See Carter Goodrich and Others, *Migration and Economic Opportunity* (New York: Columbia University Press, 1935).

[5]See further discussion of this problem in Robert W. Fogel, *Railroads and American Economic Growth: Essays in Econometric History* (Baltimore: The Johns Hopkins Press, 1964).

stretched along the railroads and left land more than about a dozen miles from the track unused, or sparsely inhabited by semi-self-sufficient, noncommercial farmers or hillbillies—even though the hills might be very low.[6] By 1890, in the area from the Ohio Valley north and west through the first tier of states beyond the Mississippi River, main and branch railroad lines made such a fine network that in fertile areas nearly all of the land was within twelve miles of a station; but in the mountains of the South and the plains of the trans-Mississippi West this was not the case.[7]

Population was literally drawn by the railroads into the "middle border" through construction work, sale of land on credit, eastern and foreign advertising, and offers of free or cheap transportation.[8] Certainly, up to 1890, the population of the states between the Mississippi River and the Rockies increased much faster than would have been possible without the railroad. Whether all of this increase was economically or sociologically desirable is another question. Instead of filling in the more easterly states, farmers were led to cultivate the staples on large acreages at immense distances from their markets, to live in difficult and uncertain climates, and to endure relative social isolation over most of the year.[9]

By making it easier to reach interior cities, the railroad undoubtedly increased the spatial mobility of labor. This was particularly true in many small to medium-sized cities that were not on navigable waterways. The effect on the major industrial concentrations, which were served by water, would be hard to estimate. Furthermore, many factors may have been as important as transportation in determining these internal migrations. The desire of the southern Negro to reach freedom before the Civil War, and thereafter to reach greater opportunity, operated in spite of nonexistent or poor public transportation. Immigrant workers were drawn to cities where they had fellow countrymen, skilled workers to where there was a special demand for their abilities, and partially employed workers to where there was full-time work, all without much apparent regard to transportation difficulties.[10] On the other hand, workers with families, who owned homes or were well located in a city, could not readily be drawn as far as Chicago from Milwaukee by wage differentials as high as 25 per cent.

The railroads sought to stimulate immigration in order to supply labor for their own construction needs, and to provide farmers and other businessmen for the development of the areas the companies served. Importation of unskilled labor under contract to work for repayment of transportation costs involved too much

[6]See Abram Kardiner, *Psychological Frontiers of Society* (New York: Columbia University Press, 1945), pp. 294–295.

[7]*Grant's Bankers and Brokers Rail Road Atlas, 1890* (New York: A. A. Grant, 1887).

[8]See James B. Hedges, *Henry Villard and the Railroads of the Northwest* (Cambridge: Harvard University Press, 1930); and Richard C. Overton, *Burlington West* (Cambridge: Harvard University Press, 1941).

[9]See Walter A. Terpenning, *Village and Open Country Neighborhoods* (New York: The Century Co., 1931), pp. 53–54.

[10]This subject requires much more research. See Charlotte Erickson, *American Industry and the European Immigrant, 1860–1885* (Cambridge: Harvard University Press, 1957).

risk.[11] The United States was too big and jobs were too plentiful. Instead, the railroads advertised for workers in Europe and in the eastern ports, and provided cheap transportation from the coast. The Castle Garden Labor Bureau, supervised by the State of New York, was probably the chief mid-century source of immigrant labor for the railroads.[12] The *Railroad Gazette* noted in 1883, the peak year of immigration up to that time and also a peak year in railroad building, that "unless the rate of immigration continued high construction would be handicapped."[13]

The shift of population in both rural and urban regions that resulted from the railroad also relocated industry and redirected trade. Many important effects of these changes were obviously economic, but from the demographic standpoint the process was often a snowballing one. The more the railroad concentrated population, the more people seeking opportunity came to these points of concentration. Thus the process of depopulating rural areas which goes far back in American history, was stimulated by the railroad.

In the nineteenth century railroad rates were not uniform, but were based on what the traffic would bear. Between points where there was competition between railroad companies, or with water carriers, rates were kept low, often below the level necessary to cover overhead costs and moderate dividends. To compensate for this the roads charged higher than average rates between points where there was no competition. As a result, manufacturing and wholesale trade were concentrated at the competitive or junction points, and the more carrier competition there was the more likely that rates would be low. If only two carriers were competing, they might form a "pool" that set rates and apportioned traffic, but between three or more carriers pools were much more difficult to maintain.[14]

Another factor tending to make big cities still bigger was that the short, early railroads frequently ended in such cities, and might fail to connect directly with other roads or to use track of similar gauge. The resulting transshipment made the points economically advantageous for processing plants.

Thus, businesses and their workers were drawn from small city locations to the major railroad junction points. In spite of lower fertility rates the major cities grew much faster than the rest of the population. By 1900 the pattern of what would now be called metropolitan areas was, in its new parts, a creation of the railroad. Chicago, the greatest of the cities built by the railroad, grew from 30,000 people in 1850 to 1,100,000 in 1890.

Among the minor effects of the increased spatial mobility brought about by railroads was the alteration of Indian relations. The railroad increased the sense of grievance of the plains tribes since it encouraged organized hunting parties which virtually exterminated the buffaloes, and brought white men into areas awarded by treaty to the Indians. But it also ended the possibility of effective Indian military resistance. The United States government abandoned the fiction of dealing with

[11] *Ibid.*, p. 73.

[12] *Ibid.*, pp. 95–96, 118.

[13] Quoted in *ibid.*, p. 76.

[14] See Julius Grodinsky, *The Iowa Pool: A Study in Railroad Competition, 1870–1884* (Chicago: University of Chicago Press, 1950); and Thomas C. Cochran, *Railroad Leaders* (Cambridge: Harvard University Press, 1953), pp. 170–172.

the Indian tribes as sovereign nations by treaty and moved to the policy of creating restricted reservations, of granting federal aid, and eventually of granting citizenship.

INSTITUTIONAL EFFECTS

Family

Initial migration from Europe, and continued migration within America, prevented the extended family, or large interdependent kinship groups, from developing in most parts of the country. Exceptions were most notable in parts of the southern seaboard and in poor rural areas. Here the more enterprising members of the family left to seek better opportunities, and those who remained sought security through emphasizing kinship. This pattern was present to a minor extent in all areas. Cities apparently had some solid residues of kinship groups that stayed in place from generation to generation while in-and-out migrants washed over them in waves.[15] But in the cities, the security afforded by the extended family was much less than in the country, and the bonds between its members correspondingly weaker. These parents, grandparents, brothers, sisters, aunts, uncles, and cousins living apart, but nearby, might be called a fringe of the nuclear family, rather than an extended family of the type prevalent in the non-European world.

While, as argued in the previous section, the railroad increased the median distance of each single move, it also made it possible to spread the nuclear fringe more widely without breaking the fabric. Before the railroad, a son who moved his nuclear family fifty miles away from the paternal homestead was living in another area. Visits taking two days in each direction would be infrequent. With rail service the trip could be made in two hours, and visiting and co-operation between relatives could be preserved. Even within a neighborhood area six or eight miles in diameter, two or more railroad stations might serve to knit a kinship group more closely together.

Education

The railroad was probably a major factor in developing higher education in engineering. At the time that Rensselaer Polytechnic Institute was founded in 1824, the Military Academy at West Point was the only college offering advanced technical studies. The founder of the Institute, Stephen van Rensselaer, thought that training in chemistry for use in agriculture would be the school's chief function, but the demand for engineers to build bridges and railroads soon moved the curriculum in that direction.

Just before the great railroad construction boom from 1849 to 1854, Amos Lawrence gave Harvard $100,000 for a scientific school. Like van Rensselaer, Lawrence was not specifically interested in developing railroad engineers; but

[15]Sidney Goldstein, *Patterns of Mobility, 1910–1950: The Norristown Study* (Philadelphia: University of Pennsylvania Press, 1958), pp. 214–215.

Lawrence Scientific School, opened in 1847, inevitably moved toward an emphasis on the civil engineering required for railroad construction. Meanwhile, Yale had established a scientific department which became Sheffield Science School in 1854. By the time the railroad boom leveled off, Dartmouth, Michigan, and other universities had opened engineering schools or departments, and Brooklyn Polytechnic Institute (1854), Cooper Union (1859), and Massachusetts Institute of Technology (1860) had been founded as independent technical schools.

As noted in the case of Rensselaer and Lawrence School, growth of railroad demand was not the sole factor in the spread of engineering education, any more than missiles and space developments were the sole cause of the great mid-twentieth-century upswing in scientific education; but without the pressure of the roads during their major building periods, from 1849 to 1857 and 1867 to 1890, the growth of such education would surely have been slower.

Business

While a few canals like the Delaware and Hudson, and the Lehigh were operated by large corporations, the railroad was the early developer of the institutions of big business. These included one or more levels of middle, or bureaucratic management; and operating policy was formulated by professional executives rather than by the principal stockholders. Inside, the railroad companies' new management forms, described by Chandler and Salsbury, developed rapidly; outside, the companies' railroad administrators added substantially to the growing upper-income group.

Inevitably, the long-range planning necessary to a railroad, and the many contacts with the public and with government, gave railroad managers an unusual awareness of their position in society. William H. Vanderbilt's, "The public be damned," was not a declaration of independence but rather the complaint of a weary man tired of giving news releases on problems of public and governmental relations. More representative is his statement: "We must be conservative and keep the public with us."[16] It took a century to develop a reasonably full conception of what the status of professional manager in a big company involved, but the process of elaborating this conception was well advanced among railroadmen by the 1880's.

Undoubtedly, this new social group put an increased emphasis on efficiency, both technologically and socially. In trade and manufacturing, wealth came from market profits, often of a chance variety; while for the professional railroad manager wise expansion and efficient operation, rather than profit, per se, were the economic aims. As President J. C. Clarke of the Illinois Central wrote in 1885: "it is better to show moderate poverty than great success," since the latter would provoke more competition and government regulation.[17] Determining the effects on society of largely increasing the number of prominent citizens with such bureaucratic or professional views will require further research. In some instances

[16]Cochran, *op. cit.*, p. 157.

[17]*Ibid.*, pp. 127–128.

the effects might be analogous to introducing five thousand space scientists into a profit-oriented community such as Houston, Texas.

The railroad popularized investment in stocks and bonds other than those of banks or governments. Every important railroad was "publicly" financed. By 1855, the securities of roads like the New York Central or the Pennsylvania were held by thousands of investors. Higher interest rates on bonds than those paid by sound governments, and the chance for appreciation in railroad stocks, led more prosperous business or professional men to buy railroads on the security exchanges. Superficially the rise of security capitalism may seem to have been an economic effect, but it probably had important social implications. The middle-class security holders formed a generally conservative and powerful social group resistant to radical experiments or change. No country in which this group is strong has moved from capitalism to socialism.

Government

Americans of the early nineteenth century did not fear their governments. The states, particularly, were seen as useful mechanisms for advancing the general welfare, and like other people in an early stage of economic development, the citizens of some states believed that for many of the needed facilities such as banks, canals, railroads, or universities, public enterprise was more practical than private enterprise. For example, a commission on canals in New York, headed by the wealthy and conservative Gouverneur Morris and composed largely of wealthy men, reported in 1811 that "by the tests of efficiency, economy and financial capacity, private enterprise was incapable of building the Erie and Champagne canals."[18]

When Massachusetts was faced with the competitive need for building a railroad, it was thought that private capital would not venture into so hazardous an enterprise "without the pledge of more exclusive privileges than it would be expedient for the legislature to grant." "The important interests of the people," Governor Lincoln held, "can only be preserved, and the honour and prosperity of the State promoted, by a system of Governmental enterprise."[19] Prompt state action was delayed in Massachusetts by fear of the cost. Some said it would be "little less than the market value of the whole territory of Massachusetts."[20] Ultimately the state paid over half the cost of the Western Railroad.

The pioneer railroads were generally either government owned or deeply involved in financing by governmental units. They also required franchises and the ability to condemn property. Thus, like canals, they were closely tied to state politics. On some roads, such as the New York Central, railroad executives were also legislators or high state officials; but, in any case, a railroad company needed professional lobbyists. The lobbyist was usually a politically active lawyer with a practice at the state capital. Robert S. Hunt writes of the Wisconsin antilobbying

[18]Nathan Miller, *The Enterprise of a Free People* (Ithaca: Cornell University Press, 1962), p. 32.

[19]Oscar and Mary F. Handlin, *Commonwealth* (New York: New York University Press, 1947), p. 186.

[20]*Ibid.*, p. 187. There is more on this theme in text and notes.

law of 1858 that it "did not prevent the roads from dominating the Wisconsin legislature throughout most of the period from 1850 to 1890."[21]

Some roads like the Pennsylvania and the Michigan Central remained under state ownership until the late 1840's. Two roads, the Western and Atlantic of the State of Georgia and the Cincinnati and Southern, although leased to private companies, were owned by the state governments until the late nineteenth and early twentieth centuries, respectively. However, there was little government construction after 1857. Of the pioneer period to 1890, Hunt says: "The notion that railroad building was a public function, widely accepted at the beginning was faint indeed at the end."[22]

In railroading, therefore, business and government were in partnership during the early period, much as in the mid-twentieth-century atomic power industry and certain defense industries. Usually state and municipal governments were bond-holders, but occasionally, as in the Camden and Amboy Railroad of New Jersey and the Western of Massachusetts, the state was a common stock-holder. Three of the nine directors of the Western were appointed by the state.[23] In return for a gift of a thousand shares of common stock, New Jersey gave the Camden and Amboy a monopoly on passenger traffic across the state; if a competitor was chartered, the state had to surrender the thousand shares to the company.[24]

Many other types of aid by the states and municipalities in both the East and the West could be described.[25] But the point to be made here is that the railroad as an innovating agency depended heavily on government encouragement. Only a few short lines between populous eastern cities were wholly financed by private capital.

The federal government became important to the railroads first for mail contracts, and after 1850 for land grants. From 1850 to 1862 the federal grants were made indirectly through the states, but from 1862 to 1870 large grants were made by the federal government directly to private corporations for potential transcontinental roads. The Central and Union Pacific companies were also aided by some $50,000,000 subscribed by the United States for second mortgage bonds, and two other roads received some minor aid of this type. In the land grants a provision was always included for reduced fares for government freight and the transportation of troops.

These policies made strong representation in Washington desirable for all mail-carrying and land-grant roads. In some cases, such as that of the Union Pacific, lobbyists dispensed railroad stock in order to gain friends in Congress. Meeting the conditions for transfer of land from the government to the railroad, locating lands, and getting rights-of-way through Indian territories kept the officers of the growing transcontinental roads in close touch with their men in Washington up to

[21]Robert S. Hunt, *Law and Locomotives* (Madison: State Historical Society of Wisconsin, 1958), p. 171.

[22]*Ibid.*, p. 167.

[23]Cochran, *op. cit.*, p. 18.

[24]John W. Cadman, Jr., *The Corporation in New Jersey* (Cambridge: Harvard University Press, 1949), pp. 54–55. No competitor was chartered before the Civil War.

[25]See F. A. Cleveland and F. W. Powell, *Railroad Promotion and Capitalization in the United States* (New York: Longmans Green, 1909); and Harry H. Pierce, *Railroads of New York: A Study of Government Aid, 1826–1875* (Cambridge: Harvard University Press, 1953).

the 1880's. Chester A. Arthur, later President of the United States, was a lobbyist for Northern Pacific in the 1870's.[26]

American courts, always guided to a degree by precedent, tend to be conservative. For many decades it is possible to evade major issues. Wisconsin law, writes Hunt, "failed to recognize the need for anything more than the time-tested techniques of an agricultural society."[27] In Massachusetts, "looseness of control, liberal incorporation and freedom of lines to use each other's tracks encouraged the tendency to allow all to proceed at their own risk without favoritism, that is, without direction from the government."[28]

Yet little by little corporations, and railroads in particular, by forcing the courts to give "specific interpretations in case after case gradually undermined and transformed the law."[29] In general, change was in the direction of more sovereignty for the corporation and less control under the assumptions of the common law.[30]

While the railroad's effect on common law and judicial process was both protean and subtle, and has had little study, its effect on statute law was obvious and important. Railroads were in fact local monopolies whether or not they were granted such privilege in their charters; and to lay out their routes they required the right of eminent domain.[31] It was inevitable that private corporations possessed of such powers would be regulated by the states that created them. New England had the largest amount of track in relation to its area, and first had to meet most of the problems of railroads as public utilities. Fortunately the history of New England railroads has been studied in detail by Edward C. Kirkland.

Eminent domain had been used by transportation companies and public utilities, such as waterworks, before the coming of the railroad. But the rapid growth of railroad companies, and public excitement about the powers of monopolistic corporations in the Jackson period brought the right into question in the Middle Atlantic and New England states. In 1833, a Massachusetts law prescribed the method by which railroads might exercise eminent domain, and in 1844, New Hampshire provided for appeal to a board of three commissioners who could fix valuations and assess damages.[32] But the more general attitude, which prevailed in the three Middle States, was that charters were self-enforcing subject only to appeal to the Courts.[33]

While for any given area the exercise of eminent domain was a passing problem, that of rates and conditions of service remained. The first state railroad commission

[26]*Billings Correspondence* (Northern Pacific Archives, St. Paul, Minnesota).

[27]Hunt, *op. cit.*, p. 171.

[28]Handlin and Handlin, *op. cit.*, p. 193.

[29]*Ibid.*, p. 149.

[30]Hunt, *op. cit.*, p. 171.

[31]The first three Massachusetts roads were granted thirty-year monopolies.

[32]Edward C. Kirkland, *Men, Cities and Transportation: A Study in New England History* (2 vols., Cambridge: Harvard University Press, 1948), Vol. I, pp. 274–275.

[33]See Lee Benson, *Merchants, Farmers and Railroads* (Cambridge: Harvard University Press, 1925), p. 2; Cadman, *op. cit.*, pp. 222–224; Louis Hartz, *Economic Policy and Democratic Thought* (Cambridge: Harvard University Press, 1948), pp. 69ff.

resulted from the question of whether a railroad, the Boston and Providence, was open to use by separately owned connecting lines. The practical issue was settled when the Boston and Providence bought the line in question, but the issue led, in 1839, to the creation of a Rhode Island Railroad Commission of five members.[34]

Gradually the New England states and New York found it impossible to handle railroad problems by charters and legislation alone. "By the fifties," writes Professor Kirkland, "it was apparent that governors, attorney-generals, legislatures and their committees, judges, and county commissioners were not able to deal continuously, expertly and successfully with a powerful, expanding, novel, railroad interest."[35] Connecticut established a railroad commission in 1853, New York in 1855, Maine in 1858, and Vermont a single commissioner in1855.[36] These early commissions were more concerned with condemnation, connections, and public safety than with rates.

By the late 1860's spread of the railroads into western areas involving long hauls and transshipment, combined with declining commodity prices and less rapidly falling railroad charges, tended to focus public discontent on rates. The first of the post-Civil War commissions, however, was an outgrowth of a number of considerations in Massachusetts, rather than of western freight-rate problems. This commission, created in 1869, had broad responsibilities for supervision, including methods of accounting and rates, but no power except to report and recommend.[37] In Illinois, a center of violent antagonism toward the railroads, the Chicago Board of Trade, aided by farm organizations, secured a law in 1871 that fixed maximum rates and prohibited discrimination but set up a commission that lacked power.[38] Before 1880, the other states of the northern Mississippi Valley and California enacted similar laws. Only Missouri gave its commission the power to adjust rates, rather than merely to enforce maximums set by law.[39]

Taking a broad view of the twenty-year struggle between midwestern railroads and public authorities, it can be said that each local group of shippers needed the railroads more than the latter needed those particular shippers, and each year the railroads gained more subtle political leverage.[40] While the United States Supreme Court upheld state regulation in *Munn* vs. *Illinois*, in 1876, the railroads had gained friendly majorities in some of the legislatures by this time and were exerting effective influence on the commissioners. As President Henry B. Ledyard of the Michigan Central wrote in 1883, "Where there are commissioners to stand between the railroads and the public much dissatisfaction can be avoided and many things made plain."[41] Another factor working in the railroads' favor was

[34]Kirkland, *op. cit.*, Vol. II, p. 234.

[35]*Ibid.*, p. 232.

[36]*Ibid.*, pp. 233–236. For New York, see Benson, *op. cit.*, pp. 6ff. The New York Commission was ended by a corrupt deal in 1857, and not recreated until 1882.

[37]Kirkland, *op cit.*, Vol. II, pp. 238–239.

[38]Benson, *op. cit.*, p. 25.

[39]Sidney F. Miller, *Inland Transportation* (New York: McGraw-Hill, 1933), pp. 79ff; Robert E. Riegel, *Story of the Western Railroads* (New York: The Macmillan Co., 1926), pp. 143ff.

[40]Cochran, *op. cit.*, pp. 189–196.

[41]*Ibid.*, p. 191.

increasing technological efficiency which made it possible to earn profits from lower and lower rates.

While the roads were adjusting to state legislation, and vice versa, there was a movement in Congress for federal action. In 1874, the Windom Report, by a congressional commission, urged federal construction of a rail and water system to provide what would later be called "yardstick" competition for the private roads.[42] From 1878 to 1885, Representative J. F. Reagan of Texas annually introduced regulatory bills. Charles E. Perkins of the Burlington Railroad was then convinced that "the public *will* regulate us to some extent—we must make up our minds to it," and other presidents struck the same note.[43] By 1886, a number of railroad executives welcomed regulation by a federal commission as a means of preventing cutthroat competition.[44]

In the course of its first half century of relations with state and federal governments the railroad had introduced most of the presentday practices of lobbying and public relations. Passes were issued to ministers, editors, and politicians; influential legislators, senators, or public executive officers were employed while in or out of office to plead the railroad's cause; editors and free-lance journalists were hired to write articles; and, on occasion, cash payments or securities were distributed to key figures.[45] While it is obviously impossible to assess the total effect of such activities, it is significant that the United States became the only major industrial nation without state-owned railroads.

SOCIAL-PSYCHOLOGICAL EFFECTS

Without special research by psychologists in historical records it is impossible to do more than note certain obvious psychological effects of the railroad.

The railroad raised the levels of expectation in relation to material success in many communities, and altered both geographic and cultural horizons. In its daily life the prerailroad rural community represented a high degree of self-sufficiency. Only in the neighborhood of big cities was community interest sufficient in New England to support the construction and maintenance of good hard-surfaced roads.[46] Lewis E. Atherton sees the prerailroad midwestern town as characterized by "a low standard of living, relative lack of specialization, and freedom from outside control."[47] When Emerson wrote of Massachusetts, "From 1790 to 1820, there was not a book, a speech, a conversation or a thought in the State," he should have added, according to Professor Kirkland, "that there was not a railroad. For the

[42]Riegel, *op. cit.*, p. 142.

[43]Cochran, *op. cit.*, p. 197.

[44]Gabriel Kolko, *Railroads and Regulation, 1877–1916* (Princeton: Princeton University Press, 1965), pp. 26ff.

[45]Cochran, *op. cit.*, pp. 184–189, 193–196.

[46]Kirkland, *op. cit.*, Vol. I, p. 37.

[47]Lewis E. Atherton, *Main Street on the Middle Border* (Bloomington: University of Indiana Press, 1954), p. 42.

railroad, even though it may not have opened wider prospects, at least revealed different ones."[48]

The initial effects of the railroad, often merely of the planning for a railroad, have been extensively discussed. Describing a meeting to promote a railroad in western Pennsylvania, the editor of the *Pittsburgh Daily Gazette and Advertiser* wrote: "We believe no other subject matter whatever, unless it was akin to a revolution when men felt their religions or civil liberties were at stake could have collected such a gathering."[49] The people of Oswego, writes Professor Pierce, were "unanimous for anything in the form of a railroad, whether it goes crooked or straight they seem to have no care."[50] In Massachusetts the influence of the church was invoked by the promoters of the Western Railroad. Ministers were asked to discourse "on the moral effects of Rail-roads."[51] One of the most interesting accounts of the feverish enthusiasm, overoptimism, and broader perspectives generated by the coming of the railroads is in R. Richard Wohl's, "Henry Noble Day."[52] The coming of a branchline railroad to Hudson, Ohio, stimulated visions of [53]

> a whole sub-system of other roads all centered on Hudson. . . . For Henry Day himself the dream had become a sacred reality. He was able to anticipate its completion while the first few spadefuls of earth were taken out for grading. He proceeded therefore to enact the logic of his expectations. He began to create a network of businesses in Hudson to service the demands arising out of the railroad building boom and to cash in on the enlarged market which would result once the roads were completed.

The resulting boom and bust was painful, but undoubtedly educational as well.

Railroads brought city newspapers to rural towns with advertisements of bargains in clothes and home fittings, and of amusements not thought of before by country people. Now on a Saturday the farm family could take a one-day cheap excursion train to the biggest nearby city and live for four or five hours in a new world.[54]

New contacts for rural people and a broader sphere of influence for urban institutions undoubtedly stimulated innovation. Bringing the farmer to the city could of itself be important. In the United States, writes H. G. Barnett, "the cultural differences between rural and urban life patterns that appear at farmers' markets are not insignificant from the standpoint of change."[55] The disruption of normal rhythms of activity by local booms, such as that of Hudson, Ohio, provided the

[48]E. W. Emerson and W. E. Forbes, eds., *Journals of Ralph Waldo Emerson* (Boston: Houghton Mifflin, 1909–1914), Vol. VIII, p. 339. Kirkland, *op. cit.*, Vol. I, pp. 92–93.

[49]Louis Hartz, *op. cit.*, pp. 11–12. From paper of March 18, 1846, cited in *Western Pennsylvania History*, Vol. XIII (1930), p. 252.

[50]Pierce, *op. cit.*, p. 42.

[51]Handlin and Handlin, *op. cit.*, p. 227.

[52]In William Miller, ed., *Men in Business* (Cambridge: Harvard University Press, 1952), pp. 153–192 (Torchbook edition, 1962).

[53]*Ibid.*, pp. 176–179.

[54]See Atherton, *op. cit.*, p. 230.

[55]H. G. Barnett, *Innovation: The Basis of Cultural Change* (New York: McGraw-Hill, 1953), p. 46.

individual with "auspicious circumstances for the emergence of new ideas, many of them predatory. . . . The microcosm in which he has lived has been destructuralized and is not habitable in that condition. He strains to give it some organization and some meaning, and in so doing he innovates or accepts definitions of the situation offered by others."[56]

To these normal incentives to innovation and change new American communities presented the added factor that, "Migrants and dispossessed populations are characteristically receptive to new ideas, whether these ideas are developed by their own members or suggested by outsiders."[57] The ideological dangers to the *status quo* inherent in the railroad were recognized by the school board of Lancaster, Ohio, which forbade the use of the schoolhouse for a meeting to promote a railroad on the basis that "such things as railroads and telegraphs are impossibilities and rank infidelity . . . if God had intended that His creatures travel at the frightful speed of fifteen miles an hour by steam He would clearly have foretold it through His holy prophets."[58] Certainly, the fundamentalist country churches had to fear the effects of increasing urbanism, cosmopolitanism, and worldly sophistication that came to town on the railroad.

An obvious social-psychological result of the railroad was a new consciousness of time. There were two distinct aspects involved in this change: first, an increasing tempo of life that had to be more closely regulated by time; and second, the substitution of standard time for individual or local time.

In prerailroad days businessmen calculated in terms of months. Inventories were replenished by semi-annual trips to wholesale markets, credit terms were six to ten months, or in some rural areas from crop to crop.[59] During the northern winters commercial activity in the interior of the country slowed nearly to a halt. The early nineteenth-century spread of industry and commerce introduced the need for punctuality in employees, but did not speed up the pace of major decision making.

The railroad and the Civil War, nearly contemporaneous in the Middle West, shortened credit terms. The railroad made it possible to do business with much smaller inventories, and the wartime uncertainties and inflation made wholesalers anxious to get their money as soon as possible. Metropolitan salesmen could now visit the shopkeepers and manufacturers of interior towns and sell them orders that could be delivered within weeks, while payment came to be expected within thirty days after delivery.

The increased tempo went deeply into social and business life. Railroad timetables became important in many ways: commutation into big cities grew from the start, children commuted to high school, newspapers and other mail arrived or left by certain trains, visitors came and went by train, and each of these activities had to be set to the minute of scheduled train times.

[56] *Ibid.*, pp. 71–72.

[57] *Ibid.*, p. 87.

[58] *Ibid.*, p. 249.

[59] Victor H. Clark, *History of Manufactures in the United States* (3 vols., New York: McGraw-Hill, 1929), Vol. II, p. 9; and Lewis E. Atherton, *The Southern Country Store, 1800–1860* (Baton Rouge: Louisiana State University Press, 1949), pp. 52ff.

In the early days of the railroad, time was still read from sundials. While such readings altered only slightly over short distances, between cities a few hundred miles east or west of each other the variation would be several minutes. A railroad junction point, such as Buffalo, would have half a dozen clocks showing the time in other major cities such as Albany, Cleveland, Detroit, or Chicago.

The railroad came to control time, first, by making the station clock the criterion for each town, and second, by establishing four time zones covering all of the United States. The latter change was the result of a decade of timetable conventions, and the ultimate agreement of 1883 superseded fifty-four regional times represented on railroad schedules.[60]

Whatever framework or model used for analysis, almost no one questions the fact that material progress has generally been welcomed in American culture. Robin M. Williams, Jr., in a recent analysis of value orientations in America outlines the major headings of achievement, success, activity, work, efficiency, practicality, progress, and material comfort. These values, all of which would be favorable to the reception of the railroad, make up half of Professor Williams' list.[61] None of the other eight, such as freedom, equality, or external conformity, appear to be of a sort that would oppose acceptance of the railroad. While other students of American culture may define the dominant values or major cultural themes in slightly different words, there seems to be no basic disagreement on the general social attitudes of middle-class Americans in the nineteenth century.[62]

Granting that value systems have, by definition, a normative quality, coincidence of the new directions of behavior introduced by an innovation and the sanctioning force of a large part of the value system should produce a strong positive reaction. Put another way, if the changes in social roles required by the innovation are in positively sanctioned directions the role players will respond with high morale and presumably some creativity.[63] These considerations reinforce what has been urged about the innovative effects of the railroad.

If, then, the railroad as an innovation is eagerly received in the culture, it should also reinforce the same traits that account for its reception. By accentuating existing cultural differentials, the railroad presumably made Americans more "American." It seems quite possible that space travel may have similar reinforcing effects.

[60]See Robert E. Riegel, "Standard Time in the United States," *American Historical Review, XXXIII* (1927), 84–89.

[61]Robin M. Williams, *American Society* (2nd ed. rev., New York: Alfred Knopf, 1960), pp. 415–470.

[62]See Clyde Kluckhohn, "Some Aspects of American National Character," *Human Factors in Military Operations,* ed. Richard Hays Williams (Baltimore: Operations Research Office, Johns Hopkins University, 1954), pp. 118–120; and Thomas C. Cochran, "The Social Scientists," *American Perspectives,* eds. Robert E. Spiller and Eric Larrabee (Cambridge: Harvard University Press, 1961), pp. 110–112.

[63]See Thomas C. Cochran, *The Inner Revolution* (New York: Harper and Row, 1964).

5.5

James B. Eads and the Great St. Louis Bridge

JOSEPH E. VOLLMAR, JR.

PART I THE EADS BRIDGE AND THE INCOMPARABLE MR. EADS

It would take a spectacular show to beat fireworks down on the levee on the second of July, but James Buchanan Eads was just the man to present it.

His great bridge spanning the Mississippi was finished at last, and he wanted to prove it wouldn't fall down. He had invited everybody in St. Louis to the show.

But by the 1870's one of every four bridges was falling; and before the decade was over, more than forty a year would collapse. Would his bridge fall this day?

Many sensible people thought so. It was almost eerie the troubles the self-proclaimed engineer had had in the seven years it had taken to build the bridge: unbelievable political and financial wrangles, shipwrecks, ice storms, tornadoes and fourteen men strangely dead from "caisson disease." Was Eads' fantasy heroic or a monumental folly?

If foolish, it was going to be catastrophic this day—July 2, 1874. Captain Eads had arranged the spectacle himself. Fourteen fifty-ton locomotives, their tenders filled with coal, water and daredevil passengers, were lined up to cross to the Illinois side. First seven were to go in a sort of warm up, then fourteen with seven on each track and finally the whole lot in one mighty, chugging pull.

Thousands stood on the banks and upper roadway of the bridge and on the waterfront. They gaped as the engineers started up, their engines noisy and eager. Suddenly, one engineer reversed his engine, and the wheels of his locomotive spun in reverse. The onlookers gasped as the balky locomotive was literally dragged across the bridge by the momentum of the other locomotives.

The bridge did not fall.

It was a miracle to the spectators. But to Eads, it was only a flourishing finish to the job he had set out to do. "Must we admit that because a thing never has been done, it never can be?" Eads had asked a convention of twenty-seven engineers on August 20, 1867.

Now, "seven years and seven million dollars later", as tongues wagged, the St. Louis Bridge was the world's first alloy steel bridge and, in fact, the first important construction of any sort in steel. It was also the biggest bridge of any type ever built up to that time. It was the first to use tubular chord members and the first to depend, in the building of the superstructure, entirely upon the use of cantilever, with no false works employed. It was the first, important use, and the deepest, of compressed air in America. And the caisson that Eads sank for his East Abutment foundations remains nearly one hundred years later the deepest at which compressed air workers ever have worked.

In his remarks before President Ulysses S. Grant on the official day of dedication two days later, Captain Eads spoke of his own calm and certainty that the bridge would endure. He said that people had commented that he must be relieved to discover his bridge was sound. "But I felt no relief," he said, "because I felt no anxiety." As much as any man who ever lived, he was the right man for the right job at the right time.

To be sure, some engineers, even great bridge builders, had proposed spanning the Mississippi at St. Louis. The late, rash Colonel Charles Ellet had planned a suspension bridge for vehicles and cattle in 1839, but the city council had rejected it as both unsafe and extravagant. Josiah Dent of St. Louis obtained charters from the states of Illinois and Missouri in 1855 to form a company to erect a railway suspension bridge with a single span of 1,500 feet. This was to cost $1,500,000, but the project was abandoned when subscriptions raised less than $40,000.

The great John A. Roebling himself, later with his son, Washington, the designer of the Brooklyn Bridge, stepped forward with a proposal, and with yet another in 1868 after Eads had already started the East Pier. His second design had suspension cables and parabolic arches.

"Railroad mania" heightened interest in a bridge to cross the river from St. Louis to Illinois Town, as East St. Louis was then called. From the East, the Baltimore & Ohio reached the Mississippi in 1856 and dumped its cargo on the riverfront for ferriage to St. Louis. Other railroads were arriving, too.

The Wiggins Ferry Company of St. Louis monopolized the lucrative business of ferriage, but all of the ferries were worthless when the river was low or stormy or laden with ice. And none moved from dusk to dawn each day.

Slight but muscular "Captain" James Buchanan Eads, whose cousin would become President, had earned his title in a lifetime of hard, onerous work on and in the river. As a thirteen-year-old, he had arrived with his parents from Indiana via Louisville, the steamboat having literally burned beneath them on the day of their arrival in 1833. While his genial but destitute father moved on to the new town of Parkhurst, Iowa, young James sold apples on the riverfront, then went to work for a drygoods merchant to help his mother out. It was the merchant, Barrett Williams, who provided books, particularly textbooks on mechanical engineering and the design of boats, that Eads studied.

At nineteen, James signed on with the "Knickerbocker," a steamboat in the Northern lead trade, as the second clerk. This boat, too, sank upon being ripped open by a snag in the river; and the young man decided that the Mississippi River bottom was a storehouse of treasures simply waiting to be salvaged. He designed a boat with twin hulls, derricks and pumps and offered a part of a business in salvaging to two young shipbuilders who would make it for him, St. Louisans Calvin Case and William Nelson. Though the boat was certainly different than other river craft, James Buchanan Eads was, nevertheless, its "captain."

The river beds were every bit as fascinating as the riverfronts and Eads, in what he called his "submarine" found sunken boats in the Cumberland, the Tennessee, the Ohio and Missouri, as well as his native Mississippi. All of this cargo he disposed of for profit and his prospects seemed good.

But neither he nor his prospects looked very promising to arrogant Colonel Dillon, whose daughter, Martha, James wanted to marry. The young suitor sold his business to venture into glassmaking but was beset with petty crises. Pots broke, whole batches of glass were ruined, his men quit, and his glassware, considered a luxury, didn't sell during the recession. In 1848, he borrowed fifteen hundred dollars, bought back his share in the "submarine" and returned to the river to dive for shipwrecked cargo.

Years later, after Martha had died of cholera, and he had married the widow of a cousin, Eunice Hagerman Eads, he grew prosperous as the captain of seven submarines. He retired to Compton Hill and a palatial house he built. During the Civil War, he used his knowledge of the Mississippi and his experience at boat design to engineer the first ironclad gunboats. Though others had also designed ironclads, it was Eads' monitors that were the best of them, opening up the Confederate-held lower Mississippi to commerce again.

A small bridge company was organized in 1866 and James Eads was chosen Chief Engineer because of his intimate understanding of the Mississippi, and particularly the river bed. Other pilots could read the surface skillfully for menacing, hidden snags and bars but only Eads knew the fluid, shifting, treacherous bed. He had seen the depth change from 20 to 100 feet at obstacles in the bottom as the result of the scouring action of currents. For more than thirty years, he had watched the river rise and recede as much as 41 feet, ice floes pile up to a depth of 20 feet and tornadoes strike like giant fists.

But not all of the obstacles were in the river. L. B. Boomer, a bridge contractor from Chicago, obtained an exclusive right to build the bridge from the Illinois legislature, called a convention of engineers to approve his truss bridge of six spans, and had printed a report in a fancy pamphlet for distribution all over the country. Eads defended his own design forcefully afterward at the Merchants Exchange, his associates backed him loyally, and he began construction on the West Abutment. He had also weathered well the condemnation of the plan by J. H. Linville, president of the Keystone Bridge Company of Pittsburgh. Linville said Eads' bridge design was "unsafe and impracticable."

The controversial blueprint was of a bridge with three spans supported by four piers. Three of these piers would rest on bedrock and the fourth on metal spikes driven into the rock. Two of the spans were to be 502 feet with the one in the center, 520 feet. Three arches of steel between the piers would be erected by cantilevering. No other falsework would be used, and therefore river traffic could move freely during the building.

The Chief Engineer and his assistants, Colonel Henry Flad and Charles Pfeiffer, sank a cofferdam on the site for the West Abutment. Of two courses of sheet piling, it was a box-like structure driven down through the sand bed of the stream to bedrock, a depth of 47 feet below "high water", the height of a memorable flood of 1826. One course of the sheet piling was inside of the other with a six-foot margin between, a box within a box, and the margin filled with clay. Men pumped out the enclosure, removed debris and excavated the sand.

But the laborers struck the grave of 29 steamers which had burned and sunk in the great fire of 1849 and the cofferdam came to rest on a part of these wrecks. Chains, old engines and the remains of aged barges were also buried there and had to be removed. Excavation was agonizingly slow and construction lagged during the winters of 1868 and 1869.

Though Congress approved his report on the bridge and Boomer was appeased, Captain Eads had to quit the work. He had contracted severe bronchitis in 1868 and decided to go abroad to convalesce, a prescription at the time for those who could afford it. What the engineer saw in Europe greatly altered his plans for the West and East Piers and the East Abutment.

Bridgebuilding in Europe was more advanced than in America. In 1851 in England, a caisson instead of a cofferdam was sunk to 61 feet for piers of the Rochester Bridge. It utilized the "plenum process" for which Lord Cochran had obtained a patent in 1850. This was a great improvement over the ages-old cofferdam, first described by Vitruvius, Roman architect to Emperer Augustus.

Eads saw the pneumatic caisson first hand at Vichy where the chief engineer for his contact, Petin, Gaudet et Cie.—a Monsieur Moreaux—was constructing a bridge over the Allier. He was utilizing an earlier invention later improved upon by Fleur St. Denis, in his bridging of the Rhine at Kehl. The pneumatic caisson was an efficient complex of four compartments, with two shafts equipped with airlocks, one for men, the other for material. To get excavated mud and rock out of the caisson without having to carry it up through the airlock, St. Denis constructed a third shaft leading downward, this one filled not with compressed air but with water. It terminated inside the caisson, and an endless chain of buckets descended through this "water shaft" and brought up the excavated sand and gravel.

Another French engineer whom Eads met, Monsieur Audernt, had built no fewer than forty piers by the pneumatic method, including one 75 feet deep at Piacenza, Italy. In the United States, the use of compressed air had been pioneered by General William Sooy Smith but only in relatively shallow foundations for railroad bridges over the Pee Dee and Santee rivers in the Carolinas.

If Eads had known of the work of the Europeans and Sooy Smith, he had apparently regarded their techniques as too dangerous for the deep and treacherous Mississippi. Now he changed his mind and decided to sink a pneumatic caisson for the East Abutment—a depth far beyond those reached by Brunel and Audernt.

Eads returned to St. Louis and his previous "headaches." At home, he reasoned that he needed experience before sinking a caisson to the great depth of the East Abutment site. He decided to construct the East Pier, where bedrock would be next deepest.

The pneumatic caissons which were built by Eads' former partner in his submarine business, William Nelson, were described by Eads as "enormous elliptical bells, 82 feet long, open at the bottom," made of wood and sheathed with iron for a cutting edge. The one for the East Pier was 60 feet wide, that for the West Pier 48 feet wide. As the workmen kept digging, the cutting edge settled further into the river bed toward rock. Steadily, they removed the stones, debris and other impediments by means of the sand pump and other workmen disposed of these away from the site of the abutment.

As the excavators, who would be called "sandhogs" in a later day, burrowed deeper, a few had stomach pains or fleeting paralysis.

Dr. A. Jaminet, called in by Eads to help the ailing men, wrote later, "A workman walking about with difficult step and a slight stoop was at first regarded as a fit object for jokes, and cases of paralysis soon became popularly known by the name of 'Grecian Bend' ". It was, of course, what is now called the "bends", and, though unknown to Eads or Jaminet at the time, referred to in Europe as "caisson disease."

When the chamber was 76 feet underwater a workman suffered such severe abdominal pains he had to be taken to a hospital. Noting that many who suffered were alcoholic and undernourished, the Chief Engineer shortened the work periods, instituted rests, served beef tea; but the men, bored, spent free time in the nearest waterfront saloons.

The East Pier caisson reached bedrock after four months at 93½ feet underwater, although high water depth would be 26 feet greater. During the course of filling the air chamber with masonry, a freshet caused the river to rise alarmingly. Ten days later, James Reilly came to the surface, said he felt fine, toppled over and died. He was the first casualty of "caisson disease" in America. Thirteen other sand excavators died during the building of the bridge, all but one on the East and West Piers. Casualties were kept at a minimum on the East Abutment because of the even more rigorous discipline Dr. Jaminet insisted upon, perhaps.

Work on the West Pier was pushed simultaneously with that on the East Pier, and the excavators struck bedrock on April 1, 1870 in a submersion of 78 feet—86 feet below the directrix (a fixed mark for the high water level of the flood of 1828).

Three of the four great supports were at bedrock, and the caisson for the East Abutment was towed into place on November 3, 1870. It was an enormous, iron-sheathed monster, requiring a boat-mounting with a ten foot draft. Three steamboats towed it across the river to the Illinois shore.

The foundation for the East Abutment progressed in fits and starts. During the winter, a huge ice gorge was deflected by a breakwater from breaking up the caisson. Spring floods threatened the abutment masonry. But the greatest peril was a raging tornado that crumpled the superstructure in minutes. The damage was repaired and the work resumed, and the East Abutment at last stood on bedrock 127½ feet below the directrix or 110 feet when the water level was "normal."

After three and a half years, the foundations stood, unshakeable, in the river bed awaiting the steel arches.

The outstanding steel company of the day, Keystone Bridge Company of Pittsburgh, was headed by J. H. Linville, who had so rudely criticized Eads' bridge as "unsafe and impracticable" in the blueprint stage. He still wanted nothing to do with the Bridge in 1870, but his brash vice-president, Andrew Carnegie, could sniff money. Keystone became the major subcontractor though Linville and Carnegie thought Eads' demands for quality were outrageous.

The financial wizard grumbled, "Nothing that would and does please engineers is good enough for this work."

Eads had asked that the new steel parts be fashioned exactly to new standards of either "elastic limit" or "modulus of elasticity," terms he then defined. He said that the elastic limit was that point to which a piece of material could be compressed under pressure or pull and still be able to resume its original form when the force was removed, and modulus of elasticity was the ratio of stress to strain.

Each rib arch was to have four ribs, each rib having two lines of tubes connected by a single system of bracing. A rib was formed by a series of straight tubes 18 inches in diameter and 10 to 12 feet long. These tubes, each made of six steel staves, would be joined by couplings, end to end, at a slight angle, which, recurring, would shape the curve of the arch. All in all, 1,036 tubes (comprising 6,000 staves), 1,024 main braces, 112 huge anchor bolts, and tension rods, steel pins, nuts and so on. Keystone subcontracted the work to Butcher Steel Company of Philadelphia, and Carnegie kept the wrought iron order for his own firm, Carnegie and Kloman. But, even so, wrought-iron skewbacks, enormous plates and sockets weighing three and a half tons each, which, bolted to the arch ends, were to be formed solid and the holes drilled—unheard of.

The assistant engineers, Flad and Pfeiffer, had devised a machine for testing strength and Eads had this installed in the Butcher plant. But after six months and 6,000 steel staves, none could be used. Two more steel firms were called in until one could make the proper alloy, and then not every stave could meet the test.

The cost of the bridge to March 1, 1872 was over $4,000,000 with another nearly three million needed for completion. A story of two men crossing on the good, old reliable ferry, one a St. Louisan and the other a stranger, has survived. The stranger had asked his acquaintance, "How much will it cost to build the bridge?"

"Seven million dollars," the acquaintance had exaggerated.

"And how long will it take to build it?"

"Seven million years."

Already five years of the seven million had gone by, and there had been the wreckage under the West Abutment, the extra costs of sinking the caisson to bedrock at the East Abutment, caisson disease, the tornado and now wrangles over the steel and wrought iron parts. Eads went to the factories to oversee the making of these parts himself.

The building of the arches was, itself, a great engineering feat. To that time, arch bridges had been supported by centering—timber falsework—in the river during raising. Eads could not obstruct the Mississippi so he projected the arches beyond the piers by cantilevering—falsework above the abutment or pier to balance the thrust of the arches.

Both Eads and Flad had solutions for closing. Eads' way was to lop off the final two steel tubes five inches and cut screw threads inside the ends of the tubes. A short wrought-iron plug fitted with two sets of threads could be screwed in, and, at the arch closing, drawn out to fit across, holding the two together. A steel band would give the joint greater strength.

On September 14, 1873, Captain Eads was in London negotiating a new loan which was contingent upon a deadline five days later for the closing of the arches. Hot weather hit, and the ribs expanded instead of contracting. Flad built long troughs under the ribs to hold, first, thirty thousand pounds of ice, then sixty. The workmen wrapped the ribs in gunny cloth which soaked up the ice water, but the temperature wouldn't quit; it rose steadily to 98 degrees by five o'clock in the afternoon on the 16th of September, and the tubes, once only a mere five-eighths of an inch too long, grew. After fifty hours on the job, the men used Eads' extensible links instead.

No sooner were the arches closed than a new political hassle loomed. Steamboaters objected to the height of the arches, only a 55-foot clearance for their smokestacks, which, in some cases, were 100 feet above the water. Unbelievably, the Secretary of War backed the steamboat demand to rid the river of this obstruction, which could mean tearing the bridge down!

Flabbergasted, Eads and Dr. William Taussig, a vice-president and general manager of the bridge company, went to see President Grant. Dr. Taussig wrote of the meeting that an angry President upbraided the Secretary: "You cannot remove this structure on your own judgment . . . I think, General, you had better drop the case."

In early 1874, Eads made yet one more trip East to reassure bondholders and bankers that the job was nearly done. While in New York, he got a telegram with truly stunning news: two tubes in the first span had ruptured.

Unless the builder could discover the cause, this would mean that the bridge, as conceived, was basically unsound. But the engineer knew his great work well, even at a distance; he reasoned that the steel cantilever cables were probably still attached to the arches and should come down. Probably the cold had contracted the cables causing the ribs to be pulled in the wrong direction. That was why the ribs broke. He wired his superintendent to loosen the cables.

When late spring came, Eads' bridge was under seige. Keystone was to supply workers as well as steel and iron parts, but Carnegie had held out for more bonuses. In February, St. Louis Bridge Company had agreed to pay Keystone officials a thousand dollars for each day before June 1st that railroads and vehicles could not use the Bridge. And Keystone was to pay a penalty of five hundred dollars for each day after June 1st required to complete their contract. At once, the arch had swarmed with workers. The upper road of the bridge—for street cars, vehicles, and pedestrians—was finished April 15th.

Carnegie, with his attorney and the manager of Keystone, rushed to St. Louis in Eads' absence to collect their bonus—in advance. Carnegie's last tantrum was to withdraw a part of the agreement to open the upper roadway for traffic and hold the bridge as security. His own workmen constituted the army. "We held the Bridge," Carnegie boasted years later. He came out well on the compromise that ensued.

On Sunday, May 24, the sidewalks were opened to pedestrians. The following third of June, Joseph Gartside drove four "spanking" bays hitched to the heaviest of the coal wagons to the other side.

Then it remained only to test the bridge for its primary purpose—providing a railroad link from West to East.

But one fearful, balky railroad engineer apparently felt too much muddy Mississippi had already gone under Eads Bridge. He reversed the wheels of his locomotive in one last, futile effort to stop the great engineer's magnificent show.

The bridge stood.

PART II JAMES BUCHANAN EADS A LEGEND OF OUR TIME

The English called him "one of the greatest engineers of all time" and gave him the Albert Medal in 1886. He was the first American to receive it.

The Hall of Fame of New York University, New York City admitted him in 1920. He was the first engineer to be so honored.

The deans of engineering schools in the 1930's declared him one of the five greatest engineers of all time—along with Da Vinci, Watt, de Lesseps and Edison.

James Buchanan Eads was not without honor in his own country and, indeed, everywhere in the world in his time. When he died on March 8, 1887, the Augusta (Georgia) Chronicle announced, "The greatest engineer in the world is dead."

But he is relatively unknown today despite his monumental achievements and the uniqueness of his character and personality. He built the Eads Bridge across the Mississippi linking eastern and western segments of the railroads, the jetties of the lower Mississippi that enabled New Orleans to grow from eleventh to second largest port in the nation and the ironclad gunboats of the Civil War that turned the tide for the successful Union effort. He knew more about the Western waters, as the Mississippi and other rivers were called then, than any man of his time. And he proposed a ship-railway across the isthmus of Tehuantepec, Mexico that was marvelously innovative in conception and lauded worldwide. His projects cost him his health and much of the profit he had made earlier in his career, but he persevered because he thought citizens were entitled to cheap transportation and easy access to all parts of the nation.

His world-wide reputation as an imaginative yet practical engineer who attempted "the impossible" was a reason he was able to keep himself financially afloat. While building the great bridge at St. Louis or the jetties at New Orleans, or the ironclads for the Union Army he was a consultant to foreign governments and was paid well for his opinion. On one occasion, at the request of the House of Lords, Mr. Eads appeared before that body and presented his opinion on the building of a ship canal from Manchester into the Mersey river; the fee he received for his three weeks study was the largest ever paid to an engineer up to that time.

Young Eads was literally cast adrift on the banks of the Mississippi when he was thirteen. His mother, sisters and he had left from Louisville by steamboat preceding his father, who intended to resettle in St. Louis. During the night before the landing in St. Louis, the boat burned and the family's possessions were lost. The boy went to work instead of to school. His employer, Barrett Williams, a drygoods merchant, lent him books, including scientific works, and James studied in his spare time.

At twenty-two Eads established a business to recover wreckage from the Mississippi, designing a special twin-hulled boat with derricks and pumps, and dived for the bounty himself. Because the salvager had the sole right to freight submerged more than five years and because he worked uncommonly hard, Eads acquired a fortune by the time he was thirty-seven. He learned the vagaries of the river intimately and understood more than anyone else the treacherous river bed. He knew that depths changed constantly because of obstacles and scouring action of the current.

His great accomplishments were the result of this understanding and his early, lonely study of the laws of physics. But the successes would not have been possible had he not had some extraordinary inner strength. He was beset with personal misfortune time and time again. He lost his first wife to a cholera epidemic while the steamboat on which they were traveling was grounded on a sandbar near Evansville, Indiana. His young son and many other members of his family succumbed to other diseases, too.

The political setbacks were nearly as devastating as his personal ones. Though the Legislatures and the Congress approved his projects, they often did not appropriate the funds allocated nor award them soon enough. Eads therefore was very nearly always his own fund raiser, and oftentimes he sacrificed profit or contributed his own money.

The bridge across the Mississippi at St. Louis dedicated July 4, 1874, was the most outstanding engineering feat of the time. Called later "the grandfather of modern bridges with steel arches," it was literally the first of its kind in many ways. It was the world's first alloy steel bridge and, in fact, the first important construction of any sort in steel. It was also the biggest bridge of any type ever built up to that time. It was the first to use tubular chord members and the first to depend, in the building of the superstructure, entirely upon the use of cantilever, with no false works employed. His patent on the pneumatic caisson was later used by other great bridge builders. Eads held seventeen patents for bridge foundation and superstructure construction.

It was the first, important use, and the deepest, of compressed air in America. And the caisson that Eads sank for his East Abutment foundations remains today the second deepest at which compressed air workers ever have worked, having been exceeded in 1963 by only a few feet.

The Eads Bridge is still in good condition and serving the railroads and vehicular traffic nearly one hundred years later.

The South Pass Jetty (now Port Eads) on the Mississippi River Delta in Louisiana was a wholly different achievement called later "the most difficult piece of engineering in river hydraulics" the world had then seen. The mouth of the Mississippi was so clogged with debris and part of over 200 million cubic yards of good topsoil (some from as far away as the Rocky Mountains) that navigation into the Gulf was impossible. Also, there was not just one mouth but many; the shallow waters spilled over the Delta. There were no channels deep enough for easy, direct navigation into the Gulf.

Army engineers would dredge the passes, managing to get a river depth of eighteen feet sometimes, but the depth would hold for no more than three years before the debris and sand would choke the passes again. Eads proposed building jetties made of pilings and "mattresses" of willow branches weighted down with concrete blocks to contain the river. The swifter current caused by the greater volume through the passage would then carry the debris and sand much farther out into the deeper Gulf of Mexico instead of accumulating at the mouths of the Mississippi. Though the easiest and deepest pass to jetty was the Southwest Pass, Eads was authorized by Congress to improve only the smallest river mouth, the south Pass (now called Port Eads) on the Mississippi River Delta in Louisiana, on a "no cure, no pay" basis. He had to prove that it would work after constructing it at his own expense before he was reimbursed for the construction costs.

By January 1878, Eads' men had created a channel twenty-two feet deep and about two hundred feet wide. Almost two and one-half million cubic yards of sediment had been removed from the bar and washed out into the sea. In July 1879, after four years, the jetties were completed and the sand bar was gone. The river was thirty-one feet deep. Immediately, river trade increased tremendously, and transportation costs dropped sharply. Savings on transportation of one year's cotton crop alone was equivalent to the cost of the entire jetty project. How well did it work? According to a Corps of Engineers representative of the U.S. Army, "Eads' work on the South Pass (now called Port Eads) on the Mississippi River Delta in Louisiana is still today self maintaining . . . A total success."

5.6

The Plan

DAVID McCULLOUGH

The shapes arise!

Walt Whitman

They met at his request on at least six different occasions, beginning in February 1869. With everyone present, there were just nine in all—the seven distinguished consultants he had selected; his oldest son, Colonel Washington Roebling, who kept the minutes; and himself, the intense, enigmatic John Augustus Roebling, wealthy wire rope manufacturer of Trenton, New Jersey, and builder of unprecedented suspension bridges.

They met at the Brooklyn Gas Light Company on Fulton Street, where the new Bridge Company had been conducting its affairs until regular offices could be arranged for. They gathered about the big plans and drawings he had on display, listening attentively as he talked and asking a great many questions. They studied his preliminary surveys and the map upon which he had drawn a strong red line cutting across the East River, indicating exactly where he intended to put the crowning work of his career.

The consultants were his idea. In view of "the magnitude of the undertaking and the large interests connected therewith," he had written, it was "only right" that his plans be "subjected to the careful scrutiny" of a board of experts. He did not want their advice or opinions, only their sanction. If everything went as he wanted and expected, they would approve his plan without reservation. They would announce that in their considered professional opinion his bridge was perfectly possible. They would put an end to the rumors, silence the critics, satisfy every last stockholder that he knew what he was about, and he could at last get on with his work.

To achieve his purpose, to wind up with an endorsement no one could challenge, or at least no one who counted for anything professionally, he had picked men of impeccable reputation. None had a failure or black mark to his name. All were sound, practical builders themselves, men not given to offhand endorsements or to overstatement. With few exceptions, each had done his own share of pioneering at one time or other, and so theoretically ought still to be sympathetic to the untried. They were, in fact, about as eminent a body of civil engineers as could have been assembled then, and seen all together, with their display of white whiskers, their expansive shirt fronts and firm handshakes, they must have appeared amply qualified to pass judgment on just about anything. The fee for their services was to be a thousand dollars each, which was exactly a thousand dollars more than Roebling himself had received for all his own efforts thus far.

Chairman of the group was the sociable Horatio Allen, whose great girth, gleaming bald head, and Benjamin Franklin spectacles gave him the look of a character from Dickens. He fancied capes and silver-handled walking sticks and probably considered his professional standing second only to that of Roebling, which was hardly so. But like Roebling he had done well in manufacturing—in his case, with New York's Novelty Iron Works—and forty years before he had made some history driving the first locomotive in America, the *Stourbridge Lion*, all alone and before a big crowd, on a test run at Honesdale, Pennsylvania. He had also, in the time since, been one of the principal engineers for New York's Croton Aqueduct and so was sometimes referred to in biographical sketches as "the man who turned the water on."

Then there was Colonel Julius Adams of Brooklyn, a former Army engineer, who was usually described as an expert on sewer construction, and who, in truth, was not quite in the same league as the others. He had, however, a number of influential friends in Brooklyn and for years he had been dabbling with designs for an East River bridge of his own. For a while it had even looked as though he might be given the chance to build it. When Roebling's proposal was first made public, he had been among those to voice sharp skepticism. That he had been included as a consultant at this stage was taken by some as a sign that Roebling was not entirely the political innocent he was reputed to be.

William Jarvis McAlpine, of Stockbridge, Massachusetts, was the president of the American Society of Civil Engineers. Kindly, genial, widely respected, he had built the enormous dry dock at the Brooklyn Navy Yard, the Albany Water Works, and a fair number of bridges. He was also the proud possessor of what must have been the most elaborate jowl whiskers in the profession and he was the one man in the group, the two Roeblings included, who had had any firsthand experience working with compressed-air foundations, or caissons, as they were called, which, in this particular case, was regarded as an attribute of major proportions.

Probably the best-known figure among them, however, was Benjamin Henry Latrobe of Baltimore, who had the face of a bank clerk, but whose endorsement alone would perhaps have been enough to settle the whole issue. He was the son and namesake of the famous English-born architect picked by Jefferson to design or remodel much of Washington, and who rebuilt the Capitol after it was burned by the British during the War of 1812. He had laid out most of the B&O Railroad and had been in charge of building a number of exceptional bridges in Maryland and Virginia.

And finally there was John J. Serrell, the only builder of suspension bridges in the group except for the Roeblings; J. Dutton Steele, chief engineer of the Reading Railroad; and James Pugh Kirkwood, a rather mournful-looking Scotsman who was an authority on hydraulics, among other things, and who, in 1848, in northeastern Pennsylvania, had built the beautiful stone-arched Starrucca Viaduct, then the most costly railroad bridge in the world.

There is no way of knowing what thoughts passed through the minds of such men as they first looked over Roebling's drawings and listened to him talk. But it is also hard to imagine any of them remaining unimpressed for very long, for all their collective experience or their own considerable accomplishments or any professional jealousies there may have been. Nor does it seem likely that any of them failed to sense the historic nature of the moment. Roebling was the recognized giant of their profession, a lesser-Leonardo he would be called, and even on paper his bridge was clearly one of the monumental works of the age. To an engineer especially that would have been obvious.

A bridge over the East River, joining the cities of New York and Brooklyn, had been talked about for nearly as long as anyone could recall. According to the best history of Brooklyn ever written, a three-volume work by a medical doctor named Henry R. Stiles, Volume II of which appeared that same year of 1869, the idea for a bridge was exactly as old as the century, the first serious proposal having been recorded in Brooklyn in 1800. Stiles wrote that an old notation, found in a scrapbook, referred to an unnamed "gentleman of acknowledged abilities and good sense" who had a plan for a bridge that would take just two years to build. Probably the gentleman was Thomas Pope of New York, an altogether fascinating character, a carpenter and landscape gardener by trade, who had designed what he called his "Flying Pendent Lever Bridge," an *invention*, as he saw it, available in all sizes and suitable for any site. His bridge to Brooklyn was to soar some two hundred feet over the water, with a tremendous cantilever fashioned entirely of wood, like "a rainbow rising on the shore," he said in the little book he published in 1811. Thomas Pope's "Rainbow Bridge" was never attempted, however, and fortunately so, for it would not have worked. But his vision of a heroic, monumental East River bridge persisted. Year after year others were proposed. Chain bridges, wire bridges, a bridge a hundred feet wide, were recommended by one engineer or another. "New York and Brooklyn must be united," Horace Greeley declared in the *Tribune* in 1849, while in Brooklyn a street running down to the river was confidently christened Bridge Street.

But nothing was done. The chief problem always was the East River, which is no river at all technically speaking, but a tidal strait and one of the most turbulent and in that day, especially, one of the busiest stretches of navigable salt water anywhere on earth. "If there is to be a bridge," wrote one man, "it must take one grand flying leap from shore to shore over the masts of the ships. There can be no piers or drawbridge. There must be only one great arch all the way across. Surely this must be a wonderful bridge."

In April 1867 a charter authorizing a private company to build and operate an East River bridge had been voted through at Albany. The charter was a most interesting and important document, for several different reasons, as time would tell. But in the things it said and left unsaid concerning the actual structure to be built, it was notable at a glance. Not a word was mentioned, for example, about the sort of bridge it was to be or to suggest that its construction might involve any significant or foreseeable problems. The cities were not required to approve the plans or the location. The charter said only that it be a toll bridge. It was important that it have a "substantial railing" and that it be "kept fully lighted through all hours of the night." I was also to be complete by January 1, 1870.

A month after the charter became law, Roebling had been named engineer of the work. By whom or by what criteria remained a puzzle for anyone trying to follow the story in the papers. In September, that same year, 1867, at a private meeting held in Brooklyn, he presented his master plan in a long formal report. But such was "the anxiety manifested on the part of the press of the two cities to present his report to the public, that it was taken and published, as an entirety . . . " The bridge had no official name at this point, and in the time since, nobody seemed able to settle on one.

At an earlier stage it had been referred to occasionally as the Empire Bridge, but the organization incorporated to build it was called the New York Bridge Company, because the Brooklyn people behind the idea saw it as just that—a bridge to New York. Roebling, on the other hand, had referred to it as the East River Bridge in his proposal and the newpapers and magazines had picked up the name. But it was also commonly called the Roebling Bridge or the Brooklyn Bridge or simply the Great Bridge, which looked the most impressive in print and to many seemed the most fitting name of all, once they grasped what exactly Roebling was planning to do.

But it was the possible future impact of such a structure on their own lives that interested people most, naturally enough, and that the press in both cities devoted the most attention to. The *Times*, for example, described the bridge as a sort of grand long-needed pressure valve that would do much to alleviate New York's two most serious problems, crime and overcrowding.

In Brooklyn, where interest was the keenest, it was said the bridge would make Brooklyn important, that it would make Brooklyn prosper. Property values would soar. Roebling the alchemist would turn vacant lots and corn patches into pure gold. Everybody would benefit. Brooklyn was already expanding like a boomtown, and the bridge was going to double the pace, the way steam ferries had. Merchants could expect untold numbers of new customers as disaffected New Yorkers flocked across the river to make Brooklyn their home. Manufacturers would have closer ties with New York markets. Long Island farmers and Brooklyn brewers could get their wares over the river more readily. The mail would move faster. Roebling had even told his eager clients how, in the event of an enemy invasion of Long Island, troops could be rushed over the bridge from New York in unprecedented numbers. In such an emergency, the old Prussian had calculated, nearly half a million men, together with artillery and baggage trains, could go over the bridge in twenty-four hours.

Most appealing of all for the Brooklyn people who went to New York to earn a living every day was the prospect of a safe, reliable alternative to the East River ferries. Winds, storms, tides, blizzards, ice jams, fog, none of these, they were told, would have the slightest effect on Mr. Roebling's bridge. There would be no more shoving crowds at the ferryhouse loading gates. There would be no more endless delays. One Christmas night a gale had caused the river to be so low the ferries ran aground and thousands of people spent the night in the Fulton Ferry house. Many winters when the river froze solid, there had been no service at all for days on end.

Some of the Brooklyn business people and Kings County politicians were even claiming that the bridge would make Brooklyn the biggest city in America, a most heady prospect indeed and not an unreasonable one either. Congressman Demas Barnes contended Brooklyn would be the biggest city in the world, once New York was "full." New York, that "human hive" John Roebling called it, was running out of space, its boundaries being forever fixed by nature. Roebling and others envisioned a day when all Manhattan Island would be built over, leaving "no decent place" to make a home, neither he nor anyone else thus far having imagined a city growing vertically. "Brooklyn happens to be one of those things that can expand," wrote the editors of the new *Brooklyn Monthly*. "The more you put into it, the more it will hold."

And such highly regarded Brooklyn residents as Walt Whitman and James S. T. Stranahan, the man behind Brooklyn's new Prospect Park, looked to the day when the bridge would make Brooklyn and New York "emphatically one," which was also generally taken to be a very good thing, since the new Union Pacific Railroad was going to make New York "the commercial emporium of the world." This was no idle speculation, "but the natural and legitimate result of natural causes," according to John Roebling. His bridge was part of a larger mission. "As the great flow of civilization has ever been from East towards the West, with the same certainty will the greatest commercial emporium be located on this continent, which links East to the West, and whose mission it is in the history of mankind to blend the most ancient civilization with the most modern." The famous engineer, it had been noticed in Brooklyn, tended to cosmic concepts, but so much the better. If there were now forty million people crossing the East River every year, as was the claim, then, he said, in ten years' time there would be a hundred million.

"Lines of steamers, such as the world never saw before, are now plowing the Atlantic in regular straight line furrows," he had written in his proposal. "The same means of communication will unite the western coast of this continent to the eastern coast of Asia. New York will remain the center where these lines meet."

This, in other words, was to be something much more than a large bridge over an important river. It was to be one of history's great connecting works, symbolic of the new age, like the Atlantic cable, the Suez Canal, and the transcontinental railroad. "Lo, Soul, seest thou not God's purpose from the first?" wrote Walt Whitman at about this time. "The earth be spann'd, connected by network . . . The lands welded together." "The shapes arise!" wrote the Brooklyn poet.

> Singing my days,
> Singing the great achievements of the present
> Singing the strong, light works of engineers . . .

But it was Roebling himself, never one to be overly modest, who had set forth the most emphatic claim for the bridge itself and the one that would be quoted most often in time to come:

> The completed work, when constructed in accordance with my designs, will not only be the greatest bridge in existence, but it will be the greatest engineering work of the continent, and of the age. Its most conspicuous features, the great towers, will serve as landmarks to the adjoining cities, and they will be entitled to be ranked as national monuments. As a great work of art, and as a successful specimen of advanced bridge engineering, this structure will forever testify to the energy, enterprise and wealth of that community which shall secure its erection.

Roebling had written that in 1867, at the very start of his formal proposal, but in all the time since, for some mysterious reason, not a spade of dirt had been turned and numbers of people, some claiming to be experts, had begun saying they were not so sure about Roebling's "advanced engineering," or whether it was worth the six to seven million dollars he had said it would cost, an estimate that did not include the price of the land required. Even if his figures were realistic, the bridge would also be about the most expensive ever built.

The editors of *Scientific American* said a tunnel would serve the purpose as well and cost less. A Navy engineer presented an alternative plan. He wanted to block off "the vexatious East River" with a dam several hundred feet wide on which he would build highways, stores, docks, and warehouses. By early 1869, when it looked as though the bridge might actually be started, the critics were sounding forth as never before. Warehouse owners along the river and others in the shipping business were calling it an obstruction to navigation and a public nuisance. The New York Polytechnic Society put on a series of lectures at Cooper Union devoted exclusively to the supposed engineering fallacies of the Roebling plan. Engineers expressed "grave apprehension." The bridge, it was stated on the best professional authority, was a monumental extravagance, "a wild experiment," nothing but an exercise in vanity. Even in Brooklyn the *Union* said another bridge and a tunnel besides would probably be built by the time everyone finished wrangling over details and questioned why, for so momentous a public work, only one engineer had been called on and no other plans ever considered.

So it had been to still such talk that Roebling had assembled his seven consultants and with total patience and candor went over everything with them point by point.

To begin with it was to be the largest suspension bridge in the world. It was to be half again the size of his bridge over the Ohio at Cincinnati, for example, and nearly twice the length of Telford's famous bridge over the Menai Strait, in Wales, the first suspension bridge of any real importance. It was to cross the East River with one uninterrupted central span, held aloft by huge cables slung from the tops of two colossal stone towers and secured on either shore to massive masonry piles called anchorages. These last structures alone, he said, would be a good seven stories tall, or taller than most buildings in New York at the time. They would each take up the better part of a city block and would be heavy enough to offset the immense pull of the cables, but hollow inside, to provide, Roebling suggested, room for cavernous treasury vaults, which he claimed would be the safest in America and ample enough to house three-quarters of all the investments and securities in the country.

The towers, the "most conspicuous features," would be identical and 268 feet high. They would stand on either side of the river, in the water but close to shore, their foundations out of sight beneath the riverbed. Their most distinguishing features would be twin Gothic arches—two in each tower—through which the roadways were to pass. These arches would rise more than a hundred feet, like majestic cathedral windows, or the portals of triumphal gateways. "In a work of such magnitude, and located as it is between two great cities, good architectural proportions should be observed," wrote the engineer. " . . . The impression of the whole will be that of massiveness and strength."

His towers would dwarf everything else in view. They would reign over the landscape like St. Peters in Rome or the Capitol dome in Washington, as one newspaper said. In fact, the towers would be higher than the Capitol dome if the dome's crowning statue of Freedom was not taken into account. So this in the year 1869—when the Washington Monument was still an ugly stone stump—meant they would be about the largest, most massive things ever built on the entire North American continent. On the New York skyline only the slim spire of Trinity Church at the head of Wall Street reached higher.

The towers were to serve two very fundamental purposes. They would bear the weight of four enormous cables and they would hold both the cables and the roadway of the bridge high enough so they would not interfere with traffic on the river. Were the two cities at higher elevations, were they set on cliffs, or palisades, such as those along the New Jersey side of the Hudson, for example, such lofty stonework would not be necessary. As it was, however, only very tall towers could make up for what nature had failed to provide, if there was to be the desired clearance for sailing ships. And as the mass of the anchorages had to be sufficient to offset the *pull* of the cables, where they were secured on land, so the mass of the towers, whatever their height, had to be sufficient to withstand the colossal *downward* pressure of the cables as they passed over the tops of the towers.

Below the water the towers were to be of limestone and each was to be set on a tremendous wooden foundation, but from the waterline up they were to be of granite. In plan each tower was essentially three shafts of solid masonry, connected below the roadway, or bridge floor, by hollow masonry walls, but left unconnected above the bridge floor until they joined high overhead to form the great Gothic arches, which, in turn, were to be topped by a heavy cornice and three huge capstones. The total weight of each tower, Roebling estimated, would be 67,850 tons, but with the weight of the roadway and its iron superstructure added on they would each weigh 72,603 tons.

The suspended roadway's great "river span" was to be held between the towers by the four immense cables, two outer ones and two near the middle of the bridge floor. These cables would be as much as fifteen inches in diameter and each would hang over the river in what is known as a catenary curve, that perfect natural form taken by any rope or cable suspended from two points, which in this case were the summits of the two stone towers. At the bottom of the curve each cable would join with the river span, at the center of the span. But all along the cables, vertical "suspenders," wire ropes about as thick as a pick handle, would be strung like harp strings down to the bridge floor. And across those would run a pattern of diagonal, or inclined, stays, hundreds of heavy wire ropes that would radiate down from the towers and secure at various points along the bridge floor, both in the direction of the land and toward the center of the river span.

The wire rope for the suspenders and stays was to be of the kind manufactured by Roebling at his Trenton works. It was to be made in the same way as ordinary hemp rope, that is, with hundreds of fine wires twisted to form a rope. The cables, however, would be made of wire about as thick as a lead pencil, with thousands of wires to a cable, all "laid up" straight, parallel to one another, and then wrapped with an outer skin of soft wire, the way the base strings of a piano are wrapped.

But most important of all, Roebling was talking about making the cables of steel, "the metal of the future," instead of using iron wire, as had always been done before. There was not a bridge in the country then, not a building in New York or in any city as yet, built of steel, but Roebling was seriously considering its use and the idea was regarded by many engineers as among the most revolutionary and therefore questionable features of his entire plan.

The way he had designed it, the enormous structure was to be a grand harmony of opposite forces—the steel of the cables in tension, the granite of the towers in compression. "A force at rest is at rest because it is balanced by some other force or by its own reaction," he had once written in the pages of *Scientific American.* He considered mathematics a spiritual perception, as well as the highest science, and since all engineering questions were governed by "simple mathematical considerations," the suspension bridge was "a spiritual or ideal conception."

His new bridge was to be "a great avenue" between the cities, he said. Its overall width was to be eighty feet, making it as spacious as Broadway itself, as he liked to tell people, and the river span would measure sixteen hundred feet, from tower to tower, making it the longest single span in the world. But of even greater import than length was the unprecedented load the bridge was designed to bear—18,700 tons.

The long river span was not to be perfectly horizontal, but would bow gracefully, gently upward. It would pass through the tower arches at an elevation of 119 feet, but at the center it would be 130 feet over the water. This, as Roebling pointed out, was thirty feet higher than the elevation fixed by the British Admiralty for Robert Stephenson's Britannia Bridge over the Menai Strait, built nearly twenty years earlier. Before long, sailing ships would be things of the past, he declared. His bridge therefore would be no obstruction to navigation, only possibly "an impediment to sailing." As it was, only the very largest sailing ships afloat would have to trim their topmasts to pass beneath the bridge.

But because of the great elevation of the river span and the relatively low-lying shores, the rest of the bridge, sloping down to ground level, would have to extend quite far inland on both sides to provide an easy grade. The bridge would have to descend back to earth rather gradually, as it were, and thus the better part of it would be over land, not water. Those inland sections of the bridge between the towers and the two anchorages were known as the land spans, and were also supported by the cables, by suspenders and diagonal stays. The ends of the bridge, from the anchorages down to ground level, were known as the approaches. In all, from one end to the other, the Great Bridge was to measure 5,862 feet, or more than a mile.

The red line Roebling had drawn on the map ran southeast from City Hall Park, in New York, crossing the river not quite at right angles, at that point where the river was returning to its essentially north-south course. At the Brooklyn Navy Yard—over to the right of the red line—the river turned sharply to the left, heading nearly due west, but then it quickly turned down the map again to merge with the harbor. And it was right there, where the river turned the second time, right about where the Fulton Ferry crossed, that Roebling had put his "Park Line" connecting New York, on the upper left of the map, with Brooklyn, on the lower right.

The precise terminating point on the New York side was at Chatham Street, opposite the park. This was *the* place for the bridge to come in, he said. For the next fifty years the park would remain "the great focus of travel, from which speedy communications will ramify in all directions." From there his red line crossed over North William Street, William, Rose, Vandewater, and half a dozen more streets, to the end of Pier 29, then over the river, straight through one of the Fulton Ferry slips, and into Brooklyn. Running parallel with Fulton Street, Brooklyn's main thoroughfare, the line cut across a patchwork of narrow cross streets—Water, Dock, Front, James—to Prospect, where it bent slightly toward Fulton, terminating finally in the block bounded by Prospect, Washington, Sands, and Fulton, or right about where St. Ann's Church stood.

Down the center of the bridge Roebling planned to run a double pair of tracks to carry specially built trains pulled by an endless cable, which would be powered by a giant stationary steam engine housed out of sight on the Brooklyn side. In time these trains would connect with a system of elevated railroads in both cities and become a lucrative source of revenue. He had worked it all out. His bridge trains would travel at speeds up to forty miles an hour. A one-way trip would take no more than five minutes. It was certain, he said, that forty million passengers a year could be accommodated by such a system, "without confusion and without crowding."

Carriages, riders on horseback, drays, farm wagons, commercial traffic of every kind, would cross on either side of the bridge trains, while directly overhead, eighteen feet above the tracks, he would build an elevated boardwalk for pedestrians, providing an uninterrupted view in every direction. This unique feature, he said, would become one of New York's most popular attractions. "This part I call the *elevated promenade*, because its principal use will be to allow people of leisure, and old and young invalids, to promenade over the bridge on fine days, in order to enjoy the beautiful views and the pure air." There was no bridge in the world with anything like it. And he added, "I need not state that in a crowded commercial city, such a promenade will be of incalculable value."

So the roadways and tracks at one level were for the everyday traffic of life, while the walkway above was for the spirit. The bridge, he had promised, was to serve the interests of the community as well as those of the New York Bridge Company. Receipts on all tolls and train fares would, he asserted, pay for the entire bridge in less than three years. To build such a bridge, he said, would take five years.

Horatio Allen and William McAlpine asked the most questions during the sessions Roebling held with the consultants. The length of the central span and the tower foundations were the chief concerns.

It had been said repeatedly by critics of the plan that a single span of such length was impossible, that the bridge trains would shake the structure to pieces and, more frequently, that no amount of calculations on paper could guarantee how it might hold up in heavy winds, but the odds were that the great river span would thrash and twist until it snapped in two and fell, the way the Wheeling Bridge had done (a spectacle some of his critics hoped to be on hand for, to judge by the tone of their attacks).

Roebling told his consultants that a span of sixteen hundred feet was not only possible with a suspension bridge, but if engineered properly, it could be double that. A big span was not a question of practicability, but cost. It was quite correct that wind could play havoc with suspension bridges of "ordinary design." But he had solved that problem long since, he assured them, in his earlier bridges, and this bridge, big as it was, would be quite as stable as the others. Like his earlier works, this was to be no "ordinary" bridge. For one thing it would be built *six times* as strong as it need be. The inclined stays, for example, would have a total strength of fifteen thousand tons, enough to hold up the floor by themselves. If all four cables were to fail, he said, the main span would not collapse. It would sag at the center, but it would not fall. His listeners were very much impressed.

There were questions about his intended use of steel and about the extraordinary weight of the bridge. Then at one long session they had discussed the foundations.

Roebling planned to sink two tremendous timber caissons deep into the riverbed and to construct his towers upon these. It was a technique with which he had had no previous experience, but the engineering had been worked out quite thoroughly, he said, in conjunction with his son, Colonel Roebling, who had spent nearly a year in Europe studying the successful use of similar foundations. McAlpine could vouch for the basic concept, since he had used it himself successfully, although on a vastly smaller scale, to sink one of the piers and the abutment for a drawbridge across the Harlem River. His caisson for the pier had been of iron and just six feet in diameter. Those Roebling was talking about would be of pine timbers and each one would cover an area of some seventeen thousand square feet, or an area big enough to accommodate four tennis courts with lots of room to spare. Nothing of the kind had ever been attempted before.

How deep did he think he would have to go to reach a firm footing, the engineers wished to know. Would he go to bedrock? And did he have any idea how far down that might be?

During the test borings on the Brooklyn side, the material encountered had been composed chiefly of compact sand and gravel, mixed with clay and interspersed with boulders of traprock, the latter of which, he allowed, had "detained this operation considerably." Gneiss had been struck at ninety-six feet. But below a depth of fifty to sixty feet, the material had been so very compact that the borehole had remained open for weeks without the customary tubing. So it was his judgment that there would be no need to go all the way to rock. A depth of fifty feet on the Brooklyn side ought to suffice and the whole operation would probably take a year.

About the prospects on the New York side, he was rather vague—but it looked, he said, as though bedrock was at 106 feet and there was a great deal of sand on the way down. Still there was a chance that rock might be found closer to the surface. An old well near Trinity Church showed gneiss at twenty-six feet, he noted, and in the well at City Hall the same rock was found at ninety feet. "The whole of Manhattan Island appears to rest upon a gneiss and granite formation," he said. The greatest depth to which similar caissons had been sunk before this was eighty-five feet. But he was willing to take his to a depth of 110 feet if that was what had to be done. His consultants said they did not think he would find that necessary.

Presently they took up the question of the timber foundations and their fate, once he left them buried forever beneath the towers, beneath the river, the rock, sand and muck of the riverbed. In his report, Roebling had explained at some length how the caissons would be packed with concrete once they were sunk to the desired position, and why, in their final resting place, well below the level where water or sea worms could reach them, they would last forever. But there were some among the consultants who wished to hear more on the subject and who had a number of questions.

That particular session on the foundations had taken place on March 9. Two days later, on the 11th, it was announced that the renowned engineers had approved the Roebling plan, "in every important particular." Their official report would come later, but in the meantime the public could rest assured that the plan was "entirely practicable."

Only Congressional authorization was needed now, since Congress had jurisdiction over all navigable waters and the bridge was to be a post road. Unlike the government in Albany, or those in either city, the government in Washington had some regulations it wished to see adhered to. Congressional legislation already drawn up stipulated that the bridge must in no way "obstruct, impair, or injuriously modify" navigation on the river. In particular, there was concern in Washington that it might interfere with traffic to and from the Navy Yard, and to be certain that every detail of the plan was fully understood, General A. A. Humphreys, Chief of the Army Engineers, decided to appoint his own review panel to give an opinion on it, irrespective of the conclusions reached by Roebling's consultants. (This was to be the only public scrutiny of the design or the location.) So at about that point it had seemed the most sensible next step would be for everyone to go take a look at some of Roebling's existing works to see how he had previously handled somewhat analogous situations. Let his work speak for itself, he had decided.

The tour was arranged almost overnight and if there was any initial intention to restrict it to a relatively small body of professionals that idea was speedily overruled. A total of twenty-one gentlemen and one lady made up the "Bridge Party," as it was referred to in subsequent accounts. In additon to the two Roeblings, the seven consultants, and three Army engineers—General Horatio Wright, General John Newton, and Major W. R. King—several prominent Brooklyn businessmen were invited, most of whom were or were about to become stockholders in the New York Bridge Company. A Brooklyn Congressman named Slocum—General Henry W. Slocum—was included, as were Hugh McLaughlin, the Democratic "Boss" of Brooklyn, and William C. Kingsley, Brooklyn's leading contractor, who was known to be the driving political force behind the bridge and the largest individual stockholder. How many of the party were aware that the tall, powerful Kingsley would also be personally covering all expenses for the tour, in addition to the seven thousand dollars in consultants' fees, is not known.

Two young engineers, C. C. Martin and Samuel Probasco, both of whom had worked for Kingsley on different Brooklyn projects, were also to go, as was the wife of one of the consultants, Mrs. Julius Adams, who is described only as an "amiable lady" in existing accounts. Why she consented to join the group, or why she was invited in the first place, no one ever explained.

Nor is there anything in the record to indicate who determined the make-up of the group. Presumably it was taken to be a representative body, having an even balance of engineering talent, business acumen, and public spirit. In any event, the editor of the Brooklyn *Eagle*, Thomas Kinsella, was also included, so that in Brooklyn at least the expedition would receive proper notice, and young Colonel Roebling appears to have been the one delegated to make the necessary arrangements.

There were, however, two very important public figures who did not make the trip, both of whom had done much to bring the project along as far as it had come and who ought to be mentioned at this point in the story.

The first was State Senator Henry Cruse Murphy, lawyer, scholar, the most respected and respectable Democrat in Brooklyn, and in Albany the leading spokesman for Brooklyn's interests. Murphy had worked harder for the bridge than anyone in Brooklyn except Kingsley, the contractor. He was the one who had written the charter for the New York Bridge Company. He had seen it through the legislature and was currently serving as the company's president. Why he failed to make the trip is not known and probably not important. But he would have added a certain tone to the group certainly and John A. Roebling, in particular, would doubtless have enjoyed his company. (The idea of Roebling keeping company with the likes of Boss McLaughlin must have raised many an eyebrow on Brooklyn Heights.)

But the absence of the second missing party was quite intentional, one can be sure; it raised no questions and required no explanation, since there had been no mention as yet, scarcely even a whisper, that he had had anything whatever to do with the bridge. He was William Marcy Tweed of New York.

The itinerary called for stops at Pittsburgh, Cincinnati, and Niagara Falls, and the announced official purpose of the expedition was to inspect four of Roebling's bridges, each of which, in one way or other, illustrated how he intended to span the East River. But a week of traveling together was also supposed to give everybody a chance to get to know one another—nothing could so cement friendships as a long train ride, Thomas Kinsella would write—and particularly, it was presumed, everyone would get to know the key man in all this, John A. Roebling.

The great engineer was still largely a mystery to the people who had hired him. Except for the times when he had expounded on his plan at the meeting in 1867, his Brooklyn clients had seen very little of him. Their ordinary day-to-day dealings had been with his son. It had been young Roebling, not his father, who had set up the makings of an office and who had taken a house on the Heights. He had been the one on hand to answer their questions and keep things moving.

The father had wanted it that way. He had remained in Trenton, showing up in Brooklyn only now and then, and staying no longer than necessary. His time was always short it seemed and even when meeting with his board of consultants he had kept each session quite formal and to the point. He had no time for anything but business, and no small talk whatever.

On occasion the two of them, father and son, would be seen walking on Hicks Street, talking intently, or down by the slate-gray river pacing about the spot where the tower was to rise, the father pointing this way and that with his good hand. They resembled each other in height and build, even trimmed their whiskers the same way. But while the son was quite handsome in the conventional sense, with strong regular features, the father's face was a composite of hard angles and deep creases, of large ears and nose and deep-sunken eyes, all of which gave the appearance of having been hewn from some substance of greater durability than mortal flesh.

Most people, later, would talk about his eyes, his fierce pale-blue eyes. But just what sort of human being there might be behind them was a puzzle. He was a man of enormous dignity, plainly enough, full of purpose and iron determination, but accustomed to deference just as plainly, somebody to be admired from a distance. His look was all-knowing and not in the least friendly. Among those who were about to stake so very much on him and his bridge, or who already had, there was not one who could honestly say he knew the man.

And so on the evening of April 14, 1869, when General Grant and his Julia were just taking up residence in the White House and the dogwood were beginning to bloom across the lowlands of New Jersey, the Bridge Party boarded a private palace car in Jersey City and started west. The only one missing from the group was the elder Roebling, who was to get on at Trenton.

5.7

Reclamation Origins

MICHAEL C. ROBINSON

Demands for a Federal Reclamation program in the West arose during the late nineteenth century. Advocates of storing water to irrigate arid wastelands stressed the need to create homes as well as regulate development of all natural resources. The movement manifested the agrarian ideals of Thomas Jefferson and sought to block usurpation of water and land by powerful interests. Various Federal and State laws fostered colonization in the arid region. Results, however, were often disappointing because of water shortages and other environmental conditions. Thus, in 1902 Federal construction of dams and canals to store and distribute irrigation water was authorized after individual, corporate, and State attempts proved inadequate.

INDIAN AND SPANISH IRRIGATION

Prior to European settlement of North America, the Southwest United States was dotted with small agricultural settlements based on the careful use of scarce water supplies. Exercising great ingenuity, Indian farmers created islands of fertility in a sea of aridity by compensating for the limitations of the natural environment. Before this sedentary social pattern evolved, essentially two subsistence economies were developed by the West's first inhabitants. Some nomadic groups survived on a diet of wild plants and small game. Others hunted large grazing animals, such as the buffalo, following the migratory herds across the plains.

The nomadic way of life set the stage for the adoption of agriculture. The small roaming bands acquired an understanding of the relationship of plant growth to moisture, temperature, soil fertility, and seasonal changes. Over many centuries, profound social and economic changes gradually took place. Groups settled in permanent villages and began the recurrent cycle of preparing the soil and planting, cultivating, and harvesting crops of maize, squash, and beans. Farming made possible the storing of winter food supplies and precluded the need for seasonal movements.

The spread of farming was circumscribed by aridity and other environmental factors. To survive the Indians made optimum use of available resources within the limits of their simple technology. Fields were established near stable water supplies and cultivated with stone, bone, and wooden implements. Villagers guarded ripening crops from birds, rodents, and other despoilers. The precious harvest was then preserved in carefully built storehouses to provide food reserves for winter and unproductive years.

Water was the only environmental factor that could be manipulated, and husbanding it became a major focus of effort. Fields were frequently located in deep mountain canyons or within the basins of large rivers. During the dry season, seeps and springs offered small sources of moisture. Indians also grew crops on river valley floodplains, where spring floods deposited silt and soaked the earth.

Simple water control measures gradually evolved. Rock check dams were built across ravines and gullies to trap alluvium and water during storm runoffs. Earthen and stone terraces erected along hill contours created vegetable plots. After cloudbursts, freshets rushing down canyons were slowed by small dams and the water spread evenly over flat areas to create productive fields. The Southwest's Puebloan peoples stored rain and spring water in natural catchment basins and small manmade reservoirs for irrigation and domestic uses. Gardens were sometimes enclosed in stone or adobe walls to abate erosion and protect plants from searing desert winds.

Along the Salt and Gila rivers in southern Arizona, the Hohokam Indians created an extensive canal system that irrigated thousands of acres and supported a complex culture. Aerial surveys disclose that they built at least 150 miles of large canals. Settlement began as early as 300 BC and by 1200 AD there were 20 or more villages. Ten distinct canal systems were constructed in the vicinity of present-day Phoenix, Tempe, and Mesa, Arizona. Many of the canals were 30 feet wide and 8 to 13 feet deep. They were excavated with wooden and stone tools. The earth and rock were carried in baskets on the backs of villagers and deposited on the canal banks.

The Hohokams constantly maintained their irrigation system. Each year new brush diversion dams, consisting of a lattice of withes and branches laced between upright poles, were placed in streams at canal inlets. Silting and bank erosion in the distribution system were countered with manual cleaning. Hohokam is a Pima Indian word which loosely translated means "the people who have gone away." The river valleys were in fact abandoned about 1400 AD, possibly because irrigation raised the water table, induced water logging, and caused alkali problems. This rendered much of the lands unfit for cultivation. In addition, the lack of holdover storage made it impossible to supply irrigation water during periods of drought.

Spanish missionaries and colonists adopted and enhanced Indian water conservation practices. In the seventeenth and eighteenth centuries, Catholic priests and settlers founded a chain of outposts stretching from San Diego to San Francisco along the Pacific Coast, and from Arizona's Santa Cruz Valley to settlements at Santa Fe, New Mexico, and as far away as El Paso and San Antonio, Texas.

Wherever they established communities, water supply systems were constructed. Spanish conservation and distribution techniques derived from observing Indians as well as implementing water storage and irrigation practices used for centuries in the semi-arid valleys of southern Spain. The ancient Romans introduced dams and aqueducts into Spain, and in the centuries following the Moorish conquest, Near Eastern storage and delivery methods were adopted. Thus, Spanish water control in America represented a merging of New and Old World technologies.

Mission and presidio sites were determined by the availability of water. No effort was spared to stabilize the community's water supply and insure its efficient use. Small storage dams on streams created reservoirs that were filled overnight and water was diverted onto fields in daylight hours. In some places, water was brought to the crops over irregular terrain through narrow channels carved out of the hard surface of hillsides. Retaining walls conserved water and inhibited erosion during rainstorms, and tile conduits carried domestic water to cisterns and patio gardens.

Furthermore, the Spanish introduced gristmills as well as various water lifting techniques such as the *noria*, a water wheel or revolving chain-and-bucket device, that raised water from wells and canals to adjacent agricultural lands.

Near the small settlements, *acequias* (community ditches) were cooperatively built and maintained to irrigate orchards and lush fields of wheat, corn, beans, melons, squash, and peppers. This system, adopted from the Moors, was introduced throughout the Southwest. The colonists and Catholic mission Indians dug and maintained irrigation ditches that carried water diverted from streams or canals by brush diversion dams.

A *mayordomo* (ditch chief) was responsible in most communities for constructing and maintaining the *acequias* and regulating water distribution to the landowners. Stone aqueducts were occasionally required to carry water over streams and arroyos (dry gulches). Over narrower expanses, the Spanish sometimes built *canoas* (dug-out log flumes supported by a wooden framework). When Americans began investigating the Far West, they encountered a thriving Spanish civilization that was based on water conservation.

MORMON ACCOMPLISHMENTS

The Mormons, who first settled Utah's Salt Lake Valley in 1847, were the first United States pioneers to establish a regional agriculture economy based on irrigation. They demonstrated to potential settlers that it was possible to create homes in an area hitherto neglected because of the bad reputation of its climate with earlier migrants.

In the 1840's, pioneers rolled their covered wagons across the plains seeking new opportunities in Oregon and California. The Mormons demonstrated that the "Great American Desert" was not a barrier to the westward march of civilization. As soon as they arrived in the Salt Lake Valley, they diverted water from small streams to soften the ground so that plows could pierce the sun-baked soil.

Relying on industry, intuition, and probably some understanding of Spanish and Indian irrigation practices, they built simple diversion dams and dug canals that transported water to fields of grain and vegetables. Fired with religious zeal and dedication, the Mormons undertook a remarkable transformation of the desert. By

1865 almost 1,000 miles of canals irrigated 1.5 million acres, and some 65,000 Latter-Day Saints lived in reasonable comfort on reclaimed land.

This success was grounded in the cohesiveness of the Mormon community and the high degree of cooperation among its members. The Utah settlement pattern—small agricultural villages surrounded by intensively cultivated 10- to 40-acre fields—maximized the social interaction needed to plan and construct common canal systems and other irrigation works. The people also exhibited a unity of purpose and constructive philosophy of life that focused their meager resources on a common cause. To the Latter-Day Saints, reclaiming the desert was part of "building up the Kingdom of God on Earth." Thus, community interest was placed above individual gain. This set of principles and attitudes fostered survival and prosperity in the face of unrelenting hardships.

Virtually every early Utah town was founded by a group selected, organized, and dispatched by church authorities. The colonists surveyed the townsite, drew for lots, built a fort, and constructed a church building that also served as a meeting house and recreation center. The adjacent fields were allotted according to family size, small fields were enclosed in a common fence, and work began on an irrigation canal. Many of the first ditches were built without surveying instruments. Makeshift levels and plumbs were used while grades were sometimes determined by sighting over pans of water. Gangs of men with shovels and a few animals pulling crude implements dug canals. Each man's share of the work was proportionate to the acreage he cultivated.

The Utah settlers worked out methods of reshaping the land and applying irrigation water evenly and in beneficial amounts over undulating fields of former desert. Through trial and error, they discovered the relationships between soil and seed types and developed irrigation techniques that promoted maximum yields on terrains of varying slope, soil texture, and mineral content.

Mormon pioneers helped establish an important legal precedent for western reclamation by departing from the English common law principle of "riparian rights" and adopting the doctrine of "prior appropriation for beneficial use." The former granted landowners bordering watercourses the right to use the water. The latter established a principle of "first in time, first in right"; the first appropriator of streams acquired rights to as much water as he could put to use for beneficial purposes. Therefore, canals could be built across the public domain without fear that water rights might be jeopardized by future development of property adjacent to canal routes.

The Mormons founded the first locally autonomous institutions to provide for equitable water distribution. Since farmers owned the canals in common, the time, place, and amount of water allocations had to be regulated and enforced. Consequently, water masters and ditch riders were appointed to ensure equality. In 1865 the Utah territorial legislature passed the first irrigation district law that authorized citizens to organize irrigation companies and levy taxes upon themselves for the building and maintenance of canals under county government supervision.

Other colonists were inspired by the accomplishments of the Utah pioneers who demonstrated that communities could flourish "under the ditch." In 1870 Nathan C. Meeker, agricultural editor for Horace Greeley's *New York Tribune*, founded the "Union Colony of Colorado." Successful irrigation colonies were also established at Riverside and Anaheim, California, in the 1870's.

Despite these accomplishments, irrigation was not a panacea for western development. Brigham Young and his followers had fortuitously populated an area of abundant small streams where the natural terrain facilitated simple diversions to nearby fields. Therefore, considerable acreage could be brought under cultivation without building large dams, tapping the waters of major rivers, and distributing water from a single canal over extensive land areas. Without water storage, however, croplands could not be expanded to take advantage of relatively abundant supplies during spring and early summer.

As the nineteenth century drew to a close, reclamationists across the country touted their cause by pointing to Mormon successes. However, they often failed to realize that the flourishing Mormon state of "Deseret" was a product of Latter-Day Saint religious zeal and beliefs, the cooperation and discipline of the sect's members, the unfettered development of innovative legal and social institutions, and a geographical setting highly favorable to irrigation. Mormon settlements were created from Mexico to Canada, but subsequent settlers of the arid region found their example difficult to emulate.

WESTERN COLONIZATION

From 1878 to 1902, irrigation expansion became a critical issue in the West. Concerned over inadequate water supply, eastern speculators and the region's residents sought to create a sound economic base by reclaiming large desert acreages. Publishers, politicians, businessmen, and farmers tried to foster settlement, but irrigation was rarely successful beyond areas where streams could be easily diverted into hand-dug canals and ditches.

Many reclamationists failed to understand the basic essentials for expanding agriculture in the arid region. Storage reservoirs were needed, but building sophisticated dams and canal systems required skills and huge capital investments often beyond the capabilities of individuals, local governments, and speculative companies. The design, construction, and operation of such facilities required employing the talents of objective, impartial engineers unswayed by "get rich quick" motives that frequently resulted in poorly built structures. Finally, there was need to resolve interstate water disputes and upgrade the operation and maintenance skills of irrigators united in ditch companies and water districts.

Institutions and ideas appropriate to agriculture in the humid East were frequently not applicable to the arid and semiarid regions beyond the hundredth meridian. Even the 1862 Homestead Act, the keystone of public land policies, did not adapt well to the arid region. Congress, viewing the success of the 160-acre freehold in more humid areas, was slow to abandon the concept that the act would continue to foster settlement.

Though the region's climate and geography differed radically from that of the East and Midwest, prospective homesteaders remained confident that the agricultural frontier offered continued opportunities to thrifty, independent, hardworking home seekers. The passion for land continued unabated. Obtaining a farm remained the paramount goal of men and women who desired either to establish a permanent home or profit from land investments. Whatever their motives, many Americans had a deep-seated desire to own a piece of the rapidly vanishing western wilderness.

Railroads, speculators, state and territorial governments, and other promoters exploited this appetite for land. Expanding agriculture was viewed as the key to stable western growth and development, and the region was portrayed to prospective immigrants as a farming paradise. Promotional pamphlets and tracts usually exaggerated crop yields, praised the climate's salubrity, and painted visions of new homes springing up on the prairie. Many agricultural pioneers, of course, met their lofty expectations for a new and abundant life. But thousands of others had their hopes shattered by the grim realities of drought and despair the boosters failed to describe.

Life on the frontier in the 1800's was often made unbearable by hostile environmental forces. Homesteaders faced droughts, blizzards, floods, and grasshoppers that destroyed crops and livestock. Even the hardiest pioneers could not withstand a long stretch of such harsh and depressing conditions.

A distraught Kansas farmer wrote that "between the drought, the chinch bugs, and the grasshoppers, we will be forced to go to Egypt or somewhere else for our corn." An Army officer recorded wintertime conditions in southwestern Nebraska during the 1870's: "it was pitiful in most instances, upon entering the poor huts to see women and children crouched shivering around the dull fires in the mist of a cloud of pulverized snow driven in upon them by the storm." Recurrent natural catastrophes forced farmers to give up and seek other opportunities. Unable to cope with western conditions, their dreams of creating a new Eden turned into nightmares.

FEDERAL LAND LAWS

Congress continued the policy of encouraging colonization of the public domain. Every land law was in a sense part of a general scheme to reclaim the wilderness. Virgin land had traditionally been granted freely or sold at low prices to individuals who would settle on and cultivate it. The fundamental assumption of these laws was that farming would proceed once the timber was cleared or the sod broken. However, in the arid West settlers often had to divert streams, dig canals, and construct weirs and flumes before crops could be planted. Other homesteaders who attempted dryland farming during wet cycles often failed because of capricious and scarce rainfall.

In the last quarter of the nineteenth century, land policies gradually evolved from disposing of the public domain in its natural state to direct Federal participation in its reclamation. Since annual rainfall would not normally support agriculture, in many areas accessibility to water rather than soil conditions usually determined land values. Husbanding scarce water supplies and insuring their efficient utilization required storing floodwaters behind dams during non-growing seasons. However, these projects involved capital expenditures far beyond the capacity of private individuals.

Congress also faced the problem of water and land monopoly, for whoever gained water rights controlled productivity on all adjacent lands. Thus, the primacy of water in western development required a new approach to the disposal of the public domain. Over time Congress was compelled to adapt public land policies to the western environment and abandon the assumption that self-reliant pioneers could easily transform raw land into verdant, productive farms.

By the end of the Civil War, westerners realized the Government must take new initiatives if the desert was to be reclaimed. They had experience with land grants to transcontinental railroads and, therefore, asked the Congress to give public land for specific irrigation projects to attract private investments. Title to the land was essential so that it could be used as security in obtaining financing.

More than 40 such requests were presented to Congress in the late 1860's and early 1870's, and in 1873 Congress appointed a special commission to investigate irrigation in California's San Joaquin and Tulare valleys. Its findings were intended to serve as the basis for action on similar requests for aid. The commission recommended Federal surveys and investigations to promote irrigation, but it did not endorse land grants to canal companies other than for rights-of-way.

Having rejected direct support, Congress sought other means of dealing with the inadequacies of existing land laws. The common argument voiced by westerners was that larger tracts of land were required in arid areas to make irrigation profitable. In March 1875, the Congress passed an act that provided for the sale of public lands in Lassen County, California. The Government was authorized to sell 640 acres at $1.25 per acre to any person who would reclaim the tract within a two-year period. It was said at the time that the bill was intended as an experiment that would serve as a precedent for a general land policy.

Concurrently, land reforms were endorsed by President Ulysses S. Grant following a 1875 trip to the territories of Wyoming, Utah, and Colorado. He stated that existing laws were "very defective," and the President recommended that land should be granted "in larger quantities to justify the expense of conducting water upon it to make fruitful, or to justify utilizing it as pasture." Grant's views were echoed by State and territorial officials and legislators who hounded Congress with various proposals for disposition of the public domain that would promote irrigation.

These efforts culminated in the 1877 Desert Land Act which applied to three States—California, Oregon, and Nevada—and eight territories. It essentially extended the Lassen County provisions to the entire area west of the hundredth meridian. Settlers could purchase one section (640 acres) of desert land if they agreed to irrigate it within three years. Desert land was defined as "all land exclusive of timber lands and mineral lands which will not, without irrigation, produce some agricultural crop." Twenty-five cents per acre was paid on application with an additional dollar per acre required upon proof of compliance. When these terms were met, the settler obtained title to the tract.

This statute marked the Federal Government's first major policy specifically for the reclamation of arid lands. It endorsed the traditional stance that land development should be left to private initiative. Though designed to aid homesteaders, the act was widely exploited by larger economic interests.

Abuses were commonplace because of huge loopholes in the statute. Residence was not required, and absentee eastern investors filed on land and held it for speculative purposes. Large cattle companies used the law to obtain control of grazing lands bordering miles of streams. The law did not specify how much water should be conveyed to the reclaimed land and many "canals" were simply plow furrows. Allegedly, one entryman splashed a can of water in a ditch on his claim and then swore he had reclaimed the tract. Hundreds of crude ditches were built solely for the purpose of obtaining title to a section which was usually sold for profit.

In the 1891 General Revision Act, the Congress initiated reforms to minimize abuses and encourage private development. The Desert Land Act was extended to Colorado and modified to permit two or more persons to associate together in construction of irrigation canals and ditches. Entrymen were required to submit plans for reclaiming their tracts and spend at least one dollar per acre in each of the three years on irrigation works and other land improvements. They had to be residents of the State in which they were filing, the maximum acreage was reduced to 320 acres, and title could not be obtained unless an eighth of the land was cultivated within three years.

The law also extended a policy previously adopted for construction of roads and railroads across public lands. It encouraged private enterprise by granting canal rights-of-way across the public domain to irrigation companies. During the 1890's, this provision was gradually extended. By 1901 the Secretary of the Interior was permitted to approve easements for almost every purpose of water conveyance.

The next major reform sought to encourage Western States to promote large reclamation projects. In 1894 Congress passed a measure introduced by Wyoming Senator Joseph M. Carey. The Carey Act granted up to 1 million acres to each of the 10 arid States if they caused the land to be irrigated, settled, and cultivated. The States were to dispose of the land to farmers in tracts no larger than 160 acres, and not less than 20 acres of each quarter section was to be cultivated within 10 years.

The general plan was for the States to contract with private construction companies for building the irrigation works. The firms would in turn sell the water rights to reimburse themselves. Thus, reclamation costs were a lien on the land. The law, however, was a dismal failure before 1900. Investors were reluctant to invest money in irrigation projects and the States were unwilling or incapable of administering the complex program. By 1902 a mere 11,321 acres had been patented under provisions of the Carey Act, although in subsequent years more than a million acres were patented under this program.

STATE AND PRIVATE IRRIGATION

Direct State attempts to support irrigation projects were often disastrous. In 1887 the California legislature passed the Wright Act that provided for irrigation districts with broad powers to undertake reclamation. However, the complex job of designing irrigation works, floating bonds, and allocating stored water was beyond the ken of the districts' directors. They often constructed poor works and printed unmarketable bonds. Fraudulent districts were created and worthless projects attempted. Managerial and technical expertise was virtually nonexistent, and districts became encumbered with huge debts.

Less than one twentieth of an intended 2 million acres were reclaimed, and the projects became entangled in litigation over water rights and bond retirements. Colorado's experiment in State irrigation was also a fiasco. The legislature spent $103,000 for five storage reservoirs from 1890 to 1894 that either held little water or had to be abandoned because of faulty dam construction.

Despite a flurry of reclamation activity in the late 1880's and early 1890's, many private irrigation projects were partial or complete failures. They were often designed, engineered, and promoted by speculators having insufficient capital and technical resources. Dams, reservoirs, and canals were sometimes poorly built and the quantity of available water frequently misjudged. Settlers were sometimes enticed onto lands with heavy alkali concentrations, short growing seasons, and inadequate drainage. Over 90 percent of the private irrigation companies were in or near bankruptcy in 1900, and it was virtually impossible to raise investment capital for new ventures.

Private irrigation projects suffered from severe handicaps. Before the 1890's, few stream gaugings were made to determine seasonal flows and variations from year to year. There were no records of water volumes diverted by upstream users, and western water rights were in such chaos that engineers and investors could rarely anticipate the amount legally available. A California canal owner when asked how to protect water rights replied that he first obtained a judicial decree given him legal title and then hired guards during periods of low flow. Between the courts and guns he got his share of the stream.

Canal projects on public lands were hampered by land laws framed in the humid East. An engineer could estimate the acreage a canal would water, but he could not judge whether the adjacent public lands would be filed on by potential irrigators or speculators. Each ditch survey, therefore, was followed by a stampede of speculative land filings under the Homestead and Desert Land Acts. By refusing to enter into water contracts, speculators could starve ditch companies into bankruptcy or force purchase of their claims to end the settlement embargo.

Irrigation companies were hastily formed. Many of the large canal projects were undertaken by promoters who obtained money from eastern investors. The basis of their operation was usually a preliminary survey and a claim to the water of a stream under the appropriation doctrine. Surveys were inadequately funded and there was little interest in accuracy since someone else's money was often at risk. In nearly every instance, engineers were pressured to reduce cost estimates so as to encourage the sale of shares. Thus, in many cases projects began without sufficient capital, work was suspended before water was furnished, and the hopes and dreams of settlers depending on canals to mature crops were dashed.

Another handicap to orderly irrigation development by private enterprise was the obstinacy of inexperienced irrigators. Many settlers from the humid East did not appreciate that irrigation required cooperation. Unlike Mormon and other cooperative ditch companies, they opposed water regulations and raised and lowered headgates when it suited them. In dry weather, farmers at the lower ends of canals discovered all the water diverted by neighbors. Thus, fights with fists, shovels, and guns frequently broke out among water users.

PIONEER CONSERVATIONISTS

Many homesteaders struggling to survive on the frontier were acutely sensitive to the domination of land and water resources by large economic interests. Widespread resentment emerged against foreign investors owning large landholdings, railroads withholding millions of acres from sale and development, and schemes by eastern monopolists to carve huge empires of timber and grazing lands from the public domain.

Farm tenancy rose rapidly in the West, causing fears that the democratic Jeffersonian ideal of America as a country of small landowners was imperiled. These frustrations found expression in various radical agrarian movements during the 1880's and 1890's. Farmers and other reform groups sought political action by banding into Farmers Alliances that evolved into the Populist Party. The movement, which had strong support in the West and South, advocated a broad spectrum of reforms to rid the country of special interest domination. They supported government ownership of railroads, free coinage of silver, election and tax reforms, and economy in government. Some Populists also called for government aid to irrigation, forest conservation, and controlled development of natural resources.

Western interests had direct and pragmatic reasons to embrace the reclamation cause. However, the crusade was also part of a national conservation movement. By the 1890's, many individuals were concerned that natural resources were being ruthlessly exploited, and they fought for regulated development of the public domain. They advocated scientific management of forest, grazing, and mineral lands through a system of classification and leasehold.

A Federal reclamation law was viewed, therefore, as the first step in the evolution of progressive programs that would insure the availability of resources for future generations. Reclamation was the first rallying point for the progressive conservation movement that emerged in the early twentieth century.

Western irrigators perceived the relationship of watersheds to stream flow and fought exploitation of timber resources. In fact, the first conservation movement was largely a coalition of irrigators and forest conservationists. Homesteaders argued that trees and other vegetation held rainfall, retarded and evened out runoff, reduced floods, stopped soil erosion and silting, and extended the irrigation season. Private power and water supply corporations as well as municipal water departments also advanced these viewpoints. Irrigators petitioned Congress to preserve watersheds by setting aside forest reserves and protecting them from timber cutting and grazing.

Frederick H. Newell, while administering hydrologic investigations for the U.S. Geological Survey (USGS), served as Secretary of the American Forestry Association. George Maxwell, master propagandist for the National Irrigation Association, advanced the reclamation cause in the 1890's with the slogan: "Save the Forests, Store the Floods, Make Homes on the Lands."

JOHN WESLEY POWELL: FATHER OF RECLAMATION

In spite of the emphasis on private development, some reclamationists advocated expanding the Federal role in western water development. "For the just half of the nineteenth century," observed Walter Prescott Webb, "there existed in the public mind a Great American Desert situated to the east of the Rocky Mountains." John Wesley Powell, perhaps more than any other individual, helped to dispel this notion and promote acceptance of the idea that the region was capable of sustaining human settlement.

In 1878 Powell made his farsighted report, *The Lands of the Arid Regions of the United States.* This study advocated a scientifically planned reclamation and settlement program to utilize water and land resources efficiently while avoiding the pitfalls of speculation and monopolistic control. Powell understood the environmental limitations of this western "district or country," as he called it, and the need to reshape eastern laws and institutions to meet its requirements. The famed scientist and explorer recognized that irrigation expansion required construction of dams and canals to either divert rivers or store floodwaters for summer use.

Furthermore, he understood that such structures required large amounts of capital and technical skills beyond the means of small farmers. Today, soil conservation districts, grazing associations, as well as irrigation and conservancy districts are common throughout the region. As early as the 1870's, Powell perceived that such cooperative institutions were essential for survival beyond the hundredth meridian.

In reality Powell proposed a regional development plan. His purpose was "not only to consider the character of the lands themselves, but also the engineering problems involved in their redemption [so that] . . . these lands may eventually be rescued from their present worthless state." He did not suggest, however, that the Federal Government become directly involved in reclamation. Instead, he envisioned farmers and ranchers organizing themselves into autonomous irrigation and pasturage districts to govern settlement, water development, and the use of other natural resources.

Powell proposed different institutions for contrasting types of land and climate. Above all he believed "the people in organized bodies can be trusted," and he recommended that "residents should have the right to make their own regulations for the division of the lands, the use of the water for irrigation and for watering the stock, and for the pasturage of the lands in common or in severalty." He proposed, therefore, the creation of dynamic social and political institutions based upon environmental factors:

> This, then, is the proposition I make: that the entire arid region be organized into natural hydrographic districts, each one to be a commonwealth within itself. . . . Each community should possess its own irrigation works; it would have to erect diverting dams, dig canals, and construct canals, and such works would have to be maintained from year to year. The plan is to establish local self-government by hydrographic basins.

Powell insisted on a comprehensive scientific approach to western development. He stressed that reservoir sites should be identified, surveyed, and withdrawn from public entry. A step-by-step plan was put forth by Powell: (1) prepare topographic maps of western lands; (2) determine the total annual volume of water in catchment areas; and (3) conduct soil surveys.

Powell's recommendations went largely unheeded but, because of the shortcomings of public land laws, support eventually developed in the Congress for a thorough inventory of western water resources and the setting aside of irrigation reservoir sites. In 1888 Powell, by then head of USGS, was authorized to survey streams and damsites in the West. To facilitate future development, USGS was empowered to withdraw from public entry all areas "necessary for the storage and utilization of water," including canals and ditches and adjacent tillable lands. By 1889, 150 canal sites were identified and 30 million arid acres were deemed irrigable.

As the survey parties progressed, however, speculators followed them into the field and filed on tracts before they could be reserved. Consequently, the Commissioner of the General Land Office issued a circular closing the public domain to further settlement until Powell's engineers finished their work. This action ignited a storm of western protests, and in 1890 Congress restricted withdrawals to actual reservoir sites and cut the survey's funds. Thus, Powell's plan for a careful inventory of water resources and controlled development was rejected in favor of identifying water storage sites followed by unrestricted settlement of adjacent lands.

Despite this setback, USGS continued to study water resources in arid areas during the 1890's. The agency's hydrographers measured streamflows, located dam and canal sites, and mapped irrigable lands. Newell became chief hydrographer for USGS in 1890 and began a lifelong commitment to Federal water development.

During those years, USGS engineers pioneered a new philosophy of water conservation. The Army Corps of Engineers, for example, viewed rivers principally as arteries of transportation and rarely considered other uses. The USGS on the other hand, investigated all aspects of precipitation, accumulation, and runoff. The USGS advanced theories on coordinating all water uses including navigation, irrigation, power, and flood control. After 1902 Newell and his team of hydrologists formed the nucleus of the Reclamation Service.

IRRIGATION MOVEMENT

William E. Smythe was the West's first spokesman to develop a common focus and ideology for the irrigation movement. In 1891 the Nebraska journalist attracted public attention with a series of articles on irrigation in the *Omaha Bee*. For the next five years, he was the West's foremost reclamation advocate—leading a broad-based popular movement to transform arid and semiarid lands into productive, small family farms. To Smythe, irrigation was not merely an economic reform but "a philosophy, a religion, and a program of practical statesmanship rolled into one."

He contended that irrigation compelled adoption of small farm units and extended cherished democratic values and agrarian institutions. Eastern public opinion was receptive to this concept of western colonization, as the traditional social and economic fabric of the Nation was challenged by the forces of urbanization, industrialism, monopoly, and the influx of large numbers of immigrants.

The journalist's philosophy was a blend of agrarian idealism and practical economic promotionalism designed to foster popular support for Federal legislation. Smythe and his followers initially sought cession of public lands to the States to help launch projects funded by private enterprise. They appealed for the enactment of State and Federal laws that would abrogate water-rights conflicts and promote efficient planning, engineering, and operation of irrigation works. Homesteads—the opening of their scenario for western development—would be followed by railroad construction, creation of new cities and industries, and development of the West's vast untapped wealth.

Smythe was made chairman of a committee to help organize the first National Irrigation Congress, which met in Salt Lake City, Utah, in September 1891. The meeting was primarily attended by real estate interests, representatives of railroad and irrigation companies, lawyers, journalists, civil engineers, businessmen, and others who had an economic stake in western expansion. The meeting's key resolution asked Congress to grant "to the States and territories needful of irrigation all lands part of the public domain . . . for the purpose of developing irrigation to render the lands now arid, fertile and capable of supporting a population." Subsequent congresses were held annually at various western cities.

As secretary of the National Irrigation Congress in 1892 and 1893, and its president from 1893 to 1895, Smythe addressed public meetings from coast to coast, advocating settlement on arid lands by small farmers. The persistence of the 160-acre provision in Reclamation law represents the homestead principle Smythe so ardently championed in the 1890's. He also exercised influence in the columns of his *Irrigation Age*, which he founded in 1891 and edited until 1896. The publication was the reclamation movement's oracle and provided a platform for the editor's views and contributions from other irrigation leaders. It monitored State and legislative developments, court decisions, irrigation projects, and offered practical farming advice to inexperienced irrigators.

The depression following the Panic of 1893 posed a threat to the irrigation movement because of its emphasis on private enterprise. The economic downturn dried up sources of investment capital, and business support for land cession to the States began to wane. New strategies for western reclamation were aired and schisms developed among advocates of different approaches. Smythe abdicated his leadership position in 1896, but the movement remained indebted to him for its structure, orientation, and organization of forces and interests that led the drive for Federal construction of irrigation works.

The movement for Federal reclamation gained considerable impetus from a 1897 report by Captain Hiram Chittenden of the Army Corps of Engineers. His investigation of dam and reservoir sites in Wyoming and Colorado resulted from Wyoming Senator Francis E. Warren's advocacy of Federal appropriations for surveys and construction of storage reservoirs in the West. Though Chittenden's report focused on flood control, he staunchly proposed the construction of storage and distribution works for irrigation. He also recommended that the Government should build, own, and operate all necessary works, holding the stored waters for public use under State and local regulations.

Chittenden formulated one of the first arguments for direct Federal reclamation and candidly discussed the shortcomings of efforts by corporations, irrigation districts, and the States to build complex irrigation works. For Chittenden, dam building was clearly a Federal responsibility. He stressed that controlling stream-flow would stabilize navigation, abate floods, provide power, and furnish water for irrigation as well as domestic and industrial use. The report crystallized support and added polemical weapons to the arsenal of the reclamation movement.

By the late 1890's, reclamation faced strong opposition from the East and Midwest. Furthermore, the West was badly divided on the best plan of action. Many irrigators continued to uphold cession of public lands to the States as the most prudent course. Reclamationists quarreled over whether irrigation should be undertaken by cooperative irrigation districts or private corporations, and many of the Plains States propagated plans for drilling artesian wells and tapping stream underflows. Even the forces upholding direct Federal reclamation promoted different strategies. One camp favored the Federal Government building the storage works and turning over the subsequent tasks of settlement and water allocation to the States. A separate wing argued that, because of past failures, only the Federal Government had the resources to implement a sound reclamation program.

RECLAMATION LEADERSHIP

In the late 1890's, George W. Maxwell emerged as the foremost leader of the irrigation movement. The National Irrigation Congress had failed to develop an ongoing institution that would promote public and congressional support for western irrigation. In 1897 Maxwell, a California lawyer with broad experience in western water development, organized the National Irrigation Association (NIA) at a meeting of the Trans-Mississippi Commercial Congress in Wichita, Kansas. The new organization's purpose was to augment, not replace, the National Irrigation Congress. It was devoted to organizing nationwide support from the general public, business, farmers, and politicians for a national reclamation act.

Maxwell faced a formidable task. The reclamation movement was split into various factions and faced strong sectional opposition. Eastern congressmen promoted their own projects such as canals and river and harbor improvements. Midwestern farmers feared competition from irrigated land, and manufacturing and laboring organizations opposed subsidizing agriculture. In addition, many people thought it unwise to tax populated areas to benefit a barren, dry frontier.

Financing was Maxwell's immediate problem. With help of the Atchison, Topeka, and Santa Fe Railroad, he was able to obtain large grants from most of the transcontinental railroads. With this support, he established NIA offices in large cities. The association's periodicals were tailored for different audiences. *The National Homemaker* in Washington, D.C., was directed toward enlisting congressional support, while the *California Advocate, Opportunity,* and *Maxwell's Talisman* conveyed the message to business, agriculture, and the general public. The NIA press bureau fed releases to newspapers throughout the country. In the Nation's capital, Maxwell received substantial support from USGS engineers including Newell.

The association's membership increased rapidly and national newspapers and powerful business and agriculture groups endorsed Federal reclamation. Public attitudes began to shift as the intersectional virtues of the proposed program became apparent. The NIA essentially argued that public land sales would fund the water projects, western agriculture would not compete with the East, and irrigation would create new markets for eastern industries.

While Maxwell marshalled public support for reclamation, Representative Francis G. Newlands of Nevada led the fight in Congress. Early in 1889, Newlands, a wealthy Californian, moved to Nevada and became deeply involved in the State's economic and political affairs. He invested enormous sums in an irrigation project on the Truckee River and was elected to Congress in 1892.

Unlike most western promoters, Newlands regarded rational planning and orderly economic development as keys to irrigation success. He employed engineers and geologists to plan his undertakings and fought for State laws that would promote effective public control and sponsorship of reclamation. Though Newlands' project in Nevada failed, he became a leading reclamation spokesman in the 1890's. He was convinced that irrigation could never be satisfactorily accomplished by private enterprise, and he became persuaded that the State governments were incapable of sponsoring successful reclamation.

Throughout 1900 and 1901, Newlands, Newell, and Maxwell worked separately and together to obtain congressional and public support for Federal reclamation. The politician, engineer, and publicist were an effective team. Newlands held a series of evening meetings at his home with western congressmen that led to congressional hearings on western irrigation. Maxwell, in addition to his nation-wide public relations campaign, lobbied vigorously for western irrigation. Newell worked to obtain support within the Federal bureaucracy.

Meanwhile, Federal irrigation became a national issue. In 1900 all of the major parties inserted pro-irrigation planks in their platforms. The Ninth National Irrigation Congress added its approval and, owing largely to Maxwell's efforts, organizations throughout the country passed resolutions favoring Federal reclamation. Backers included the United Mine Workers, National Association of Manufacturers, National Board of Trade, and the National Business League.

By 1901 Newlands with the assistance of Maxwell and Newell had formulated a prospective national reclamation law. The measure was introduced in the House on January 26, 1901. Since Newlands was a Democrat, Republican Senator Henry Hansbrough of North Dakota agreed to submit an identical measure to the Senate five days later.

The first session of the Fifty-Sixth Congress was stormy. A myriad of other irrigation bills was offered which the East and Midwest opposed. Some westerners led by Wyoming Senator Francis E. Warren tried to have direct appropriations for storage dams included in rivers and harbors legislation. Opposition to Federal reclamation mellowed somewhat when Montana Senator Thomas Carter filibustered to death a rivers and harbors bill with its many pet projects for eastern congressmen. Though Newland's bill was not passed, by the close of the session eastern support for western irrigation had grown, hastened by a severe drought in the Middle West.

Newland's proposal for Federal reclamation disavowed past efforts to encourage irrigation by private enterprise and the States. His position was that the Government build and finance irrigation works from a revolving fund composed of proceeds from public land sales. Therefore, the West would pay for the public works.

Hoping to forestall disagreement among western congressmen over project locations, he suggested that the Secretary of the Interior have complete discretion in selecting projects and apportioning funds. Congressional control of annual appropriations frequently produced confusion, poor coordination and efficiency in rivers and harbors projects. The Nevada congressman proposed that expert planning based on thorough technical investigations, instead of logrolling, would direct the course of western irrigation. Newlands was adamant in his belief that families, not corporation farmers, should own reclaimed desert tracts; and he insisted that no individual should acquire title to more than 160 acres. To protect the program from speculation, entry on project lands would be tightly regulated.

FEDERAL RECLAMATION

Strong opposition arose in Congress from eastern and midwestern representatives who advanced anti-reclamation arguments later echoed throughout the twentieth century: farmers were already producing surpluses; it would be unfair to aid western irrigators at the expense of eastern farmers; future generations should not be burdened by debts for water projects; opening lands was a government subsidy to the few; and Federal reclamation violated State rights and was therefore unconstitutional.

Newlands and supporters of his bill developed social and financial arguments in favor of enactment: it would extend the frontier and relieve population pressures and overcrowded cities; reservoirs would enhance conservation; irrigation crops would be consumed locally; the program would be self-liquidating; new markets would be created for eastern manufactured goods; and the dams and canals would be no more paternalistic than Federal rivers and harbors projects built throughout the rest of the country.

The reclamation movement received an unexpected boost as a result of the assassination of President William McKinley. When Theodore Roosevelt became President on September 14, 1901, he was already a champion of western reclamation. Having lived in the dry region, he had firsthand knowledge of its vital need for water conservation. Both arch-conservationist Gifford Pinchot and Newell personally influenced Roosevelt's views before 1900. They were asked by the new President to prepare portions of his first annual message to Congress that emphasized forestry and irrigation. Maxwell was invited to help draft the presentation which was a ringing manifesto for a Federal reclamation act and multiple-purpose water development.

Roosevelt opened his address by stating that the forests were natural reservoirs which helped conserve water resources and abate floods and droughts. "Forest conservation," he said, "is therefore an essential condition of water conservation." After illustrating the interrelationship of forests and water, he remarked that reliance upon timberlands could not fully regulate water in arid regions. Large reservoirs were required to regulate streamflows and store floodwaters. The Federal Government, he argued, should:

> Make the streams ... of the arid regions useful by irrigation works for water storage. ... The storage of floods in reservoirs at the headwaters of our rivers is but the enlargement of our present policy of river control, under which levees are

> built in the lower reaches of the same streams. The Government should construct and maintain these reservoirs as it does other public works.

Sweeping aside constitutional objections, the President simply cited the General Welfare Clause of the Constitution as authority. The work he maintained could not be done by the States. Far-reaching interstate problems were involved and the resources of individual States were often inadequate. Water development was, he concluded, "properly a national function. . . . "

Newlands made some modifications in his bill, and from January to June 1902 it again ran the gauntlet of congressional debate. However, strong public support and the President's endorsement eroded the opposition. The bill passed both houses of Congress by wide margins and was signed into law by Roosevelt on June 17, 1902.

The law launched a progressive program that employed Federal engineering as a tool for social progress. Its authors envisioned barren lands being transformed into productive farms that would support new homes and communities. The statute dedicated the proceeds from public land sales to a revolving Reclamation Fund that financed water storage and diversion works in 16 western States and territories. (Texas was added in 1906). Irrigable public lands were subject to entry under the 1862 Homestead Act, with each settler obligated to repay his proportionate share of construction costs in 10 years. Local water user associations would assume operation and maintenance responsibilities once half of the repayment obligations were met, but the Federal Government retained title to major works. State and territorial laws governed the appropriation, use, and distribution of project water; and existing Federal, State, and private rights on interstate streams remained undisturbed. Although in theory the act applied to reclaiming the public domain, many acres of project land were already privately owned. Therefore, the sale of water rights was provided for a maximum of 160 acres to existing landowners.

The 1902 Reclamation Act was, therefore, the culmination of efforts to adapt national public land policies to western conditions. The funding of surveys and reserving from entry potential dam and reservoir sites reflected an expanding recognition of the need for water storage. Various rights-of-way acts sanctioned the doctrine of beneficial use, affirmed State control over water appropriations, and assured development of streams flowing through the public domain. Congress also resolved the question of financing extensive reclamation when it became evident that irrigation projects were not attractive to private capital. As the new century dawned, the Federal Government embarked on a socially progressive public works program designed to create new family farms.

5.8

Power by Persuasion

R. A. KINGERY, R. D. BERG, AND E. H. SCHILLINGER

Ninety-four-year-old John Anderson stood before a group of his neighbors and local dignitaries on a farm near Ludlow, Illinois, and solemnly smashed a kerosene lamp against an electric light pole. The ceremony marked the passing of the kerosene lamp era in Champaign County; more significantly, it heralded the advent of electricity on the farms. It was 1938—almost 20 years after Illinois agricultural engineers began to promote rural electrification. It had been a long and frustrating battle because of the very people it was to help—the farmers.

The transition of the United States from an agricultural country to an industrial one had not been easy. While society was still largely agrarian, the farmer was often overcharged by the railroads that hauled his produce and the warehouses that stored it. He often could not get credit. He faced steadily falling prices. He went into debt.

Not until just before the First World War had the farmer's economic situation improved. Then in 1920, '21, and '22, he was hit by a severe depression that forced him to sell his crops at less than it had cost to produce them.

By the late 1920's, the farmer was nearly destitute. Never a strong consumer of manufactured goods, he now tried to reduce expenses by becoming self-sufficient. He tried to grow his own food, build his own fences and buildings, make his own clothes, and conduct his life so that if he should get any return for his labors, he could keep all of it. He was hardly inclined to believe that a small investment in modernization could bring him a better, happier life—particularly when the modernization involved the use of electricity, possibly the last thing he would have felt important to his life.

Rural electrification in Champaign County and in Illinois had begun long before the federal government and the bulk of the nation's farmers had realized the need for electricity on the farms. Much of the state's progress and the influence it had on the rest of the nation may be traced back to the early 1920's and to one man—a man with the foresight to recognize the plight of a major industry and a major segment of the population without power.

Emil Wilhelm Lehmann came to the University of Illinois in 1921 as head of the new Department of Farm Mechanics. He would occupy that position for 34 years. The youngest of ten children, Lehmann had grown up on a cotton farm in Mississippi where his German-born parents had settled. After earning bachelor's and master's degrees in electrical engineering from Mississippi and Texas A & M, Lehmann decided to combine his interests of farming and engineering. In 1914 he completed a degree in agricultural engineering at Iowa State University, and received a professional degree in agricultural engineering from the same institution in 1919.

As Lehmann saw it, the farm families' failure to recognize the importance of proper power and mechanization was the main barrier to farm modernization. He told the members of his department at Illinois in the first departmental meeting: "It isn't that we don't know how to help the farmers; it's just that they don't want to be helped. It's as if they're afraid of us, or maybe they're just afraid of changing."

The farmers' resistance to progress was nothing new to agricultural engineers. Problems which had plagued farmers for generations continued to plague those of the twentieth century. The annual race to harvest corn crops by hand before the weather and the market broke; mules and horses laid up for days with shoulders rubbed raw under loose harnesses; weeks of isolation following spring rains on dirt roads—these were the grim and timeless realities of rural life as it had been for centuries.

By 1920 a few daring farmers were beginning to experiment with motorized tractors, radical departures from the dependable plow horses which had served their agrarian masters since early times. But such innovators were in the minority. With the exception of gasoline engines, which were used sparingly about the farmsteads as energy sources for water pumps and belt-driven machinery, farm living had not changed appreciably for centuries. Limited production and small scale farming made it impossible for individual farmers to amass enough capital to bring progress to their operations; on the other hand, there could be no production increases until agriculture moved into the twentieth century. Small wonder the U of I agricultural engineers adopted as their slogan, "The Profession with a Future." Agriculture had only one way to go—forward.

Lehmann was convinced that standards of living and work within the rural community could be and had to be improved. This was a novel concept to most farmers, who had accepted their sunup-to-sundown work day and the back-breaking drudgery of their labor when they chose their profession. More important than Lehmann's conviction were the tireless energy and dogged persistence he possessed and demanded from all who worked with him. Lehmann found that if he could not get what he wanted in any other way, he could talk most opponents into agreement. Lehmann's energy and tenacity may have caused some unhappiness in the Department, but they also produced results. One University official remarked that if he sent out a letter asking for suggestions concerning any problem, he could always count on at least one reply—Lehmann's.

Lehmann had first recognized the potential of electricity for farm labor at Texas A & M. His thesis topic there had been the use of electricity in agriculture.

Measuring the electrical energy consumed by washing machines, ranges, refrigerators, and almost every other piece of equipment he could find in the community, Lehmann produced data which indicated that the farm household's potential electrical appetite surpassed that of a city family because of the necessity of refrigerating and preserving foods. He speculated that there was an untapped potential for electrical labor-saving devices around the farmstead, too.

Early in his career as head of the Agricultural Engineering Department at the University of Missouri, Lehmann became noted as a pioneer in rural home improvements. He organized traveling demonstrations of home improvements which he took around the countryside. The exhibition truck hauled an open two-story frame structure with kitchen plumbing on the first level and bathroom fixtures on the second. Groups of farm families would gather at the home of a neighbor to hear Lehmann expound on the convenience and value of indoor pumps, bathtubs, and toilets. For a rural population with few outside distractions, the demonstrations were festive outings for the entire family. But response to the lectures, though polite and interested, was disappointing. These were audiences that prided themselves on their frugality and practicality.

On one occasion, when Lehmann was having particular difficulty convincing the farmers that they could afford home improvements and that appliances were worth their cost, he caught sight of a pickup truck with a crate in the back of it. In the crate was a dog. "Mind telling me, sir, how much you paid for that hound dog?" Lehmann called to a man leaning against the truck.

"Thirty-eight dollars," the man called back, "plus two dollars for shipping."

"Don't you realize that for that amount of money you could have a pump right in your kitchen, so that your wife wouldn't have to go outside every time she wanted water?"

"That may be," the man replied, "but I bet that sink can't tree a coon."

Unlike engineers in other fields, whose research findings were directed to industry, the agricultural engineers dealt directly with their consumers—the farmers. Individuals were not likely to gamble their incomes on new ideas when the old ones had proved workable over a period of years. Thus the agricultural engineer was forced to do more than teach and do research—he also had to sell.

In 1920, less than 10 per cent of rural America was served by electricity. By comparison almost all of Holland's farms, 90 percent of Germany's, and 50 percent of Sweden's were electrified. Although population density on the European continent made electrification there easier and cheaper, the absence of electricity on American farms was especially deplorable to men like Lehmann because of the nation's generally high standard of living. For every farm family with electric power there were four with automobiles and two with telephones.

Lehmann wanted to show the farmers that adequate electrical equipment could save them money as well as provide conveniences and comfort. The power companies, too, needed to be convinced that there was a profitable market for their service in the corn-covered land as yet unbroken by transmission lines.

The utilities could profitably provide service to city dwellers at 35 customers per mile of line, but they balked at the idea of extending this service to the more sparsely populated rural areas. Even small towns had to rely on local power plants. There was almost no provision for the 230,000 Illinois farm homes. Individual generator units producing up to one kilowatt of electricity were available to those few farmers who could afford them. These were intended for lighting, the one innovation which appeared useful to most farmers. In the words of Professor Lehmann, lighting made it possible to "read the current events of the day, study various farm problems, and plan for the next year's work" after it was too dark to work outside. Concerning any other use of electric power, however, skeptical farmers asked, "Even if we could afford it, what would we use it for?" Lehmann reasoned that a lower rate for increased service would spur farmers on to find new uses for electric power, a proposition that could mean profit for the farmers—and for the utilities.

In 1923 his group at Illinois conducted an initial study on northern Illinois farmsteads served by power lines. The results of the investigation were disappointing. They emphasized rather than disproved the risk involved for power companies in rural neighborhoods. Of the 93 farm homes having electric service, all were lighted by electricity. However, only half of the consumers used electric motors for limited power operations like pumping, grain grinding and elevating, and household operations. Most of their power was supplied by windmills; gasoline engines made up the difference.

The following year Lehmann, in cooperation with the local power companies, set up an experiment of his own—a three-year practical study of electricity in rural Tolono. Ten farms were wired to determine whether sufficient electricity would be used to make the installation profitable for both the farmers and the utility companies. More important, from Lehmann's point of view, was the opportunity to use these farms just as he had the plumbing exhibitions—as demonstrations that electricity and agriculture worked well together.

During the first year of the experiment, electrical equipment was provided free of charge by manufacturers to the families with an option to purchase after the first year. In addition to lights, all ten families were given refrigerators, vacuum cleaners, and cream separators. Utility motors, dishwashers, food mixers, ranges, grain elevators, and buttermakers were also made available. Although manufactured commercially, most of the household appliances were considered luxury items even in the urban areas.

During the study, the farm women often used their appliances to make cookies and cakes for the scores of curious farmers who dropped in to see the electrified farms. Professor Lehmann was a frequent visitor, and became a close personal friend of the cooperating farm families. Though he was serious and a bit formal in his relationships with the agricultural engineering faculty, Lehmann relaxed and enjoyed his Tolono friends. Once, after an evening meeting at the home of a cooperator, Lehmann led the group in a lively jig. One of the men recalled, "When I first saw him I thought, 'Here's a city fellow from the University going to tell me how to run my farm.' But after I got to know him, I knew he was just a farmer like me—only educated."

He became particularly fond of Mr. and Mrs. Harry Reifsteck. Mrs. Reifsteck, a young bride who had come to her husband's farm less than a year before, was used to the greater comforts and conveniences of city living. She felt that the heavy labor associated with farm household tasks deprived a woman of the intellectual and spiritual activities which made her life interesting. "It is possible for a farm wife to be a good housekeeper, but not a good homemaker. There is quite a difference in the meaning of these two words. I would much rather that people said I was a good homemaker then merely a good housekeeper." She found that electricity gave her time for both.

So enthusiastic was her acceptance of electrical living (she and her husband tried every piece of equipment available and eventually bought all but the electric ironer) that Lehmann decided to enlist her help in convincing other farm women of the advantages to be gained with electricity.

The small woman soon became a familiar sight on the rural roads in her blue Ford coupe. At Home Bureau meetings and other gatherings she told other farm wives her experiences with electricity on the farm. Farm and electrical magazines published her articles and radio stations began to broadcast interviews with her.

Lehmann's group kept detailed data on the way the women on the ten farms used their time. Soon it became clear that electric washing machines and vacuum cleaners alone were saving each housewife as much as ten hours a week, leaving more time for gardening, poultry raising, and other profit-making activities.

But the data revealed an interesting side effect: the number of hours the women spent on personal grooming increased in every case, and almost doubled in one instance. This unexpected development resulted from increased leisure time and social contact, the latter brought on partly by the study itself, which required frequent visits to the farm homes by the investigators. "You need to look prettier for a man from the University than you do for the chickens," one of the Tolono women said.

Lehmann also used the farms as testing areas for experimental electrical equipment that had come out of the farm mechanics, agronomy, and other laboratories. Tests at various research centers had already established that egg production could be increased with controlled lighting by as much as 21 per cent during the winter months when prices were high. Poultry house lighting with automatic time switches could lengthen the chicken's day to 12 or 13 hours; the 40-watt bulbs served as artificial suns to rouse the sleeping chickens while it was still dark outside. The Tolono experiments bore out the earlier findings and introduced the system to Illinois.

The Tolono experiment indicated beyond any doubt that there was a place for electricity on the farm. On the ten test farms, only draft horses and tractors played a greater role than electricity in supplying energy. Energy and time tables compiled for farmstead equipment showed that the highest monthly total averaged 278 kilowatt-hours; the lowest, 42. This compared with an average monthly household consumption in the cities of 50 kilowatt-hours.

The task of expanding the applications of electricity on the farms now lay before the researchers, but there could be no electrical appliances without electricity. A means had to be found to make electricity available to the farmers at a price they would be willing and able to pay. Lehmann devoted a six-month sabbatical leave from the University to working out equitable electric rates with the Central Illinois Public Service Company in Springfield. Armed with the indisputable results of his Tolono experiment, he was eventually able to convince the power companies that there was a tremendous potential for electric power on the farms—in fact, a much larger per capita market than in the cities.

Lehmann won a number of rate concessions from the major utilities, but it soon became apparent that these would not be sufficient to do the job he wanted. If most Illinois farmers were to have electric power, a broader, more positive program, geared especially to the farms and their problems of distance and scattered population, would have to be provided.

At the end of the Tolono experiment, rural electrification seemed almost as far away as it had at the beginning. The great value of the experiment lay in the fact that many farmers had become aware of the potential of electric power for their work. County Farm and Home Bureau representatives and members of other farm organizations worked with the University Extension representatives to bring the story of the Tolono successes to every rural community in the state. Farmers themselves began seeking information. Studies of a similar nature in other state universities were making the issue of farm electrification a national concern. It became evident that, if the rural segment of the population were ever to have electrical power, the issue would have to be supported by someone other than the power companies. The farm interest groups turned from the utilities in the early 1930's and headed toward Washington, where they hoped their arguments would receive more attention.

In the fall of 1935, six years after the Tolono test ended, President Franklin D. Roosevelt issued an Executive Order establishing the Rural Electrification Administration. Set up to subsidize the utilities and thereby encourage their extension into the rural areas, the program proved to be ineffective and unpopular with the power companies, and participation was negligible. Nevertheless, the people who had struggled for so many years for a comprehensive program of rural electrification took heart.

The farm lobbies regrouped for a battle with Congress the following year. In the legislative hopper was a bill entitled the Rural Electrification Act of 1936. It had been drawn up by the Vice-President of the United States in consultation with the president of the National Federation of Farm Bureaus and the president of the Illinois Agriculture Association. Illinois, one of the most progressive and prosperous of the farm states, had long been recognized in Washington as a leader in the fight for rural electrification legislation. Spearheading the Illinois effort was Lehmann's own Champaign County Farm Bureau.

Victory came early in the session. On May 20, 1936, the Rural Electrification Act was signed into law by President Roosevelt. It would be one of the most permanent and successful acts to come out of the New Deal. Under the program, the Rural Electrification Administration would make available 35-year loans at two per cent interest to farmers wishing to work together to extend lines to their property. These systems would be controlled, operated, and jointly owned by the participating farmers. In the spring of 1936, when the REA began, approximately 16 per cent of rural America had access to electrical power. Thirty years and hundreds of millions of dollars later the number had climbed to over 98 per cent.

The coming of electric power to the farms was the catalyst which triggered the agricultural revolution. Radical changes and improvements in machinery, crop production, soil and water conservation, farm buildings, and home improvements followed almost immediately. Agricultural engineers suddenly saw their world become a universe. At Illinois the Department of Farm Mechanics was incorporated into the College of Engineering as the Department of Agricultural Engineering.

In 1938, as the first tangible results of the Rural Electrification Act were manifesting themselves over rural America, President Roosevelt delivered a speech at

Barnesville, Georgia. In the course of his address he spoke of his first visit to his retreat in Warm Springs, Georgia:

> There was only one discordant note in that first stay of mine at Warm Springs. When the first-of-the-month bill came in for electric light for my little cottage, I found that the charge was 18 cents a kilowatt hour—about four times what I pay at Hyde Park, New York. That started my long study of public utility charges for electric current and the whole subject of getting electricity into farm homes. . . . So it can be said that a little cottage at Warm Springs, Georgia, was the birthplace of the Rural Electrification Administration.

An important and farsighted piece of legislation may well have been born in the little cottage at Warm Springs, as President Roosevelt stated; but many of the ideas and much of the inspiration behind it had their origins elsewhere—such as in the minds of men like Emil Lehmann.

5.9

Women and Technology in American Life

RUTH SCHWARTZ COWAN

Critics of women's history contend that it is a nonsubject. The history of women, they say, is inseparable from the history of men; therefore women's history, as a category unto itself, is analytically unjustified. That point of view has some merit, but only insofar as the history of women is, in fact, indistinguishable from the history of men. There is a sense in which political history was the same for both men and women. Abraham Lincoln was the president of all of us even though some of us didn't get to vote for him. In that sense our histories are indistinguishable. In many other senses the experiences of men and women have been profoundly different. One wonders, for example, why our history books have neglected to tell us how our founding mothers reacted to the news that in the new republic they had helped to nurture they would not be allowed to vote. Similar examples could be adduced but the point is clear: the history of women as a group has been different from that of men as a group, and that history should be told.

Thus, investigation of the relation between women and technology in American life is analytically justifiable only insofar as that relation has been different from men's. A careful reading of most histories of American technology (or for that matter Western technology) would not lead one to suspect that important differences existed, but that is a property of histories of technology and not of historical reality. There are at least four significant senses in which the relation between women and technology has diverged from that of men.

WOMEN AS BEARERS AND REARERS OF CHILDREN

Women menstruate, parturate, and lactate; men do not. Therefore any technology that impinges on those processes affects women more than it affects men. There are many such technologies, some of them with long histories, such as pessaries, sanitary napkins, tampons, intrauterine devices, childbirth anesthesia, artificial nipples, bottle sterilizers and pasteurized and condensed milks. Many psychological theories suggest that those three processes are fundamentally important in the psycho-social development of individuals. Thus a reasonable person might suppose that the history of technological intervention with those processes would be known. That reasonable person would be wrong, of course. The indexes of the standard histories of technology—Singer's, Kranzberg and Pursell's, Daumas's, Giedion's, even Ferguson's otherwise magnificent bibliography—do not contain a single reference to such a significant cultural artifact as the baby bottle. Along with its attendant delivery systems, this simple implement has revolutionized a basic biological process, transformed a fundamental human experience for vast numbers of infants and mothers, and been one of the more controversial exports of Western technology to less-developed countries—yet it finds no place in our histories of technology.[1]

There are many questions that scholars might reasonably ask about the baby bottle. How long has it been part of Western culture? When a mother's milk could not be provided, which classes of people used the bottle and which the wetnurse, and for what reasons? Which was a more crucial determinant for widespread use of the bottle: changes in milk technology or changes in bottle technology? Who marketed the bottles, at what price, to whom? How did mothers of different social classes and ethnic groups react to them? Can the phenomenon of "not enough milk," which was widely reported by American pediatricians and obstetricians in the 1920s and 1930s, be connected with the advent of the safe baby bottle? Which was cause and which effect?

In short, the history of the uniquely female technologies, with the single exception of the technologies of contraception, is yet to be written.[2] The same is true of the technologies of childrearing, a process not anatomically confined to females but more or less effectively limited to them by many unspoken social contracts. We know a great deal more about the bicycle than we do about the baby carriage, despite the fact that the carriage has played a more lasting role in the transport of infants than the bicycle in the transport of adults. We recognize the importance of toilet training in personality formation, but we have not the faintest idea whether toilet training practices have been affected by the various technologies that impinge upon them, such as inexpensive absorbent fabrics, upholstered furniture, diaper services, wall-to-wall carpeting and paper diapers. The crib, the playpen, the teething ring, and the cradle are as much a part of our culture and our sense of ourselves as harvesting machines and power looms, yet we know almost nothing of their history. Only an historian of technology who is concerned with and versed in women's history is likely to take the trouble to search it out.

WOMEN AS WORKERS

Women have been part of America's economy from its earliest days. In the colonial period they tended cows, delivered babies, kept taverns, published newspapers, and stitched fancy clothes, among other things. During the period of industrialization they tended looms, folded paper bags, packed cigars, helped with harvests, washed laundry, and stitched fancy clothes, again among other things. With the advent of automation they punch cards, handle switchboards, pack cookies, teach school, tend the sick, and stitch fancy clothes—again among other things. All along they have been paid for their work—in land, in produce, or in cash.

But any analysis of the relation between women and the American technological order will show that, as workers, women find their circumstances much different from those of men. The economic facts of life for women are so pervasive as to be almost universal truisms. As with the aforementioned anatomic facts of life there are three economic facts of life: when doing the same work women are almost always paid less than men; considered in the aggregate women rarely do the same work as men (jobs are sex-typed); and women almost always consider themselves, and are considered by others, to be transient participants in the work force.[3]

These characteristics of women as workers predate industrialization; they were economic facts even before our economy was dominated by cash. Sex typing of jobs occurred in the Jamestown settlements before the household economy had completely replaced the communal economy. Unmarried or poor women worked as laundresses in return for a portion from the communal store; men who were not entitled to grants of land worked as cooks and bakers.[4] Unequal pay for equal work was also characteristic of the early settlements. Adventurers who came to settle in Maryland were allotted 100 acres of land for every manservant they brought with them and 60 acres of land for every womanservant. Unmarried free men in Salem and Plymouth were given allotments of land when they requested them, but (after the first few years of settlement) unmarried women were not. The first American effort to obtain equal economic rights for women may well have been the request made in 1619 by the Virginia House of Burgesses that husbands and wives be granted equal shares of land on the grounds that the work of each was equally crucial to the establishment of a plantation.[5] That women were regarded as transitory members of the workforce even then is shown by many things for example, the fact that when girl children were "put out" for indenture or apprenticeship the persons who received their work were rarely required to teach them a trade,[6] or the fact that women who owned and operated businesses in the colonial period were almost always widows of the men who had first established the business; they consciously advertised themselves as worthy of patronage on those grounds alone.[7] Parents of daughters did not expect that their girl children would need to know any occupation other than housework. Although young women expected that they might need to support themselves while unmarried, they thought gainful employment would become unnecessary and undesirable after marriage; married women expected not to be gainfully employed unless their husbands died or were disabled.

So it was, and so it continues to be—despite industrialization, unionization, and automation. The Equal Pay Act of 1964 attempted to legislate equal pay for equal work for women but in 1968 it was still true that women were earning from 40.5 to 65.9 per cent of what men in the same job classifications were earning.[8] Power technologies have eased and simplified thousands of jobs, yet the labor market is still dominated by sex-typed occupations, despite the fact that the worker's "strength" is no longer a relevant criterion. In the garment industry in New York, for example, men cut and women sew. Thus when a manufacturer goes into the labor market to find a skilled cutter, he enters one labor market, with its own price structure and its own supply-demand pattern. If he is looking for a skilled sewing-machine operator, he enters a different labor market.[9] A sophisticated statistical analysis of sex typing in the labor market has demonstrated that although some job classifications shifted from being male dominated to being female dominated between 1900 and 1960 (ironically, none except obstetrics has gone the other way), the total amount of sex typing has not changed appreciably. In 1900 66 per cent of all employed women would have had to shift their jobs into male-dominated fields in order for the distribution of women and men in all fields to resemble chance; in 1960 that figure was 68.4 per cent. (The corresponding figure for racial typing of jobs in 1960 was 46.8 per cent.[10])

And of course it is still true, as it was in colonial days, that women are not regarded, by themselves and by others, as permanent members of the workforce. Most young women do not invest in expensive training for themselves because they anticipate leaving the workforce when they marry and have children. Employers are equally unwilling to invest in training women because they anticipate the very same thing—and with some statistical basis for their suspicion; the labor-force participation rates of females fall off sharply between the ages of 18 and 25, the years when most women marry and begin their families.[11] The cumulative result of these attitudes is that women prefer to place themselves in fairly unskilled, unresponsible, and therefore lower-paying positions and employers are content to have them remain there.

These three characteristics of women as workers—the fact that they work for less, that many jobs are not open to them because of sex typing, and that they are transient members of the workforce (and therefore difficult to organize and unionize)—should be of signal importance in any discussion of rates of technological change, yet they are rarely considered in that context. We are accustomed to thinking about the price and availability of labor as one of the key determinants of rates of change in any given industry or any given locality, but we are not accustomed to thinking of the price and availability of labor as determined by the sex of the laborers.

The ways in which the sex of workers interacts with technological change can be illustrated by two somewhat different cases.

The first is the cigar industry in the second half of the 19th Century, a case in which technological change was accelerated by the availability of female workers.[12] During the middle decades of the century cigarmaking was localized in factories but the product was entirely handmade by skilled male workers, most of them Spanish, Cuban, and German. In 1869 the New York cigarmakers went on strike and in retaliation several manufacturers arranged for the immigration of Bohemian women who worked in the cigar trade in their native land. These women were not as skilled as the men they replaced; they used a simple molding tool to shape the cigar. They were also accustomed to working at home, which meant that they were amenable to the tenement system of manufacture, which was much cheaper for the employers. The women were effective in breaking the strike. In subsequent years more women cigarmakers immigrated when the cigar trade in Bohemia was disrupted by the Franco-Prussian War. There were other male cigarmakers' strikes in New York and elsewhere during the 1870s and 1880s; their net effect was that some manufacturers converted entirely to the tenement system and others were induced to try some simple pieces of machinery (also of European origin) that could be operated by women. The women were laborers of choice because they knew the cigar trade yet were willing to work for less and had not been organized. As one New York cigar manufacturer put it in 1895, "... the handwork has almost entirely disappeared. The suction tables, which are in reality nothing else than wrapper cutting machines, are used as price cutters. More so, because there are only girls employed on them."[13]

A somewhat contrary case is that of the ladies' garment industry in the 20th Century. Here technological change seems to have been slowed by the presence of female workers. Since sewing machines were first connected to central power supplies (steam or electricity) in the last quarter of the 19th Century, there has been little technological change in the sewing process, despite the high competitiveness of the industry and despite substantial changes in the technology of the processes that are auxiliary to sewing—namely cutting and pressing. The sewing process could potentially be automated but there appears to be little incentive for manufacturers to do so, partly because the expense would be very great and many of the companies are very small. Yet another reason stems from the fact that sewers are women and sewing is work that women from traditional cultures find congenial. The ladies' garment industry has been mainly populated by successive waves of fairly skilled immigrant women of various sorts: American farm girls who came to the cities to escape rural life in the middle of the 19th Century, immigrant women from Italy and eastern Europe, Black women from the south, Puerto Rican women, and now Chinese women. Although these women are skilled and although their trade has been unionized successfully for many years, the wages paid to sewing-machine operators are, as one would expect, significantly lower than the wages paid to other skilled machine operators and the technology of sewing has remained fairly static.[14]

There is yet another sense in which the characteristics of women as workers have interacted with the American technological order. Here we confront one of the most firmly grounded shibboleths about the relation between women's work and technology. It is true that there has been a vast increase in the number of women in the workforce in the past century, during which some occupations (such as clerical work) have almost completely changed sex. It is also true that during the same period power-driven machinery has entered many fields, requiring much less human energy to do work that was once very hard to do. Acknowledging that these facts are true, historians and others have concluded that technological change has drawn women into the workforce by opening to them fields of work that were previously closed because they were presumably the weaker sex. That conclusion appears to be almost entirely unwarranted. Women have replaced men in several occupations in which hard physical labor was not required before industrialization (for example, cigar making), in some occupations in which no significant technological change occurred (for example, school teaching), and in some occupations in which technological change made no difference to the physical labor involved (for example, bookkeeping). In all these cases the crucial factor is not physical labor but price; women replaced men because they worked for less.[15]

Alternatively there are many trades in which work has been transformed by the introduction of new machines, but in which women have not replaced men. Typesetting is a perfect example.[16] From the colonial period to the present there have always been a few women typesetters, but they worked in the smallest shops, often shops that were family owned. Typesetting was generally an apprenticed trade in the 19th Century and women were not set to apprenticeships. In any event it was widely believed that women could not do the work of typesetting efficiently because they were not able to carry the heavy type cases from the composing tables to the press. Women typesetters were occasionally used to break strikes or to cut wages, a practice that did not endear them to the typesetters' unions that were formed during the early decades of the century. In 1887 the linotype machine was introduced and after that time the work of typesetting was not terribly much more difficult than the work of typewriting. Various modifications of that machine, and the more recent introduction of photographic processes, have made the work easier still—but men still dominate the trade. There are several reasons. After women were admitted to the typesetters' unions, in the latter part of the 19th Century, they had to agree to work for scale, which meant, of course, that employers had little interest in hiring them; following that the advent of protective labor laws meant that night work for women was very carefully regulated, which made it unfeasible for women to become typesetters since so much of the work is on newspapers.

Thus technological change has been, at best, a mixed blessing for women who have wished, for whatever reason, to find gainful employment. More jobs are open to them that they are fit (either biologically or socially) to perform, but all too many of those jobs are at the very lowest skill and salary levels.

WOMEN AS HOMEMAKERS

Both men and women live in homes, but only women have their "place" there—another one of those salient facts about women's lives that make their interaction with technology somewhat different from men's. The homes in which we live, the household implements with which we work, and the ways in which that work is organized have changed greatly over the years, but the character of that change and its effect on the people who work in homes (predominantly women) has proved to be difficult to gauge.

Some tasks have disappeared (for example, maintaining a coal fire) but other tasks have replaced them (waxing a linoleum floor). Some tasks have been made easier (ironing) but other tasks have been made harder (compare shopping in a supermarket to shopping over the telephone). It takes less time and effort to wash and iron a sheet than it once did, but there are now vastly more sheets to be washed in each household. If there is less work to be done at home because so many aspects of home production have been absorbed into the marketplace (for example, canning of vegetables), there are now fewer hands to do the work, as there are fewer servants, fewer unmarried females living at home, and fewer children per family; an equivocal picture at best.[17]

But one point is worth making. Despite all the changes that have been wrought in housework, and they have been many, the household has resisted industrialization with greater success than any other productive locale in our culture.[18] The work of men has become centralized, but the work of women remains decentralized. Several million American women cook supper each night in several million separate homes over several million separate stoves—a specter that should be sufficient to drive any rational technocrat into the loony bin, but that does not do so for reasons we shall consider presently. Out there in the land of household work there are small industrial plants that sit idle for the better part of every working day; there are expensive pieces of highly mechanized equipment that are used only once or twice a month; there are consumption units that trundle out to their markets weekly to buy eight ounces of this nonperishable product and twelve ounces of that one. There are also workers with no job descriptions, time clocks, or even paychecks. Cottage industry is alive and well and living in suburbia.

Why? there is no simple answer to that question, but here is an attempted rough list of what some of the components of the answer might be, presented in no particular order and with no pretense of knowing either the relative weight that should be attached to each or the relative likelihood that some are causes and others effects. To start with, since the middle of the 19th Century Americans have idealized the household as the place where men could retreat from the technological order—and so a place that had to remain apart from that order. Increased efficiency in the home has occasionally been advocated by advanced thinkers (many of whom have been women, such as Catharine Beecher, Ellen Richards, Charlotte Perkins Gilman, or Lillian Gilbreth) but the general public has been more or less hostile to the notion.[19] The farm kitchen has been the American mythic dream, not the cafeteria. The maintenance of an individual home, with the latest and showiest equipment, has been an indication of social status for a good part of our national history, but that equipment was not intended to make the total job of the housewife less time consuming; as soon as one job was made easier another was introduced to take its place. In fact the maintenance of an inefficient mode of operation, requiring the full attention of the housewife, has been almost the *sine qua non* of the symbolic quality of the individual home.

Related to that is the fact that since the middle of the 19th Century most Americans have also regarded communalization of households as socialistic and therefore un-American. There have been repeated attempts at communalizing some of the household functions, especially during the first two decades of the 20th Century—communal canneries, laundries, kitchens, even nursery schools appeared in many communities—but they have all failed for want of a supportive community attitude.

The implements invented or developed for the home have very special features that may set them apart from other implements. Many of them were initially developed, for example, not for home use, but for commercial use: the automatic washing machine, the vacuum cleaner, the small electric motor, the refrigerator immediately come to mind.[20] Most of them were not developed by persons intimately connected with the work involved; inventors tend to be men and homemakers tend to be women. Moreover, many of the implements were marketed through the use of selling techniques that also had little relation to the work performed. These three factors lead to the hypothesis that the implements that have transformed housework may not have been the implements that housewives would have wanted.

Thus, for reasons that may have been alternately economic, ideological, and structural, American homes have not become part of the industrial order in the same sense that American business has, because few Americans have wished it so.

WOMEN AS ANTITECHNOCRATS

That brings us to a final and somewhat related point. For the better part of its cultural life the United States has been idealized as the land of practicality, the land of knowhow, the land of Yankee ingenuity. No country on earth has been so much in the sway of the technological order nor so proud of its involvement in it.

If practicality, knowhow, and willingness to get one's hands dirty down there with the least of them are signatures of the true American, then we have been systematically training slightly more than half our population to be un-American: the women. We socialize our men to aspire to feats of mastery over nature, yet we also socialize our women to aspire to feats of submission to nature. Men are hard; women are soft. Men are meant to conquer nature; women are meant to commune with it. Men are rational, women irrational; men are practical, women impractical. Boys play with blocks; girls play with dolls. Men build; women inhabit. Men are active; women are passive. Men are good at mathematics; women are good at literature. If something is broken, daddy will fix it. If feelings are hurt, mommy will salve them. We have trained our women to opt out of the technological order as much as we have trained our men to opt into it.

All that is probably just as much true today as it was in the heyday of the archetypically passive, romantic Victorian female. An interesting survey of American college girl's attitudes towards science and technology in the 1960s revealed that the girls could not assimilate the notion of themselves as future engineers or inventors. There was no single occupation that they regarded as less attractive to themselves than the occupation of engineer and—what is equally revealing—no single occupation that they thought their male contemporaries and their parents would be less pleased to have them pursue.[21]

Thus women who might wish to become engineers, inventors, mechanics, or jackhammer operators would have to suppress some deeply engrained notions about their own sexual identity in order to fulfill their wishes. Very few people have the courage to take up such a fight. It is no wonder that women have played such minor roles in creating technological change; in fact it is a wonder that there have been any female engineers and inventors at all.

The recent upsurge in "antiscience" and "antitechnology" attitudes may be thus correlated very strongly with the concurrent upsurge in women's political consciousness. As more and more women begin to play active and powerful roles in political life they will be carrying with them a set of unspoken assumptions about the technological order that may become increasingly surprising to men.

References

1. On this last point, see Nicholas Wade, "Bottle feeding: Adverse effects of Western technology," *Science* 184: 45–48, 5 April 1974.
2. N. E. Hines, *A Medical History of Contraception* (Williams and Wilkins, Baltimore, 1936). Does not, of course, cover more recent developments.
3. On women workers see R. W. Smuts, *Women and Work in America* (Columbia University Press, New York, 1959); Juanita Kreps, *Sex in the Marketplace: American Women at Work* (Johns Hopkins University Press, Baltimore, 1971).
4. Julia Cherry Spruill, *Women's Life and Work in the Southern Colonies* (University of North Carolina Press, Chapel Hill, N.C., 1938), p. 6.

5. Ref. 4; also Edith Abbott, *Women in Industry: A Study in American Economic History* (Appleton, New York, 1910), p. 11.
6. Ref. 5 (Abbott), pp. 30–32.
7. Ref. 4, *passim.*
8. Ref. 3 (Kreps), pp. 2–3.
9. On sex typing of jobs see Valerie Kincaid Oppenheimer, *The Female Labor Force in the United States,* Population Monograph Series, No. 5 (University of California Press, Berkeley, 1970).
10. Edward Gross, "*Plus ça change* . . . The sexual structure of occupations over time," *Social Problems* 16: 202, 1968.
11. Ref. 3 (Kreps), pp. 28–29.
12. Ref. 5 (Abbott), chap. 9.
13. *Ibid.*, p. 263.
14. This account is based on Elizabeth Faulkner Baker, *Technology and Women's Work* (Columbia University Press, New York, 1964).
15. *Ibid.,* Chaps. 13 and 17.
16. *Ibid.*, pp. 170–177; Ref. 5 (Abbott), chap. 11.
17. This picture is drawn in more detail in Ruth Schwartz Cowan, "The 'Industrial Revolution' in the home: Household technology and social change in the 20th Century," *Technology and Culture* 17: 1–23, 1976.
18. On this point see Allison Ravetz, "Modern technology and an ancient occupation: Housework in present-day society," *Technology and Culture* 6: 256–260, 1965.
19. See Kathryn Kish Sklar, *Catharine Beecher: A Study in American Domesticity* (Yale University Press, New Haven, Conn., 1973); Caroline Hunt, *The Life of Ellen H. Richards* (Whitcombe and Barros, Boston, Mass., 1912); Waida Gerhardt, "The pros and cons of efficiency in the household," *J. Home Economics* 20: 337–339, 1928.
20. Siegfried Giedion, *Mechanization Takes Command* (Oxford University Press, New York, 1948), pp. 556–606.
21. Alice Rossi, "Barriers to the career choice of engineering, medicine or science among American women." In Jacqueline Mattfield and Carol VanAken, Eds., *Women and the Scientific Professions* (MIT Press, Cambridge, Mass., 1965), pp. 51–127.

Selected Additional Readings

A. Briggs, *Iron Bridge to Crystal Palace*, Thames and Hudson, London, Eng., 1979.

D. S. L. Cardwell, *Turning Points in Western Technology*, Science History Publications, New York, 1972.

C. M. Cipolla, *Clocks and Culture, 1300–1700*, Walker and Company, New York, 1967.

C. M. Cipolla, *Guns, Sails, and Empires,* Pantheon Books, New York, 1965.

R.S. Cowan, "A Case Study of Technology and Social Change: The Washing Machine and the Working Wife," *Clio's Consiousness Raised: New Perspectives on the History of Women*, M. Hartman and L. W. Banner (editors), Harper & Row, New York, 1974, pp. 245–253.

R. S. Cowan, "The Industrial Revolution in the Home: Household Technology and Social Change in the 20th Century," *Technology and Culture*, January 1976, pp. 1–23.

F. E. C. Cullick, "The Origins of the First Powered, Man-Carrying Airplane," *Scientific American*, July 1979, pp. 86–100.

G. Dellaire, "The Engineer: What Role in the Development of Civilization," *Civil Engineering*, October 1977, pp. 64–70.

R. J. Forbes, *Man the Maker*, Abelard-Schuman, New York, 1958.

J. J. Flink, *The Car Culture*, The M.I.T. Press, Cambridge, Mass., 1975.

S. Gideon, *Mechanization Takes Command*, Oxford University Press, New York, 1948.

J. Gimpel, *The Medieval Machine*, Holt, Rinehart & Winston, New York, 1976.

J. G. Landels, *Engineering in the Ancient World*, University of California Press, Berkeley, 1978.

D. S. Landes, *The Unbound Prometheus*, Cambridge University Press, London, 1972.

E. T. Layton (editor), *Technology and Social Change in America*, Harper & Row, New York, 1973.

D. McCullough, *The Path Between The Seas*, Simon and Schuster, New York, 1977.

R. H. Merritt, *Engineering in American Society, 1850–1875*, University of Kentucky Press, Lexington, 1969.

E. T. Morison, *From Know-How to Nowhere*, Basic Books, New York, 1974.

E. T. Morison, *Man, Machines and Modern Times*, The M.I.T. Press, Cambridge, Mass., 1966.

L. Mumford, *The City in History*, Harcourt, Brace & World, New York, 1961.

L. Mumford, *Technics and Civilization*, Harcourt, Brace & World, New York, 1934.

D. F. Noble, *America By Design*, Alfred A. Knopf, New York, 1977.

A. Pacey, *The Maze of Ingenuity*, Holmes & Meier, New York, 1974.

I. deS. Pool (editor), *The Social Impact of the Telephone*, The M.I.T. Press, Cambridge, Mass., 1977.

N. A. F. Smith, *Man & Water*, Peter Davies, London, 1975.

N. A. F. Smith, "Roman Hydraulic Technology," *Scientific American*, May 1978, pp. 154–161.

A. Trachtenberg, *Brooklyn Bridge*, Oxford University Press, New York, 1965.

J. Vanek, "Household Technology and Social Status and Residence Differences in Housework," *Technology and Culture*, July 1978, pp. 361–375.

L. White, Jr., *Medieval Religion and Technology, Collected Essays*, University of California Press, Berkeley, 1978.

R. M. Wik, *Henry Ford and Grass-roots America*, The University of Michigan Press, Ann Arbor, 1972.

PART 6

Technology and Developing Nations

By and large, our present problem is one of attitudes and implements. We are remodeling the Alhambra with a steam-shovel, and are proud of our yardage. We shall hardly relinquish the shovel, which after all has many good points, but we are in need of gentler and more objective criteria for its successful use.

Aldo Leopold, A Sand County Almanac

Let fanciful men do as they will, depend on it, it is difficult to disturb the system of life.

James Boswell, The Life of Samuel Johnson

This book is based on the idea that engineering and technology are devoted to the service of humanity and to the growth of the culture and benefits we enjoy today. This interplay between engineering and culture begets a relationship between the humanities, as we know them, and engineering. This is a relationship that forms an inseparable tie between technology and the "good life"; examples of this tie have filled the preceding chapters. It would be difficult to deny that technology has served to aid the progress of our civilization for the benefit of all.

For the last several decades (particularly those since World War II), there has been a growing concern that technology has been diverted from service to humanity as a whole and directed to the benefit of the few without regard to the effect upon the many. Differences have developed that are expressed in the deep-seated suspicions of those in the technologically unsophisticated "have-not" countries toward those in the more technologically advanced nations. Some of these suspicions are founded in the intent of technological exploitation, while others are based on the errors and mistakes of well-intentioned efforts to aid the world's less developed nations.

This concern over the proper use of technology to benefit those for whom it is intended by those who have made great use of its potential to improve the lot of humanity has generated a movement that is variously called "appropriate technology" or "intermediate technology." The terms are generally interchangeable and recognize that a technology suitable to one culture cannot be transferred to another without disruptive effects on the second. These effects tend to tear down existing social, economic, and cultural systems to the detriment of a society. Thus we find a failure of the technology to achieve its intended aim to benefit the receivers. Kamenetzky (1) expressed it well when he said, "Technology has been an instrument of liberation, but it must be admitted, an instrument of oppression as well." At a later point in the same paper he noted that though chemical engineering (and perhaps all engineering) has contributed to the betterment of human life in many countries, yet "there still remains famine, toil, and repression among two-thirds of humanity."

The terms "appropriate technology" and "intermediate technology" need some definition. Though there are many, the one presented in *Science* (2) appears to be suitable:

> . . . appropriate technology generally means technology that is small, easy to understand and maintain, cheap, dependent on local resources, and fitted to local needs. It makes heavy use of renewable resources . . . and makes minimal demands on capital and on non-renewable environmental and energy resources.

The significance of intermediate technology is indicated by some criteria that must be satisfied to be considered in this classification. Kenkare (3) gives such criteria:

(a) Is the best use made of human and material resources?
(b) Does the technology bring social benefits to the mass of the population or does it benefit only a few?
(c) Does the technology reflect an improvement on traditional methods or is it progress according to some arbitrary but irrelevant standard?
(d) Is there a measure of flexibility in the technology to allow for change and growth in subsequent years?

The concern for appropriate technology is primarily as it relates to the emerging or less developed nations, though certainly all lands can gain from technology that emphasizes renewable resources and "makes minimal demands on capital and on non-renewable environmental and energy resources" (2). But what characterizes developing nations? In an attempt to answer this complex question, Crosthwaite and Bergan (4) group developing nations into three broad categories to facilitate discussion:

- Those with economies in the early stages of development almost wholly dependent upon the export of raw materials, primarily the surplus unprocessed production of peasant farming.
- Partially developed countries, often with substantial reserves of professional expertise and skilled labor but frequently deficient in financial resources.

> Wealthy nations, especially those who have achieved this position in recent years. They may be underpopulated and in addition have not had the time to produce their own cadres of professionals and associated skilled manpower.

Kenkare (3) also emphasizes that developing nations frequently have a dual economy prevailing in the urban and in the rural areas; that unemployment or seasonal employment creates poorly distributed incomes, a relative lack of capital, and a growing population.

Early attempts at applying technology to aid the development and growth of emerging nations took the form, in most instances, of the transfer of highly sophisticated technology designed for mass production by skilled laborers to these nations. This concept is expressed in a United Nations report of the early 1960s as quoted by Myrdal (5), "The reason for emphasizing industrialization is that industrial development would absorb rural underemployed persons." The result was often a mass migration of unskilled labor from the rural areas to the site of a new industry, with the resulting development of extensive slums and a shift in the unemployment problem from rural to urban areas. A companion problem was the breakdown of the social and economic structure of the country that was often based on the close ties among people in a village society. These problems are discussed in detail in the selections of this part.

The concern for appropriate technology for developing nations has been expressed by many authors. Gunnar Myrdal (6), for example, states "it had to be learned that scientific-industrial technology, to be maximally useful in the underdeveloped countries, cannot simply be transferred but must be adapted to the conditions prevailing there." Probably the most widely read proponent of appropriate technology, E. F. Schumacher (7), emphasized in his book *Small Is Beautiful* the primary need for technology to "maximize work opportunities for the unemployed and the under-employed. For a poor man, the chance to work is the greatest of all needs, and even poorly paid and relatively unproductive work is better than idleness."

The need for appropriate technology can be illustrated by many examples. Each serves to emphasize the concept of appropriate technology to those of us whose thinking has been conditioned by our association with an advanced level of technology enjoyed in the developed nations. The "Nova" program *India: Machinery of Hope* (8) presents many illustrations of the need for, and the nature of, appropriate technology in India. Weiss (9) presents the problem through considering the alternatives available for supplying water from a near-surface aquifer to small farmers, while Bliss (10) has used the introduction of the wagon to the Papago Indians to show a positive, yet unexpected, effect of the introduction of a new technology to a technologically unsophisticated people. The adverse effect of modern technology on employment is indicated by Kothari and Mendlovitz (11) in a discussion of the impact of mechanized trawlers on the fishers of South Asia. Lederer and Burdick (12) present a fictionalized example in *The Ugly American*. In it, one of the more likable Americans develops a bicycle-powered pump for local villagers in Sarkhan. The power source, the bicycle, was readily available, and the modification to permit it to be used as a pump without destroying its utility as a bicycle was made easily with village materials and tools. This was appropriate technology at work.

The selections in this part were chosen to show the problems of technology transfer to the developing nations. E. F. Schumacher's essay "Education for Good Work" presents an excellent introduction to the "problem" as a whole. Schumacher's work is particularly appropriate as the opening for this part since it was this farsighted, erudite economist who, more than anyone else, stimulated the widespread current interest in appropriate technology with his popular work *Small Is Beautiful.* That effort contributed to our motivation to compile this volume.

D. K. Mukerji's paper "Mahatma Gandhi's Views on Machines and Technology" provides an insight to the thinking of one of the most influential world leaders of the twentieth century. Gandhi was concerned with the improvement of the quality of life for all India's peoples, but he saw technology as being destructive to that end. The particular problems facing India before it gained independence (and to a great extent today) reflect the impact of modern technology on the culture and values of a people whose society was greatly different from that of the Western world. Gandhi saw India's future in cottage industries, typified in his examples by the spinning wheel. That device gave all an opportunity to work, produce, and earn without the expenditure of extensive capital. Gandhi did not oppose machinery per se, but he felt it should not be allowed to displace human labor. What he objected to was "[a] the introduction of machines which would create more unemployment, [b] exploitation of poor laborers by rich capitalists, and exploitation of one country by another, [c] the moral evils attending the mad craze for large-scale industries, centralization, and monopolies, and [d] man's limbs being atrophied by too much dependence on machines" (13). Gandhi felt (and stated many times) that "the supreme consideration was man."

"Constraints on Engineering in Third World" was written by an Australian engineer, D. S. Mansell. His concerns are with those problems that engineers encounter in the practice of their profession within cultures other than their own. The physical constraints imposed by nature are accepted as one of the many factors entering into any engineering project. What is not so easily recognized are the constraints that are imposed by human considerations and the need for engineers to understand the culture of the country in which they work. These aspects are discussed with empathy, based on the author's experiences in New Guinea.

This part concludes with a short extract from the contemporary novel *Wheels* by the popular author Arthur Hailey. Many of Hailey's books have technological settings; *Wheels*, for example, is set in the automotive industry. The extract we use is a side issue in the novel. It illustrates the response of "big business" (represented by the central character, Adam Trenton) when faced with a proposal to produce what we would call appropriate technology. Using manpower and simple technology, the device in this case is a bicycle-powered farm machine suitable for the threshing of any kind of grain. The selection illustrates what the people of many developing nations unfortunately believe to be the unconcerned attitude, of those familiar with advanced technology, to their problems and needs.

The theme of this part can be summarized effectively by Schumacher's (14) urging that we expend greater effort to the development of "technology with a human face."

References

1. Kamenetzky, Mario, "The Role of Chemical Engineering in the Countries of Latin America," a paper presented at the World Congress on Chemical Engineering in the Service of Man, Paris, September 1972.
2. Holden, Constance, "NCAT: Appropriate Technology with a Mission," *Science*, 195, March 4, 1977 p. 857.
3. Kenkare, A. S., "Technology for the Developing World," *Chartered Mechanical Engineer*, March 1975, Institute of Mechanical Engineers, pp. 87–90.
4. Crosthwaite, C. D., and A. Bergan, "The Needs of Developing Countries in Particular so far as Design and Construction Are Involved," *Problems Associated with Design and Construction in Developing Countries*, International Association for Bridge and Structural Engineering, Zurich, 1977, p. 28.
5. Myrdal, G., "The Transfer of Technology to Underdeveloped Countries," *Scientific American*, September 1974, p. 176.
6. Myrdal, ibid., p. 173.
7. Schumacher, E. F., *Small is Beautiful*, Perennial Library, Harper & Row, New York, 1973, p. 173.
8. *India: Machinery of Hope*, "Nova," WGBH Educational Foundation, 1979.
9. Weiss, C., Jr., "Mobilizing Technology for Developing Nations," *Science*, 16 March 1979, pp. 1083–1089.
10. Bliss, W. L., "In the Wake of the Wheel: Introduction of the Wagon to the Papago Indians of Southern Arizona, *Human Problems in Technological Change—A Casebook*, H. Spicer (editor), Russell Sage Foundation, New York, 1952, pp. 23–32.
11. Kothari, R., and S. Mendlovitz, "The Perversion of Science and Technology," *Current*, October 1978, pp. 54–58.
12. Lederer, W. J., and E. Burdick, *The Ugly American*, W. W. Norton, New York, 1958, pp. 214–231.
13. Datta, D. M., *The Philosophy of Mahatma Gandhi*, The University of Wisconsin Press, Madison, Wisc., 1953, p. 114.
14. Schumacher, E. F., *Small Is Beautiful*, Perennial Library, Harper & Row, New York, 1973, p. 146.

6.1

Education for Good Work

E. F. SCHUMACHER

In the beautiful hills of Surrey, where I live, you can go for walks, and even on a Sunday you never meet anybody, though you may hear the distant roar of traffic down to the coast. And the story goes that an economist went there for a Sunday-afternoon walk, and he met none else than God Almighty, which gave him a bit of a shock, and he didn't know what to say. He remembered that as a little boy he had been told that what is a thousand years to us is but a minute to the Lord. And he asked Him "Is this so?" and the Lord said, "Yes, this is quite so." By that time he had recovered his composure, and he said to Him, "Then perhaps it may also be true that what is a million pounds to us is only a penny for you." And the Lord said, "Yes, that's quite true." And he said, "Well, Lord, give me one of those pennies." The Lord said, "Certainly, my dear chap. I don't happen to have it on me, but just wait a minute while I fetch it."

And so it is with the salvation expected from the growth of GNP: just wait another minute and we shall all be rich and happy. This notion is becoming increasingly questionable. It's an awesome view one takes from outside the United States, GNP-wise the most successful country of the world, unbelievably high gross national product, and, to put it mildly, a country with problems. A country with poverty, indeed a country with degrees of poverty that many other countries at a much lower GNP are not experiencing. I'm not saying this in a spirit of criticism, but simply in analysis. After all, the Western Europeans are pursuing the same aims; they're only less successful Americans.

To me, this concept of GNP means nothing at all. I know, for certain technical reasons of managing certain money flows in the economy, it may be quite useful; but as a measurement of any kind of achievement to me it's meaningless. Because it's a purely quantitative concept. Statistics don't have to be accurate; they have to be significant. My theory has always been that figures don't mean anything if you can't make them sing. How can anybody assert that "growth" is a good thing? If my children grow, this is a very good thing; if I should suddenly start growing, it would be a disaster. Therefore, the qualitative discrimination is the main thing; it's far more important than some mysterious adding-up of everything. We've all learned at school that you must add together only things of essentially the same quality. So you can't add together apples and the number of evenings spent watching television; this is meaningless. Let's look at the qualitative aspect of the whole matter. Surely our job is always to make up our minds what's good and do our best to let the good grow; and to make up our minds what it evil or not so good and try to diminish it. Whether the two processes, put together, mean a new growth or a net diminution shouldn't interest anyone. The quality of life—not the quantity—yes, that's what matters. GNP, being a purely quantitative concept, bypasses the real question: how to enhance the quality of life.

If you take the world distribution of income, you get very, very many people extremely poor, very many people extremely rich, hardly anybody in the middle. This is a pathological distribution. That is to say, there is not one world anymore. Even on that humble level any semblance of unity has been lost. This applies not only between the rich countries and the poor countries, where it applies in a horrifying form; it also applies inside many societies. All the developing countries are evermore becoming dual societies with a lot of people immensely poor, a few people immensely rich, and nothing in the middle.

Some people don't observe the idea that they have purposes in life other than the material. To put it in the shorthand of Christian language, their job is really to save their souls and they are using far more of the material means than is necessary for this purpose, and it's most likely that therefore they are greatly hindered. So we see that the rich society, that is, the very few people at the rich end, cannot possibly be a model for the world as a whole, and unless we find a new model we are just going to drift deeper and deeper into disaster. The question is, Is there another pattern that would make life really worth living? These patterns have existed and they have produced some of the greatest works of art and culture that we know of. And they were produced in a culture of poverty, not in a culture of affluence.

What is the culture of poverty? It seems to me that it is a culture in which a very firm distinction is made between two categories of goods, which I shall call the ephemeral goods on the one hand, goods produced in order that they should be destroyed, and eternal goods—perhaps eternal is too big a term—goods which are not to be destroyed, on the other hand. They may be destroyed by accident, but that's not their purpose. In all real cultures the eternal goods were outside the economic cultures, because, after all, how can you calculate eternity? When they built a cathedral, they didn't budget it, they said, "Only the best is good enough. Only that which can be offered to the glory of God is worthy of the dignity of man." But when they came to ephemeral goods they lived frugally.

I visited Florence a year ago for the first time. Opposite this fantastic cathedral there is the statue of the architect, and on the pedestal is the Latin inscription, which, with the greatest difficulty I deciphered. It said, "This is Arnolfo, who, instructed by the municipality of Florence to build a cathedral of such splendor that no human genius can ever surpass it. On account of the superlative endowment of his mind he proved equal to this gigantic task." Instructed, you see. Outside the cathedral are the famous bronze doors by Ghiberti, and you learn that it took him twenty-eight years to do them. There was no patronage behind it, no economist calculated that if these bronze doors were made in twenty-five years instead of twenty-eight, the tourist trade would be increased by three years. The tourist trade had increased by 3000 percent; you can hardly see the doors now, but that is another matter.

The moment we allow the economic calculus to invade everything, then nothing becomes worthwhile anymore. I tried to imagine what it would look like if I found the statue of the architect opposite one of the high-rise office-block buildings in London. It would probably read, "This is Mr. R. W. Smith, member of the Royal Institute of British Architects, who, instructed by the Greater London Council to create an office block of such superlative cheapness per square yard, per square foot, that no human genius can ever underbid it, on account of his superb endowment with computers proved equal to this mean task."

Our problems are so serious that the best way to talk about them is light-heartedly. All we can hope is that, with the many improvements that are really beginning and trying to get us straight, we should not dissipate our powers by neglecting the strength of science and technology. What is needed is a redirection of them, making use of the best stock, and not wasting too much time in being heroic and reinventing the wheel. Get organized, make the best use of people, even those who are not like-minded, who are just as jittery about the future as anybody else, but who can help you.

There are three ways to develop a new intermediate technology. One way is to take the existing very low-level technology and say simply, look, this can be upgraded. It can be done much better, and if one has the eye for it, one finds it can be enormously improved with virtually no capital investment. That's one way, upward.

Another way is to say, Well, now here we make this marvelous equipment but our production processes are designed mainly to save labor. But this machinery is not suitable for poor countries. Let's, in a sense, downgrade it, to fit it into the condition of the developing country.

The third way is to conceive it anew, to lay on design studies after one has conceived the right question. To get the peasants of the developing countries to do this themselves is very difficult; in most cases they just do not have the resources to do it. They have fewer scientists. That's the tragedy. Where a country does swing around and say, Now we want to work on our own problems, which has happened to some extent in a country like Ghana, all things become possible. But too many are looking to Imperial College, London, or Cornell or MIT, instead of attending to the problems of their own country. You don't get much innovation, much real reconsideration of technology, from the developing countries. You don't get *much*, but there is still quite a lot, taking the world as a whole, and therefore one of the most important activities to be undertaken—and we are undertaking it with our slender resources—is to survey the world and see what already exists and evaluate it and perhaps test it. The Filipinos have some absolutely excellent, very simple equipment for rice culture, and so have the Japanese. We see if that equipment is suitable for other conditions and, if it is, make it available.

Now we come to a definition of the whole thing. It's a funny thing how the word, the concept, which is merely a label attached to some intuition, some knowledge, then becomes the thing itself. The Buddhists have a nice expression, that Buddhism is the finger pointing at the moon. The moon is the thing, Buddhism simply directs your attention to it. So a phrase like "appropriate technology" or "intermediate technology" is simply a finger pointing at the moon. And the moon can't be fully described; it can be pointed out in terms of specific situations.

I was asked in the early 1960s to go to India. I didn't know what to do about rural India, but I traveled the whole country, by every conceivable conveyance or nonconveyance, and finally I came back to Delhi and wrote them a little memorandum about it and they assembled all the planners for a two-day seminar. I said, The question that must be raised is, what is the appropriate technology for rural India? The answer, to the best of my insight, is an intermediate technology, something much better than the hoe and much cheaper and easier to maintain than the tractor plow. So, if I make myself understood, intermediate is an answer. At first I got a very bad reception and I left in a huff. But fifteen months later they arranged an all-India conference on intermediate technology. (They still didn't like the term so they called it appropriate technology. That's all right, it's a purely formal term, it makes my life very easy: when I come to India and somebody, as happened quite recently, comes to me and says, Oh, Mr. Schumacher, I don't believe in appropriate technology, I just look him straight in the face and say, Oh, that's splendid. Do you believe in inappropriate technology?)

To determine what precisely is appropriate is pretty difficult, but it has to fit the particular situation. Normally in the rural areas markets are small and the big mass-production units don't fit, so one of the subsidiary definitions is probably that it will be a small unit. I've recently been to Sri Lanka and they told me, We have little sugar estates all over the place but have only one big sugar refinery. To take all the cane into that one place now, with oil costs, transport costs, is uneconomic. Where can we get mini-plants? They had inquired all over the world. They decided that no one had such a plant; the "big system" is not interested. In fact, the mini-plant for sugar refining, just the thing Sri Lanka needs, has been developed in India against the opposition of the large-scale sugar industry. Now that it exists, everybody says, Thank God it exists; now the little people of Sri Lanka and other places have a chance of making a living.

The second point, perfectly obvious, is that in rural areas—poor countries—we do not have an intellectual infrastructure so that we have only to whistle and there comes your manager, top engineer, lawyer, etc. The appropriate technology at the intermediate level will be simple enough that you don't have to have these specialists.

To illustrate, let's say you set up a factory, making anything, in the desert. You find it very much more difficult than setting up a factory in, say Denver. Why? Because in the desert there is no infrastructure, that is to say, you'll have to build roads, you'll have to bring the people nearer the factory. You'll have to build the houses for these people, you'll have to build schools for the people, and hospitals, etc., etc. You can't walk out the front door of the factory because you have an accountancy problem and around the next corner you find a first-class accountant. You've got to maybe fly him from thousands of miles away. If you need a new personnel manager you can't, as you can in Denver, just put an advertisement in the local paper and you immediately get twelve applicants. So all those things that are givens when you do it in Denver, that you don't have to bother about, but not givens out in the desert, that is called infrastructure. The infrastructure in many places has accumulated over the years, several generations, and in the European countries for hundreds of years. It includes an immense number of wires and tubes and pipes beneath the roads. If they have to be created in a poor country, in a hurry, they're immensely expensive. Each mile of road can run into millions of dollars. So if you have to have elaborate infrastructure first and then a mass-production industry, you have to wait a long, long time and I don't know where the wealth would come from to build this elaborate infrastructure. But if you have small-scale it doesn't require an elaborate infrastructure and you can become productive quickly and then perhaps later you will have the wealth to make the infrastructure a bit more elaborate. That's all I mean.

In Burma I found the Buddhists, Burmans like it if one speaks in parables. And I said, Well, you know, there's a road and from the road there's a path and by the path there's a shed. And in the shed is a hen that lays an egg. Well, now, all that—the road, the path, the shed, the hen—is not what you want. You want only that one egg. If you spend all your money on the road, the path, and the shed, you're then broke and can't even have a hen to lay an egg; it's not very good business. So you want to minimize the infrastructure requirement; you want to produce in such a way and at such a location that you don't have to spend all your money on roads, paths, sheds, and the like, and you can spend more money on getting more hens to lay more eggs.

Third, since we are dealing with poverty and not with rich people, it's useless to go to them with some extremely expensive apparatus, wonderful as it may be, and say, Please get out of my way and I'll show you how to do things, which means I'll show you what you can do when you're rich. The poor watch it with increasing astonishment and say, Well, we're not rich so it means nothing to us. It just proves to us that we are nothing. It destroys people. It's what we call, in high-falutin language, a negative demonstration effect. Then we can go home and talk about the culture gap and how we can close it. Intermediate technology says, No, if I want to help I must be genuine about it and become a poor man myself to understand it, and then I quickly realize what fits. Certainly it must be something which can be done without this great capital intensity. Please notice I am not saying that it can be done labor intensively, I am not aiming at that, but rather at capital *saving*. They are very different things. I have no one who enters the situation imaginatively and feels that he wants to do useless labor, so labor intensity as such is not desirable. But capital saving is desirable.

These are some of the criteria that come only gradually into one's mind, and because the child has to have a name we call it intermediate technology. It's a difficult approach because the aid givers, who are they? Basically they are townies, but the problem can be solved only in the rural areas and the gap between townie and country folk is immense. They are, in a manner of speaking, educated and they try to help people who, according to their criteria, are less educated. The gap between the educated and the uneducated is enormous. And thirdly they are rich and they are trying to help the poor and the gap between rich and poor is also enormous. So there are three big chasms to be bridged, which requires extreme effort of the imagination, at which we, the rich countries, the aid-giving countries, have been very poor. We said, We have the answer, now get out of the way and we'll implement it.

Here is a really symbolic example of simple technology that can serve as a symbol for the whole thing, about a different way of thinking. Much of the third world is arid country, not because it doesn't rain, but because the rainwater flows off into the rivers and down into the sea for certain reasons. The Lord has arranged it so that by and large it rains everywhere; there are few places that are exceptions. In some places, like England, it rains all the time; in other cases it comes in great big blobs, and when it comes in great big blobs it runs off the soil very quickly unless you are very careful. Now don't let it run down to the river and out to sea, and say, "Oh, we've got that marvelous technology, we'll build a desalination plant." Even if you do it, you still have fresh water only in one place when it's wanted all over the country. You want the water when you want it—*that* is the thinking. Don't be misled by the marvelous things we do; do the necessary things. Give your intelligence to the construction, for instance, of underground rainwater tanks. Of course we know how to do it expensively. How can we do it cheaply? How can we do it so that the actual expenditure the villager has to make for materials he doesn't have is minimized? Maybe he has to put in a lot of labor. But in many of these countries there's a great surplus of labor. That is appropriate technology, as an example.

I'll give you another example. In order to have efficient oxcarts, the wheels ought to have steel rims. We've forgotten how to bend steel accurately except with big machines in Pittsburgh or Sheffield. How do you do it in a small rural community? Is it beyond the wit of man to do this on a small scale? No, we remember that our forefathers knew how to do it before James Watt, and they had a most ingenious tool. We found one of those tools in a French village, more than two hundred years old—brilliantly conceived, clumsily made. We took this to the National College of Agricultural Engineering in England and said, "Come on, boys, you can do better than that. Upgrade it, use your best mathematics to work out the required curvature and whathave you." The upshot of it is that while hitherto in the modern world the smallest instrument to do this bending job would cost something on the order of £700, and require outside power and electricity to operate, this tool upgraded to the level of knowledge of 1974 can be made by the village blacksmith. It costs £7, it doesn't require electricity, and anyone can do it. Now this is something quite different from going back into the preindustrial era. It is using our knowledge in a different way, and we know it can be done.

I emphasize our interest in nonagricultural activities in rural areas. There is only a given amount of land and if the population increases they have to subdivide the land more and more and more. The urgent need is to *bring industry into the rural areas.*

A couple of years ago I went to a place about ninety miles out of Delhi, called Khurja. In the late 1940s the government of India said Khurja is the ideal place where we can make our own crockery. We don't have to import it. They set up a pottery development station but nothing happened. The center attracted several first-class technical people. One was Mr. T. N. Sharma, and he realized that nobody had got down to the problem of technology. He persuaded the government to send him around the world to study the technology. He went out as Mr Sharma but he came back as Dr. Sharma, and immediately he was offered the managership of a big mass-production firm in Calcutta, which he accepted. Suddenly he remembered himself; Good Lord, I was going to help these poor fellows in Khurja. And so he went to Khurja and selected out of all that he had seen the appropriate technology for the little man. Within twelve years, three hundred pottery factories have grown out of the ground like mushrooms.

I was there and found thirty thousand people that are now earning a living from pottery and hospital porcelain and electrical porcelain. It is magic. I was fascinated. Three hundred firms. I moved around and I got hold of a chap who was employing a couple of hundred people. And I said, Where do you come from? He said, I come from the Punjab. Why did you come here? Are you a traditional potter? Oh, no. I come from a farming village. Why did you leave there? No land. My brother's got the land. And where did you learn this? You are running a factory now. I learned it from Dr. Sharma. Well, yes, Dr. Sharma brought you the implements and so on, and trained you, from a technical point of view; but where did you learn management? He said, Learn what? You just manage, he said. Well, I said, you have three hundred people here who come off the farms from little villages and are entrepreneurs employing altogether thirty thousand people. How many go bankrupt here? He said, I've never heard of one. One has to think about this. I have no reason to believe that this totally frank chap was telling lies. I asked other people. We assume that all of this is so extremely difficult because we have created this extremely complex technology. But when it is back to reality, to real simplicity, then management ceased to be a great problem.

Recently I was seated in a restaurant, next to a family of three, a father and mother and a very bright little boy, I would think between eight and ten years of age. They studied the menu, and the boy said, "Oh, I want liver and bacon." The waitress was there; the father studied the menu, the mother studied the menu, and then the father ordered three steaks. The waitress said, "Two steaks, one liver and bacon," and went off. The boy looked at his mother and said, "Mummy, she thinks I'm real!"

The question, Are we real? arises particularly in this vast country. I know these situations in industry, when certain people who have been doing the same boring job for a long time are, perhaps by chance, perhaps by way of policy, treated as *people*: and they experience something similar to this little boy. Good Lord, they think, in this place I'm real; I'm not just a means to an end.

Yes, I think you are put into this life with the task of learning to distinguish between that which is really real and really important and permanent and of true value on the one hand, and things trivial, amusing, ephemeral, and of no real value on the other hand. Your intellect has to make that distinction. The world has to attach itself to the things that really matter and not to those ephemeral trivialities which make the most noise. That is the message of religion. I know it is normally handed down in all sorts of other ways, but unless you do that you are an unhappy, messed-up person. It has taken me a long time to discover why Religion has split up into so many different religions; it's so you can choose the one that is the most practical for you. The most practical to me was the Roman Catholic version of Christianity, and now I am relieved of such totally offbeat questions as: How could something incredible, like the human being, have come about by an accidental combination of atoms? So I say, Come off it, this is just stupid. I don't know how it is, but I believe there is a Creator. The moment I believe in this higher level, it would be most improbable that the Creator could have put into life such loquacious beings as you and I and never say a word to us. He has actually communicated to us.

This is called by the simple word "revelation." We have the sacred books of mankind, and having spent many, many years studying them, not only in the Christian tradition, I find that it's the same spirit that is communicating to all of us. By various means, in a subtle way, an educative way. You always have to stretch yourself to understand; it is not meant to be automatic. This is the great education we can receive in life, and once we get hold of that, then suddenly we find that we are no longer worried, we have energy actually to act in this life, we can distinguish between the phony and the real questions, and we're happy.

Our system has been extremely destructive and I think we must get down to the metaphysical or, if you like, the philosophical or religious causes of this situation. We have been hearing lately a great deal of the tragedy of the commons, which shows the metaphysical deterioration from which we are suffering, because as a matter of historical fact, the commons in Western Europe were well looked after. But of course, when this feeling that man is the servant of this world in a sense, or at least a trustee, was organized out of our thinking, then the commons were no longer maintainable. So it is those causes that we have to become aware of. It is quite true that tinkering with the present system has very little effect, except where the tinkering represents a return to the true values. It is of little use to call for a very big change in values without at least incorporating these values in some new structures, no matter how small. Because as we know, or as some of us know, we have been told that in the beginning was the word but then the word must come and become flesh and dwell among us. We are now locked in a crisis of this modern civilization which has given up, more or less, the two teachers that mankind had.

What are these two teachers? One is the marvelous system of living nature: the town civilization is not in touch with that system. And the other one is the traditional values, the traditional wisdom of mankind, which we have also rejected and replaced by some extraordinary structure which we call objective science.

Let us try to utilize this opportunity to recall to our minds what a human person really is. I am saying nothing orginal, I just remind myself of it. First of all, in some way or another, whatever way you try to express it, he comes from the divine level onto this earth. He is a son or daughter of the divine. Second, he is a social being; he doesn't come alone. He is put into this social context. And third, he or she is an incomplete being. He has been sent here to complete himself. From this insight have been formulated all the ethics and all the instructions to the human race. As a divinely arrived being he is called upon to love God in traditional language. As a social being he is called upon to love his neighbor. And as an incomplete individual being he is called upon to love himself. The social organization ought to reflect these three absolute needs. If these needs are not fulfilled, if he can't do it, he becomes unhappy, destructive, a vandal, a suicidal maniac. The social, political, and economic organizations ought to reflect these needs.

But they do not. Let's take the lowest of the three, namely the need to complete himself with joyful, constructive labor, work, to feel that he is, as a child of God, created. Well, in most jobs he can't do it. Most jobs have been organized to be as dull as anything can possibly be. This is now one of the critical points in many parts of the world because people are no longer prepared to accept it. This is a situation that is still, as it were, cushioned: in some plaes where you have migrant labor they can still be pushed around. But all that is now finishing. So the kind of industrial trouble that we have in England arises because we don't have a cushion of recent immigrants who will allow themselves to be pushed around. We don't have foreign workers as Germany and many other European countries have and the British are no longer prepared to tolerate certain industrial conditions. They say, You put your thinking cap on and make something that is humane.

The other need, to serve one's neighbor, and the first need, namely to act as a divinely derived being in accordance with our moral impulses, are also frustrated to a very large extent by the size of our organizations. It is only in a small organization that we can meet people face to face and make decisions face to face.

In large organizations all sorts of little people have to implement rules and regulations, and rules and regulations can never be fair. As Thomas Aquinas said, Justice, untempered by mercy, is brutality. Mercy, uncontrolled by justice, is the mother of dissolution. So if you have a big structure you have to have a lot of rules and regulations and the little people who have to deal with the clients and customers and so on. They can't act in accordance with their moral impulses: they say, Oh, I'm sorry, I am here only to execute the rules I am given. I agree, it is most unfair, but I can't do anything about it. So his impulses are frustrated and you get this extraordinary possibility, namely at the top and throughout the ideology of the big thing, the good intentions and the moral people create, because of wrong structuring, an immoral, wicked effect.

I could put it also in another way, which has certain advantages because it embraces the whole of our life, that if the size of the unit is wrong, this means inevitably the eradication to a large extent of T.L.C. Now T.L.C. is the only fertilizer that really works. Don't inquire where you can buy it—you can make it yourself. T.L.C. means tender loving care, and this is what our life needs. When it is organized out of the system, then everything becomes unproductive and madly expensive. Agriculture is a good example of this: through the technological development, the withdrawal of human labor from agriculture, the substitution for T.L.C. of chemicals and big mechanization, we now have a system that has no real future because it is based totally on nonrenewable fossil fuels, and instead of working with nature it bullies nature.

It is very interesting in Britain, time and again our planners will take good agricultural land and build a new council housing estate upon it and then we environmentalists protest that this is an alienation of land and we don't have enough land anyhow. It is criminal to withdraw this land from food production. But three or four years later we find that fifty families have settled on this land, they all have their gardens, and the land produces more than the farm has ever produced. All because the T.L.C. factor has been brought back into it.

Everything is just a matter of the carrying capacity of this earth. There are studies which show that when food production is done on a sort of family scale, intensively, then the output per acre is on the average five times as high as that of a well-run farm. In Britain, where we produce only about half the food that we need, all our modern people—I call them the people of the forward stampede—say, With ever increasing application of chemicals we must double food output. Then they say it is impossible anyhow, yet they now have a plan to increase food output by 3 1/2 percent a year. Actually it is falling by about 7 percent a year. These people are almost unteachable. The question has to be raised, Can we afford agriculture in Britain? I don't think we can really afford agriculture, we can afford only horticulture, where we get really high output per acre, and this requires new structure.

This is the sort of constructive work that we now have to do. Getting down to details. We must realize that we are making a bad job today, a very bad job, and we must ask ourselves, How can we make a better job tomorrow? Let us see the abnormality in the things we consider most normal.

This system is wrong; it doesn't work properly. And the further intensification along the same lines is just tripling the case of fanaticism. Fanaticism has been defined as an attitude of redoubling one's efforts when one knows one is on the wrong track. People are worrying their heads about their standard of living when the world around them, where it really matters, is collapsing. I mean, it is not a thing to be taken lightly that in many, many places in the so-called rich world it is really hazardous, even for a man, upstanding, big man, to try to walk home after dusk alone. These are the real things, the gadgets are not real: if I get mugged, that is a real thing.

I will tell you a moment in my life when I almost missed learning something. It was during the war and I was a farm laborer and my task was before breakfast to go to yonder hill and to a field there and count the cattle. I went and I counted the cattle—there were always thirty-two—and then I went back to the bailiff, touched my cap, and said, "Thirty-two, sir," and went and had my breakfast. One day when I arrived at the field an old farmer was standing at the gate and he said, "Young man, what do you do here every morning?" I said, "Nothing much. I just count the cattle." He shook his head and said, "If you count them every day they won't flourish." I went back, I reported thirty-two, and on the way back I thought, Well, after all, I am a professional statistician, this is only a country yokel, how stupid can he get. One day I went back, I counted and counted again, there were only thirty-one. Well, I didn't want to spend all day there so I went back and reported thirty-one. The bailiff was very angry. He said, "Have your breakfast and then we'll go up there together." And we went together and we searched the place and indeed, under a bush, was a dead beast. I thought to myself, Why have I been counting them all the time? I haven't prevented this beast dying. Perhaps that's what the farmer meant. They won't flourish if you don't look and watch the quality of each individual beast. Look him in the eye. Study the sheen on his coat. Then I might have gone back and said, "Well, I don't know how many I saw but one looks mimsey." Then they would have saved the life of this beast.

There I learned something. I then also found that in all human traditions there has been a very great antagonism against all this counting business. I don't know how many of you still know your Bible, but you can find it in two places, Chronicles and Kings.The first chap who arranged the census was King David, and when he arranged the census the Lord was utterly furious. He gave him a choice between three penance punishments. And David, said, Yes, yes, I know I have sinned. He didn't argue back. You know, those old Jews used to argue back very freely. He immediately understood there was something wrong in having a census which treats people as if they were units, whereas they are not. Each is a universe.

6.2

Mahatma Gandhi's Views on Machines and Technology

D. P. MUKERJI

This paper comes out of the conviction that any study of the social changes brought about by technology should be based primarily on an understanding of the conditions of both the society which introduces technology and the society to which it is introduced. If these conditions are crystallized in two systems of values, one which has accepted, and the other which has not accepted "technical progress" as desirable or technological advance as "a self-evident good" (R. K. Merton, *Social Theory and Social Structure*, ch. xiii, p. 317); and further, if the strains of technological advance, which any scheme of technological assistance brings in its wake, must be reduced or eliminated so as to render the assistance genuinely effective, then one of the chief concerns of the sociologist as well as of the technical administrator is the discovery of the terms of the normative system of both countries concerned. This does not preclude the various types of specific researches on the social effects of technological change; on the contrary, the normative study may be fortified by their scientific conclusions. What is stressed here is the methodological issue. For example, with regard to the impact of machines or technology (associated with Western Europe) on Asian countries (believed to belong to a pre-technical stage) the underlying assumption that technological advance is a "self-evident" good should be brought to the surface so as to clear the way for the comparative normative approach. It is heartening to know that eminent Western sociologists have recognized the importance of this approach.

But it is disheartening that no formulation of the Eastern value-system has been made by an Eastern sociologist. One reason for this may be that Eastern people are still too deeply involved in their system and, therefore, the technological impact on their basic values is as yet superficial. Another reason may be that those who could formulate and compare the value-systems are the very people who believe in technical advance as a "self-evident good" and, therefore, do not worry about the problem beyond the stage of annoyance with temporary maladjustments which, in their view, a welfare state, or a similar agency or agencies, would benevolently remove sooner or later. Certain economic interests in India, in particular, also seem to be far too committed to techological advance to be anxious to study the conflict of value-systems involved in the resultant strain. Though one hears about cottage industries and their place in Indian life and notices earnest administrative encouragements, the importance attached to them seems to be mainly on the score of their being able to provide 'some' employment to those who are being thrown out of 'employment' by technological advance. Gandhiji, however, was deeply and primarily concerned with the value-systems. There were others too, but they are less known. Gandhiji put his views very sharply indeed. One may not like his manner of posing the problem, one may consider it as partial, one may dismiss it, if one chooses, as many 'educated' men and industrialists of India have chosen. But his statements remain a challenge to the entire problem of technological change and schemes of technical assistance. They should be taken seriously because many new disturbing features of Indian life cannot be explained or removed otherwise. From them one might also infer that the term 'underdeveloped economy', which is the excuse of technical assistance, was inappropriate insofar as it confused the co-existence of two different value systems by placing them on the assembly-line of historical development in which economic growth, being the supreme value, was subservient to and dependent only upon technological advance. Perhaps, Gandhiji was unfair to the European civilization; it may also be that he did not subscribe to the unilinear concept of history. But it is certain that he had other values and his understanding of India, at least, was unerring. We Indians love to think that Gandhiji's views correctly represent the unformulated values of the vast majority of Indian population towards social changes. They, as he would say, would welcome change on their own human terms.

Let us know exactly what he said in regard to machines and when. (He never used the word 'technology', but it is clear that he meant it.) Much misunderstanding of his position prevails everywhere. But he did not want to be misunderstood; in fact, the very clarity of his style leaves no room for misunderstanding. The first unequivocal statement of his position occurs in *Hind Swaraj* or *Indian Home Rule*, written originally in Gujerati, on the return voyage from London to South Africa in 1908 'in answer to the Indian school of violence and its prototype in South Africa'. It was first published in the columns of *Indian Opinion of South Africa*, in the form of the editor's answers to readers' queries. In this period Gandhiji was finding himself through many a personal and a few social experiments. Violence had become the desperate creed of Indian nationalists. He was analyzing this creed all the time, digging its base, so to say, and reaching towards an alternative moral creed. *Hind Swaraj* was the first formulation and bears all the marks of religious conversion. The subsequent history of the book is interesting, but not quite relevant here. But it is on record that in January 1921 Gandhiji would 'withdraw nothing except one word of it, and that in deference to a lady friend'. As Mahadev Desai wrote in his preface to the 1938 edition, 'Even in 1938 he would alter nothing in the book, except perhaps the language in some parts'. We will see, however, that certain alterations were made, but they were more in the nature of elaboration in terms of reality, e.g., the relative unpreparedness of Indian people for the practice of 'a higher simplicity and renunciation', which he knew to be India's values, than as deviations from a fundamental position.

His writings show (letter to a friend quoted in *Mahatma*, vol. I, p. 129-30) that his opposition to England or to the Western or even the European civilization was not merely on the limited ground of political and economic subjection, but on the much wider issue of conflict between the values of different civilizations. *Hind Swaraj* discusses this issue with eloquence and poses the conflict between Indian values and Western values in pure white and pure black. Many historians and sociologists would be more cautious about India's ancestry, about her deliberate wisdom in rejecting machinery, city-life and the evils thereof. They would question the latent virtuous assumptions about India's past and her society. They would attribute them mostly to lack of opportunities and the incurable human habit of making a virtue of necessity. Nor would they fail to detect a high order of spiritual values in the Western, European, or modern civilization and a low order of social values prevailing in the East, in India old and new.But here at last was the Indian positing of a felt contrast in the fierce clarity of exaggeration. The exaggeration was similar to that of a rebel slave who would assert with vehemence his own human dignity and clothe it in historical terms. At the same time, it was not a metaphysical rebellion, so typical of India and the East. It was not a protest against the universal condition of man whose life is interrupted by disease, old age, or death, as was that of Buddha. It was essentially a moral rebellion, couched in the social terms of civilization, which Gandhiji defined as good conduct with the entire weight on performance of duty and observance of morality. The supreme duty was to attain mastery over mind and passions in the performance of which we know ourselves, that is, knowledge accrues. The performance implied proper use of hands and feet and the porcess led to the limitation of indulgences, reduction of wants, and simplification of life. All these ideas formed a whole pattern of thought, beliefs, attitude and action which placed Indian civilization in sharp opposition to what he sometimes called the Western, at other times the European, but what was really modern civilization centered on material values.

Let us follow the development of Gandhiji's ideas. In October 1924, soon after he had broken one of his famous fasts, he gave an interview to a student from Santiniketan, Sir Ramchandran, who questioned him on his views on the place of Art in national regeneration and on machinery. Ramchandran asked (*Mahatma*, vol. 2, p. 212).

R: 'Are you against all machinery?'

G: 'How can I be when I know that even this body is a delicate piece of machinery? The spinning wheel itself is a machine; a little tooth-pick is a machine. What I object to is the craze for machinery, not machinery as such. The craze is for what they call labour-saving machinery. Men go on 'saving labour' till thousands are without work and thrown on the streets to die of starvation. I want to save time and labour, not for a fraction of mankind, but for all. I want the concentration of wealth, not in the hands of a few, but in the hands of all. Today machinery helps a few to ride on the backs of millions. The impetus behind it is not the philanthropy to save labour, but greed. It is against this constitution of things that I am fighting with all my might.'

R: 'Then you are fighting not against machinery as such, but against its abuses, which are so much in evidence today?'

G: 'I would unhesitatingly say yes; but I would add that scientific truths and discoveries should first cease to be mere instruments of greed. Then labourers will not be overworked and machinery instead of becoming a hindrance will be a help. I am aiming not at eradication of all machinery, but limitation.

R: 'When logically argued out, that would imply that all complicated power-driven machinery should go.'

G: 'It might have to go, but I must make one thing clear. The supreme consideration is man. The machine should not tend to keep atrophied the limbs of man. For instance, I would make intelligent exceptions. Take the case of the Singer Sewing Machine.'

R: 'But, in that case, there would have to be a factory for making these Singer Sewing Machines and it would have to contain power-driven machinery of ordinary type.'

G: 'Yes, surely. But I am Socialist enough to say that such factories should be nationalized or state-controlled. They ought only to be working under the most attractive and ideal conditions, not for profit, but for the benefit of humanity, love taking on the place of greed as the motive. It is an alteration in the conditions of labour that I want. This mad rush for wealth must cease and the labourer must be assured not only of a living wage but a daily task that is not a mere drudgery. The machine will, under these conditions, be as much a help to the man working it as to the state, or the man who owns it. The present mad rush will cease and the labourer will work, as I have said, under attractive and ideal conditions. This is but one of the exceptions I have in mind. The sewing machine had love at its back. The individual is the one supreme consideration. The saving of labour of the individual should be the object and honest humanitarian considerations, and not greed the motive . . . Therefore, replace greed by love and everything will come right.'

Next morning the interview was continued. Ramchandran persisted:

R: 'If you make an exception of the Singer Sewing Machine and your spindle, where would these exceptions end?'

G: 'Just where they cease to help the individual and encroach upon his individuality. The machine should not be allowed to cripple the limbs of man.'

R: 'But I was not thinking just now of the practical side. Ideally would you not rule out all machinery? When you except the sewing machine, you will have to make exceptions of the bicycle, the motor-car, etc.'

G: 'No, I don't because they do not satisfy any of the primary wants of man; . . . Ideally, however, I would rule out all machinery, even as I would reject this very body, which is not helpful to salvation and seek the absolute liberation of the soul. From that point of view, I would reject all machinery, but machines will remain because, like the body, they are inevitable. The body itself, as I told you, is the *purest piece* of mechanism; but if it is a hindrance to the highest flights of the soul, it has to be rejected.'

The quotation is important from many points of view. The views expressed in 1924 seem to mark a departure from those of 1908. Gandhiji was a realist in the best sense of the term. The seeds that had been sown in the *Hind Swaraj* grew into a plant in the upturned soil of India. Gandhiji had started the non-co-operation movement and the Khilafat movement, but he realized that the people of India were not yet ripe for the supreme renunciation his values demanded. Non-violence and Truth could not be the bread of the masses. That was the bitter lesson of Chauri-Chaura (1922). The masses wanted bread, and for them 'God was bread and bread was God'. A terrible famine raged in Orissa, and it haunted his dreams. In various cities of India strikes occurred; and their lesson was not lost on him. His sense of limits, a gift which every moral genius must possess, came into play. And in that process he realized a few historical truths. It was clear to him that the impetus behind the large-scale use of machinery was profit or 'greed'—which was 'in the constitution of things' as they were, and not philanthropy or love, that 'today machinery helps a few to ride on the backs of millions', that the labourer must get a living wage and a secure daily task and his labour should not be drudgery, and above all, that man, that is, the labourer was the supreme consideration. In Gandhiji's view man was the producer, the bread-worker by hand, an idea which he had imbibed early in his career in South Africa from Ruskin's *Unto This Last*, which he had translated as *Sarvodaya*. Gandhiji, be it underlined, would have nationalized or state-controlled factories of power-driven machinery to control the profit and produce for 'the benefit of humanity', love taking the place of greed as the motive. At this point Gandhiji presumably believed that the State was, and would be, an agency for transforming greed into a love for humanity, though elsewhere he was less hopeful. All this appears to be a move away from the uncompromising position taken up in the *Hind Swaraj*. Be that as it may, it was not a sacrifice of his basic, or he called it, the 'ideal' position. 'Ideally, however, I would rule out all machinery, even as I would reject this very body (the purest piece of machinery) which is not helpful to salvation and seek the absolute liberation of the soul.' This idea of salvation and absolute liberation from the body is, in my view, the key-note of Gandhian values, or Indian values, as Gandhiji and many others would interpret them. In short, Gandhiji would have welcomed the 'natural destruction' of machinery and mills, together with law-courts, railways and hospitals, but not a violent destruction. By 'natural' he appeared to have meant the potential nature of the scholastics—the nature that sprang from love, purity, simplicity, and flowered in fulfilment and renunciation. His exceptions arose from the 'actual' nature of man in India, his unpreparedness. (*Young India*, January, 1921 'A Word of Explanation'.) Truly could Ghandhi say: 'Ripeness is all.'

These values, it is obvious, centered in renunciation and non-possession. The Hindu idea of renunciation is not *vairagya*, which is probably a Buddhist concept incorporated in Hinduism. Renunciation in the Hindu sense is *aparigraha*(non-possession) of Geta, as Acharaya Vinoba has pointed out more than once in the pages of the *Harijan* (10 April 1949). The concept of love, or altruism, the good of all, as opposed to the hedonistic calculus of the greatest good of the greatest number (*Autobiography*, part iv, ch. viii; *Young India*, 9 December 1926), was probably a mixture of Vaishnava, Christian and later Buddhist ideas. Whatever its origins, it worked very well indeed, particularly as a means of propagation of Gandhiji's ideas among the masses. In other words, the 'ideal' pattern of Hindu values was never forsaken by Gandhiji. It was woven round 'wantlessness'. How could technology and machines geared to the production of goods for the satisfaction of wants, which created more wants, joint wants, derived wants, the infinite hyperbola of wants, be consonant with the pattern of Indian norms? How could such norms square for that matter, with Economics, grounded as it was on wants and their satisfaction? If absolute liberation of the soul from the body be the utter sum of existence, then Gandhiji, and with him, every Hindu who was aware of his ancestry, would raise the eternal query: Why this craze for machinery? Why machine-civilization at all? Other problems such as machines helping a few to ride on the back of millions, the concentration of power and wealth, of justice for the labourer as man, securing attractive conditions of life for him and of giving him security of employment, etc., etc., would be subsidiary. These latter, in Gandhiji's opinion, ultimately hinged upon non-possession, *aparigraha*, wantlessness, subordination of body and bodily wants to the need of the soul's liberation from its physical encasement, which was *the* end.

These subsidiary problems, however, were far from socially insignificant. In the Gandhian view of life, they were related to the means. To many 'educated' Indians they were the ends, or the primary values. Gandhiji had many opportunities of discussing them with those who were more sensitive to the needs and ideas of the day. Rabindranath Tagore's objection was of a different type, but it was met by the logic of means. The poet had written in 1925 against the charkha (the spinning wheel), because he felt that it would bring about a deathlike sameness in the nation. Gandhiji met his argument on the highest level. Taking his cue from the Hindu philosophical conception of oneness, identity, or sameness, which Sankar had carried to its logical extreme, Gandhiji held that 'behind a variety of occupations there is an indispensable sameness also of occupation'. After inveighing for a while against exploitation both by European and Indian manufacturers, Gandhiji conceded:

> Machinery has its place; it has come to stay. But it must not be allowed to displace the necessary human labour. An important plough is a good thing. But if by some chance one man could plough up by some mechanical invention of his the whole of the land of India and control all the agricultural produce and if the millions had no other occupation, they would starve, and being idle, they would become dunces, as many have already become. There is hourly danger of many more being reduced to that unenviable state. I would welcome every improvement in the cottage machine but I know that it is criminal to displace the hand labour by the introduction of power-driven spindles unless one is at the same time ready to give millions of farmers some other occupation in their homes. (*Young India*, 5 Nov. 1925; *Mahatma*, vol. 2, p. 283)

Here was argument on the purely economic level of means, that is, of employment and unemployment. (Also, vol. 4, 34, p. 238–9) To a modern Western economist it may appear to be old-fashioned. He thinks that he has devised excellent measures against various types of unemployment, cyclical, structural, frictional, seasonal, and all that, and he is not worried if even under full employment two to five per cent are unemployed. But Gandhiji was not to be deluded by such theories and measures emerging out of the practices of countries that had been wedded to industrial and technical civilization, that had colonies to exploit and accepted competitive values in production, trade and commerce. The essence of Gandhiji's concession in this open letter to Tagore is, however, historical. That is to say, so long as 'some other occupation in their homes' was not available—and it was not likely to be available in that historical context, or in the near future, because imperialist exploitation would not allow India to create alternative occupations—Gandhiji would stick to the spinning wheel and be against the displacement of labour by machinery. In other words, India in her present context should have labour-intensive economy for the sake of full employment.

Gandhiji was very respectful towards Tagore, and as we have seen met the poet's cultural charge with economic arguments. But he was not so soft towards the Indian Communist M.P. Mr. Saklatwala, in his rely to the latter's appeal, which he duly published in *Young India*. The differences with Tagore were not vital, but with Mr. Saklatwala they were. Gandhiji wrote in *Young India* under the caption 'No and Yes':

> His facts are fiction and his deductions based upon fiction are necessarily baseless. And where these are true, my whole energy is concentrated upon nullifying their, to me, poisonous results. I am sorry, but we do stand at opposite poles. There is, however, one great thing common between us. Both claim to have the good of the country and humanity as our only goal. Though we may for the moment seem to be going in opposite directions, I expect we shall meet some day. I promise to make amends when I discover my error. Meanwhile, my error, since I do not recognize it as such, must be my child and my solace.

Having said this in true humility—and there is not the slightest reason to doubt it—Gandhiji reveals himself in the full panoply of his original, uncompromising, absolute, non-historical faith. He wrote:

> For unlike Saklatwala, I do not believe that multiplication of wants and machinery contrived to supply them is taking the world a single step nearer its goal. Comrade Saklatwala swears by the modern rush. I whole-heartedly detest this mad desire to destroy distance and time, to increase animal appetites and go to the ends of the earth in search of their satisfaction. If modern civilization stands for all this, and I have understood it to do so, I call it satanic and with it the present system of government, its best exponent.

Then follow some sentences which would remind one of the wrath of prophets, but with a difference.

> I would destroy that system today, if I had the power. I would use the most deadly weapons, if I believed that they would destroy it. I refrain only because the use of such weapons could only perpetuate the system though it may destroy its present

> administration. Those, who seek to destroy men rather than their manners, adopt the latter and become worse than those whom they destroy under the mistaken belief that the manners will die with the men. They do not know the root of the evil.

The last paragraph seems to retain its pertinence today in spheres wider than the Indian. The points to be noticed in Gandhiji's reply to Saklatwala's appeal are Gandhiji's firm faith in wantlessness as a cardinal human virtue, and his objection only to the 'modern rush', in which he included both the Western and the Indian values of the day. We may further note his association of modernism with the British Government in India. Strictly from the point of view of the propagation of an unswerving faith, this mixture of basic values with nationalism was excellent. A sociologist would not cavil at it. Technological values are usually associated with the nationalist values, particularly in the Eastern countries in the period of their anti-imperialist struggle which centers in their opposition to the obstacles that imperialism places in their economic growth, and also in the period of economic advance which is held to be possible only with the help of technology. But it is equally understandable that nationalist, anti-subjectionist motives and attitudes should be integrated with the basic values which are, or are interpreted to be, specific to the culture of the nation. While it is true that in this extract no reference is made to the Indian-ness of the objection to the modern rush and the argument appears to rest on the personal level, it is apparent, as it certainly was to the Indian of 1925, that it was a typically Indian argument securely grounded on the Indian philosophy of life, aparigraha, non-possession, enjoyment through giving away, *tyaktena bhunjitha.*

We must repeat that Gandhiji collected other arguments round this basic objection to machines. They were mainly what we would call sociological arguments on the score of excessive population on land, of idle labour, of bad distribution of wealth, benefit to the nation, that is to say, the welfare of the people. He referred to unemployment again and again, to bad health, unwholesome food, and to the decay of art. He categorically stated that labour had a unique place in swaraj, or independence, and formed its content. Each argument depended on another and the whole formed a pattern of positive values. In 1935, on 23 April, he said after opening the first All-India Village Industries Exhibition at Indore:

> The reason why our average life-rate is deplorably low, the reason why we are getting more and more impoverished is that we have neglected our 100,000 villages. We have indeed thought of them, but only to the extent of exploiting them. We read thrilling accounts of the 'glory that was Ind', of the land that was flowing with milk and honey; but today it is a land of starving millions. We are sitting in this fine pandal under a blaze of electric lights, but we do not know that we are burning these lights at the expense of the poor. We have no right to use the lights if we forget that we owe these to them.

Gandhiji would seldom forget to remind his audience of their responsibility towards the people. The word 'owe these to them' is an English rendering of the Hindu concept of rina or debts, contracted by every individual at his birth to his ancestors, his gods, sages, and to nature, the universe or society, debts which must be duly discharged in the course of existence. Hindu social organization is built on the principle of obligations which Gandhiji insisted on resuscitating in all social and economic spheres of activity, *vide* his concept of rich men holding their riches in trusteeship for the poor. Only in the sphere of political struggle against foreign rule would he allow the exercise of the Western sense of rights, and these too hedged in by the sense of social obligations, which was one vital significance of non-violence. Gandhiji continued:

> There is a difference between the civilization of the East—the civilization of India—and that of the West.

Formerly he had contrasted only the modern civilization with the earlier, pre-technical one and minimized, if not ignored this distinction between the East and the West. However, this again is not quite a shift in position; in fact, he was also speaking to the West. And he was only referring to the defects of the tendencies of the Western civilization, and not to its inherent nature, as Mahadev Desai pointed out in his preface to the 1938 edition of *Hind Swaraj:*

> It is not generally realized wherein the difference lies. Our geography is different, our history is different, our ways of life are different. Our continent, though vast is a speck on the globe, but it is the most thickly populated, barring China. Well, the economics and the civilization of a country where the pressure of population on land is greatest (the conjunction 'and' should not be made much of), are and must be different from those of a country where the pressure is the least. Sparsely populated America may have need of the machinery. India may not need it at all. When there are millions upon millions of units of idle labour, it is no use thinking of labour-saving devices. . . . The reason of our poverty is the extinction of [cottage] industries, and our consequent unemployment.

There follow certain figures about the increasing dependence on land as a result of the destruction of cottage industries and the loss of health through the elimination of vitamins in grain after being ground in machines. Gandhiji's context was the first exhibition of Village Industries, which he was building up as the base of his constructive programme. He was an anti-machinist with a purpose, and the language of his argument was suitable to the masses hearing him.

Gandhiji seemed to have been perpetually alive to this problem of unemployment. On 22 October 1937 he inaugurated the Educational Conference at Wardha and developed his ideas of education through handicrafts. It was a new setting for his constructive programme—an integration of living in love, with efficiency and independence, without exploitation, conflict and poverty, and with education of the body and mind. In expounding his thesis, he said:

> Then, take the question of machinery. I think that machinery is not necessary for us at all. We should use khadi (home-spun cloth); and, therefore, we do not require mills. We should try to produce all the necessary cloth in villages, and we need not be the slaves of machines. I am afraid, by working with machines we have become machines ourselves, having lost all sense of art and handwork. If you still think that we cannot do without machines, the scheme (of new education) I have placed before you will be futile. You wish to keep our village alive by means of machines and think of imparting education to the village children through them. Machines will only help in

> making all the thirty-five crores of people unemployed. If you think that machines are really indispensable, you must reject the scheme and suggest a new one (*Mahatma*, vol. 4, p. 238–9).

In 1946 India was passing through a severe food crisis. Gandhiji offered advice to those who sought it. There were suggestions and counter-suggestions. Even in those dire days he would not move from his fundamental ground. In one of his replies he said:

> I regard the existence of power wheels for the grinding of corn in thousands of villages as the limit of our helplessness. I suppose India does not produce all the engines and grinding machines. . . . The planting of such machinery and engines on a large scale in villages [a suggestion made by a correspondent for resolving the food crisis] is also a sign of greed. Is it proper to fill one's pockets in this manner at the expense of the poor? Every such machinery puts thousands of hand *chakkis* [grindstones for corn] out of work, and takes away employment from thousands of housewives and artisans who make these *chakkis*. Moreover, the process is infective and spreads to every village industry. The decay of the latter spells too the decay of art. If it meant the replacement of the old crafts by the new ones, one might not have much to say against it. But this is not what is happening. In the thousands of villages, where power-machinery exists, one misses the sweet music in the early morning of the grinders at work. But to come to the main point. Whilst I hold that these power engines are, at present, being put to wrong use, it would be some compensation if these engines, in addition to the present use, were also used to pump water out of the rivers, tanks and wells for irrigation. (*Mahatma*, vol. 7, p. 71–2)

Probably the most comprehensive and yet succinct account of the place of machinery in the context of Independence was written by Gandhiji in the *Harijan* of 15 July 1946. It was a clarification for the benefit of Congressmen of his concept of independence. He painted a glorious picture of self-sufficient villages, giving free and voluntary play to mutual forces, highly cultured in the sense that there every man and woman knows what he or she wants and, what is more, knows that no one should want anything that the others cannot have with equal labour, a society based on the living force of truth and non-violence, a society not like a pyramid but like 'an oceanic circle.'[1]

On 31 July 1946 Gandhiji addressed a conference of Ministers of Industries of different states in Poona. There he clarified his conception of village industries and referred to the imbalance between town and village economics and the need for establishing justice in their relation. In that connexion he came to a fairly elaborate explanation of what he meant by machines. Sir D. G. Tendulkar, the author of *Mahatma*, gives the following summary:

[1]'I may be taunted with the retort that this is all Utopian and, therefore, not worth a single thought. If Euclid's point, though incapable of being drawn by any human agency, has an imperishable value, my picture has its own for mankind to live. Let India live for this picture, though never realizable in its completeness. . . . In it, there is no room for machines that would displace human labour and that would concentrate power in few hands. Labour has its unique place in a cultured human family. Every machine that helps every individual has a place. But I must confess that I have never sat down to think out what the machine can be. I have thought of Singer's Sewing Machine. But even that is perfunctory. I do not need it to fill in my picture.' (*Mahatma*, vol. 7, p. 201–2.)

'Ours has been described as the machine age,' remarked Gandhi, 'because the machine dominates our economy. What is a machine? One may ask. In a sense, man is the most wonderful machine in creation. It can neither be duplicated nor copied.' He had, however, used the word not in its wider sense, but in the sense of an appliance that tended to displace the human or animal labour instead of supplementing it, or merely to increase its efficiency. That was the first differential characteristic of the machine. The second characteristic was that there was no limit to its growth or evolution. That could not be said of the human labour. There was a limit beyond which its capacity or mechanical efficiency could not go. Out of this circumstance arose the third characteristic of the machine. It seemed to be possessed of a will or genius of its own. Machine was antagonistic to man's labour. Thus, it tended more to displace man, one machine doing the work of a hundred, if not a thousand, who went to swell the army of unemployed and underemployed, not because that was desirable, but because that was its law. In America it had perhaps reached the extreme limit. He had been opposed to it not from today, but even before 1909, when he was in South Africa surrounded by machines. Their onward march had not only impressed him,. but had repelled him.

'It then dawned upon me that to suppress and exploit the millions, the machine was the device par excellence; it had no place in man's economy if, as social units, all men were to be equal. It is my belief that machine has not added to man's stature and it will not serve the world, but disrupt it, unless it is put in its proper place. Then I read Ruskin's *Unto This Last*. . . . I saw clearly that if mankind was to progress and to realize the idea of equality and brotherhood, it must act on the principle of *Unto This Last*.'

In the machine age, this principle had no place. Under it the fittest alone survived, to the exclusion and at the cost of the weak. 'But that is not my picture of independence, in which there is room even for the weakest," observed Ghandhiji. 'That requires that we must realize all available human labour, before we entertain the idea of employing mechanical power.' (*Mahatma*, vol. 17, p. 214–15.)

We have traced the development of Gandhiji's ideas on machines, and on machine-civilization and found that despite many concessions to the 'proper' use of machines his values were definitely opposed to those which make for technological civilization and are made by it. By 'proper' he meant, positively, that which was prompted by love and good for humanity, and negatively, what did not lead to concentration of wealth in a few hands and inequality, to centralization of power, to urbanization, to unemployment, to political, economic and social exploitation. These evils, which in his view, were the characteristics of the modern society, with its American apogee, were the consequences of the large-scale use of machines, and they had to be fought with vigour. To that extent he was placing Indian (Eastern) against Western (Euro-American, modern) values. It was certainly not a case of revivalism, but a clear statement of a principle of social organization which was different from the one that had succeeded in imposing itself on the strength of political suzerainty. His minimal idea was to establish co-existence of different social systems, on the basis of equality, though the prophetic strain that came to him in the course of his experiments with Truth led him to think that the values he propagated would also be good for the Western world. We will leave it to the Western sociologist to ponder over this issue. An Indian sociologist can only mention that Gandhiji's protagonism of Indian values was not a manifestation of the romantic agony of nationalist historians, nor was it a reactionary, obscurantist throw-back. It is submitted here that he was a revolutionary and what a revolutionary in India should be, viz., an Indian revolutionary, that is, one who would first be steeped in Indian realities and then evaluate the nature of changes in social realities in order to create fresh norms. Gandhiji did not go to the past: in fact, he was not an Indologist; he only went to the roots and the sources. And 'the deeper you go to the root the more radical you become'.

At this stage it will be relevant to note the revolutionary elements in his views. Their importance arises from the problem before us, that we here must consider the conditions under which technological change can take place without causing the various tensions and frustrations which bring about aggression, violence, and war, which lead to mental unhealth.

In other words, we must ask ourselves whether technology cannot but generate these evils; whether technology should always depend upon wants and their increase and bring about a state of culture in which material wants are king. The inner significance of Gandhiji's concessions to the use of machines, that is to say, the logical meaning of the 'proper' use, is that they do not, and need not, go together. It is perfectly possible, in Gandhiji's opinion; and it is also logical to assume under stated conditions, that technology can be introduced into India without upsetting the Indian pattern of values. These conditions are non-possession, aparigraha, the 'oceanic' constitution of Indian independence in terms of self-sufficient villages with their group-existence fulfilled through the panchayats, bread-labour, sharir-srama, with its concomitant of the dignity of labour symbolized in *khadi*, *sarvodaya*, that is, total awakening or uplift, and of course, non-violence and Truth, that is satyagraha. Of these, *aparigraha*, non-possession, or wantlessness and *srama* or labour, alone are selected for discussion. (Gandhiji would have emphasized Truth.) Now, non-possession in the context of human history has been an individual value, and at best, a value for the elite-group, known as the Brahmin caste, to be perpetually practised by it. Others practised it, but the Brahmin was the specialist. Gandhiji would institutionalize it in the State which would own, and not possess for greed or profit. To this extent he was a Socialist—or even a Communist as he called himself before Mr. Louis Fischer (*Mahatma*, vol. 7, p. 190), but with this difference that his Socialism did not grow out of industrial civilization, technological values, class-conflict, or according to the operations of the laws of dialectics. (That it could only be social expression of Truth, non-violence and faith in God, is important, but not relevant to our selected purpose.) It was to grow out of agriculture, cottage industries, and 'oceanic' rural organization, into a non-possessive State which would be wedded to public good and be a guarantee of non-exploitation by large machinery in the hands of the rich. Meanwhile, Gandhiji would ask the rich and the fortunate few to hold their fortunes in trusteeship, practising aparigraha themselves. Trusteeship, however, for Gandhiji, was an interim measure sanctioned by Indian traditions.

A second revolutionary element in Gandhiji's prescription is the concept of *srama* or manual work. So far as this writer knows *srama*, or the value of the dignity of labour, was not quite an Indian value. In a hierarchical society, types of work are defined and relegated to different strata on the two assumptions: that spiritual 'work', that is, pure contemplation, is the highest type; that each stratum, or caste, which is fixed by birth, has its own *swadharma* ('bond', 'religion') the practice of which means fulfilment of personality, and the departure of which means 'destruction' of self. But Gandhiji had a different conception of labour. He writes from *Yeravada Mandir*, ch. IX):

> The law, that to live man must work, first came home to me upon reading Tolstoy's writing on bread-labour. But even before that I had begun to pay homage to it after reading Ruskin's *Unto This Last*. The divine law that man must earn his bread by labouring with his own hands, was first stressed by a Russian writer named T. M. Boudaref. Tolstoy advertised it and gave it wider publicity. In my view, the same principle has been set forth in the third chapter of the Gita, where we are told he who eats without offering sacrifice eats stolen food. Sacrifice here can only mean bread-labour. There is a world-wide conflict between capital and labour, and the poor envy the rich. If all worked for their bread, distinctions of rank would be obliterated; the rich would still be there, but they would deem themselves only trustees of their property, and would use it mainly in the public interest. The ideal body-labour is agriculture, but the next best would be spinning, weaving, carpentry and smithery; and the logical, common body-labour is scavenging. . . .

Surely, this is not Indian value.The dissociation of bodily labour from mental and spiritual labour has had a long history, which Gandhiji did not take into account. His attitude towards what is known as the caste-system cannot be discussed here. But on the matter of bread-labour bringing about economic equality he was anti-caste and, therefore, a revolutionary—almost a Socialist. In other words, if the Socialist gave up the usual Western assumption, i.e., no high technology, no Socialism, and remained content with the use of certain special types of machinery which would not displace labour or exploit human beings for greed or concentrate power, then Gandhiji would bless him. If further, the Socialist accepted this idea of bread-labour and built on the revolutionary content of this view, the ending of the separation of physical labour—now the only duty of a whole class of people who form the majority—from mental labour which is now the monopoly of the few, then the difference between him and Gandhiji would not exist, except in the matter of wantlessness. One could argue here that there was a danger in this concept of bread-labour, that it involved the possibility of lowering the impulse and the level of intellectual work by making the intellectual workers work physically for bread without raising the intellectual level and stimulating the impulse of physical workers. But Gandhiji would reply that this could be averted.

Thus the Gandhian conclusion in regard to machines and technology is logical if one accepts the postulates that India has a separate norm of values—with the hidden assumption that values determine conduct—that she has a separate principle of social organization which would be disturbed and even destroyed by large-scale use of machinery for greed and profit, and that a proper use would presuppose certain attitudes, some traditional and others not, but all working in alliance, and also a type of State that would own and control large machineries, if they were indispensable for defined purposes. Otherwise, the machineries to be used would be of a special type suitable for removing the drudgery of handicraft and improving its quality, if possible. They would operate in the general context of de-centralized economy, in close alliance with agriculture. Gandhiji would thus remove the strings of Capitalism and Socialism alike.

In this author's opinion, Gandhiji's views have to be carefully studied before any scheme of technical assistance and large-scale technological development is initiated. While it is very true that among certain strata and sections of the Indian people these views appear strange, even though lip-service is paid to them, and that such people would want to initiate rapid technological change in the name of economic advance, evolution and progress, it is also clear that an unintelligent injection of technology would so disturb the existing social pattern of human relations that work would cease to be associated with joy and workmanship, that skill would be replaced by efficiency, that 'the public identity of the job' would be lost, that scientific management and discipline would squeeze the labourer of all humaneness, and that a new instrument of social power would 'teach docility' or 'break the intransigence of workers', all the time keeping greed, profit and more profit in the background, beyond the sight of those who are to be immediately benefited by higher wages, better conditions of living, welfare-measures and the like (*Social Theory and Social Structure*, R. K. Merton, p. 318–22).

In other words, the sociologist would do well to study the sociology of the demand for technology in India and the sociology of its supply. Fortunately, the Indian masses are not yet fully taken in by the technological values. They are being acted upon by technology, and they are showing healthy, normal reactions to the injection by absenteeism, inattentiveness, a sort of lackadaisical attitude towards work in the factories, unpunctuality, the so-called absence of pride in work and workmanship, but unfortunately, quite often in accidents and ill-health. Even strikes, which are bemoaned as symptoms of industrial conflict, can be traced to the search of the soil-bound peasantry of India for mental and social peace in holiday, away from the scene of work, the din of factories and technology, to the villages where the pattern which they know and instinctively feel to be right, rules even today.

An Indian sociologist cannot thus help questioning the manner and the possible hidden motivations of a technological advance, of those who want it and those who supply it. He can, at least, categorically state, with Gandhiji, that if change is inevitable, let it come in the shape of certain types of machines, at the proper time, in the proper context; and that if large-scale use of machinery is unavoidable, let it be owned and managed by the State, a new form of State. Judging from experience, he cannot share Gandhiji's hope of an interim trusteeship by the fortunate few. In short, the whole problem of technological advance in India, which, let it be repeated, is undeveloped only in the purely technological sense, and, therefore, in the sense of being more socially integrated and less fragmentized, has to be studied from the points of view of both the types of machinery to be introduced and of the types of people who would accept them with due regard to the motivations of introduction. In India, human beings are not yet atomistic, so that no functional specification in the most common productive processes, which are those of agriculture, is possible. Nor has it occurred in many industries. In India, no productive section of society is 'universalistic', that is, only very few 'criteria possibly present in any segment of the population without regard to previous social relationships or memberships in irrelevant groups' are available. In India, human relations are affective rather than rational and impersonal. The normative system suitable for the industrial mode of production through large-scale use of machines is thus not the normative system of India. A matter of additional importance is the fact that this normative system has combined with nationalism, anti-imperialism, and Gandhiji's interpretation of independence to convince some people that it is still valuable. So change must take account of these facts in order not to produce the same evils to which the West has been an heir.[1]

[1]Most of the quotations from Gandhiji's writings have been taken from D. G. Tendulkar's monumental biography *Mahatma*, only seven volumes of which are available at the time of writing, the eighth being in print. The 1946 edition of Gandhiji's *Hind Swaraj* has been used here, and the 1951 edition of *Sarvadaya, Its Principles and Programme.* Both are published by Navajivan Publishing House, Ahmedabad, Gujerat, India. An autobiography: *The Story of My Experiment with Truth*, translated from Gujerati by Mahadev Desai, 1939 edition, Phoenix Press, London has also been cited.

6.3

Constraints on Engineering in Third World

DONALD S. MANSELL

INTRODUCTION

The underlying principles of any science-based discipline such as engineering are universal, which leads the practitioner to believe that the application of those principles is also universal. An engineer who moves across international boundaries to practice his profession may make this assumption either knowingly or unwittingly. Also in engineering there is a considerable amount of prior experience behind any professional judgment made, and it is very difficult to bring this background component of one's thinking into the open to question its relevance and to detach oneself from the particular environment in which it has been gained. The engineer therefore tends to transport this part of the decision-making process with him wherever he goes as though it is of equal validity in all situations, usually being conscious of it in only a very vague way.

This paper considers the subject of geographical variation in factors that constrain engineering decisions, paying particular attention to those that are of human rather than natural origin.

THE THIRD WORLD

The rate of development of nations has not been uniform. Doubtless even in very early times great differences in economic, cultural, and political development existed, but these differences have increased as the centuries have passed, since development tends to be a self-propagating phenomenon, although individual nations have waxed and waned. Consequently, despite the probability that all humans have common ancestors, it is now found that there are differences in many aspects of life in different communities, including inequalities of economic and educational kinds, and the reflection of these in economic and political stratification to the point of the dependency of some nations on the support of others.

The term "The Third World" had a political origin, distinguishing those states that were outside the political polarization of the capitalist and socialist blocs after World War II. But it now has an economic significance in current usage.

It does not follow that there is a simple economic criterion by which the "Third World" may be identified and distinguished from the remaining wealthier groups of nations. It is perhaps not even particularly profitable to attempt a neat categorization, and the present paper does not depend on this being done. In fact, the writer's thesis is that there is a technological uniqueness in all societies whether developed or not. The reasons for making specific reference to the Third World are that its members are usually purchasers of foreign technology and that the particular constraints to which attention is drawn here are highlighted more easily in developing nations.

Although there is no need to define the Third World precisely it is helpful to recognize the existence of different types of country within this grouping. Crosthwaite and Bergan (1) have drawn the following distinctions between countries among the Third World group, the parenthetical examples being the writer's opinions:

1. "Countries with economies in the early stages of development almost wholly dependent upon the export of raw materials, primarily the surplus unprocessed production of peasant farming" (many African and Pacific Island states).
2. "Partially developed countries, often with substantial reserves of professional expertise and skilled labour but frequently deficient in financial resources" (India and the People's Republic of China).
3. "Wealthy nations, especially those which have achieved this position in recent years. They may be underpopulated and in addition have not had the time to produce their own cadres of professionals and associated skilled manpower" (Arabian oil-exporting states).

Third World countries share a recipient status in technological exchange and share the characteristic features resulting from being either in the throes of rapid social and political evolution or poised on the brink of it. They are almost all recently independent of colonial powers with consequent sensitivities towards outsiders and with consequent infrastructure weaknesses that frequently include gaps in the supply of technological and managerial skills.

CONSTRAINTS COSTS AND BENEFITS

Engineers are now becoming familiar with increasingly formal descriptions of the decision-making process. Two types of mathematics that are useful in describing the decision-making process are optimization theory and decision theory. In the first it is asserted that decisions such as designs may not only be checked as to whether they are permissible or not permissible, but they may be compared to find which is optimal or "best." Decision theory introduces the notion of uncertainty, asserting the possibility of rationally adjusting assessments of uncertainty so that decisions can be compared in order to minimize the risk of loss or maximize the chance of gain. In both types of analysis there is acceptance of the facts that decisions are constrained, that they incur costs, and that they confer benefits. The engineer's task in making a decision is to ensure that significant constraints are not omitted, and to place a proper valuation on all the costs and benefits. As has been claimed previously herein, a large part of the mental processes used in doing this are not formalized and are almost subconscious. They are akin to those of a gambler deciding whether to place a bet.

It follows that since constraints, costs, and benefits have local values, decisions must be expected to vary to reflect the setting in which they are applied.

It is obvious that many natural factors will guarantee this variation in optimal decisions including climate and topography. Those who have worked in the Third World have no difficulty citing examples of equipment and techniques imported from temperate climates that proved to be most inappropriate in the different environment of the purchasing country. One ought to be entitled to expect that constraints of this type are obvious to all, although continuing failures show that they are not. Undoubtedly, these problems will continue until the Third World constitutes a sufficiently large and discriminating market to insist on designs being prepared to meet its particular needs. But since these natural constraints should be obvious attention will be devoted here to factors of a more human kind that are easy to overlook and difficult to take into account.

It is not difficult to imagine the pleasure of an engineer who has designed, installed, and commissioned a small hydro-electric plant to serve an isolated village, and who has been gratified by the enthusiasm of the village people for the new facility. Evidence of the benefits in increased social intercourse in the evenings, opportunities for night-time reading by school children, and the feasibility of the use of small electrical appliances would have a dramatic effect on the engineer's motivation and job satisfaction. Imagine then the dismay caused when the reception from another village to the suggested building of a similar supply is one of either sullen noncooperation or of outright resentment and opposition. This happens when the village concerned harbors suspicions of outsiders and external influences, cherishing its independence and self-sufficiency. Influential elders may be aware of the threat to their status and authority from the young people having contact with schools and towns. Their evaluation of the benefits received from the facility is therefore much lower, and the costs incurred in building it seem to them to be much higher.

Or it may be that the village welcomes the installation of the scheme but does nothing to maintain it, perhaps even vandalizing it. Examples abound of village pumps not working, for lack of a simple washer, or of incompleted penstocks in microhydro schemes jammed with rocks and rubble, because the equipment was seen as belonging to "government" or "mission" and being of only incidental benefit to the villagers. The local value placed on the costs and benefits of the schemes in these situations obviously differs from what had been expected by the engineer responsible for them. It could be said that cultural factors had intruded to place unexpected constraints on the practice of the technology involved.

CULTURE

It is a pity that the term "culture" has gained such a restricted meaning in popular usage of the English language, for its wider meaning is very useful in the present context. For people of European origin "our" culture is perceived as comprising baroque music, cathedral architecture, poetry and folk dances, while "their culture" is caricatured in terms of primitive tools, quaint myths, and rather distasteful social practices. Since technologists in general and engineers in particular frequently consider their professions as outside the bounds of their national culture, and possibly as an internationalized or supranational noncultural activity, it is worthwhile reminding oneself of what culture really is. For present purposes, it is intended to reject common usage that regards "cultivated" and "cultured" as synonyms for attributes conferred by culture, and to embrace the more comprehensive definitions preferred by anthropologists, of which the following is an example (4): " . . . historically created designs for living, explicit and implicit, rational, irrational and non-rational, which may exist at any given time as potential guides for the behaviour of men." Alternatively a very simple definition is " . . . everything in a human society which is learned."

Without attempting to be definitive in a listing of what is covered by these comprehensive definitions one can easily identify the following aspects of culture that differ from the usual illustrations of cultivated pursuits. In every instance it is possible to see an impact on the determination of engineering decisions.

Obligations.—The obligations and tacitly prescribed roles of males, females, castes, and other identifiable groups are of great importance in social infrastructure, and may be relevant to an engineer. They may be important constraints on the cost and "trainability" of labor, or determinants of market demand. In many developing countries the young have a deeply entrenched obligation of deference to their elders, creating problems for the educated young indigenous engineer or supervisor required to exercise authority over subordinates to whom he should normally defer. Is it therefore wise to use local engineers in positions of responsibility in their own area among their own people? This is a question that can be answered only on the spot, where the benefits of better communication and greater trust can be weighed against the costs of satisfying familial and community obligations.

Loyalties.—Most Third World communities have evolved systems of loyalties that have successfully protected their well-being. Their survival in the struggles with environment and human enemies is proof of the success of the systems that are extant, and community leaders who are aware of this are frequently jealous guardians of the traditions that reinforce them. An alien engineer working with these people brings a different set of loyalties, and must adapt them to succeed. His own culture has an established system of loyalties between employer and employee, or between merchant and customer that is embodied in the systems of business ethics and law to which he is accustomed. If in a foreign setting these loyalties are much lower in status than loyalty to an ethnic group of an extended family a new and unfamiliar set of management problems are created, requiring new decision criteria.

Engineers from a nuclear family culture may disparage the extended family relationship, regarding it as antiquated and resenting its competition for the loyalties of their subordinates. It makes great demands on the employee's wages, on his time, and on his hospitality, bringing him into conflict with the employer. However, it is as well to recognize its virtues. It provides the worker with an awareness of his worth, to the benefit of his motivation, particularly if the kinship group is a potential beneficiary of the engineering project concerned. It can provide an effective communication link, particularly when members of the group with which the engineer must negotiate are on his staff, and have an understanding of the language, taboos, and personalities involved. It may even provide a means of discipline. There are occasions when there is an advantage in bringing a kinship group in as a disciplinary body to judge one of its members who is guilty of misdemeanor. Instead of the group instinctively coming to the defense of the culprit when his employer disciplines him, the employer and the group may combine to protect the reputation of the group, and the guilty party is left with no grounds for resentment. Customary "courts" are frequently quick, fair, and effective.

Finally, it should also be remembered that if a worker is encouraged to maintain his links with his extended family he usually has certain rights in his own village to support him when the contract is over and his employment ceases.

Beliefs.—Not only are people's beliefs varied but they are sometimes unexpected. We mostly do not fully comprehend our own beliefs, much less those of other people, for they go much deeper than mere intellectual assent or committment. They have a profound effect on the relationships between people, finding their expression in complex systems of loyalties and obligations. They also influence the relationship between people and nature.

One important outcome is sometimes a rather mystical relationship between an individual and his kin, and another is a similar quality in his relationship with the land that his people have traditionally occupied. An engineer wishing to use this land for passage or develoment ignores this special relationship only at the risk of unexpected and significant costs to the project. Australians of European origin have never understood the Aborigines' attitude to land, which differs radically from the commercial concept embedded in European culture, and a similar cultural gap occurred in Papua New Guinea.

When a Commission of Enquiry into Land Matters traveled around Papua New Guinea taking evidence of the complaints and opinions of the people shortly after Self-Government (1973) a Papua New Guinean engineering student protested to the Commissioner (who was a Papua New Guinean himself) ". . . my grandfather had no right to sell our land to the white man." As that particular discussion unfolded it became obvious that the complaint was not against the price in trade goods that had been paid, nor against one particular person or generation receiving the benefit. Although these were criticized, the main force of the complaint was against the very concept of land as a saleable commodity. The student believed that his people had a form of relationship with the land that was not of an ownership kind, and that his grandfather must have had a different understanding of the contract from what the purchaser thought.

It would be presumptuous of the writer to define what that relationship might be, and no alien engineer is likely to comprehend it fully. But it is essential to the successful responsible practice of engineering to respect emotional links between people and land when these links have developed over many generations and are part of the system of beliefs of the people concerned.

Similarly, the relationship of an animist with nature may introduce unexpected constraints upon the evaluations made in an environment impact study, since individual trees, animal species, or topographic features may be invested with properties or rights that are not accounted for in the engineer's conventional assessment of costs and benefits.

A third illustration of a belief that is potentially important in the application of a technology such as engineering is the belief that every disaster is willed to occur, and that it cannot be accidental. In some societies all instances of accidental injury are seen as the consequence of malice on the part of someone who may be unknown or who may be detected by some means. Retribution is called for, and it is not difficult to imagine the implications when the parties to the dispute are involved in an engineering project in any capacity. An incident of this kind occurred in Papua New Guinea when a construction bus was involved in an accident. Several workers were killed, and the driver belonged to a different ethnic group. Although the European supervisors would have considered the driver not responsible, all of the people on the site from his province felt themselves to be in danger of indiscriminate retribution. They asked to be driven out of the area, but the construction authority refused to do so. They therefore commandeered the authority's transport and drove their entire group to safety in a major town. Charges were laid against them but their action was upheld in court as a justifiable precaution for self-preservation. That they were refused sanctuary in the first instance and then charged with misuse of vehicles is an example of a lack of sensitivity to local beliefs that affect engineering construction work.

Skills.—The skills component of a culture is one that is widely recognized and therefore needs little emphasis. However, it is relevant to point out that a nation's stock of skills is almost unique, having been established by the chance combination of local materials, immigration, markets, climate, and other aspects of culture. In one place there are significant reserves of skill in masonry, in another it is carpentry or metal working. Where a certain skill exists there is the possibility of rapid and effective transfer of technology from an alien source to build up local skills to a level that will serve engineers well in most projects undertaken. Where there is no traditional relevant skill the time required for training may be much longer, even to the extent that it influences basic decisions at the design or project planning stage. The mining industry, which in recent years has been obliged to pay heed to the strictures of newly independent governments on counterpart training and indigenization of its work forces has obtained much useful experience of the problems and techniques in trade training with trainees who have no relevant skills at the outset. The sophistication of the training aids and training programs created by large mining companies and the relevant equipment manufacturers should be examined by all engineers planning to work in the Third World.

Education.—For the successful application of technology education should go much deeper than mere literacy, or mere numeracy and should be more fundamental than the mere transfer of trade skills. If it is inadequate, the quality of engineering work suffers markedly, because in engineering there is a dependence on analytic, synthetic, and symbolic thinking of which alphabetic and numeric systems are only a small component. Classification, analysis, and synthesis are intellectual skills that must be learned. Without them clerks cannot be relied upon to keep accurate records, tradesmen cannot interpret drawings or exercise initiative, and communication becomes imprecise and unreliable. In some Third World countries the tradition of education and of symbolism is very ancient and technological training therefore becomes a natural extension of the existing culture. This paper classifies its subsections using a number system devised in a culture that is now considered part of the Third World, and is reproduced by a process that had its origins in another developing country. In other places the local language has no satisfactory number system at all, and graphic symbols are used primarily as an art form with little regard for representational precision. Symbols may abound, and are frequently of great cultural importance, but they are unlikely to be created to facilitate analysis, even of simple fundamental geometrical properties such as length, area, and volume. Consequently, algebra and geometry have no local cultural basis from which to develop, while the concepts of proportions, rates, and linear combinations that are essential to much engineering decision seem to be difficult to learn.

These higher level intellectual skills are crucial for supporting engineering staff and the extent of their development is therefore important to the alien engineer, but his greatest difficulty is likely to be restricted literacy and numeracy among manual workers and their immediate supervisors. In some countries there is no universal lingua franca, so communication barriers become even more difficult, affecting efficiency and thus influencing costs and safety. One intriguing aspect of this is that bilingual indigenous staff may be literate in English but not in the local tongue despite conversational fluency in both.

Expectations.—Engineering projects are always performed with specified standards of quality as objectives, both in design and in execution. These standards have a profound influence on costs and on benefits. There are occasions in the Third World when standards should be higher than in developed nations because of greater rigor in service conditions. There are also many applications that call for standards appropriate in any country. But it is more common for optimal standards to be lower because the savings in reduced costs outweigh the penalties in reduced benefits. This effect is related to expectations. In many cultures time is less important than it is in the frenetic bustle of European culture. Road standards that reduce construction costs at the expense of traffic delays are then more readily accepted as optimal. The collapse of the Erap No. 1 bridge on the main arterial highway in Papua New Guinea in 1976 forced traffic to divert to a ford. The level of expectations of the highway users was such that the collapse was not even reported in the newspapers!

Similarly the people's expectations in standards of public hygiene vary between cultures, influencing decisions on water treatment and waste disposal. Feachem (3) argues that optimal water resources goals in the less developed countries are a reduction in the effort required of people for fetching their water and first steps in the control of water-related infections. The responsible water engineer must therefore be prepared to consider the possibility that an optimal decision that will match the expectations of the people he serves is to provide each of several villages with filtered water in a single well or standpipe, rather than to provide only one community with a treated and reticulated supply.

Effective engineering in the Third World should reflect the client's expectations and not those of the alien engineer. This is a difficult adjustment to make since one's own expectations are created in the formative years and reinforced in professional training. There is a natural reluctance to use this training to create something that is felt to be inferior. Furthermore, it is difficult to assess the expectations of an alien client. Not only do the normal communication difficulties interfere, but people's ambitions are not static and not uniform within the community. Nevertheless the attempt at adjustment must be made.

As an example of the possible differences between an engineer's experience of standards and those of the people he serves, one might consider villagers in remote mountainous terrain in Papua New Guinea who sometimes have become seasoned air travelers before ever seeing a motor car. Their scales of values in transportation matters are therefore rather different, with their education lacking a complete stage in the evolution of automotion. Instead of progressing from foot (5 km/hr) to animal (10 km/hr) to road vehicle (80 km/hr) to flight (200 km/hr) there are Papua New Guineans who have never experienced the second stage and who have moved from the first to the fourth and then back to the third. They lack awareness of traffic hazards and thus have one of the highest traffic mortality rates in the world (5).

It is also necessary to recognize the low expectations of privacy in many communities and the low status of certain groups within their societies, both of which affect the costs and benefits in housing design decisions.

It should not be concluded that a low standard is always preferred. Large urban centers have much the same needs in most countries, and to extrapolate simple rural technological solutions to a metropolitan area could have disastrous consequences for health, social order and the operational efficiency of industry, commerce, and administration. For example expensive capital assets such as manufacturing machines, computers, and libraries require the same environmental protection in any country.

In the process of gaging appropriate standards it can be frustrating for engineers to discover that sometimes expectations are not consistent with income—perhaps a universal phenomenon. A building development group went to some trouble in Papua New Guinea to develop a house plan making effective use of natural round poles from commerical timber plantation thinnings. The possibility seemed to exist to bring the cost of this house down to a level that a local wage earner could afford. When prototype houses were constructed, the civil servants to whom they were offered rejected them outright, regarding this form of construction as a retrograde design suitable only for remote and unsophisticated populations. The European insistence on sawn timber has established that as the only acceptable type of construction for houses in towns and has damned natural round poles to the same disrespect as indigenous thatching, despite its technological adequacy and its probable economy.

Similarly there is a striking contrast between the acceptance in China of the single-axle, single-cylinder "Gong Nong" tractor as a means of transportation in communes (capable of drawing not only implements but also small passenger trailers) and the reluctance of Papua New Guinea village people to concede that any solution to their transport problems is acceptable unless it is in common use among the relatively wealthy ex-patriate population. So a high proportion of communal savings are used on the purchase of six-cylinder trucks and four-wheel drive utilities, which are less versatile than the Gong Nong and are expensive to maintain.

Tradition.—Communities that are small, isolated, and self-sufficient have probably survived through the evolution of a successful culture protected by the authority of a universal acceptance of the importance of tradition. This creates a conservatism that ensures that proven solutions are used which are assumed to be at least successful if not necessarily efficient. But an alien engineer necessarily challenges tradition. He brings a knowledge of alternative solutions that the community may use. In most instances he also brings with him a new authority, perhaps because he represents government, or perhaps merely because he is educated and carries evidence of successful innovation in the form of motor transport, radios, and other devices.

Although the community may accept much of what the engineer has to offer it is as well to recognize that traditions and their authority do not vanish overnight but remain in tension with innovation. It has been known for people in areas devastated by earthquake to insist on rebuilding in heavy nonductile materials contrary to expert technical advice, despite the availability of safer forms of contruction better suited to their self-interest. Thus tradition dies hard, and must be reckoned with in design and construction. The challenge for the engineer is to marry modern technology to tradition wherever possible in order to minimize social "costs" and to increase its effectiveness.

Economic System.—The economic environment in the Third World varies from subsistence agriculture with a relatively underfinanced central government responsible for most engineering work, through underdeveloped cash economies with limited trade established around market towns, to the larger scale economic models of socialism and industrialized capitalism. Exchange may be based on simple barter, a cash economy, or on a sophisticated banking system, while the range of goods and services involved in that exchange varies to a similar extent.

A consequence is that an engineer servicing industry in a capital town usually conducts his professional affairs in a different manner from a colleague engaged in provincial construction work. In the provinces work is bedevilled by the absence of on-the-spot supplies and services that the normal economic infrastructure provides in a more developed commercial setting. Subcontractors may not be available, accommodation for employees may be nonexistent, and a host of similar constraints may occur, generally leading to the conclusion that the project must be largely self-sufficient.

The degree of organization of labor into unions, and the role of government in the enterprise frequently reflect the economic system as a whole and combine with the other factors mentioned to require rather different decisions in project planning.

Legal System.—Engineering projects involve relationships between people, including commercial contractual relationships between parties, industrial relationships between workers and managers, and the legal requirements of a duty of care that every person owes to others whom his work may affect. If different countries the legal systems in which these relationships are embedded vary in significant ways.

Discriminatory laws exist which reserve certain occupations and land ownership to nationals. Examples are found easily enough in Fiji, Papua New Guinea, Malaysia, and many parts of Africa. In the past job reservation has usually been restricted to those occupations for which local training is relatively well developed, including most trades. But these restrictions now affect many levels of management and the experience of personnel available for these managerial tasks may be limited. In particular, there may be a legal obligation to incorporate into the professional engineering team nationals of the country where the work is being done, because of legal requirements for "counterpart" training. In some instances it is obligatory to include local equity in the company that undertakes the project, special company tax regulations are enforced and constraints are placed on the repatriation of profits.

But in addition to these legislative factors, customary law affects the operation of engineering projects if the traditional rights of landowners or individual workers are injured or jeopardized. Customary law frequently sanctions remedies that appear to an alien to be unusual and possibly disproportionate, and comes into effect if the injury for which remedy or redress is sought is not quickly identified and attended to. At the same time there is much to be gained by making use of the available processes of customary law to the fullest extent in the settlement of disputes because they have the authority of ancient usage, they are locally available and can be called upon quickly and cheaply.

Reference to the legal system, even in a brief mention such as this, should not overlook law enforcement. Police forces in developing countries are often inadequately staffed in terms of experience and skill. However, it seems that even in the most affluent societies security of engineering works must be treated very seriously, so perhaps this is one of the few truly universal constraints on technology.

APPROPRIATE TECHNOLOGY

Given that for climatic, topographic, and cultural reasons technological solutions should vary from one environment to another, it follows that a particular solution may be inappropriate in a particular place. This is one way of interpreting the message of Schumacher (6), Dickson (2), and others who have attacked conventional engineering value judgments, although they go further to claim that some technology is inappropriate anywhere. Schumacher was particularly concerned about the diversion of effort into what he called "stunt" technology and pressed for greater priority for the development of "technology with a human face."

The present paper has attempted to highlight the need for sensitivity in all technological work, even when the scale of operations is not small enough to be classed with Schumacher's "Intermediate Technology." This latter term has fallen out of favor, losing place to "Appropriate Technology." Dickson chose "Alternative Technology" for the title of his book, which is perhaps a better reflection of the actual polarization of technologists into two groups, with the radical element epitomizing its objectives by reference to windmills and self-sufficient production. The majority of engineers are at the conservative pole of this contrast, following the historically established path of development of techniques, devices and systems that are capital intensive and sophisticated. This polarization is both unnecessary and undesirable. Engineers with a sensitivity to all of the needs of society feel a compulsion to develop appropriate technology free of ties to a particular social or political attitude. An appropriate technology is by definition an optimal one, giving due consideration to the constraints already mentioned. It will sometimes mean small watermills constructed from local materials with maximum utilization of local skills. But it will also mean sophisticated water and wastewater treatment systems, horizontal cement kilns, and heavy-duty trucks when concentrations of population and demand favor such technologies.

At the present time the fact that the Appropriate Technology movement is based in the wealthiest capitalist societies presents a danger that it will degenerate into a new form of cultural imperialism or at least a technological paternalism. To the extent that it is a disaffected reaction to the environmental consequences of industrialized and urbanized capitalism it could become an attempt, however well-meaning, to prevent the Third World from experiencing the high-consumption wealth that has recently turned sour in the countries with high standards of living. If this were to happen one could not blame the Third World for resenting it. If Third World nations have something to learn from the social and ecological disorders in their more affluent industrialized neighbors in our "global village" it would be best for them to make the discoveries themselves rather than to be instructed by the rich to avoid the discomforts of wealth.

The other side of this coin is that technologists must be broad-minded in reviewing alternatives for decision. In the Third World, and perhaps elsewhere, an optimal decision may mean adoption of designs or processes that have been previously considered superseded or substandard. Simple inefficient devices may justify preference over more modern designs because of higher local content in materials and labor, or because of easier maintenance. Despite the professional bias that resists it, engineers will need to admit that sometimes a crude design is nevertheless the best choice. The bias stems from the store of conventional wisdom that is part of the engineer's subconscious thought in making decisions using judgment to chose between competing alternatives.

CONCLUSIONS

It has been emphasized that engineers earn their living by evaluating needs and solutions to those needs, and by choosing wisely so that the best solution is put into effect. This is true of the design of a facility or a system, and the planning and execution of its construction.

In searching for the best solution the decision maker is subject to many influences, some of which are explicit and even quantitative, while some are implicit and call upon a background of education and experience of which the individual concerned is scarcely conscious. These latter influences are sometimes referred to as "engineering judgment."

One's judgment is conditioned to a considerable extent by the culture within which experience and education have been gained. Consequently the influence of judgment on decisions made in a different culture may be far from optimal if all costs and benefits are given proper evaluation.

The responsibility of an engineer is to learn to understand not only the inanimate elements with which he works, but the culture of the people he serves so that his decisions are sound, successful, and optimal.

Appendix.—References

1. Crosthwaite, C. D., and Bergan, A., "The Needs of Developing Countries in Particular So Far As Design and Construction Are Involved," *Problems Associated with Design and Construction in Developing Countries*, International Association for Bridge and Structural Engineering Zurich, 1977, p. 27.
2. Dickson, D., *Alternative Technology and the Politics of Technical Change*, Fontana/Collins, Glasgow, Scotland, 1974.
3. Feachem, R., "The Rational Allocation of Water Resources for the Domestic Needs of Rural Communities in Developing Countries," *Proceedings*, 2nd World Congress, International Water Resources Association, New Delhi, India, Vol. II, Dec., 1975, p. 539.
4. Kluckholn, C., and Kelly, W., "The Concept of Culture, " *The Science of Man in the World Crisis*, R. Linton ed., Columbia University Press, New York, N.Y., 1945, p. 78.
5. Lynch, R. L., "Traffic Accident Study, Western Highlands Province, July 1974–June 1975," *Research Report R33/76*, Papua New Guinea University of Technology, Lae, Papua New Guinea May 1976.

Schumacher, E. F. *Small Is Beautiful*, Blond and Briggs, London, England, 1973.

6.4

Wheels (excerpt)

ARTHUR HAILEY

They walked along a basement passageway, then, selecting a key from several on a ring, Hank Kreisel opened a gray metal door. As they entered the room beyond, bright fluorescent lighting flooded on.

They were, Adam saw, in an engineering experimental workshop. It was spacious, organized, among the best-equipped of its kind that he had seen.

"Spend a lot of time in this place. Do pilot stuff," Kreisel explained. "When new work comes up for my plants, bring it down here. Then figure out best way of production at cheapest unit cost. Pays off."

Adam remembered something which Brett DeLosanto had told him: that Hank Kreisel had no engineering degree, and his only training before beginning business for himself was as a machinist and plant foreman.

"Over here." Kreisel led the way to a low, wide work table. An object on it was covered by a cloth which he removed. Adam looked curiously at the metal structure underneath—an assemblage of steel rods, sheet metal, and connected internal parts, the size about equal to two bicycles. On the outside was a handle. As Adam turned it, experimentally, parts within the structure moved.

Adam shrugged. "Hank, I give up. What the hell is it?"

"Obviously," Erica said, "it's something he's submitting to the Museum of Modern Art."

"Maybe that's it. What I ought to do." Kreisel grinned, then asked, "Know much about farm machinery, Adam?"

"Not really." He turned the handle once again.

Hank Kreisel said quietly, "It's a threshing machine, Adam. Never been one like it, or this small. And it works." His voice took on an enthusiasm which neither Adam nor Erica had heard before. "This machine'll thresh any kind of grain—wheat, rice, barley. Three to five bushels an hour. Got pictures proving it . . . "

"I know enough about you," Adam said. "If you say it works, it works."

"Something else works, too. Cost. Mass-produced, it'd sell for a hundred dollars."

Adam looked doubtful. As a product planner, he knew costs the way a football coach knows standard plays. "Surely not including your power source." He stopped. "What *is* your power source? Batteries? A small gas motor?"

"Thought you'd get around to that," Hank Kreisel said. "So I'll tell you. Power source isn't any of those things. It's some guy turning a handle. Same way *you* did just now. Same handle. Except the guy I'm thinking of is an old Eastern geezer in a jungle village. Wearing a slope hat. When his arms get tired, a women or a kid'll do it. They'll sit there, hours on end, just turn the handle. That's how we'll build this for a hundred bucks."

"No power source. Too bad we can't build cars that way." Adam laughed.

Kreisel told him, "Whatever else you do. Do me a favor now. Don't laugh."

"Okay, I won't. But I still can't see mass-producing, in Detroit of all places, a piece of farm machinery"—Adam nodded toward the thresher—"where you *turn a handle, for hours on end,* to make it work."

Hank Kreisel said earnestly, "If you'd been to places where I have, Adam, maybe you would. Parts of this world are a long way from Detroit. That's half our trouble in this town: we forget those other places. Forget that people don't think like we do. We figure everywhere else is like Detroit, or ought to be, so whatever happens should be our way: the way we see it. If others see different, they have to be wrong *because we're Detroit!* We've been like that about other things. Pollution. Safety. Those got so hot we had to change. But there's a lot more thinking left that's like religion."

"With high priests," Erica put in, "who don't like old beliefs challenged."

Adam shot her an annoyed glance which said: *Leave this to me.*

He pointed out, "A good many who are moving up in the industry believe in rethinking old ideas and the effect is showing. But when you talk about a hand-operated machine—any kind of machine—*that* isn't a forward change; it's going backward to the way things were before the first Henry Ford." He added, "Anyway, I'm a car and truck man. This is farm machinery."

"Your company has a farm products division."

"I'm not involved with it, and don't expect to be."

"Your people at the top are. And you're involved with them. They listen to you."

"Tell me something," Adam said. "Did you put this up to our farm products people? Did they turn you down?"

The parts manufacturer nodded affirmatively. "Them and others. Need someone now to get me in a board room. So I can raise interest there. Hoped you'd see it."

At last it was clear precisely what Hank Kreisel wanted: Adam's help in gaining access to the corporate summit of his company, and presumably the ear of the president or chairman of the board.

Erica said, "Can't you do it for him?"

Adam shook his head, but it was Hank Kreisel who told her, "He'd have to believe in the idea first."

They stood looking at the contraption with its handle, so alien to everything in Adam's own experience.

And yet, Adam knew, auto companies often did become involved in projects having little or nothing to do with their principal activity of producing cars. General Motors had pioneered a mechanical heart for use in surgery, and other medical devices. Ford was working on space satellite communication, Chrysler dabbling in planned communities. There were other examples, and the reason for such programs—as Hank Kreisel shrewdly knew—was that someone high in each company had taken a personal interest to begin with.

"Been down to Washington about this thresher," Kreisel said. "Sounded out a lot of guys in State. They go for this. Talk of ordering two hundred thousand machines a year for foreign aid. It'd mean a start. But State Department can't do manufacturing."

"Hank," Adam said, "why work through another company at all? If you're convinced, why not build and market this yourself?"

"Two reasons. One's prestige. I don't have the name. Big company like yours does. Has the marketing setup, too. I don't."

Adam nodded. That much made sense.

"Other reason is finance. I couldn't raise the dough. Not for big production."

"Surely, with your track record, the banks . . ."

Hank Kreisel chuckled. "I'm into the banks already. So deep, some days they think I held 'em up. Never had much cash of my own. Surprising what you can do without it."

Adam understood that, too. Plenty of individuals and companies operated that way, and almost certainly Hank Kreisel's plants, their equipment, inventories, this house, his place at Higgins Lake, were mortgaged heavily. If Kreisel ever sold his business, or a part of it, he could reap millions in cash. Until he did, like others he would continue month by month with cash flow problems.

Again the parts manufacturer turned the thresher handle. Inside, the mechanism moved, though accomplishing nothing now; what it needed was grain to bite on, fed into a quart-size hopper at the top.

"Sure this is offbeat. Could say it's been a dream with me. Had it a long time." Hank Kreisel hesitated, seeming embarrassed by the admission, but went on, "Got the idea in Korea. Watched guys 'n dames in villages, pounding grain with rocks. Primitive: lots of muscle, small results. Saw a need, so started figuring this gizmo. Worked on it, on and off, ever since."

Erica was watching Hank Kreisel's face intently. She, too, knew something of his background, having learned it partly from Adam, partly elsewhere. Suddenly a picture took shape in her mind: of a tough, hard-fighting United States Marine in an alien, hostile land, yet observing native villagers with such understanding and compassion that, years afterward, an idea born at that time could stay with him like a flame.

"Tell you something, Adam," Kreisel said. "You too, Erica. This country's not selling farm machinery overseas. Leastways, not much. Ours is too fancy, too sophisticated. It's like a religion with us—the way I said: everything has to be powered. Must be electric, or use an engine, or whatever. What's forgotten is, Eastern countries have unending labor. You call for a guy to turn a handle, fifty come hurrying like flies—or ants. But we don't like that idea. Don't like to see dams built by coolies carrying stones. Idea offends us. We figure it's inefficient, not American; we say it's the way the pyramids were built. *So what?* Fact is: situation's there. Won't change for a long time, if ever. Another thing: out there, not many places to repair fancy machinery. So machines need to be simple." He took his hand away from the thresher whose handle he had continued turning. "This is."

Adam thought: Strangely, while Hank Kreisel had been speaking—eloquently for him—and demonstrating what he had built and believed in, he had a Lincolnesque quality which his tall, lean figure emphasized.

Would the idea work, Adam wondered? *Was* there a need, the way Hank Kreisel claimed? *Was* it a worthwhile project to which one of the Big Three auto companies might lend its world prestige?

Adam began firing questions based on his product planner's training in critical analysis. The questions embraced marketing, expected sales, distribution, local assembly, costs, parts, techniques for shipping, servicing, repair. Each point Adam raised, Kreisel seemed to have thought of and been prepared for, with the needed figures in his brain, and the responses showed why the parts manufacturer's own business had become the success it was.

Selected Additional Readings

"Technology Transfer: Toward a Redirection," *Science News*, November 27, 1976, pp. 342–343.

Appropriate Technologies in Civil Engineering Works in Developing Nations: An Exploratory Appraisal of the State of the Art, prepared by the Department of Economics and Social Affairs, United Nations, New York, 1976.

E. K. Bryant, "Environmental Impact of the Aswan High Dam," *The Military Engineer*, May–June 1980, pp. 181–183.

P. H. DeForest, "The Appropriate Technology Debate in the United States: Social Values and Public Policy," A paper presented at the Annual Meeting of The American Society of Mechanical Engineers, New York, December 2–7, 1979.

D. Dickson, *The Politics of Appropriate Technology*, Universal Books, 1975.

R. C. Dorf and Y. L. Hunter (editors), *Appropriate Visions*, Boyd & Fraser Publishing Company, San Francisco, 1978.

A. Holmberg, "The Wells That Failed," *Human Problems in Technological Change—A Casebook*, H. Spicer (editor), Russell Sage Foundation, New York, 1952, pp. 113–123.

C. Pursell, "The History of Technology as a Source of Appropriate Technology," *Public Historian*, winter 1979, pp. 15–22.

N. Rubin and W. M. Warren (editors), *Dams in Africa*, Augustus M. Kelley, New York, 1968.

C. Shirley, "Superdams," *The Atlantic*, June 1972, pp. 35–41.

PART 7

The Techno-Human World

That which did all this was mind, because man did it with intelligence; it was not fate, because they did it by choice.

Giambattista Vico, The New Science

All things must change to something new, to something strange.

Henry Wadsworth Longfellow

John Dewey's maxim that the true purpose of all knowledge lies in making choices among alternatives, based on future consequences (1), is an appropriate theme for this concluding part. These "choices," so important to the engineer, will be our major concern.

In the last two decades some of the negative effects of technology have stirred public reaction: air and water pollution and the question of the safety of nuclear energy have all been at the focus of national attention. In this part we shall consider several aspects of the impact of technology on our world: humanity's place within the Earth's ecosystems; direct and indirect effects of technology; and ways to assess technology's future impact.

The first article, "The Human Ecosystem," by Marston Bates, a well-known zoologist and author of popular works on science, sets the stage for the part's discussion by defining our world from a "systems" point of view: all living organisms and their physical environment form interacting systems (2). Thanks to technology, humans have developed the ability to work on a larger scale than any other animal. As a result, the human capacity for influence on other delicate systems is immense. Bates closes by asking whether our biosphere can withstand "the constant growth of populations and new technology" the human species imposes on it.

The eminent medieval historian, Lynn White, Jr., puts the problem of ecological crisis into a historical perspective in the second selection. He examines what human values (specifically Christian beliefs) tell us about people's relationship with the environment. White finds the Christian view ("nature has no reason for existence save to serve man") has been largely responsible for our own current ecological problems (3).

Ecologists and nonecologists alike can agree that it is frequently difficult to foresee all the ramifications of a technological innovation. The newspaper article by Isaac Asimov (author of over 200 books about science, religion, science fiction, and popular history) contains many examples to show that even the most beneficial invention may have unpleasant side effects. Still he argues that we cannot abandon some of our industrial technology, "since it all hangs together": new discovery must help us solve old problems.

In a sophisticated discussion, J. Weizenbaum, professor of computer science at M.I.T., uses the example of the impact of the computer on society to illustrate that the indirect, subtle effects of technological innovation are often more significant than the direct, obvious effects. An important part of Weizenbaum's argument revolves around the phenomenon of the "technological mentality" that he sees as a dangerous by-product of our technological world: our thought processes are so thoroughly pervaded by technological metaphors that we allow technology to formulate the important questions, thus obscuring the real issue—"the behavior of the whole man."

Public concern about the dangerous side effects of technology has been a feature of American life since at least the first quarter of the nineteenth century (4). In response to a long outcry over steam boiler explosions on steamboats, the federal government established its first regulatory agency in 1852. Since then it has become an accepted fact that the government has a duty to act in the interest of public safety. The culmination of government interest in regulating technological development came in 1972 with the establishment of the Office of Technology Assessment (a method of anticipating the impact of a product of technology before going ahead with producing it) (5).

The next three selections by Lindbergh, Harrison, and Ramo are drawn from a compendium of papers published by the House Committee on Science and Technology (1979). They deal with science and technology policy. In the Charles A. Lindbergh letter, the great aviator and conservationist expresses, to a former chairman of the House committee, his belief that we need regulations and laws to ensure that science and technology work for the "basic benefit of mankind." Anna Harrison, a professor of chemistry and former president of the American Chemical Society, argues for the use of risk-benefit analysis to determine the "net value to a society of a given technological development," and views the Committee on Science and Technology as the public's surrogate in making important decisions regarding the role of science and technology in society. Finally, S. Ramo, author of many technical books and articles (and from 1954–1958 chief scientist of the U.S. intercontinental ballistic missile program), makes the case for creative cooperation between the government and the free-enterprise system, so technology can benefit society to the utmost.

We have been discussing the problem of technological impact on later twentieth-century earth. Perhaps the most urgent problem our rampant technology poses for us is whether technological possibility necessarily means desirability. A provocative article by J. C. Wallace, "The Engineering Use of Human Beings," raises the question and asks what the role of the engineer is in deciding what technology should come to pass (6). His conclusion is challenging:

> I would say that if we can educate the engineer to the critical need for directing the technology in the interests of preserving and enhancing certain values within the society, and if we can discourage him from altering the society at the expense of those values, then we will have accomplished a great thing. And we will have moved to a higher plateau of human endeavor.

Morris (7) ties the efforts of the engineer to those of several other service professions in the second half of this century.

In closing, we return for a moment to the computer because this invention brought a new dimension to the study of technological impact. Weizenbaum (8) reminds us that those who work on artificial intelligence can now contemplate

> nothing less than to build a machine on the model of man, a robot that is to have its childhood, to learn language as a child does, to gain its knowledge of the world by sensing the world through its own organs, and ultimately to contemplate the whole domain of human thought.

The choice open to the technologist is this: to learn everything he can about humans from the study of the computer or to allow the computer to define a human as an information-processing system and no more. Is a human only a machine? Now, more than ever, it is important for engineers to ask what it means to be a human being.

The final selection in the work, a public lecture by M. Kranzberg, the noted science historian, provides a useful review of questions we raise in this book about the interface between the humanities and engineering.

References

1. We owe the John Dewey maxim to Wenk, E., "Engineering in the 80's," *The Bridge* (National Academy of Engineering), spring 1980, p. 21.
2. See Vaissière, R., "A Model Marine Ecosystem: 24 Hours in the Life of a Rock-Pool," in J. Benthall (editor), *Ecology in Theory and Practice*, Viking, New York, 1972, pp. 101–113.
3. For a review of antitechnological thinking see Florman, S. C., *The Existential Pleasures of Engineering*, St. Martin's, New York, 1976, pp. 45–56.
4. See Burke, J. G., "Bursting Boilers and the Federal Power," *Technology and Culture*, winter 1966, pp. 1–23.
5. On the history of technological assessment see Pytlik, E. C., D. P. Lauda, and D. L. Johnson, *Technology, Change and Society*, Davis Publications, Worcester, Mass., 1978, pp. 266–280. For a negative evaluation see Drucker, P. F., "The Futility and Dangers of Technological Assessment," in J. G. Burke and M. C. Eakin (editors), *Technology and Change*, Boyd and Fraser, San Francisco, 1979, pp. 449–453.
6. J. C. Wallace, "The Engineering Use of Human Beings," *Technology and Culture*, January 1969, p. 19.
7. For an overview of how other professionals see engineering relating to their work, vide Morris, M. D., et al., *Civil Engineers in the World around Us*, American Society of Civil Engineers, New York, 1974.
8. For a thorough discussion of this important problem, see Weizenbaum, Joseph, *Computer Power and Human Reason*, W. H. Freeman, San Francisco, 1976. The quote is from pp. 202–203.

7.1

The Human Ecosystem

MARSTON BATES

In the West, our desire to conquer nature often means simply that we diminish the probability of small inconveniences at the cost of increasing the probability of very large disasters.

Kenneth E. Boulding, 1966, p. 14

Our planet has been aptly called "Spaceship Earth." It forms, overwhelmingly, a closed system as far as materials are concerned. Science fiction to the contrary, we have no present basis for believing that this essential isolation will be altered—that we can colonize other parts of the solar system or import from outer worlds any appreciable quantities of materials. This earth is our habitat and probably will be as long as our species survives. We would do well, then, to treat it carefully and to take thought in planning our actions.

Early men—the food-gathering and hunting peoples of the Old Stone Age—were closely interacting parts of the biological communities in which they lived. Tools and language made them unusually efficient hunters but not really different in their impact on the community from other kinds of social carnivores or omnivores. The first major change in man's relations with nature came with the deliberate making of fire. No other animal starts fires. Their frequency must have increased greatly when man began setting them, with far-reaching ecological effects.

Man's relations with the rest of nature underwent a much greater change with what anthropologists call the "Neolithic Revolution"—although "revolution" is perhaps a poor word for changes that may have taken millenia and that occurred at different times and in different sequences in various parts of the world. The important changes, which were at least in part interrelated, involved the cultivation of plants, the domestication of animals, the settlement of villages, the making of pottery, and a series of improvements in toolmaking. Man began to alter the biological community in which he lived by removing vegetation he did not want or could not use and replacing it with crops.

Gordon Childe (1941), the British anthropologist, looked at post-Neolithic developments in terms of two further "revolutions"—the Urban, turning on the transport and storage of food, which made possible the formation of cities and the specialization of labor and knowledge, and the Industrial, based on the harnessing of power other than the muscles of men or beasts. C. P. Snow (1959) would add a recent revolution, the Scientific, resulting from the union of science and technology for the solution of practical problems—the revolution that gave rise to the dramatic and drastic increase in power available to man in the twentieth century, including the development of nuclear and electronic techniques. The Scientific Revolution in particular has the potential to bring about even more sweeping worldwide ecological changes than we have seen in the past.

The idea of looking at cultural development in terms of a few relatively abrupt "revolutions" is obviously an oversimplification. Yet equally obviously, man's history does not show a smooth and gradual increase in knowledge and power. Shifts in techniques and ideas have resulted in periods of rapid change; and there have been long periods in the history of every culture in which neither ideas nor ways of life have changed much—periods of temporary near-equilibrium, or "stagnation" as some would say. The charting and description of these developments are matters for historians: and, although they may never reach agreement, their efforts are nonetheless thought-provoking in the attempts that each of us make to understand ourselves and our world.

Whatever the history, whatever the interplay of cause and effect, the result is the curious paradox of man as a part of nature, and man as apart from nature—a force of geological magnitude changing the face of the earth. Many of the activities that we think of as peculiarly human have parallels in other animals, but man has come to work on a different scale. Compare, for instance, Hoover Dam with the work of beavers, or Manhattan Island with a gopher town. The difference in scale is so great that we can only regard it as a difference in kind. In a somewhat comparable way, the human ecosystem has become a different sort of phenomenon from anything else that we know.

The world of life—the biosphere—can be looked at as a single, interconnected though endlessly diversified system. There is no denying the diversity. A desert and a forest are different kinds of places however difficult it may be to delineate their boundaries. Yet the many different kinds of biological communities that make up the biosphere display a very real underlying unity. Life everywhere is organized on the same basic principles.

Essentially all life as we know it depends on the transformation of radiant energy from the sun into the chemical energy of hydrocarbons through the process of photosynthesis based on chlorophyll. The chlorophyll-bearing organisms of the seas are mostly microscopic floating algae of the surface waters where sunlight penetrates. Such algae are probably also more important in fresh water than are the fixed plants of the shallows. But on land the "vascular" plants, the herbs, grasses, shrubs, and trees, are overwhelmingly important. These organisms that carry on the photosynthetic process are called *producers* by ecologists. The animals that live off them directly or indirectly are the *consumers.*

The British ecologist Charles Elton (1936) has called the animals that live directly off the plants "key industry animals," because all the rest of the animal system depends on them. They can also be called "first-order consumers," which in turn are eaten by "second-order consumers," and so on. This leads to the idea of a food chain—grass, grasshoppers, frogs, snakes, hawks—which is a neat idea, but vastly oversimplified, because any attempt to diagram who eats whom in a biological community results in a complicated network of lines more appropriately called a "food web."

Most ecologists would add a third category of organisms, the *decomposers:* bacteria, fungi, and the like, that cause rot and decay, reducing the corpses of dead animals and plants to dust (or mud). The component chemicals are then available to be used again—carbon from the carbon dioxide of the air combined with oxygen from water through photosynthesis and the other necessary chemicals through absorption by plants or digestion by animals.

Most vegetable material never passes through the animal system at all, as is obvious enough in a glance at the uneaten leaves of any forest or prairie. This economy of abundance is necessary for the functioning of the biological community. When, because of some upset in the balance of the system, the forest is defoliated or the prairie overgrazed, the consequence is catastrophe for both the plants and the animals. In general, throughout nature, animal populations do not multiply up to the limit of their food supply—other controlling factors intervene. Plants, however, under favorable circumstances, may occupy all available space.

Most of the key industry animals (first-order consumers) are small, and often individuals of a given species are very abundant—insects on land, and minute crustaceans in water. Yet some of the largest of animals are herbivores—the elephants of today and the giant mammals, reptiles, and even birds of times past. Generally, however, the herbivores are more numerous than the predators that feed on them; and various studies have shown that, with each step away from the producer plants, only six to ten percent of the available energy is transmitted. It is not possible for animals, either in terms of energy or numbers, to operate beyond the fourth- or fifth-order consumer level.

What limits the number of individuals of a given species of animal? In 1798, Thomas Robert Malthus published his little book, *An Essay on the Principle of Population as It Affects the Future Improvement of Mankind.* Concern with the human population problem can be neatly dated from the publication of this book. An enlarged and much revised edition, published in 1803, states his basic propositions thus:

1. Population is necessarily limited by the means of subsistence.
2. Population invariably increases where the means of subsistence increase, unless prevented by some very powerful and obvious checks.
3. These checks, and the checks which repress the superior power of population and keep its effects on a level with the means of subsistence, are all resolvable into moral restraint, vice and misery.

This has been called the "dismal theorem" of Malthus, and there are those who would exorcise it with ridicule. But, dismal though it may seem, its basic validity remains unaffected either by man's ingenuity in finding means for the increase of his own subsistence or the likelihood that he will continue to achieve such increases up to some limit not yet reached. The theorem, of course, was intended to apply to people: "moral restraint" and "vice" can hardly refer to any animal except man. Under "misery," however, Malthus included such things as disease and starvation, which are not restricted to humans. In any event, it is instructive to consider his propositions in the context of the behavior of animal populations in general and not just that of *Homo sapiens*. Food is certainly the ultimate limit on the population of any animal, whether we think in terms of local or global populations. Yet, as noted above, populations in nature very rarely multiply up to the limit of the food supply. What, then, are the checks that normally limit animal populations? What prevents a population from multiplying until it reaches the point at which its food supply is exhausted? The search for the answers to such questions has led to a great deal of experiment, observation, and speculation. It is clear that there is no single kind of check, no simple answer—that we are dealing with a complex system of checks and balances that we cannot yet describe completely, let alone fully understand (e.g., Dawson, 1968).

Perhaps the most general check is the so-called balance-of-nature resulting from the fact that animals not only eat but are eaten. Herbivores tend to be controlled by carnivores before they reach the point of exhausting food supply—as has been amply suggested by the damaging multiplication of such animals as deer or rabbits when man has removed their "enemies" such as pumas, wolves, or foxes. The carnivores may be limited in turn by other carnivores that live on them. The young of all animals are particularly susceptible to predation and to death through accident. And the most lordly carnivore may fall victim to parasitic disease. Disease particularly tends to be "density-dependent," becoming more prevalent as individuals of a particular host species become more common.

The whole system has a great deal of flexibility, of "play." A given predator in a particular region may concentrate largely on some one species of prey until this becomes rare and hard to find, then shift its attention to other species. More commonly, a predator may eat a variety of animals: but since the most abundant kinds would be most often caught, there is a sort of automatic limit on abundance for any given species.

The availability of food, however, is not the only factor limiting population size. A population may be limited by the availability of suitable breeding and rearing sites or by other special requirements of habitat. We are gradually learning about a variety of mechanisms that operate within a population to limit its growth. These include such things as cannibalism and failure to reproduce because of crowding, endocrine stress, and so on. Perhaps the most extensively studied of these limiting factors is territoriality. With many kinds of animals, especially fishes, lizards, birds, and mammals, an individual, a pair, or a social group may inhabit a particular area and defend it against intrusion by other individuals of the same species. Different kinds of territoriality have been described. All serve to space individuals; but where the defended area is large enough to include more food than can be utilized by the defenders, territorial behavior keeps the number of individuals below the theoretical limit imposed by the means of subsistence. The limit thus can be availability of appropriate space as well as food or other controls.

There is thus a constant turnover of individuals within a biological community, but with mortality balanced by reproduction so that the result, over a period of time, is a fairly steady state—a kind of dynamic equilibrium in the functioning of the community. This is comparable to the equilibrium that is maintained by the body of an individual: cells dying and being replaced, various organ systems working together harmoniously through an elaborate system of nerve and endocrine controls. This tendency to maintain a steady state despite environmental changes, stresses, and shocks (called *homeostasis*) is expressed in various natural population controls and the resulting balance of nature (Slobodkin, 1962; Slobodkin, Smith, and Hairston, 1967).

Ecologists find it convenient to use, as their unit of study, the *ecosystem* rather than only the biological community, thus taking into account both the living organisms and their physical environment, which together form an interacting system. Every organism is affected by the conditions of the world in which it lives, but every organism also has some effect on these conditions, however trivial. The kind of forest growing in a particular region is in part a consequence of the soil, climate, and water supply of that region; but the kind of soil is also in part a consequence of the type of forest—coniferous, hardwood, or other—and the presence of the forest has a measurable influence on local climatic conditions and water supply.

The reciprocal interactions between organisms and environment are particularly striking in the case of the human species. Our activities are influenced in many ways by the nature of the physical setting in which we live—coastal, inland, desert, mountain, or forest. But we are also capable of altering the environment with unprecedented speed and effect. We would do well, then, to think in terms not of modern man in dominant relationship with other biological communities but of the human ecosystem—of the man-altered landscape and its biological components.

Man's actions can be looked at as efforts to simplify the biological relationships within the ecosystem to his own advantage. By clearing land and planting crops or orchards a complex of mixed species of wild plants may be replaced by a single kind of plant, a monoculture, which may extend over a wide area. In living off grain or fruit or tubers, man functions as a first-order consumer. In largely vegetarian societies he is, thus, a "key industry animal," which means that a large population can be supported—but he is also a dead end, not giving support in turn to the usual predators and scavengers. With modern medicine he has even largely defeated the parasites.

Man's food web, in such societies, is thus reduced to a simple producer-consumer interaction, often with the decomposer system greatly modified as plant growth is maintained by adding fertilizers to the soil. Man also tries as far as possible to reduce or eliminate competition, controlling the insect pests of the crops and attempting to eliminate vertebrate competitors, whether crows, rats, or raccoons.

As a meat-eater, man becomes a second-order consumer, growing grain for his chickens or hogs or pasture for his cattle and then eating the animals. Again there is an attempt to eliminate competition from hawks, weasels, big cats, or wolves; again there is a vastly simplified food relationship, unique to the human ecosystem.

These simplified food relationships are efficient as long as they can be sustained, and they now support the large and expanding human population. In fact, they could be regarded (in conjunction with medical reduction of the death rate) as a cause of this population growth, in terms of the second of the Malthusian propositions. There is no doubt about the effectiveness of the system to support an exploding human population *up to a point*, but it carries the danger of inherent imbalance and some large questions of ethics and aesthetics.

The danger in the simplified ecosystem is in its liability to catastrophe (e.g., Ehrlich, 1968, p. 4). The most complex of natural ecosystems (in the sense of those including the largest numbers of different kinds of organisms) are the most stable in that they are the least liable to great fluctuations in number of individuals of a particular species from year to year. Greatest contrast observed is between complex biotas such as those of the tropical rain forest or coral reef and the relatively simple biotas of the arctic tundra or northern forests. Records of fluctuations in various tundra animals suggest either a cyclic Malthusian relationship with food supply or catastrophic physiological or psychological controls. Such cyclic fluctuations are unknown in the rain forest, where the complexity of relations makes for flexibility and results in a relatively steady state among limited populations of the many species there.

Charles Elton (1958) has summarized six lines of evidence for the relative stability of complex communities. First there is the mathematical argument: "models" of simple predator-prey population relations show conspicuous fluctuations. As Elton says, "Put in ordinary language, this means that an animal community with only two such species in it would never have constant population levels, but would be subject to periodic 'outbreaks' of each species." The second line of evidence comes from attempts to test these simple relationships under laboratory conditions. "One thing stands out from the results: it is very difficult to keep small populations of this simple mixture in balance, for not only do they fluctuate but one or both of the species is liable to become extinct." Greater stability can be achieved by arranging for complex experimental conditions that provide cover for the prey to hide and dodge about in; but it is still difficult to maintain prey-predator populations for any length of time.

"The third piece of evidence," to quote Elton again, "is that natural habitats on small islands seem to be much more vulnerable to invading species than those of the continents." Animals and plants, accidentally or intentionally brought into simple island communities by man, often undergo catastrophic population explosions. Hawaii has seen many examples of this. The efforts at cure have involved attempts to restore a steady state by increasing the complexity of the system through the introduction of parasites and predators of the invading aliens.

"The fourth point is that invasions and outbreaks most often happen on cultivated or planted land—that is, in habitats and communities very much simplified by man." Our crops are notoriously subject to damage by pests, both native and foreign, which must be kept in control by chemical means—the chemicals sometimes producing unexpected side effects.

Elton's fifth line of evidence is based on the contrast in stability between tropical and arctic communities, as mentioned above. His final point involves recent research on orchard pest control. "Orchards are especially good for testing the effects of ecological variety, because they are half-way between a natural woodland and an arable field crop—less complex than the wood but more complex than the crop, and more permanent." Attempts to control orchard pests have disclosed many unexpected relationships among the populations involved. Elton quotes an observer as remarking: "We move from crisis to crisis, merely trading one problem for another."

Analogy can be made between the value of diversified ecosystems and the value of diversified national economies: where the economy depends largely on a single crop or a single industry, the danger of disruption is always greater than in situations where the economy has a diversified base. Large areas devoted to a single crop, covered by a single species of plant, or inhabited by concentrated numbers of a particular kind of animal, always present conditions conducive to the development of epidemics. History affords many examples. One of the most dramatic and best documented involves the Irish potato blight of 1845 (Bates, 1955). The Irish population provided a neat demonstration of the validity of the Malthusian propositions within something like a closed system by growing rapidly, following introduction of the potato, to more than eight million at the time of the census of 1841. When the crop failed in 1845 and again in 1846 because of the sudden appearance of the blight, more than a million people died directly of starvation, and as many as could emigrated. The shock in this case resulted in an eventual stabilization of the Irish population, which has remained at a level of about four million for the last fifty years—half of the preblight figure. In Malthusian terms, "moral restraint" (aided by migration) has checked the tendency to multiply up to the limit of the means of subsistence again.

Modern man, thus, depends on the simplified food relations of intensive agriculture, and modern technology has been efficient in developing methods of pest control. But the danger is always there—of a new pest, of immunity to chemical methods of control, and of environmental pollution from the pesticides. It would seem most prudent, therefore, to preserve as much of the natural diversity as possible, as is done in densely settled Europe with the hedgerows—at least until we understand the system well enough to foresee, guard against, and repair the consequences of possible oversimplification.

The aesthetic argument depends on the impression that a varied landscape is more pleasing, more satisfactory for living, than a monotonous one. This may be a matter for debate; and it carries over from landscapes to the general question of the value of diversity in styles of life, in ways of thinking and acting. From diversity comes the possibility of change, of adaptive response to new conditions, of development and evolution.

There would thus seem to be a need, in planning, to maintain as much diversity as possible within the human habitat. There is also a need to preserve substantial areas in which disturbance by man is kept at a minimum—the concept of nature reservations and national parks. Such parks serve at least four different and at times conflicting purposes: the recreation function, providing the possibility of escaping temporarily from the all-encompassing man-altered landscape; the research function, allowing the continued study of natural ecosystems; the museum function, preserving for future generations adequate samples of the diversified biosphere that has developed on our planet; and, perhaps most important of all, a reservoir function, whereby are sustained the organisms from which destructively altered environments may be restored.

The ethical question has been little explored by our philosophers. We have the power to alter the landscape and to exterminate species of animals and plants that we do not like or that we do not find convenient to our purposes. But do we have the right? Geological history is, of course, in part a record of the disappearance and replacement of organic types. But the rate of extinction increased enormously following the evolution of man as an ecologic dominant—and the rate is accelerating with the growth of human numbers, human power, and human intervention. Elton mentions a remark from Albert Schweitzer's book *My Life and Thought*, pointing out that a fault of ethical philosophy has been that it has dealt only with the relations of man to man. And Aldo Leopold has decried the same oversight in terms of our need to develop an "ecological conscience."

Problems of immediate practicality and utility may override such aesthetic and ethical considerations locally or in emergencies; but the three in large measure coincide. There is need periodically to take stock of and to reconsider our behavior in terms of our own survival. As the demographers have shown, we are rapidly multiplying toward the spatial limits of our earth and toward a possibly tragic final test of the truth of the dismal propositions developed by Malthus. Our industrial society is not sufficiently aware of its needs to recycle the very materials on which its survival depends. We are changing the composition of air, water, and soil in ways that are to some extent irreversible and that have already in places reached disastrous proportions.

In very large measure, we have the knowledge we need to live harmoniously within the confines of our "Spaceship Earth" if we do not insist on the constant growth of populations and new technology. The question that looms threateningly over the future of our species is, will we gain the wisdom to use this knowledge and to generate the new knowledge needed to evaluate and to cope with the environmental consequences of advancing technology?

References

Bates, M. 1955. *The prevalence of people*. New York: Scribner.

Boulding, K. 1966. *Human values on the spaceship earth*. New York: National Council of Churches of Christ in the U.S.A.

Childe, V. G. 1941. *Man makes himself*. London: Watts & Co.

Dawson, P. S. 1968. Xenocide, suicide and cannibalism in flour beetles. *Am. Naturalist* 102(924): 97–105.

Ehrlich, P. R. 1968. World population: a battle lost? *Stanford Today* ser. 1, no. 22, pp. 2–7.

Elton, C. 1936. *Animal ecology*. New York: Macmillan.

Malthus, T. R. 1798. *An essay on the principle of population as it affects the future improvement of mankind*. Facsimile reprint in 1926 for J. Johnson. London: Macmillan & Co.

Slobodkin, L. B. 1962. *Growth and regulation of animal populations*. New York: Holt, Rinehart & Winston.

Slobodkin, L. B. 1968. Aspects of the future of ecology. *BioScience* 18(1): 16–23.

Slobodkin, L. B., F. E. Smith, and N. G. Hairston. 1967. Regulation in terrestrial ecosystems, and the implied balance of nature. *Am. Naturalist* 101: 109–124.

Snow, C. P. 1959. *The two cultures and the scientific revolution*. New York: Cambridge Univ. Press.

7.2

The Historical Roots of Our Ecologic Crisis

LYNN WHITE, JR.

A conversation with Aldous Huxley not infrequently put one at the receiving end of an unforgettable monologue. About a year before his lamented death he was discoursing on a favorite topic: Man's unnatural treatment of nature and its sad results. To illustrate his point he told how, during the previous summer, he had returned to a little valley in England where he had spent many happy months as a child. Once it had been composed of delightful grassy glades; now it was becoming overgrown with unsightly brush because the rabbits that formerly kept such growth under control had largely succumbed to a disease, myxomatosis, that was deliberately introduced by the local farmers to reduce the rabbits' destruction of crops. Being something of a Philistine, I could be silent no longer, even in the interests of great rhetoric. I interrupted to point out that the rabbit itself had been brought as a domestic animal to England in 1176, presumably to improve the protein diet of the peasantry.

All forms of life modify their contexts. The most spectacular and benign instance is doubtless the coral polyp. By serving its own ends, it has created a vast undersea world favorable to thousands of other kinds of animals and plants. Ever since man became a numerous species he has effected his environment notably. The hypothesis is that his fire-drive method of hunting created the world's great grasslands and helped to exterminate the monster mammals of the Pleistocene from much of the globe is plausible, if not proved. For 6 millennia at least, the banks of the lower Nile have been a human artifact rather than the swampy African jungle which nature apart from man, would have made it. The Aswan Dam, flooding 5000 square miles, is only the latest stage in the long process. In many regions terracing or irrigation, overgrazing, the cutting of forests by Romans to build ships to fight Carthaginians or by Crusaders to solve the logistics problems of their expeditions, have profoundly changed some ecologies. Observation that the

French landscape falls into two basic types, the open fields of the north and the *bocage* of the south and west, inspired Marc Bloch to undertake his classic study of medieval agricultural methods. Quite unintentionally, changes in human ways often affect nonhuman nature. It has been noted, for example, that the advent of the automobile eliminated huge flocks of sparrows that once fed on the horse manure littering every street.

The history of ecologic change is still so rudimentary that we know little about what really happened, or what the results were. The extinction of the European aurochs as late as 1627 would seem to have been a simple case of overenthusiastic hunting. On more intricate matters it often is impossible to find solid information. For a thousand years or more the Frisians and Hollanders have been pushing back the North Sea, and the process is culminating in our own time in the reclamation of the Zuider Zee. What, if any, species of animals, birds, fish, shore life, or plants have died out in the process? In their epic combat with Neptune have the Netherlanders overlooked ecological values in such a way that the quality of human life in the Netherlands has suffered? I cannot discover that the questions have ever been asked, much less answered.

People, then, have often been a dynamic element in their own environment, but in the present state of historical scholarship we usually do not know exactly when, where, or with what effects man-induced changes came. As we enter the last third of the 20th century, however, concern for the problem of ecologic backlash is mounting feverishly. Natural science, conceived as the effort to understand the nature of things, had flourished in several eras and among several peoples. Similarly there had been an age-old accumulation of technological skills, sometimes growing rapidly, sometimes slowly. But it was not until about four generations ago that Western Europe and North America arranged a marriage between science and technology, a union of the theoretical and the empirical approaches to our natural environment. The emergence in widespread practice of the Baconian creed that scientific knowledge means technological power over nature can scarcely be dated before about 1850, save in the chemical industries, where it is anticipated in the 18th century. Its acceptance as a normal pattern of action may mark the greatest event in human history since the invention of agriculture, and perhaps in nonhuman terrestrial history as well.

Almost at once the new situation forced the crystallization of the novel concept of ecology; indeed, the word *ecology* first appeared in the English language in 1873. Today, less than a century later, the impact of our race upon the environment has so increased in force that it has changed in essence. When the first cannons were fired, in the early 14th century, they affected ecology by sending workers scrambling to the forests and mountains for more potash, sulfur, iron ore, and charcoal, with some resulting erosion and deforestation. Hydrogen bombs are of a different order: a war fought with them might alter the genetics of all life on this planet. By 1285 London had a smog problem arising from the burning of soft coal, but our present combustion of fossil fuels threatens to change the chemistry of the globe's atmosphere as a whole, with consequences which we are only beginning to guess. With the population explosion, the carcinoma of planless urbanism, the new geological deposits of sewage and garbage, surely no creature other than man has ever managed to foul its nest in such short order.

There are many calls to action, but specific proposals, however worthy as individual items, seem too partial, palliative, negative: ban the bomb, tear down the billboards, give the Hindus contraceptives and tell them to eat their sacred cows. The simplest solution to any suspect change is, of course, to stop it, or, better yet, to revert to a romanticized past: make those ugly gasoline stations look like Anne Hathaway's cottage or (in the Far West) like ghost-town saloons. The "wilderness area" mentality invariably advocates deep-freezing an ecology, whether San Gimignano or the High Sierra, as it was before the first Kleenex was dropped. But neither atavism nor prettification will cope with the ecologic crisis of our time.

What shall we do? No one yet knows. Unless we think about fundamentals, our specific measures may produce new backlashes more serious than those they are designed to remedy.

As a beginning we should try to clarify our thinking by looking, in some historical depth, at the presuppositions that underlie modern technology and science. Science was traditionally aristocratic, speculative, intellectual in intent; technology was lower-class, empirical, action-oriented. The quite sudden fusion of these two, towards the middle of the 19th century, is surely related to the slightly prior and contemporary democratic revolutions which, by reducing social barriers, tended to assert a functional unity of brain and hand. Our ecologic crisis is the product of an emerging, entirely novel, democratic culture. The issue is whether a democratized world can survive its own implications. Presumably we cannot unless we rethink our axioms.

THE WESTERN TRADITIONS OF TECHNOLOGY AND SCIENCE

One thing is so certain that it seems stupid to verbalize it: both modern technology and modern science are distinctively *Occidental*. Our technology has absorbed elements from all over the world, notably from China; yet everywhere today, whether in Japan or in Nigeria, successful technology is Western. Our science is the heir to all the sciences of the past, especially perhaps to the work of the great Islamic scientists of the Middle Ages, who so often outdid the ancient Greeks in skill and perspicacity: al-Rāzī in medicine, for example; or ibn-al-Haytham in optics; or Omar Khayyám in mathematics. Indeed, not a few works of such geniuses seem to have vanished in the original Arabic and to survive only in medieval Latin translations that helped to lay the foundations for later Western developments. Today, around the globe, all significant science is Western in style and method, whatever the pigmentation or language of the scientists.

A second pair of facts is less well recognized because they result from quite recent historical scholarship. The leadership of the West, both in technology and in science, is far older than the so-called Scientific Revolution of the 17th century or the so-called Industrial Revolution of the 18th century. These terms are in fact outmoded and obscure the true nature of what they try to describe—significant stages in two long and separate developments. By A.D. 1000 at the latest—and perhaps, feebly as much as 200 years earlier—the West began to apply water power to industrial processes other than milling grain. This was followed in the late 12th century by the harnessing of wind power. From simple beginnings, but with remarkable consistency of style, the West rapidly expanded its skills in the development of power machinery, labor-saving devices, and automation. Those who doubt should contemplate that most monumental achievement in the history of automation: the weight-driven mechanical clock, which appeared in two forms in the early 14th century. Not in craftsmanship but in basic technological capacity, the Latin West of the later Middle Ages far outstripped its elaborate, sophisticated, and esthetically magnificent sister cultures, Byzantium and Islam. In 1444 a great Greek ecclesiastic, Bessarion, who had gone to Italy, wrote a letter to a prince in Greece. He is amazed by the superiority of Western ships, arms, textiles, glass. But above all he is astonished by the spectacle of waterwheels sawing timbers and pumping the bellows of blast furnaces. Clearly, he had seen nothing of the sort in the Near East.

By the end of the 15th century the technological superiority of Europe was such that its small, mutually hostile nations could spill out over all the rest of the world, conquering, looting, and colonizing. The symbol of this technological superiority is the fact that Portugal, one of the weakest states of the Occident, was able to become, and to remain for a century, mistress of the East Indies. And we must remember that the technology of Vasco da Gama and Albuquerque was built by pure empiricism, drawing remarkably little support or inspiration from science.

In the present-day vernacular understanding, modern science is supposed to have begun in 1543, when both Copernicus and Vesalius published their great works. It is no derogation of their accomplishments, however, to point out that such structures as the *Fabrica* and the *De revolutionibus* do not appear overnight. The distinctive Western tradition of science, in fact, began in the late 11th century with a massive movement of translation of Arabic and Greek scientific works into Latin. A few notable books—Theophrastus, for example—escaped the West's avid new appetite for science, but within less than 200 years effectively the entire corpus of Greek and Muslim science was available in Latin, and was being eagerly read and criticized in the new European universities. Out of criticism arose new observation, speculation, and increasing distrust of ancient authorities. By the late 13th century Europe had seized global scientific leadership from the faltering hands of Islam. It would be as absurd to deny the profound originality of Newton, Galileo, or Copernicus as to deny that of the 14th century scholastic scientists like Buridan or Oresme on whose work they built. Before the 11th century, science scarcely existed in the Latin West, even in Roman times. From the 11th century onward, the scientific sector of Occidental culture has increased in a steady crescendo.

Since both our technological and our scientific movements got their start, acquired their character, and achieved world dominance in the Middle Ages, it would seem that we cannot understand their nature or their present impact upon ecology without examining fundamental medieval assumptions and developments.

MEDIEVAL VIEW OF MAN AND NATURE

Until recently, agriculture has been the chief occupation even in "advanced" societies; hence, any change in methods of tillage has much importance. Early plows, drawn by two oxen, did not normally turn the sod but merely scratched it. Thus, cross-plowing was needed and fields tended to be squarish. In the fairly light soils and semiarid climates of the Near East and Mediterranean, this worked well.

But such a plow was inappropriate to the wet climate and often sticky soils of northern Europe. By the latter part of the 7th century after Christ, however, following obscure beginnings, certain northern peasants were using an entirely new kind of plow, equipped with a vertical knife to cut the line of the furrow, a horizontal share to slice under the sod, and a moldboard to turn it over. The friction of this plow with the soil was so great that it normally required not two but eight oxen. It attacked the land with such violence that cross-plowing was not needed, and fields tended to be shaped in long strips.

In the days of the scratch-plow, fields were distributed generally in units capable of supporting a single family. Subsistence farming was the presupposition. But no peasant owned eight oxen: to use the new and more efficient plow, peasants pooled their oxen to form large plow-teams, originally receiving (it would appear) plowed strips in proportion to their contribution. Thus, distribution of land was based no longer on the needs of the family but, rather, on the capacity of a power machine to till the earth. Man's relation to the soil was profoundly changed. Formerly man had been part of nature; now he was the exploiter of nature. Nowhere else in the world did farmers develop any analogous agricultural implement. Is it coincidence that modern technology, with its ruthlessness toward nature, has so largely been produced by descendants of these peasants of northern Europe?

This same exploitive attitude appears slightly before A.D. 830 in Western illustrated calendars. In older calendars the months were shown as passive personificatons. The new Frankish calendars, which set the style for the Middle Ages, are very different: they show men coercing the world around them—plowing, harvesting, chopping trees, butchering pigs. Man and nature are two things, and man is master.

These novelties seem to be in harmony with larger intellectual patterns. What people do about their ecology depends on what they think about themselves in relation to things around them. Human ecology is deeply conditioned by beliefs about our nature and destiny—that is, by religion. To Western eyes this is very evident in, say, India or Ceylon. It is equally true of ourselves and of our medieval ancestors.

The victory of Christianity over paganism was the greatest psychic revolution in the history of our culture. It has become fashionable today to say that, for better or worse, we live in "the post-Christian age." Certainly the forms of our thinking and language have largely ceased to be Christian, but to my eye the substance often remains amazingly akin to that of the past. Our daily habits of action, for example, are dominated by an implicit faith in perpetual progress which was unknown either to Greco-Roman antiquity or to the Orient. It is rooted in, and is indefensible apart from, Judeo-Christian teleology. The fact that Communists share it merely helps to show what can be demonstrated on many other grounds: that Marxism, like Islam, is a Judeo-Christian heresy. We continue today to live, as we have lived for about 1700 years, very largely in a context of Christian axioms.

What did Christianity tell people about their relations with the environment?

While many of the world's mythologies provide stories of creation, Greco-Roman mythology was singularly incoherent in this respect. Like Aristotle, the intellectuals of the ancient West denied that the visible world had had a beginning. Indeed, the idea of a beginning was impossible in the framework of their cyclical notion of time. In sharp contrast, Christianity inherited from Judaism not only a concept of time as nonrepetitive and linear but also a striking story of creation. By gradual stages a loving and all-powerful God had created light and darkness, the heavenly bodies, the earth and all its plants, animals, birds, and fishes. Finally, God had created Adam and, as an afterthought, Eve to keep man from being lonely. Man named all the animals, thus establishing his dominance over them. God planned all of this explicitly for man's benefit and rule: no item in the physical creation had any purpose save to serve man's purposes. And, although man's body is made of clay, he is not simply part of nature: he is made in God's image.

Especially in its Western form, Christianity is the most anthropocentric religion the world has seen. As early as the 2nd century both Tertullian and Saint Irenaeus of Lyons were insisting that when God shaped Adam he was foreshadowing the image of the incarnate Christ, the Second Adam. Man shares, in great measure, God's transcendence of nature. Christianity, in absolute contrast to ancient paganism and Asia's religions (except, perhaps Zoroastrianism), not only established a dualism of man and nature but also insisted that it is God's will that man exploit nature for his proper ends.

At the level of the common people this worked out in an interesting way. In antiquity every tree, every spring, every stream, every hill had its own *genius loci*, its guardian spirit. These spirits were accessible to men, but were very unlike men; centaurs, fauns, and mermaids show their ambivalence. Before one cut a tree, mined a mountain, or dammed a brook, it was important to placate the spirit in charge of that particular situation, and to keep it placated. By destroying pagan animism, Christianity made it possible to exploit nature in a mood of indifference to the feelings of natural objects.

It is often said that for animism the Church substituted the cult of saints. True; but the cult of saints is functionally quite different from animism. The saint is not *in* natural objects; he may have special shrines, but his citizenship is in heaven. Moreover, a saint is entirely a man; he can be approached in human terms. In addition to saints, Christianity of course also had angels and demons inherited from Judaism and perhaps, at one remove, from Zoroastrianism. But these were all as mobile as the saints themselves. The spirits *in* natural objects, which formerly had protected nature from man, evaporated. Man's effective monopoly on spirit in this world was confirmed, and the old inhibitions to the exploitation of nature crumbled.

When one speaks in such sweeping terms, a note of caution is in order. Christianity is a complex faith, and its consequences differ in differing contexts. What I have said may well apply to the medieval West, where in fact technology made spectacular advances. But the Greek East, a highly civilized realm of equal Christian devotion seems to have produced no marked technological innovation after the late 7th century, when Greek fire was invented. The key to the contrast may perhaps be found in a difference in the tonality of piety and thought which students of comparative theology find between the Greek and the Latin Churches. The Greeks believed that sin was intellectual blindness, and that salvation was found in illumination, orthodoxy—that is, clear thinking. The Latins, on the other hand, felt that sin was moral evil, and that salvation was to be found in right conduct. Eastern theology has been intellectualist. Western theology has been voluntarist. The Greek saint contemplates; the Western saint acts. The implications of Christianity for the conquest of nature would emerge more easily in the Western atmosphere.

The Christian dogma of creation, which is found in the first clause of all the Creeds, has another meaning for our comprehension of today's ecologic crisis. By revelation, God had given man the Bible, the Book of Scripture. But since God had made nature, nature also must reveal the divine mentality. The religious study of nature for the better understanding of God was known as natural theology. In the early Church, and always in the Greek East, nature was conceived primarily as a symbolic system through which God speaks to men: the ant is a sermon to sluggards; rising flames are the symbol of the soul's aspiration. This view of nature was essentially artistic rather than scientific. While Byzantium preserved and copied great numbers of ancient Greek scientific texts, science as we conceive it could scarcely flourish in such an ambiance.

However, in the Latin West by the early 13th century natural theology was following a very different bent. It was ceasing to be the decoding of the physical symbols of God's communication with man and was becoming the effort to understand God's mind by discovering how his creation operates. The rainbow was no longer simply a symbol of hope first sent to Noah after the Deluge: Robert Grosseteste, Friar Roger Bacon, and Theodoric of Freiberg produced startlingly sophisticated work on the optics of the rainbow, but they did it as a venture in religious understanding. From the 13th century onward, up to and including Leibnitz and Newton, every major scientist, in effect, explained his motivations in religious terms. Indeed, if Galileo had not been so expert an amateur theologian he would have got into far less trouble: the professionals resented his intrusion. And Newton seems to have regarded himself more as a theologian than as a scientist. It was not until the late 18th century that the hypothesis of God became unnecessary to many scientists.

It is often hard for the historian to judge, when men explain why they are doing what they want to do, whether they are offering real reasons or merely culturally acceptable reasons. The consistency with which scientists during the long formative centuries of Western science said that the task and the reward of the scientist was "to think God's thoughts after him" leads one to believe that this was their real motivation. If so, then modern Western science was cast in a matrix of Christian theology. The dynamism of religious devotion, shaped by the Judeo-Christian dogma of creation, gave it impetus.

AN ALTERNATE CHRISTIAN VIEW

We would seem to be headed toward conclusions unpalatable to many Christians. Since both *science* and *technology* are blessed words in our contemporary vocabulary, some may be happy at the notions, first, that, viewed historically, modern science is an extrapolation of natural theology and, second, that modern technology is at least partly to be explained as an Occidental, voluntarist realization of the Christian dogma of man's transcendence of, and rightful mastery over, nature. But, as we now recognize, somewhat over a century ago science and technology—hitherto quite separate activities—joined to give mankind powers which, to judge by many of the ecologic effects, are out of control. If so, Christianity bears a huge burden of guilt.

I personally doubt that disastrous ecologic backlash can be avoided simply by applying to our problems more science and more technology. Our science and technology have grown out of Christian attitudes toward man's relation to nature which are almost universally held not only by Christians and neo-Christians but also by those who fondly regard themselves as post-Christians. Despite Copernicus, all the cosmos rotates around our little globe. Despite Darwin, we are *not*, in our hearts, part of the natural process. We are superior to nature, contemptuous of it, willing to use it for our slightest whim. The newly elected Governor of California, like myself a churchman but less troubled than I, spoke for the Christian tradition when he said (as is alleged), "when you've seen one redwood tree, you've seen them all." To a Christian a tree can be no more than a physical fact. The whole concept of the sacred grove is alien to Christianity and to the ethos of the West. For nearly 2 millennia Christian missionaries have been chopping down sacred groves, which are idolatrous because they assume spirit in nature.

What we do about ecology depends on our ideas of the man-nature relationship. More science and more technology are not going to get us out of the present ecologic crisis until we find a new religion, or rethink our old one. The beatniks, who are the basic revolutionaries of our time, show a sound instinct in their affinity for Zen Buddhism, which conceives of the man-nature relationship as very nearly the mirror image of the Christian view. Zen, however, is as deeply conditioned by Asian history as Christianity is by the experience of the West, and I am dubious of its viability among us.

Possibly we should ponder the greatest radical in Christian history since Christ: Saint Francis of Assisi. The prime miracle of Saint Francis is the fact that he did not end at the stake, as many of his left-wing followers did. He was so clearly heretical that a General of the Franciscan Order, Saint Bonaventura, a great and perceptive Christian, tried to suppress the early accounts of Franciscanism. The key to an understanding of Francis is his belief in the virtue of humility—not merely for the individual but for man as a species. Francis tried to depose man from his monarchy over creation and set up a democracy of all God's creatures. With him the ant is no longer simply a homily for the lazy, flames a sign of the thrust of the soul toward union with God; now they are Brother Ant and Sister Fire, praising the Creator in their own ways as Brother Man does in his.

Later commentators have said that Francis preached to the birds as a rebuke to men who would not listen. The records do not read so: he urged the little birds to praise God, and in spiritual ecstasy they flapped their wings and chirped rejoicing. Legends of saints, especially the Irish saints, had long told of their dealings with animals but always, I believe, to show their human dominance over creatures. With Francis it is different. The land around Gubbio in the Apennines was being ravaged by a fierce wolf. Saint Francis, says the legend, talked to the wolf and persuaded him of the error of his ways. The wolf repented, died in the odor of sanctity, and was buried in consecrated ground.

What Sir Steven Ruciman calls "the Franciscan doctrine of the animal soul" was quickly stamped out. Quite possibly it was in part inspired, consciously or unconsciously, by the belief in reincarnation held by the Cathar heretics who at that time teemed in Italy and southern France, and who presumably had got it originally from India. It is significant that at just the same moment, about 1200, traces of metempsychosis are found also in western Judaism, in the Provençal *Cabbala*. But Francis held neither to transmigration of souls nor to pantheism. His view of nature and of man rested on a unique sort of pan-psychism of all things animate and inanimate, designed for the glorification of their transcendent Creater, who, in the ultimate gesture of cosmic humility, assumed flesh, lay-helpless in a manger, and hung dying on a scaffold.

I am not suggesting that many contemporary Americans who are concerned about our ecologic crisis will be either able or willing to counsel with wolves or exhort birds. However, the present increasing disruption of the global movement is the product of a dynamic technology and science which were originating in the Western medieval world against which Saint Francis was rebelling in so original a way. Their growth cannot be understood historically apart from distinctive attitudes toward nature which are deeply grounded in Christian dogma. The fact that most people do not think of these attitudes as Christian is irrelevant. No new set of basic values has been accepted in our society to displace those of Christianity. Hence we shall continue to have a worsening ecologic crisis until we reject the Christian axiom that nature has no reason for existence save to serve man.

The greatest spiritual revolutionary in Western history, Saint Francis, proposed what he thought was an alternative Christian view of nature and man's relation to it: he tried to substitute the idea of equality of all creatures, including man, for the idea of man's limitless rule of creation. He failed. Both our present science and our present technology are so tinctured with orthodox Christian arrogance toward nature that no solution for our ecologic crisis can be expected from them alone. Since the roots of our trouble are so largely religious, the remedy must also be essentially religious, whether we call it that or not. We must rethink and refeel our nature and destiny. The profoundly religious, but heretical, sense of the primitive Franciscans for the spiritual autonomy of all parts of nature may point a direction. I propose Francis as a patron saint for ecologists.

7.3

The Polemics of Progress

ISAAC ASIMOV

Pity the poor scientist—he's damned if he does and damned if he doesn't.

Scientists, usually considered cold and felt to be unconcerned with the welfare of humanity in their avid search for knowledge, are, indeed, regularly perturbed at the possible consequences of their work.

Last July, a group of molecular biologists proclaimed they would stop experiments involving the transfer of genes into micro-organisms. They fear the creation of a type of bacteria resistant to antibiotic control, or possibly a new strain of virus capable of causing cancer. If something of this sort should escape the laboratory, the results might be devastating.

Earlier, the work on the supersonic transport plane (SST) had been stopped in the United States partly because some scientists wondered if the exhaust chemicals released into the upper atmosphere might not weaken the ozone layer. If it did, more ultraviolet from the sun would then reach the Earth's surface and there might be an increase in skin cancer. Alternatively, the exhaust chemicals might have a long-range effect for the worse on the earth's weather.

In general, science and the public are now aware that some new scientific or technological advance which may seem to be interesting or useful may also have unpleasant side effects. More and more, the tendency is to exert caution and to consider the consequences before committing the world to something that may not be reversible if we subsequently change our minds.

The trouble is that it is not always easy to tell what the side effects will be, or in which direction. In 1847, the Italian chemist Ascanio Sobrero produced nitroglycerine for the first time. He heated a drop of it and it exploded shatteringly. Sobrero realized in horror its possible application to warfare, and stopped all research in that direction at once.

It didn't help, of course. Others followed up the work, and it and other high explosives were indeed used in warfare by the close of the 19th Century.

Did that make high explosives entirely bad? In 1866, the Swedish inventor Alfred B. Nobel learned how to mix nitroglycerine with diatomaceous earth to produce a mixture which was completely safe to handle and which he called "dynamite." With dynamite, earth could be moved at a rate far beyond the pick-and-shovel efforts of all the ages before and without brutalizing men at hard labor.

It was dynamite that helped forge the way for the railroads in the final decades of the 19th Century, that helped build dams, subways, building foundations, bridges and a thousand other grand-scale constructions of the industrial age.

It was, after all, man's choice to use explosives to kill men and to destroy his works. It can as easily be used to help man and construct what he needs.

It is this double-edged sword of good-and-evil that has hung over human technology from the beginning. The invention of knives and spears increased man's food supply—and improved the art of murder. Go farther back and the discovery of fire gave warmth and light, safety and cooked food—and also the possibility of arson. Go still farther back, and the invention of speech made man human—and very often a liar.

Even the most fearsome discoveries hold within them the capacity for great good, and the most noble discoveries the capacity for great evil.

In the first class is the discovery of nuclear energy which now places all the earth under the threat of quick destruction in a thermonuclear holocaust or of slower destruction by radioactive pollution—yet also offers the possibility of fusion power as an ultimate solution to man's energy problems.

In the second class, let us think back to 1798 and the discovery of vaccination by the English physician Edward Jenner. It put an end almost at once to the terror of smallpox, and was seized upon with wild jubilation by the people of Europe and America. Jenner became a folk hero almost at once. Even Napoleon, who was holding prisoners of war against all diplomatic representations, released them at once when Jenner made the request.

For over half a century, this first victory over infection stood alone. Then, in the 1860's, the French chemist Louis Pasteur advanced the germ theory of disease. Once physicians understood the cause of infection, they learned how to prevent, ameliorate and cure, so that modern medicine begins with Pasteur's discovery.

Did anyone refuse to take advantage of advances in medicine—of vaccines and antitoxins, of anesthesia and asepsis, of chemical specifics and antibiotics? Virtually no one.

Did anyone view medical advance as dangerous to humanity? Virtually no one. Even today, when so many people are concerned about the abuses of technology, who has a bad word for research into the cure of arthritis, circulatory disease, birth defects, or cancer?

And yet the side effects of the various medical discoveries of the last century done more to assure human destruction than anything the nuclear physicists have done. The population explosion today is not caused by any rise in the rate at which babies are produced. Rather, the average birth rate in the world is lower than it has ever been. The trouble is that, thanks to medicine, the death rate has dropped precipitously. It is for that reason, and for that reason only that mankind is now swarming on Earth in numbers that make it impossible to live decently and is increasing at a rate that may destroy civilization by the end of the century.

Could not scientists foresee this dangerous side effect of these most humanitarian of all discoveries? Some could. As far back as 1798, the pioneer sociologist Thomas R. Malthus pointed out the dire consequence of overpopulation initiated by any improvements in man's lot.

Does that mean that science should have avoided bringing about such an improvement? Does it mean we should have kept mankind a short-lived race?

Or does it mean we should have used science to correct the possibly deleterious side effect? Should scientists have devised methods that would make it simpler to reduce the birth rate and keep it matching the falling death rate? Obviously!

That could have been done, too, but there was a great deal of social resistance to that. Many people—not scientists—who seized with eagerness and avidity on everything that lowered the death rate, held fast against any attempt to lower the birth rate, considering that immoral. To this very day, there are strong forces—not scientists—that stand in the way of a lowered birth rate.

It is almost always the decision of men of business, of government, of the military, that determine whether a scientific advance is to be used for good or for evil, to improve man's lot or to destroy it. Very often, these men choose evil in some shortsighted lunge for the immediate advantage; and then, all too frequently, it is the scientists who are blamed for the greed and blindness of those others who are given Paradise, and turn it into Hell.

Science and technology are getting a bad press these days. Our young people are increasingly scornful of the materialism of our culture, and speak about returning to an older and simpler day in a kind of nostalgia for the prescientific world.

In the question-and-answer period that followed one of my talks, a college student asked me if I honestly believed science had done anything to increase man's happiness.

"Do you think you would be just as happy if you had lived in the days of ancient Greece?" I asked.

"Yes," he said.

"How would you enjoy being a slave?" I asked.

That's it, you see. People have a peculiar and romantic notion of the "good old days" that were really the horribly bad old days. They fancy themselves in Athens, talking to Socrates and listening to the latest play by Sophocles—never as a slave brutalized in the Athenian silver mines.

They think of themselves as Roman senators talking to Cicero and Cato—never as slaves, whipped and scarred in the wheatfields of Sicily. They imagine themselves to be medieval knights on armored chargers—never as starving peasants mercilessly slaughtered by those same knights.

Throughout the history of civilization, right down to modern times, the wealth and prosperity of a relatively small number of people has been built on the animal-like labor and wretched existence of a large number of peasants, serfs and slaves. What's more, in a world without science, nothing could be done about it. Slavery and peonage were taken for granted.

Look through the Bible from the first verse of Genesis to the last verse of Revelation and you will find not one word of condemnation of slavery as an institution. There will be suggestions that slaves be treated humanely but not a syllable to the effect that slavery is morally wrong. Look through all the writings of the great philosophers of the past and you will find no whisper of condemnation against slavery as an institution. To Aristotle it seemed quite clear that there were people who seemed to be fitted by temperament to be slaves.

Isn't it strange that there was never any moral outcry against slavery before the days of science, but that it came to be recognized as a dreadful wrong, and to be abolished in civilized nations once science became prominent?

It was the scientist, supposedly amoral, supposedly cold and aloof, supposedly divorced from value judgments, who brought about this great victory for morality and humanity. His investigations just happened to make possible the harnessing of the energy of the inanimate world. With steam, electricity and radio beams to do our work for us, there was no longer the need for the comparatively weak and fumbling human muscle—and slavery began to vanish.

The Quakers were the first to cry out against slavery in the 18th Century, just as the Industrial Revolution was beginning, and without the latter, the former would have come to nothing. Indeed, if we were to turn away from science now, if a noble young generation abandoned the materialism of an industry that seems to be concerned with things rather than with ideals, what would happen? Without the machinery of that materialistic industry, we would inevitably drift back to slavery.

Nor is this the only example of the manner in which it is the scientist who is the humanitarian, and not those who talk ideals but who can do nothing to implement them.

In the preindustrial age, mankind was subject to the constant onslaught of infectious disease, and life expectancy varied from an average of 25 to 35, depending on the time and place. All the love of parents, all the prayers of the congregations, all the lofty generalizations of philosophers, could not prevent a child dying of diptheria in the bad old days, or prevent half a nation of dying of the plague.

It was the cold curiosity of men of science, magnifying and studying forms of life invisible to the naked eyes, that worked out the cause of infectious disease, demonstrated the importance of hygiene, of clean food and water, of efficient sewage systems. It was they who worked out vaccines, antitoxins, chemical specifics and antibiotics. It was they who saved hundreds of millions of lives.

It was scientists, too, who won the victory against pain, and who discovered how to soothe anguish when prayer could not. The discovery of anesthesia lulled the pangs of childbirth, for one thing; it did so despite the thunders from the pulpit against this advance on the ground that the Biblical curse on Eve condemned all women to the dreadful pain. Who were the humanitarians then?

Who can argue against the glories of art, literature, the manner in which the mind of man has wrested beauty from the universe and distilled it into shape, form, sound, pattern, words, images? Surely that existed before science did, and what does science give us that can uplift the soul in this fashion?

For one thing, it is possible to point out that the vision of the universe made apparent by the careful labor of three centuries of scientists far outweighs, in beauty and majesty, all the creations of all human artists put together.

Beyond that, it is also a fact that before the days of modern technology, the full flower of art and of the human intellect was reserved for the few who were aristocratic and rich. What chance did the peasants and slaves have to immerse themselves in art and literature, or in anything but endless sweat and toil?

It was the technical advances that went into the development of printing that scattered books widely, brought them within the reach (eventually) of all, made universal literacy practical. It was the movies, radio, the record-player, and television that brought much of the marvels of mankind (along with much of the refuse) to even the meanest.

It was the technical advances in communication and transportation that placed all the world within reach of so many of us, enlarged our diets, increased our knowledge of each other.

And yes, science has helped create problems, too, very serious ones, and we must labor to solve those problems in the only way that history tells us serious problems in the past have been solved—by science.

But would you perhaps prefer to avoid those problems by avoiding science in the first place, by doing without it? Would you prefer, then, to find happiness by going back to ignorance, pain, disease, slavery, death?

Because that's what the bad old days of the prescientific world were all about.

In these days of urban decay and energy crisis, there is a constant longing to return to the land and flee back to a simpler day. Too bad, but it can't be done. You have a tiger by the tail and you can't go home again.

You never could. The history of mankind is one of a constant move onward toward greater economic and social complexity, and an inability ever to move back again willingly. One can only be thrust back by catastrophe.

This goes all the way back. When mankind learned how to make use of fire some 50,000 years ago or so, it meant protection against predators, and it meant more and better food. It also meant that mankind could venture out of the tropics and into those climates where it grew cold at night and very cold in winter.

Do you suppose that this didn't bring problems? When the fire went out in the cave on a winter night and, for any reason, could not be relit, there was the danger of freezing. And if one continually huddled over the fire through all the cold stretches, the smoke would ruin one's lungs.

Why not give up fire, then, go back to the tropics and live on uncooked food in a simple, carefree ways of one's ancestors? Ah, one could not. Man had extended his range and increased his numbers thanks to fire.

If he got rid of fire, he would have to contract his range and decrease his numbers. The people from the cold areas returning to the tropics would find the tropics full. There would be a catastrophic struggle for the smaller supply of fire-free food.

The fact is that no people in history, as far as we know, ever deliberately gave up fire, having once learned to use it. They either endured its discomforts, or they learned to do away with them by further technological advances. They learned better ways of making fire, better ways of heating dwelling places, better ways of handling smoke.

This has remained true ever since. Not only fire, but no other fundamental technological advance has ever been given up willingly by any society. There has been no way to do it.

About 8000 B.C., mankind discovered agriculture. Again it made possible an increase in numbers, and even with increased numbers, people had never eaten so well.

But it meant that people had to give up the free, nomadic life of earlier generations, and remain bound to the soil where the crops were growing. It meant hard labor, sowing, and reaping, and getting rid of weeds, and killing off animal pests. It meant banding together and trying to fight off surrounding tribes who were still food-gathering and had no objection to helping themselves to your crops.

It also meant that insect pests multiplied and sometimes destroyed the crops, or that the rains failed and sometimes withered your crops.

Where irrigation was introduced to make harvest more dependable, it meant the formation of a large political unit and the social tyranny of a king, an aristocracy, and a priesthood. Even if everything worked well, and the land grew prosperous and populous, any infectious disease that got started ran through the crowded population like wildfire.

Who needs it all, then? Why not abandon agriculture and go back to the wilder, freer ways of hunting and food-gathering? Wouldn't that mean less work, less worry, less war, less pestilence?

But you can't! Abandon agriculture and out of every 10,000 people living in an agricultural area only 100 survive, and no one ever wants to gamble on being one of those 100. For that reason, no agricultural community in history ever willingly gave up farming out of a desire to return to a simpler day.

The solution to the problems to which agriculture gave rise could only be solved by moving forward with additional advances in technology—the use of oxen in place of men, the use of horses in place of oxen, the use of crop-rotation, fertilizer, contour-plowing and so on.

So it is with our industrial age now. The use of new sources of energy, of new methods of using that energy, of new devices and gadgets of all kinds, has once again increased man's numbers and his range. (This time it has taken him all the way to the moon.)

With industrialization, too, there are problems, as we all know. Can we solve them by undoing industrialization and by returning to a way of life that existed only two centuries ago?

If there is a shortage of gasoline, can't we abandon our automobiles in a pinch and go back to the horse-and-wagon that satisfied our sturdy ancestors? Can't we give up our oil furnace, if we must, for wood in the fireplace? Can't we give up our electric lights and use candles?

No, we can't. There are far more of us now than there were in the time of our sturdy ancestors. There aren't enough horses any longer to move us about, nor enough wood to warm us, nor enough candles to light our way. Besides which, if we try to travel by horse-and-wagon, keep warm by means of a fireplace, or read by the light of a candle, we will quickly find that the simple life is rotten.

Nor are our industrial techniques used only to keep our effete city dwellers in luxury. They do a lot more than that.

In 1800, when the earth was still supported almost entirely by nonindustrial methods, the population of the planet was 900,000,000, and they didn't live very well, either—worse than we do now, on the average.

Nowadays, the population is pushing four billion, more than four times the population in 1800. Where does the food come from that is needed to support the extra 3 billion? It comes from the industrialization of the farm; from the use of high-energy machinery to plow and seed and harrow and weed and reap and bind and everything else. It comes from the chemical fertilizers and chemical insecticides produced by sophisticated high-energy chemical factories.

We can't abandon industrialization, if only because our food supply depends on it.

You can talk about "natural food" all you want, but if everyone decided to grow food without chemical fertilizers or insecticides or machinery, it would mean that only one-quarter of the world population could be fed and the other three-quarters would have to die.

Can we abandon some of our industrial technology and hold on to the rest? That would be very difficult, since it all hangs together.

We can save, we can conserve, we can cut out waste, but we must keep what we have. The only solution, as has always been true in the history of mankind, is to solve problems by still further advances in technology, advances that will help us increase our energy supply, recycle our resources, restore our soil, and humanely help us save the ecology.

7.4

On the Impact of the Computer on Society

JOSEPH WEIZENBAUM

The structure of the typical essay on "The impact of computers on society" is as follows: First there is an "on the one hand" statement. It tells all the good things computers have already done for society and often even attempts to argue that the social order would already have collapsed were it not for the "computer revolution." This is usually followed by an "on the other hand" caution which tells of certain problems the introduction of computers brings in its wake. The threat posed to individual privacy by large data banks and the danger of large-scale unemployment induced by industrial automation are usually mentioned. Finally, the glorious present and prospective achievements of the computer are applauded, while the dangers alluded to in the second part are shown to be capable of being alleviated by sophisticated technological fixes. The closing paragraph consists of a plea for generous societal support for more, and more large-scale, computer research and development. This is usually coupled to the more or less subtle assertion that only computer science, hence only the computer scientist, can guard the world against the admittedly hazardous fallout of applied computer technology.

In fact, the computer has had very considerably less societal impact than the mass media would lead us to believe. Certainly, there are enterprises like space travel that could not have been undertaken without computers. Certainly the computer industry, and with it the computer education industry, has grown to enormous proportions. But much of the industry is self-serving. It is rather like an island economy in which the natives make a living by taking in each other's laundry. The part that is not self-serving is largely supported by government agencies and other gigantic enterprises that know the value of everything but the price of nothing, that is, that know the short-range utility of computer systems but have no idea of their ultimate social cost. In any case, airline reservation systems and computerized hospitals serve only a tiny, largely the most affluent, fraction of society. Such things cannot be said to have an impact on society generally.

SIDE EFFECTS OF TECHNOLOGY

The more important reason that I dismiss the argument which I have caricatured is that the direct societal effects of any pervasive new technology are as nothing compared to its much more subtle and ultimately much more important side effects. In that sense, the societal impact of the computer has not yet been felt.

To help firmly fix the idea of the importance of subtle indirect effects of technology, consider the impact on society of the invention of the microscope. When it was invented in the middle of the 17th century, the dominant commonsense theory of disease was fundamentally that disease was a punishment visited upon an individual by God. The sinner's body was thought to be inhabited by various so-called humors brought into disequilibrium in accordance with divine justice. The cure for disease was therefore to be found first in penance and second in the balancing of humors as, for example, by bleeding. Bleeding was, after all, both painful, hence punishment and penance, and potentially balancing in that it actually removed substance from the body. The microscope enabled man to see microorganisms and thus paved the way for the germ theory of disease. The enormously surprising discovery of extremely small living organisms also induced the idea of a continuous chain of life which, in turn, was a necessary intellectual precondition for the emergence of Darwinism. Both the germ theory of disease and the theory of evolution profoundly altered man's conception of his contract with God and consequently his self-image. Politically these ideas served to help diminish the power of the Church and, more generally, to legitimize the questioning of the basis of hitherto unchallenged authority. I do not say that the microscope alone was responsible for the enormous social changes that followed its invention. Only that it made possible the kind of paradigm shift, even on the commonsense level, without which these changes might have been impossible.

Is it reasonable to ask whether the computer will induce similar changes in man's image of himself and whether that influence will prove to be its most important effect on society? I think so, although I hasten to add that I don't believe the computer has yet told us much about man and his nature. To come to grips with the question, we must first ask in what way the computer is different from man's many other machines. Man has built two fundamentally different kinds of machines, nonautonomous and autonomous. An autonomous machine is one that operates for long periods of time, not on the basis of inputs from the real world, for example from sensors or from human drivers, but on the basis of internalized models of some aspect of the real world. Clocks are examples of autonomous machines in that they operate on the basis of an internalized model of the planetary system. The computer is, of course, the example par excellence. It is able to internalize models of essentially unlimited complexity and of a fidelity limited only by the genius of man.

It is the autonomy of the computer we value. When, for example, we speak of the power of computers as increasing with each new hardware and software development, we mean that, because of their increasing speed and storage capacity, and possibly thanks to new programming tricks, the new computers can internalize ever more complex and ever more faithful models of ever larger slices of reality. It seems strange then that, just when we exhibit virtually an idolatry of autonomy with respect to machines, serious thinkers in respected academies [I have in mind B. F. Skinner of Harvard University (1)] can rise to question autonomy as a fact for man. I do not think that the appearance of this paradox at this time is accidental. To understand it, we must realize that man's commitment to science has always had a masochistic component.

Time after time science has led us to insights that, at least when seen superficially, diminish man. Thus Galileo removed man from the center of the universe, Darwin removed him from his place separate from the animals, and Freud showed his rationality to be an illusion. Yet man pushes his inquiries further and deeper. I cannot help but think that there is an analogy between man's pursuit of scientific knowledge and an individual's commitment to psychoanalytic therapy. Both are undertaken in the full realization that what the inquirer may find may well damage his self-esteem. Both may reflect his determination to find meaning in his existence through struggle in truth, however painful that may be, rather than to live without meaning in a world of ill-disguised illusion. However, I am also aware that sometimes people enter psychoanalysis unwilling to put their illusions at risk, not searching for a deeper reality but in order to convert the insights they hope to gain to personal power. The analogy to man's pursuit of science does not break down with that observation.

Each time a scientific discovery shatters a hitherto fundamental cornerstone of the edifice on which man's self-esteem is built, there is an enormous reaction, just as is the case under similar circumstances in psychoanalytic therapy. Powerful defense mechanisms, beginning with denial and usually terminating in rationalization, are brought to bear. Indeed, the psychoanalyst suspects that, when a patient appears to accept a soul-shattering insight without resistance, his very casualness may well mask his refusal to allow that insight truly operational status in his self-image. But what is the psychoanalyst to think about the patient who positively embraces tentatively proffered, profoundly humiliating self-knowledge, when he embraces it and instantly converts it to a new foundation of his life? Surely such an event is symptomatic of a major crisis in the mental life of the patient.

I believe we are now at the beginning of just such a crisis in the mental life of our civilization. The microscope, I have argued, brought in its train a revision of man's image of himself. But no one in the mid-17th century could have foreseen that. The possibility that the computer will, one way or another, demonstrate that, in the inimitable phrase of one of my esteemed colleagues, "the brain is merely a meat machine" is one that engages academicians, industrialists, and journalists in the here and now. How has the computer contributed to bringing about this very sad state of affairs? It must be said right away that the computer alone is not the chief causative agent. It is merely an extreme extrapolation of technology. When seen as an inducer of philosophical dogma, it is merely the reductio ad absurdum of a technological ideology. But how does it come to be regarded as a source of philosophic dogma?

THEORY VERSUS PERFORMANCE

We must be clear about the fact that a computer is nothing without a program. A program is fundamentally a transformation of one computer into another that has autonomy and that, in a very real sense, behaves. Programming languages describe dynamic processes. And, most importantly, the processes they describe can be actually carried out. Thus we can build models of any aspect of the real world that interests us and that we understand. And we can make our models work. But we must be careful to remember that a computer model is a description that works. Ordinarily, when we speak of A being a model of B, we mean that a theory about some aspects of the behavior of B is also a theory of the same aspects of the behavior of A. It follows that when, for example, we consider a computer model of paranoia, like that published by Colby *et al.*, (2), we must not be persuaded that it tells us anything about paranoia on the grounds that it, in some sense, mirrors the behavior of a paranoiac. After all, a plain typewriter in some sense mirrors the behavior of an autistic child (one types a question and gets no response whatever), but it does not help us to understand autism. A model must be made to stand or fall on the basis of its theory. Thus, while programming languages may have put a new power in the hands of social scientists in that this new notation may have freed them from the vagueness of discursive descriptions, their obligation to build defensible theories is in no way diminished. Even errors can be pronounced with utmost formality and eloquence. But they are not thereby transmuted to truth.

The failure to make distinctions between descriptions, even those that "work," and theories accounts in large part for the fact that those who refuse to accept the view of man as machine have been put on the defensive. Recent advances in computer understanding of natural language offer an excellent case in point. Halle and Chomsky, to mention only the two with whom I am most familiar, have long labored on a theory of language which any model of language behavior must satisfy (3). Their aim is like that of the physicist who writes a set of differential equations that anyone riding a bicycle must satisfy. No physicist claims that a person need know, let alone be able to solve, such differential equations in order to become a competent cyclist. Neither do Halle and Chomsky claim that humans know or knowingly obey the rules they believe to govern language behavior. Halle and Chomsky also strive, as do physical theorists, to identify the constants and parameters of their theories with components of reality. They hypothesize that their rules constitute a kind of projective description of certain aspects of the structure of the human mind. Their problem is thus not merely to discover economical rules to account for language behavior, but also to infer economic mechanisms which determine that precisely those rules are to be preferred over all others. Since they are in this way forced to attend to the human mind, not only that of speakers of English, they must necessarily be concerned with all human language behavior—not just that related to the understanding of English.

The enormous scope of their task is illustrated by their observation that in all human languages declarative sentences are often transformed into questions by a permutation of two of their words. (John is here → Is John here?) It is one thing to describe rules that transform declarative sentences into questions—a simple permutation rule is clearly insufficient—but another thing to describe a "machine" that necessitates those rules when others would, all else being equal, be simpler. Why, for example, is it not so that declarative sentences read backward transform those sentences into questions? The answer must be that other constraints on the "machine" combine against this local simplicity in favor of a more nearly global economy. Such examples illustrate the depth of the level of explanation that Halle and Chomsky are trying to achieve. No wonder that they stand in awe of their subject matter.

Workers in computer comprehension of natural language operate in what is usually called performance mode. It is as if they are building machines that can ride bicycles by following heuristics like "if you feel a displacement to the left, move your weight to the left." There can be, and often is, a strong interaction between the development of theory and the empirical task of engineering systems whose theory is not yet thoroughly understood. Witness the synergistic cooperation between aerodynamics and aircraft design in the first quarter of the present century. Still, what counts in performance mode is not the elaboration of theory but the performance of systems. And the systems being hammered together by the new crop of computer semanticists are beginning (just beginning) to perform.

Since computer scientists have recognized the importance of the interplay of syntax, semantics, and pragmatics, and with it the importance of computer-manipulable knowledge, they have made progress. Perhaps by the end of the present decade, computer systems will exist with which specialists, such as physicians and chemists and mathematicians, will converse in natural language. And surely some part of such achievements will have been based on other successes in, for example, computer simulation of cognitive processes. It is understandable that any success in this area, even if won empirically and without accompanying enrichments of theory, can easily lead to certain delusions being planted. Is it, after all, not terribly tempting to believe that a computer that understands natural language at all, however narrow the context, has captured something of the essence of man? Descartes himself might have believed it. Indeed, by way of this very understandable seduction, the computer comes to be a source of philosophical dogma.

I am tempted to recite how performance programs are composed and how things that don't work quite correctly are made to work via all sorts of strategems which do not even pretend to have any theoretical foundation. But the very asking of the question, "Has the computer captured the essence of man?" is a diversion and, in that sense, a trap. For the real question, "Does man understand the essence of man?" cannot be answered by technology and hence certainly not by any technological instrument.

THE TECHNOLOGICAL METAPHOR

I asked earlier what the psychoanalyst is to think when a patient grasps a tentatively proffered deeply humiliating interpretation and attempts to convert it immediately to a new foundation of his life. I now think I phrased that question too weakly. What if the psychoanalyst merely coughed and the cough entrained the consequences of which I speak? That is our situation today. Computer science, particularly its artificial intelligence branch, has coughed. Perhaps the press has unduly amplified that cough—but it is only a cough nevertheless. I cannot help but think that the eagerness to believe that man's whole nature has suddenly been exposed by that cough, and that it has been shown to be a clockwork, is a symptom of something terribly wrong.

What is wrong, I think, is that we have permitted technological metaphors, what Mumford (4) calls the "Myth of the Machine," and technique itself to so thoroughly pervade our thought processes that we have finally abdicated to technology the very duty to formulate questions. Thus sensible men correctly perceive that large data banks and enormous networks of computers threaten man. But they leave it to technology to formulate the corresponding question. Where a simple man might ask: "Do we need these things?," technology asks "what electronic wizardry will make them safe?" Where a simple man will ask "is it good?", technology asks "will it work?" Thus science, even wisdom, becomes what technology and most of all computers can handle. Lest this be thought to be an exaggeration, I quote from the work of H. A. Simon, one of the most senior of American computer scientists (5):

> As we succeed in broadening and deepening our knowledge—theoretical and empirical—about computers, we shall discover that in large part their behavior is governed by simple general laws, that what appeared as complexity in the computer program was, to a considerable extent, complexity of the environment to which the program was seeking to adapt its behavior.
>
> To the extent that this prospect can be realized, it opens up an exceedingly important role for computer simulation as a tool for achieving a deeper understanding of human behavior. For if it is the organization of components, and not their physical properties, that largely determines behavior, and if computers are organized somewhat in the image of man, then the computer becomes an obvious device for exploring the consequences of alternative organizational assumptions for human behavior.

and

> A man, viewed as a behaving system, is quite simple. The apparent complexity of his behavior over time is largely a reflection of the complexity of the environment in which he finds himself. . . . I believe that this hypothesis holds even for the whole man.

We already know that those aspects of the behavior of computers which cannot be attributed to the complexity of their programs is governed by simple general laws—ultimately by the laws of Boolean algebra. And of course the physical properties of the computer's components are nearly irrelevant to its behavior. Mechanical relays are logically equivalent to tubes and to transistors and to artificial neurons. And of course the complexity of computer programs is due to the complexity of the environments, including the computing environments themselves, with which they were designed to deal. To what else could it possibly be due? So, what Simon sees as prospective is already realized. But does this collection of obvious and simple facts lead to the conclusion that man is as simple as are computers? When Simon leaps to that conclusion and then formulates the issue as he has done here, that is, when he suggests that the behavior of *the whole man* may be understood in terms of the behavior of computers as governed by simple general laws, then the very possibility of understanding man as an autonomous being, as an individual with deeply internalized values, that very possibility is excluded. How does one insult a machine?

The question "Is the brain merely a meat machine?", which Simon puts in a so much more sophisticated form, is typical of the kind of question formulated by, indeed formulatable only by, a technological mentality. Once it is accepted as legitimate, arguments as to what a computer can or cannot do "in principle" begin to rage and themselves become legitimate. But the legitimacy of the technological question—for example, is human behavior to be understood either in terms of the organization or of the physical properties of "components"—need not be admitted in the first instance. A human question can be asked instead. Indeed, we might begin by asking what has already become of "the whole man" when he can conceive of computers organized in his own image.

The success of technique and of some technological explanations has, as I've suggested, tricked us into permitting technology to formulate important questions for us—questions whose very forms severely diminish the number of degrees of freedom in our range of decision making. Whoever dictates the questions in large part determines the answers. In that sense, technology, and especially computer technology, has become a self-fulfilling nightmare reminiscent of that of the lady who dreams of being raped and begs her attacker to be kind to her. He answers "it's your dream, lady." We must come to see that technology is our dream and that we must ultimately decide how it is to end.

I have suggested that the computer revolution need not and ought not to call man's dignity and autonomy into question, that it is a kind of pathology that moves men to wring from it unwarranted, enormously damaging interpretations. Is then the computer less threatening than we might have thought? Once we realize that our visions, possibly nightmarish visions, determine the effect of our own creations on us and on our society, their threat to us is surely diminished. But that is not to say that this realization alone will wipe out all danger. For example, apart from the erosive effect of a technological mentality on man's self-image, there are practical attacks on the freedom and dignity of man in which computer technology plays a critical role.

I mentioned earlier that computer science has come to recognize the importance of building knowledge into machines. We already have a machine—Dentral—(6) that commands more chemistry than do many Ph.D. chemists, and another—Mathlab—(7) that commands more applied mathematics than do many applied mathematicians. Both Dendral and Mathlab contain knowledge that can be evaluated in terms of the explicit theories from which it was derived. If the user believes that a result Mathlab delivers is wrong, then, apart from possible program errors, he must be in disagreement, not with the machine or its programmer, but with a specific mathematical theory. But what about the many programs on which management, most particularly the government and the military, rely, programs which can in no sense be said to rest on explicable theories but are instead enormous patchworks of programming techniques strung together to make them work?

INCOMPREHENSIBLE SYSTEMS

In our eagerness to exploit every advance in technique we quickly incorporate the lessons learned from machine manipulation of knowledge in theory-based systems into such patchworks. They then "work" better. I have in mind systems like target selection systems used in Vietnam and war games used in the Pentagon, and so on. These often gigantic systems are put together by teams of programmers, often working over a time span of many years. But by the time the systems come into use, most of the original programmers have left or turned their attention to other pursuits. It is precisely when gigantic systems begin to be used that their inner workings can no longer be understood by any single person or by a small team of individuals. Norbert Wiener, the father of cybernetics, foretold this phenomenon in a remarkably prescient article (8) published more than a decade ago. He said there:

> It may well be that in principle we cannot make any machine the elements of whose behavior we cannot comprehend sooner or later. This does not mean in any way that we shall be able to comprehend these elements in substantially less time than the time required for operation of the machine, or even within any given number of years or generations.
>
> An intelligent understanding of [machines'] mode of performance may be delayed until long after the task which they have been set has been completed. This means that though machines are theoretically subject to human criticism, such criticism may be ineffective until long after it is relevant.

This situation, which is now upon us, has two consequences: first that decisions are made on the basis of rules and criteria no one knows explicitly, and second that the system of rules and criteria becomes immune to change. This is so because, in the absence of detailed understanding of the inner workings of a system, any substantial modification is very likely to render the system altogether inoperable. The threshold of complexity beyond which this phenomenon occurs has already been crossed by many existing systems, including some compiling and computer operating systems. For example, no one likes the operating systems for certain large computers, but they cannot be substantially changed nor can they be done away with. Too many people have become dependent on them.

An awkward operating system is inconvenient. That is not too bad. But the growing reliance on supersystems that were perhaps designed to help people make analyses and decisions, but which have since surpassed the understanding of their users while at the same time becoming indispensible to them, is another matter. In modern war it is common for the soldier, say the bomber pilot, to operate at an enormous psychological distance from his victims. He is not responsible for burned children because he never sees their village, his bombs, and certainly not the flaming children themselves. Modern technological rationalizations of war, diplomacy, politics, and commerce such as computer games have an even more insidious effect on the making of policy. Not only have policy makers abdicated their decision-making responsibility to a technology they don't understand, all the while maintaining the illusion that they, the policy makers, are formulating policy questions and answering them, but responsibility has altogether evaporated. No human is any longer responsible for "what the machine says." Thus there can be neither right nor wrong, no question of justice, no theory with which one can agree or disagree, and finally no basis on which one can challenge "what the machine says." My father used to invoke the ultimate authority by saying to me, "it is written." But then I could read what was written, imagine a human author, infer his values, and finally agree or disagree. The systems in the Pentagon, and their counterparts elsewhere in our culture, have in a very real sense no authors. They therefore do not admit of exercises of imagination that may ultimately lead to human judgment. No wonder that men who live day in and out with such machines and become dependent on them begin to believe that men are merely machines. They are reflecting what they themselves have become.

The potentially tragic impact on society that may ensue from the use of systems such as I have just discussed is greater than might at first be imagined. Again it is side effects, not direct effects, that matter most. First, of course, there is the psychological impact on individuals living in a society in which anonymous, hence irresponsible, forces formulate the large questions of the day and circumscribe the range of possible answers. It cannot be surprising that large numbers of perceptive individuals living in such a society experience a kind of impotence and fall victim to the mindless rage that often accompanies such experiences. But even worse, since computer-based knowledge systems become essentially unmodifiable except in that they can grow, and since they induce dependence and cannot, after a certain threshold is crossed, be abandoned, there is an enormous risk that they will be passed from one generation to another, always growing. Man too passes knowledge from one generation to another. But because man is mortal, his transmission of knowledge over the generations is at once a process of filtering and accrual. Man doesn't merely pass knowledge, he rather regenerates it continuously. Much as we may mourn the crumbling of ancient civilizations, we know nevertheless that the glory of man resides as much in the evolution of his cultures as in that of his brain. The unwise use of ever larger and ever more complex computer systems may well bring this process to a halt. It could well replace the ebb and flow of culture with a world without values, a world in which what counts for a fact has long ago been determined and forever fixed.

POSITIVE EFFECTS

I've spoken of some potentially dangerous effects of present computing trends. Is there nothing positive to be said? Yes, but it must be said with caution. Again, side effects are more important than direct effects. In particular, the idea of computation and of programming languages is beginning to become an important metaphor which, in the long run, may well prove to be responsible for paradigm shifts in many fields. Most of the commonsense paradigms in terms of which much of mankind interprets the phenomena of the everyday world, both physical and social, are still deeply rooted in fundamentally mechanistic metaphors. Marx's dynamics as well as those of Freud are, for example, basically equilibrium systems. Any hydrodynamicist could come to understand them without leaving the jargon of his field. Languages capable of describing ongoing processes, particularly in terms of modular subprocesses, have already had an enormous effect on the way computer people think of every aspect of their worlds, not merely those directly related to their work. The information-processing view of the world so engendered qualifies as a genuine metaphor. This is attested to by the fact that it (i) constitutes an intellectual framework that permits new questions to be asked about a wide-ranging set of phenomena, and (ii) that it itself provides criteria for the adequacy of proffered answers. A new metaphor is important not in that it may be better than existing ones, but rather in that it may enlarge man's vision by giving him yet another perspective on his world. Indeed, the very effectiveness of a new metaphor may seduce lazy minds to adopt it as a basis for universal explanations and as a source of panaceas. Computer simulation of social processes has already been advanced by single-minded generalists as leading to general solutions of all mankind's problems.

The metaphors given us by religion, the poets, and by thinkers like Darwin, Newton, Freud, and Einstein have rather quickly penetrated to the language of ordinary people. These metaphors have thus been instrumental in shaping our entire civilization's imaginative reconstruction of our world. The computing metaphor is as yet available to only an extremely small set of people. Its acquisition and internalization, hopefully as only one of many ways to see the world, seems to require experience in program composition, a kind of computing literacy. Perhaps such literacy will become very widespread in the advanced societal sectors of the advanced countries. But, should it become a dominant mode of thinking and be restricted to certain social classes, it will prove not merely repressive in the ordinary sense, but an enormously divisive societal force. For then classes which do and do not have access to the metaphor will, in an important sense, lose their ability to communicate with one another. We know already how difficult it is for the poor and the oppressed to communicate with the rest of the society in which they are embedded. We know how difficult it is for the world of science to communicate with that of the arts and of the humanities. In both instances the communication difficulties, which have grave consequences, are very largely due to the fact that the respective communities have unsharable experiences out of which unsharable metaphors have grown.

RESPONSIBILITY

Given these dismal possibilities, what is the responsibility of the computer scientist? First I should say that most of the harm computers can potentially entrain is much more a function of properties people attribute to computers than of what a computer can or cannot actually be made to do. The nonprofessional has little choice but to make his attributions of properties to computers on the basis of the propaganda emanating from the computer community and amplified by the press. The computer professional therefore has an enormously important responsibility to be modest in his claims. This advice would not even have to be voiced if computer science had a tradition of scholarship and of self-criticism such as that which characterizes the established sciences. The mature scientist stands in awe before the depth of his subject matter. His very humility is the wellspring of his strength. I regard the instilling of just this kind of humility, chiefly by the example set by teachers, to be one of the most important missions of every university department of computer science.

The computer scientist must be aware constantly that his instruments are capable of having gigantic direct and indirect amplifying effects. An error in a program, for example, could have grievous direct results, including most certainly the loss of much human life. On 11 September 1971, to cite just one example, a computer programming error caused the simultaneous destruction of 117 high-altitude weather balloons whose instruments were being monitored by an earth satellite (9). A similar error in a military command and control system could launch a fleet of nuclear tipped missiles. Only censorship prevents us from knowing how many such events involving nonnuclear weapons have already occurred. Clearly then, limitations of the systems he is capable of designing brilliantly clear. The very power of his systems should serve to inhibit the advice he is ready to give and to constrain the range of work he is willing to undertake.

Of course, the computer scientist, like everyone else, is responsible for his actions and their consequences. Sometimes that responsibility is hard to accept because the corresponding authority to decide what is and what is not to be done appears to rest with distant and anonymous forces. That technology itself determines what is to be done by a process of extrapolation and that individuals are powerless to intervene in that determination is precisely the kind of self-fulfilling dream from which we must awaken.

Consider gigantic computer systems. They are, of course, natural extrapolations of the large systems we already have. Computer networks are another point on the same curve extrapolated once more. One may ask whether such systems can be used by anybody except by governments and very large corporations and whether such organizations will not use them mainly for antihuman purposes. Or consider speech recognition systems. Will they not be used primarily to spy on private communications? To answer such questions by saying that big computer systems, computer networks, and speech recognition systems are inevitable is to surrender one's humanity. For such an answer must be based either on one's profound conviction that society has already lost control over its technology or on the thoroughly immoral position that "if I don't do it, someone else will."

I don't say that systems such as I have mentioned are necessarily evil—only that they may be and, what is most important, that their inevitability cannot be accepted by individuals claiming autonomy, freedom, and dignity. The individual computer scientist can and must decide. The determination of what the impact of computers on society is to be is, at least in part, in his hands.

Finally, the fundamental question the computer scientist must ask himself is the one that every scientist, indeed every human, must ask. It is not "what shall I do?" but rather "what shall I be?" I cannot answer that for anyone save myself. But I will say again that if technology is a nightmare that appears to have its own inevitable logic, it is our nightmare. It is possible, given courage and insight, for man to deny technology the prerogative to formulate man's questions. It is possible to ask human questions and to find humane answers.

References and Notes

1. B. F. Skinner, *Beyond Freedom and Dignity* (Knopf, New York, 1971).
2. K. M. Colby, S. Weber, and F. D. Hilf, *Artif. Intell.* **1**, 1 (1971).
3. N. Chomsky, *Aspects of the Theory of Syntax* (M.I.T. Press, Cambridge, Mass., 1965); ——— and M. Halle, *The Sound Pattern of English* (Harper & Row, New York, 1968).
4. L. Mumford, *The Pentagon of Power* (Harcourt, Brace, Jovanovich, New York, 1970).
5. H. A. Simon, *The Sciences of the Artificial* (M. I. T. Press, Cambridge, Mass., 1969), pp. 22–25.
6. B. Buchanan, G. Sutherland, and E. A. Feigenbaum, in *Machine Intelligence*, B. Meltzer, Ed. (American Elsevier, New York, 1969).
7. W. A. Martin and R. J. Fateman, "The Macsyma system," in *Proceedings of the 2nd Symposium of Symbolic and Algebraic Manipulation* (Association for Computer Machines, New York, 1971); J. Moses, *Commun. Assoc. Computer Mach.* **14** (No. 8), 548 (1971).
8. N. Winer, *Science* **131**, 1355 (1960).
9. R. Gillette, *ibid*, **174**, 477 (1971).

7.5

Letter from Charles A. Lindbergh

Switzerland
July 1, 1970

Hon. Emilio Q. Daddario (1),
Chairman, Subcommittee on Science, Research and Development
Rayburn House Office Building
Washington, D.C.

Dear Congressman Daddario:

I have been thinking a great deal about our correspondence, and about the breakfast table discussion at your home in Washington last month. I become constantly more impressed with the wisdom of your new approach to scientific research and development through consideration of its effect on the future welfare of mankind.

In the success of the ideas you advance here may lie, at least symbolically, the answer to whether our human species will continue to progress (even to exist) far into the future—whether we are the trunk or only an overspecialized branch of evolutionary life.

The critically important fact that man is beginning, vaguely, to realize in this 20th-century A.D. is that the impact of the human mind on life's evolution has been negative, and that major changes in his thought and action are essential. We see our surface-of-the-earth environment breaking down at the same time our genetic defects are increasing—both rapidly.

Our science and technology are at once responsible for the rapidity of breakdown and our realization that a breakdown is taking place. Speaking fundamentally, the human intellect is becoming aware of the vulnerabilities that accompany its power, that a crisis exists, and that to avoid self-destruction it must exercise control over its accumulating knowledge.

In this new sense of humility impacting on intellectual arrogance, I find cause for hope. It is practically exemplified in the attempt you and your committee are making to estimate the effect of research and development projects on human welfare, and to relate your findings to the support these projects are given.

What you are trying to do, as I see it, and what man has not yet accomplished, is to relate intellectual action to human progress and evolution in the basic sense. Here, I think, it is necessary to bear in mind that life and awareness were developed by instinctive evolution working over aeons, that the human mind impacted on this instinctive or "natural" evolution only a short time ago, and that the titanic problems we now face have been caused by a still much shorter period of intellectual domination. (I am tempted to say intellectual tyranny.)

Has our human intellect the wisdom to replace instinct or—since I think replacement is neither possible nor desirable—to work with it in basic evolutionary progress? We have no acceptable alternative to making the attempt. Ways must be found, through approaches such as you are planning, to give intellectual support to the fantastically complicated requirements for basic progress—requirements that are here obvious, and there so infinitely subtle that no mortal will ever be able to fully recognize and control them.

It seems to me essential for us to accept the fact that intellectual support of basic progress will necessitate reconsideration of long cherished ideals, laws, and customs. In guiding such reconsideration, we can lay down certain principles, one of them being that man must place more value on the human lifestream than on himself as an individual—realizing that he is only a temporary manifestation of that lifestream, that his salvation and immortality lie in it rather than in himself. Possibly this will involve an intellectual religion rooting into intuitive religions of the past. I do not think it will involve sacrifice when its full significance is understood.

Since all projects must relate to our environment, I believe your studies will quickly bring out the primary and pressing needs of protecting that environment both for our lifestream and for existing individuals. I do not see how adequate protection can be achieved other than by government supervision of natural resources, based on the proposition that each generation is their custodian for generations that will follow and that no one has a right to despoil or waste mankind's aeon-formed inheritance.

Such supervision will not be a curtailment of freedom but an emphasis of it, for we have reached a time when a framework of regulation is unavoidable if we are to maintain an environment in which the freedoms we have known are meaningful or can continue even to exist.

In hindsight, realizing the damage that has been done through our science and technology, I do not see how it could have been prevented except through laws, regulations, and accompanying education. It seems to me the same situation will apply to most future research and development projects. The study of probable effects may well indicate that support should be increased for one project and reduced or eliminated for another. I should think that major savings in talent and money could be made in this way. But I believe that sufficiently effective results can be achieved only by paralleling research and development programs with the laws, regulations, and education needed to turn the final results of these programs to the basic benefit of man.

Here we return to the ultimate question of whether the human intellect has enough wisdom to regulate its own creations, to use its science and technology for real rather than apparent progress. If it has, our future seems unlimited, with man's present day awareness only an indication of what an immortal human lifestream may achieve. If it has not, then we may be close in time to our own extermination.

Let me say again that I believe you and your committee's approach to the problems involved gives cause for encouragement.

Sincerely,
CHARLES A. LINDBERGH

Annotations

1. The letter serves as the preface to a "Compendium of Papers on National Science and Technology Policy Issues" (1979), submitted to the House Committee on Science and Technology (Ninety-sixth Congress). Chairman Fuqua's statement which follows puts the letter in context:

The [preceding] letter, not previously made public, was written to former Representative Emilio Q. Daddario when he was chairman of the Subcommittee on Science, Research and Development. The Subcommittee was then completing Congress' first comprehensive survey into "science policy" as a discrete government interest—a study resulting in the subcommittee's report "Toward a Science Policy for the United States," released in October 1970.

Mr. Lindbergh's observations in this area represent a distinct and fundamental point of view. Their inclusion in this compendium is with the permission of Mr. Daddario who subsequently served as the first Director of the Office of Technology Assessment as well as president and board chairman of the American Association for the Advancement of Science.

We are indebted to Mr. Daddario for making this thoughtful letter available to the Committee and his permission to reprint it as part of this compendium.

DON FUQUA, CHAIRMAN

7.6

Science, Technology and Public Benefits

ANNA J. HARRISON

The benefits to be derived by society through the use of science and technology are determined by the value judgments made by the public and by the productivity of the scientific and technological community. The two are closely coupled.

As an academic chemist, my commitment to the value of basic research has been and is essentially an article of faith. As a member of the National Science Board (1972–1978) and as a current officer of the American Chemical Society, I have been and am deeply concerned about the social, economic and political factors which impact the productivity of the scientific and technological community. These flow from the value judgments made by the public and it is this aspect of the total picture that I shall seek to address in this letter. The word "public" will be used to refer to all of those who have the right to participate in the democratic process through the exercise of the right to speak and the right to vote.

All technological change, regardless of how great its positive contribution to society, has associated with it negative impacts. This is a characteristic of change, all change, and is in no way a unique characteristic of technological change. One example: developments in medical technology have served society well; these same developments have also contributed to our unprecedented population growth and all of the problems associated with greater numbers of peoples.

Herein lies the basis of the concept of benefit/risk analysis. This is a useful concept; its evaluation, in the quantitative sense, eludes us. Our capacity to perceive the future seldom, if ever, encompasses either the total benefits or the total risks. If both benefits and risks can be evaluated in dollars, economists can, at least in principle, arrive at a numerical value for the benefit/risk ratio. Similarly, if the benefits and risks associated with a particular drug can be evaluated in terms of lives saved and lives lost, the medical profession can, at least in principle, assess a numerical value for the benefit/risk ratio. If benefits are in terms of lives and risks in terms of dollars, or vice versa, the ratio can only be evaluated by assigning a dollar equivalent to life. To do so is a value judgment par excellence. In most cases, benefits and risks are both complicated functions involving short term and long term social, economic and political consequences.

The concept of benefit/risk analysis is a useful concept and every effort should be made to understand the nature and the magnitude of all benefits and all risks in whatever units they can be assessed. Such information, inadequate as it may be, is the only rational approach to the value judgments that must be made as to the net value to society of a given technological development.

The public is, for the most part, not prepared either psychologically or philosophically to capitalize upon the opportunities and to accept the responsibilities inherent in the use of modern science and technology. Much of the public does not understand the nature of and the consequence of either scientific capabilities or technological options and much of the scientific and technological community does not understand the nature of the social, economic and political matrix of which science and technology are a part. This limitation of understanding is a significant barrier to meaningful communication and a significant barrier to the intelligent use of science and technology in the service of society. I believe it to be the greatest barrier and that we must re-evaluate our educational programs in science for all students, those who are to become professional scientists and those who are not.

It is too early to assess, but it is timely to pursue, the factors that have led to the frequency and the vehemence with which the implementation of technological options are opposed by segments of the public. Four such factors are readily identifiable:

For a time, the rate of introduction of technological options was unprecedented in terms of total numbers and in terms of diversity.

A great deal is being learned about toxic responses, particularly chronic toxic responses, to exposures of radiation and to exposures of specific chemical substances and combinations of chemical substances.

An expanding population has placed unprecedented demands upon renewable and nonrenewable resources and placed an unprecedented burden of waste products upon the environment.

That which is socially acceptable in one period may become unacceptable in another.

I suspect that historians will attribute an abrupt change in attitude toward the conservation of the world's resources and the protection of the world's environment to the emotional impact of those magnificent space photographs of the earth as it receded into the distance. It was not that the photographs provided us with new information but that the photographs so dramatically portrayed the finite nature of the planet and its beautiful but fragile environment.

I also suspect that for a time the scientific and technological community sold, or more probably the public bought, an unrealistic concept of that which science and technology can deliver. So many wonderful things had been produced, it was perhaps easy to assume that science and technology can produce anything, including a riskless society. This, for a time, was compounded by confusion as to who is to make those inescapable value judgments in regard to the use of science and technology—the total public or the scientific and technological community. This conflict is now clearly resolved in favor of the public as defined earlier but apprehension lingers on.

The Committee on Science and Technology is a surrogate for the public in making value judgments as to the appropriate role for science and technology in our democratic society. The Committee on Science and Technology is also the designated leader of Congress to develop appropriate mechanisms to facilitate the fulfillment of the role for science and technology. The decisions made and the actions taken are powerful forces in determining the future of the nation and the quality of life of its people.

The scientific and technological capabilities of this nation are a living legacy of unprecedented value.

7.7

Science, Technology and the Economy

SIMON RAMO

Are we in the United States using science and technology to the fullest on behalf of our society? This is not to ask whether we are following up every clue to nature's undiscovered secrets and are building every machine it is technically possible to build. These latter are very different and less serious questions. We seek here rather to inquire if our scientific and technological know-how is being applied adequately where there is strong evidence of high economic and social rewards for the effort.

To this important question the answer is "no." While our nation today has more capability in science and technology than ever before, we are using it less—less, that is, as a fraction of what the full, beneficial potential could be. Scientific research, if more avidly pursued, can teach us how to apply our resources, natural and human, more effectively than we now do. Advanced technology can be employed more broadly to develop additional economically advantageous products whose production would create new jobs to fight unemployment. Further efforts can lead us to superior methods for increasing supply and lowering costs as a counter to inflation, to substitutes for materials in short supply and to ways of acquiring raw materials and manufacturing for our needs with less harm to the environment. But we are becoming slower, more timid and less innovative in applying science and technology toward such possible ends.

PENALTIES OF TECHNOLOGICAL LAG

Lethargy and negativism regarding research and development are especially penalizing to the U.S. society because our values and habits are so strongly based on a generous availability of the fruits of these disciplines. Our high standard of living is rooted in many decades of technological advance. Political experts assure us that no approach to our social and economic problems is politically viable if it contemplates the average citizen's accepting a materially reduced personal supply of goods and services. It is equally unrealistic politically to expect those now disadvantaged to abandon their aspirations for a higher living standard. If these are political truths, then the tools of science and technology must be sharpened and applied vigorously, because such action is indispensable for a feasible approach to national problems.

Of course, when we speak here of employing technology fully, we specifically do not mean the unthinking application of it, the misuse of technology on projects and products the public does not in the end really want and that are more harmful than beneficial. Also, even if we were to attain perfection in the selection of technological programs and the implementation were one hundred per cent efficient, we still would not be guaranteed a healthy economy and a happy society. If we handle badly numerous nontechnological decisions, we can have inflation, recession, high unemployment and other ills all at once. However, without a strong technological foundation, our needs will not be satisfied. Science and technology are not sufficient. But they are necessary.

They are essential to the United States for more than domestic tranquility. International economic competitiveness by the United States and our contribution to world stability and progress are both dependent on the status of our science and technology. If world economic health is fostered by each nation's doing for the rest what it can do best and trading with the others for their most suitable items in return, [we are—Ed.] doing our part. In this regard the situation is again disappointing. The areas of science and technology in which we lead are becoming fewer and in many of those areas the leadership gaps are being closed by others. In still other high technology fields where we are close to unique in our capabilities and opportunities, we are moving ahead slowly.

High technology product exports (such as aircraft, computers, petroleum equipment, chemicals) are presently the important contributors to the positive portion of the U.S. trade balance. The only other strong such contribution comes from agriculture and that is in substantial part because our fertilizer, farm machinery and other farming and food technology leads the world. To maintain our position in these fields requires continuing technological advances to increase productivity steadily. Unfortunately, the rate of productivity increase for the United States is now lower than that of every other major industrial country. We used to be ahead of others in expenditures for R & D as a fraction of the GNP. We no longer are, despite the known dependence of productivity increase on application of new technology.

"ANTI-TECHNOLOGY"

Now why is this? One reason is the public opinion wave of "anti-technology" in the United States. A substantial fraction of our nation's citizens seem to equate technology with the devil. In attaining our high production of goods and services, as they perceive it, we have been crowded into congested cities, this before we have learned how to live together. TV means to them vapid programs loaded with violence that miseducate our children. The automobile, they note, kills 50,000 people a year, fouls the air and forces us to spend hours each day in traffic snarls. We waste money going to the moon, the atom bomb may destroy civilization and the nuclear reactor may poison our environment. We make our soil more productive, but insecticides may do us in.

It can be argued, of course, that those who make these statements obviously fail to distinguish between the tools of man, valuable when properly used, and man's own misuse of them. However, the presence in the nation of a broad anti-technology bias, applied indiscriminately, impairs our overall ability to reach objective, sound, nonemotional decisions on the development and employment of technology. This prejudice, like all prejudices, stands in the way of our arriving at meaningful value judgments.

GOVERNMENT-PRIVATE RELATIONS

An even more serious limitation to our wise utilization of science and technology is the public confusion as to the right roles for free, private enterprise, on the one hand, and government control, sponsorship and regulation, on the other. Admittedly, a muddle of inconsistent opinions by the citizenry about all manner of things is to be expected in a democracy, but the situation is particularly chaotic as regards science and technology matters. Thus, for example, many people are convinced "big business" is socially irresponsible. Whenever any kind of a problem surfaces, they are certain private enterprise can be counted on solely to over-exploit it, seeking monopolies and "unconscionable" profits at the expense of the consumer. They thus look to government alone, and the private sector not at all, to provide solutions. Since technological advance is most often identified with big business, they reason that such advance means unneeded product changes, or, just as harmful, withheld product advances, and higher prices in either instance.

Another large fraction of the voters are fed up with government spending, and big government generally. They see the government as a huge and increasingly incompetent and inefficient bureaucracy. As to government sponsored scientific research and technological advance, the government is believed by these citizens to be engaging in wasting dollars into the billions as it engages in boondoggles or tries to do what should be left to free enterprise.

As a capping indication of the degree of confusion, it seems much of the population holds both of these extreme views at once. They distrust both the private sector's and the government's involvement with science and technology and can be counted on for a strictly negative stance.

AN UNSOLVED SYSTEMS PROBLEM

We see in the United States today a severe mismatch between the high potential of technological advance and the low potential of the country's social-political progress. The problem is not technology per se. We are simply not organized to use science and technology to the fullest. It is the interface of technological and non-technological factors that is critical and controlling. The whole is an unsolved "systems" problem. For instance, in choosing where and how to apply science and technology, it would be helpful to have clearer national goals, because for a "systems approach" it is required that we know what we are after before we can get very far. When we find it hard to articulate what kind of a society we want, it is naturally difficult to pinpoint the effective use of science and technology as tools to help us build it. Similarly, we cannot make satisfactory decisions without an understanding of trade-offs and options. We need to be in a position to compare the "good" that can come from specific technological advances against the "bad" and the cost. As to our employment of technological change, we can be likened to a bunch of carpenters, sawing and hammering away, often getting fingers in the saws, sometimes hitting each other's heads as they bring their hammers back, and hitting their own thumbs as the hammers come down. They don't quite know what they are trying to build, sense an unsatisfactory situation, and meanwhile blame the saws and hammers.

LIMITATIONS OF FREE ENTERPRISE

We have two routes for initiatives to get things done in the United States: free enterprise and government. The approaches are not mutually exclusive. In fact, we have always had a hybrid system and the need for a balance as we allocate assignments to private or government action. The free enterprise system—private capital at risk seeking a return—has served well for two centuries to connect the needs and desires of the citizens with the capabilities of technology. This wedding has continually blessed us with an immense stream of products and services. Why does not the free market automatically solve the system and organizational bottlenecks just referred to? Why is the free enterprise route not still the practical answer to the challenge of putting science and technology to work sensibly and completely?

For one thing, inflation and government spending, tax policies, and over-regulation have impaired earnings and lowered available funds in industry for plowback to create technological advances. Furthermore, the increasing likelihood of inadequate return militates against investing funds still available to exploit technological opportunities. At the same time, an increasing fraction of promising technological projects calls for investments beyond the net worth of most corporations, and large corporations in the U.S. now are essentially not allowed to team up to share the risk. For such projects, the "risk-to-return" ratio is too great. The start-up losses are too high, the time to "turn-around" to an eventual profit phase is too long, and the dependence of success on unpredictable political decisions is forbiddingly severe. The market for these large-scale projects is far from being a free one and is instead a cacophony of semi-autonomous and conflicting interests. Finally, many of the potentially most beneficial programs, like development of ocean resources and many space systems, require a high degree of international cooperation, something difficult to arrange.

For most technological innovation, the straightforward economic limitation is now the most important. Technological advance occurs mainly in industry and technological corporations, large and small, which are not now on the average earning enough to provide adequate funds for plowback to extend their technologies. After the small difference between selling price and costs is used to pay taxes, the interest on borrowed money, dividends to the shareholders and the inflated costs of replacing depreciating facilities, too little is left to improve methods and develop new products. So dislocating has been the effect of inflation and recession that many corporations have overborrowed. They now find themselves with too high a debt in relation to their equity capital and their net earnings. Since this condition is paralleled by high interest rates, their debt expense is extraordinary. Many companies' stocks are selling on the market at less than the book values of their assets even though because of inflation those assets could not be replaced at those stated book figures. This means new equity capital is as hard to come by as are earnings for reinvestment. It also explains why few new, small corporations are being formed now to develop positions in new technological products.

The important drag on innovative effort in U.S. industry is certainly not a lack of creative, inventive talent, but rather the incentive to provide funds to back it, which condition goes with the other present characteristics of the U.S. economy: inflation, high taxes, high costs in dealing with government regulatory requirements, high consumption to investment ratios, and high ratio of government to private sector spending. Given the state of the economy, no government program to stimulate more innovative effort and R & D in private industry could be nearly so effective as, for instance, a successful government program to lower inflation or one to cause more discretionary cash flow to be available to technological coporations by a lowering of taxes. If the government could create these changes, the seeds of beneficial science and technology advance would quickly blossom into copious crops.

Let us now consider a number of examples that illustrate the relation of the rate of application of technological advance to economic-political constraints.

Oil Well Technology

If existing U.S. continental oil wells were equipped with submergible pumps for secondary and tertiary recovery wherever warranted, our oil reserves would be greatly increased, perhaps even doubled, the experts tell us. Technologically, such installations are thoroughly practical. Oil can be brought to the surface that otherwise would have to be left below, counted as unavailable. Most often, the installations would be economically feasible as well, that is, if the price per barrel were higher than present controls permit for old oil, more in the range of prices the nation is paying now for imported oil. Billions of dollars invested in this way would bring us tens of billions in equivalent additional petroleum. But the critical pricing parameter, to make the private investment a sound one, lies outside the free market category and is a political matter. This pricing involves a contest among numerous influential constituents with wide differences of opinion on the issue. The private sector thus cannot justify investing the required funds. The "risk-to-return" ratio is too high.

Improved Urban Transportation

Almost every large U.S. city could gain by installing a first class public transportation system. But it has to be the right system. The technology has to be carefully selected and matched to each city's layout, industry pattern, employment distribution, health care and educational facilities—i.e., to that city's economy, social conditions and life style. The economic gain, assuming a sound application of the correct technology to the cities' needs, could be prodigious. Perhaps half our population would be benefited. A typical resident could save hours each week over present means for getting to and from work, travel at less expense, consume less energy, cut air pollution, and decrease the accident rate.

But is it intelligent for American technological corporations to invest their resources with the objective of developing, then selling, and finally earning a return from, mass transit systems for American cities? Take Los Angeles as an example. Imagine that some private company a few years ago actually decided to devote several years of effort, using a necessarily large team of systems engineers and specialists, to matching technology with L.A.'s social and economic needs and designing the multi-billion dollar transit system required. Then, further assume, as we must to be complete, that all the sub-systems and components (vehicles, controls, energy distributing systems, maintenance and reliability procedures and organizations, legal permissions and approvals) were worked out and debugged.

All this would have involved an expenditure of a hundred million dollars or more and would be only a beginning. Much of this effort would have had to be repeated for other cities because of the importance of bringing prices down by realizing as much commonality as possible in equipment and procedures for multiple installations. Then the private concern would have to design, build and tool up several plants to start producing the specifically required hardware, refine it by full-scale testing, and manufacture parts in some quantity to be sure of price and quality before taking on contracts for shipment and installation. If the company had gone this far (it would by now be about a billion dollar investment) and had done everything perfectly, would it then have succeeded in selling a system to Los Angeles? That is extremely doubtful. Who exactly is this customer we have labeled Los Angeles? What does Los Angeles truly want and how are its requirements being decided? Can Los Angeles afford and raise the purchasing price? Hundreds of separate (and often quarrelling, apathetic, self-seeking and uncooperative) groups, both private and governmental, are involved in the answers to these questions. Critically important and potentially beneficial as the proper application of transportation technology appears to be in the city, the market in Los Angeles is not clear. The "risk-to-return" ratio would be absurdly high for the private corporation we imagined electing to go after this field. The start-up cost would be huge. Even assuming success eventually, the time to pay off would be too long and the city or county government's eventual approach to the pricing of the service, fundamental to the promise of return on the private investment, would be too unpredictable.

Environmental Technology

Done suitably, with a mixture of creativity and common sense, the selective depolluting of major lakes, rivers, harbors and coastal waters stands to yield a good return on the investment in health, quality of life, and economic gain through preservation and maintenance of vital natural resources. To achieve optimum results, much scientific research and technological work is needed to understand pollution phenomena more deeply, develop superior non-polluting approaches to the use of the waters, and produce a myriad of specialized equipments. Of course, such technical effort would be meaningless or misguided without attention to all the complex interface problems. Learning quantitatively and qualitatively about city waste, industrial effluents, power utility outputs, transportation pollutants; developing techniques and assessing the costs of minimizing, limiting and altering these environmental impairments; puzzling out the economic value to the society of cleaner waters (how clean?)—these and numerous other factors and trade-off issues must be analyzed. The decision makers, in the end the public, must be able to see the options and compare the benefits with the price to be paid.

But what company, or group of companies, would invest in a private overall systems solution to, say, depolluting Lake Erie, working out with their own funds the many trade-offs, developing the new chemical processes, superior purification and fuel burning methods, optimum energy use techniques and the rest? Many specific and difficult science and engineering accomplishments would be vital for an intelligent, broad changeover to an integrated lower pollution system. But in what context should the individual technological efforts be chosen? What criteria would be used? Even if a combine could be created of, say five of our largest corporations that among them possess all the technical expertise required, how could they presume to design and install a system that would affect the economy and social make-up of the industrialized communities of millions of citizens distributed around the Lake?

If that private combine invests billions to develop a sound solution (a moon landing involved some tens of billions, but depolluting a large industrialized geographical area entails much more), it is not clear who would be their customer. To be sure, they might offer a proposal to install superior waste disposal equipment to a city, one of the many contributing sewage to the Lake, but why should that city pay to pollute the Lake less unless all other polluters agree to do their part in a balanced program? The polluters would not only have to agree; all would need to possess the funds to buy their piece of the system. Finally, suppose the numerous independent segments of the entire market come together by some magic. That is, they decide to buy, stand ready with the funds and are set to contract with the entrepreneurial syndicate to install the entire developed system. Would not the Justice Department be expected to stop this combine of private corporations, claiming an illegal joining of large corporations into a group possessing domination of the market? And the government would be right, would it not? If some such partnership sank billions into the creation of a depollution system market, they probably would possess a near monopoly since a second such team (equally resourceful as to finance and technology) would be unlikely to choose to enter as a competitor.

Let us look quickly at a few quite different examples.

Electronic Brainpower

Electronic data systems can now absorb, store, categorize, process, ponder, move and present information in vastly higher quantities, with greater speed, yet with radically reduced cost and more reliability and accuracy than ever before attainable. This synthetic "electronic brainpower" can make each human being smarter at his job. We all spend much of our time doing something with information because information makes the world go 'round. By a new man-technology partnership in information handling, the potential for increasing the value of every hour of a person's time, and hence the nation's productivity, is tremendous—in business and industry, banks, the professions, airlines, hospitals, educational institutions, as well as in all levels of government.

As with the telephone, then the automobile, and later, TV, such electronic information handling advances become economical only if the users are numbered in the many millions. The United States is close to unique in the world in possessing the combination of characteristics needed for leadership in this information technology advance: a science and technology lead; a single, integrated large market; a clear need; opportunity for high return on invested resources. In fact, virtually every man, woman, and child in the United States could be aided—in his work, schooling, traveling, shopping, etc.—by the ready availability of such technology. If every worker's output were to be improved to an extent equivalent to a mere few hundred dollars a year by the application of innovative electronic information technology, this saving would justify an investment of a thousand dollars per worker for the purchase of hardware and software. However, such a mass application of electronic information technology would sum up to hundreds of billions of dollars, too big for any one corporation. The progress thus will come through a hybrid of free market and government involvement—the latter through direct government sponsorship and regulation and with the government as a principal customer—not through free enterprise alone.

To be more specific, take, for instance, Electronics Funds Transfer, the use of electronic signals to replace the present flow of billions of pieces of paper a year for control and accounting of transactions. In the future, bills will be paid, orders placed, deposits to accounts made, interest charged or credited, funds assigned and moved from individuals, companies and governments to one another, accounting completed—all in accordance with electronic signals. However, the full blossoming of this low paperwork society requires a revolution in the rules governing the flow of money. Look what the government will have to do: redefine banking; decide who can be allowed to profit from the holding and moving of money; standardize interest rates on funds involved; create new privacy and security regulations; set standards for the form of the electronic signals that carry the funds transfer information and for the many equipments making up the systems so they will all interconnect compatibly. We see that to achieve electronic funds transfer we face a challenging organization problem, one requiring the teaming of government and private industry with science and engineering. In electronic information handling, as in many other frontier areas, the speed with which society can obtain the productivity and other benefits is not limited by the technology, but rather by the rate of progress in organization.

Space Technology

The United States is ahead of other nations in space technology. In the coming decades, economic cost-effectiveness exists for numerous large-scale satellite systems: increased telephonic communication between continents and over large continental areas; the same for television broadcoasting; airline navigation and traffic control; navigation and communication with ships at sea; computer-to-computer communication of the data essential to the operations of business and industry; educational broadcasting to homes and schools in a vast uplifting of education by video programming; military communications, command, control, intelligence and reconnaissance for military alertness or arms control; earth scanning to improve discovery, analysis, and utilization of mineral, water, agricultural and other earth resources; weather prediction, and ultimately the beginnings of weather control; library networks to provide research and operating data to medical, legal and other professional activities. The rate of such developments will depend on the handling of a maze of government-industry arrangements on a world basis. Consider specifically a system to plot and examine the earth's resources by satellites working with a ground-based network of communication equipment, computers, and data analyzers. How could a private entrepreneur, in the form of one or more American companies, create the system, operate it and go about obtaining rewards to cover its investment? The private entity could not scan our nation and others and offer the information for sale to realize income without a government sanctioned position—which would carry with it government regulation—which means the price of the information made available would have to be controlled and it would not be solely a "private enterprise" project.

Fusion Energy

Controlled thermonuclear fusion has the possibility of providing us with cheap and plentiful energy. Many millions of dollars have already been spent trying to understand the underlying physics, and it is estimated by most experts that decades will be required before all of the remaining physical science and engineering details are worked out sufficiently that this approach to energy supply could be ready for application. It may turn out that it never will prove feasible. The total costs in the start-up phase are seen to be in the billions of dollars, maybe in the tens of billions. When we have to speak of funding at such levels and time in the decades, we are beyond the boundaries of areas suitable for speculative investment by private entities.

Coal

By all-out use of science and technology and the full cooperation of government and private industry (neither of which is in the cards today) we could have a plentiful supply of energy for hundreds of years by properly utilizing our huge reserves of coal. Technology can be advanced to desulphurize coal, produce liquid and gaseous fuel from it, mine it more safely, obtain a good deal of it without mining (in situ), burn it more cleanly and fully, and limit environmental harm. The resulting energy supply would be higher priced than today's oil but the funds to pay for the fuel would stay in the United States.

Success requires both the know-how of private industry and enormous capital placed at risk. A corporation choosing to make a major entry into new coal technology would have to commit funds in the range of billions of dollars and decades might pass before a return on that investment would be realized. Indeed such return might never come, so great are the risks associated with unpredictable decisions on critical aspects that it is clear will be made by the government. The complete "system" needed for a much greater utilization of coal in the United States involves a host of private and public organizations that are semi-autonomous and not readily directed by any one body: land owners, mine operators, labor unions, railroads, power generating and water supply utilities, numerous engineering and manufacturing organizations, and many, many agencies in the federal government that deal with prices, environment, labor, and transport, to name only a portion. It is not difficult to appreciate why we are not moving very rapidly in developing and applying new coal technology.

Food and Nutrition

Almost every nation of the world faces food insufficiency. The United States of America, with its resources and know-how, has a special opportunity here. Science and engineering advances could improve every aspect of food and nutrition. From seed to mouth, potentials for greater economy, efficiency, nutrition, productivity, preservation and waste reduction exist. However, consider the interactive problems of land utilization, energy and water supply, environmental and health control (use of pesticides, fertilizers, food additives and preservatives), fundamental research in plant growth and nutrition, storage and packaging, distribution methods, price controls, and foreign sales policies—all pieces of a grandly complex system. Working this systems problem requires innovative teaming of government and private enterprise. Today the government is mainly in the regulatory business as regards food. No entity is providing integrated leadership. Only slow progress in the utilization of science can result.

If these organization problems could be solved, in a decade or two the U.S. position in the world in food could be like that of the OPEC nations in oil. Even now agricultural exports make a substantial favorable contribution to our balance of payments. If the U.S. technological superiority on every food resource front were exploited, the United States would not have a balance of payments problem for a century ahead.

Ocean Resources

As a final example, today we have the technological means to advance rapidly our understanding of ocean sciences. New sources of minerals, food and energy and improved transportation and weather forecasting are among the possibilities that loom over the horizon. Full exploitation of ocean resources could change the world economy. This potential is more than matched, however, by the difficulties of world cooperation. To cite only one example, who owns the raw materials that may be extracted from the deep ocean bottom? Private know-how and private capital will not be available for such efforts until there is a positive answer to this question. The governments of the world are meeting in "Law of the Sea" negotiations, and unfortunately, getting nowhere.

THE GOVERNMENT'S ROLE

Free enterprise acting alone, we observe, is a way to realize at most only a portion of all the good that science and technology can bring us. The preceding examples suggest that government is a necessary participant to cope with the remainder. But what about the government's role in direct science and technology advance? What should the government try to accomplish all on its own, not necessarily using private enterprise as a co-contributor to the advance?

Of course, the public, speaking through the government, must control and regulate against impairment of the citizen's welfare that could arise from technological projects performed for the narrow benefit of a private group. Also, to advance some aspects of science and technology for the general society's good, the government must be relied upon for strong and essential, perhaps even solitary, sponsorship. This is perhaps most obvious in the development of military weapons systems, or for sending a man to the moon, and for similar projects considered as necessary in the national interest but where there is no consumer product, no market involving the general public. But why not have the government do more, beyond those necessary projects free enterprise cannot or will not perform alone? Why now embark on a major effort to increase the innovation and the productivity of America by all-out development, under government funding and sponsorship, of new technology? This, it could be argued, would lead to new methods and products, cutting costs, increasing supply, getting around shortages, and making for more jobs.

To go this far, to consider having the government move into direct technological advance, even into those segments where private enterprise is still strong and willing, or potentially so, takes the government right into the heart of the function of a free market. As a step in pondering this, consider that, low as is the level to which free enterprise may have sunk in public favor, there remains today a bottom limit to this disfavor. Despite the diminishing popularity of the concept of capitalism, most of those who talk or think against it in the United States are far from ready to accept as the alternative a totally government controlled economy. Most do not want a more massive government bureaucracy. They will rebel against their take-home pay's being cut further, the government's taking a bigger chunk of the gross for more spending while they lose the freedom to spend their earnings as they choose. They do not want the government allocating the resources of the nation, deciding what technology to further, and what products should be manufactured. All this is to say that most have an interest in keeping free enterprise alive and healthy, even though many may not know what this requires and may often espouse views that hurt free enterprise's ability to make the very contribution to the society these people expect and desire.

The federal government employs scientists and engineers in policy-making and management of government sponsored technical effort. Also, the government maintains laboratories in certain highly specialized fields and these contain their share of experts. However, in the United States today, most of the expertise in science and technology lies in private industry and the universities. If the government were to launch a program of research and development with the serious objective of raising the innovative technological level of the nation it would have to set out to influence, augment and direct the efforts of thousands of technically trained inviduals in the private sector and deal in a detailed way with the enormous infrastructure in which they now operate. The technological industry sees the government as particularly ineffective when it comes to matching research and development with the market's real needs, that is with the end consumer's true desires. Technology advance is infinite in its variety and dimensions. The new products that could be developed, the changes in methods and procedures that are technically feasible and the possible alterations of virtually every system for producing, distributing and using technical projects and services are endless in number. The nation has only the resources, whether it be technically trained people or funding to back them up, to carry through a tiny percentage of all that can be conceived. The trick is to choose the right research and development, to assign the limited sources so as to best meet the citizen's sense of values. Even those who are convinced that the free market is an inadequate means for obtaining the proper match of resources to goals, probably would grant that if this balancing is to be attempted by the federal government instead, it could only be worse. It would require the creation of a huge bureaucratic government system for detailed analysis, selection of projects, and assignment of resources. It is submitted that today only a small fraction of the electorate really believes that the government is suited to such control of scientific research and technological development.

Concern exists today that U.S. industrial technological advance is insufficient and some believe government programs are needed to enhance industrial R & D. In industrial technological innovation, the government can help most, not by funding new efforts, but by improving the economy, as already stated, so the private sector will have the financial means and incentives to perform more R & D of its own choice. Lowering capital gains taxes and decreasing the taxation (presently "double") on dividends would especially encourage investment in innovation. (Germany's and Japan's investment ratios are much higher than ours and so are their productivity increases.)

On the other hand, the federal government must continue to be the main sponsor of basic research in the nation's universities. Unfortunately, even here, in this necessary role, the government is not performing well. The low funding leaves good research talent unemployed. The government's bureaucratic approaches are causing too large a chunk out of every dollar granted for research in the universities to go for administrative expenses. Also, the growing emphasis by the government on proven relevance for research underestimates the high eventual payoff in probing the unknown secrets of nature. Admittedly, the government must spend its money wisely, but we have gone past common sense in the control process. It costs less to accept that some research will prove useless, than to try to guarantee that all research will pay off.

SOME RECOMMENDATIONS

Now what does all this add up to? This high potential for science and technology to benefit the society more and at the same time this difficulty in obtaining these benefits. This inability of free enterprise to encompass more than a fraction of the sponsorship. This limitation on the capability of the government to take over where the private sector cannot provide the desirable advances. What views and policies should we promote as steps to improve the situation?

Let us offer some specific recommendations separately to the three concerned entities: the free enterprise sector, the government, and the public.

Those interested in preserving the free enterprise system (such as leaders of big business and others who see the free market and the private enterprise concept as valuable routes towards utilization of advanced technology) should accept that ours is a hybrid society. It is not helpful and oftentimes harmful to press so hard, as these supporters of free enterprise often do, for it to be labeled as the all encompassing answer to all problems. Free enterprise can only provide for a part of our needs and a creative, cooperative relationship between the private sector and the government should be sought.

Government leaders should recognize the government's involvement in science and technology matters to be permanent and set a long-range policy for that involvement. It should answer such questions as: Which research and development areas must the government sponsor, and which are best left to free enterprise? How can the required regulatory duties of the government, the setting of the rules for free enterprise, be handled so as to enhance and not impair national innovative effort? The government should appreciate the broad impact of science and technology on virtually every important government decision making area, and count a healthy scientific and technological effort in the country as vital to the overall national economy. The government should itself be innovative in creating a national environment for innovation. It should direct its sponsorship in science and technology matters towards the generation of information to remove risks and thus to enhance the contribution that can be made by free enterprise. The government should resist the creating of new government R & D efforts and expenditures to enhance innovation in private industry. Above all, the government should act on the premise that improved use of science and technology will quickly follow an improvement in the economy, this improvement characterized by less inflation, taxes and government spending, with discretionary funds for investment in innovation more available to the private sector.

The public would do well to foster neither extreme view: that the free enterprise system alone can do everything best; nor that free enterprise is bad and passé and the government must do everything. If the public is interested in the flexibility, freedom of opportunity, motivation and incentives of free enterprise, then it should recognize that free enterprise can only contribute to the nation's needed research and development if industry is allowed to generate the necessary funds for investment. "Profits" should be applauded if they are reinvested boldly and wisely for research and development

Finally, the public should get over its anti-technology prejudices. It should decide once and for all that science and technology are not independent, external enemies but the public's own tools, and get on with seeing that these tools are used to achieve the goals the citizens choose.

7.8

The Social Ecology of Technology

MELVIN KRANZBERG

One of my most vivid childhood memories is attending the Saturday matinee and watching the melodramatic movie serials. Near the end of virtually every episode of "The Perils of Pauline," the heroine, played by Pearl White, lay lashed to a railroad track while a speeding locomotive bore down upon her. To many social critics, this flickering image of the silver screen provides an apt metaphor for our contemporary situation: we fear that technology is rushing headlong to destroy us; we struggle, but naught avails. Technology, symbolized by the locomotive, the man-made juggernaut, is out of control and is bent on our destruction.

TECHNOLOGY IN LITERATURE

Today's serious literary figures say much the same thing, and in just as melodramatic a fashion. The modern novel, the contemporary drama, and today's poetry have as one of their most insistent themes the fear that technology is taking over from man. And, in this current literature of antitechnology, two metaphors constantly appear; Frankenstein's monster from the 19th-century thriller by Mary Wollstonecraft Shelley, and the robots from Karel Čapek's play of the 1920s, *R.U.R.* Both these metaphors engulf us in a bloodchilling speculation that man has created the mechanical means of his own annihilation, either literally or figuratively.

These metaphors are more apt than the critics who use them realize. For one thing, the monster of Frankenstein and Rossum's Universal Robots were created by men who sought to do good for mankind. Dr. Frankenstein said, "I thought that if I could bestow animation upon lifeless matter, I might in the process of time . . . renew life where death had apparently devoted the body to corruption."

Frankenstein thus conceives of himself as man's benefactor, as one who will "pour a torrent of light into our dark world." The engineers who manufactured Rossum's Universal Robots were also motivated by altruism. As Harry Domin, general manager of the robot plant, states,

> In ten years time Rossum's Robots will produce so much corn, so much cloth, so much everything, that things will cost practically nothing. There will be no poverty. All work will be done by living machines. Everybody will be free from worry and liberated from the degradation of labor. Everybody will live only to perfect himself.

At this point, our antitechnologists go off the track in utilizing these metaphors. While willing to admit that technologists are motivated by altruism, they imply that the creations turned against their creators because there is something wrong with technology. If they would remain true to the letter and spirit of their metaphors, however, they would discover that it was man himself who was responsible for turning these technical creations against humanity. The monster of Frankenstein—and I am speaking here about the original novel—sought only warmth, understanding, and sympathy. Yet his gruesome form and appearance turned people against him, until finally, out of despair, he turned against mankind. In the case of Frankenstein's monster, it was not the machine which failed, but human beings, including Dr. Frankenstein himself. By indicating the monster's responsiveness to music and human friendship, Mel Brooks' recent movie, "Young Frankenstein," is truer to Mrs. Shelley's original novel than was Boris Karloff in the older movie version which you watch on the TV late, late movies.

In the play *R.U.R.*, it is also a human rather than a technological defect which creates problems. The trouble is caused, we learn, by Dr. Gall, head of the physiological and experimental department of R.U.R., who changed the composition of the robots and sought to make them more human by making them more irritable. Only when the robots ceased to be machines and became more human did they become more dangerous.

The story of Frankenstein is a psychological thriller. The fault is not in the monster created by man but in man himself. *R.U.R.* is a modern parable, an indictment of selfish materialism, not of the machine. It is important that we remember this distinction, that we do not allow our modern literary Luddites to stampede us into unthinking criticism of technology by literary allusions that are incorrect.

TECHNOLOGY IN HISTORY

Yet there is something even more disturbing about such literary swipes at technology that have become almost a ritual among today's bleeding-heart humanists. For they posit some sort of conflict between man and his technology, as if the two were somehow opposed to each other and locked in mortal combat. I find this somewhat strange because the historical facts indicate quite the opposite: that man and his technology are inextricably intertwined, that technology has served as one of man's chief tools as he sought to achieve his goals,both glorious and ignoble. Indeed, man and his technology form part of a dynamic ecological system.

Although some people think of ecology in narrow terms of the physical environment, it is much broader in scope than that. According to its Greek root, ecology means the study of "houses," so that Xenophon entitled his study of the management of households the *Oeconomicus*. Since the household is regarded as a "living environment," ecology thus became a science concerned with the dynamics of interrelationships in the physical environment. Here, I extend ecology to include the social environment, for man with the aid of technology has created his own physical and social environment which has thus become part of the larger ecological system.

Our ecologists and environmentalists tend to think only of the natural environment, and then only in terms of a static state of nature. But that is scientifically inaccurate, as any climatologist or geologist can attest. It is also historically inaccurate. The natural habitat of civilized man is the city. The physical environment of civilized man includes his own artifacts as well as the handiwork of nature.

The environment of 20th-century *homo Americanus* includes elevators, telephones, electric ovens, air-conditioners, TV sets, automobiles, vacuum cleaners, central heating, refrigerators, and umpteen other objects. If we are concerned about protecting our environment—and I think we should be—it is equally important that we protect our civilized, man-made environment as well as our natural environment. Indeed, a good portion of our social environment has come about because we used our technology to protect us against a natural environment which, except on the island of Tonga-Tonga or a few other remote spots, is pretty inhospitable and cruel to man—and let's not forget it.

THE SYMBIOSIS OF MAN AND TECHNOLOGY

We must go back to the very beginning of our species to find the origins of the symbiotic relationship between man and his technology. Anthropoligsts seeking the origin of mankind have attempted to differentiate between what constitutes "almost man" and the genus "man." One distinction they find is that man employed tools. Man as we know him probably could not have evolved or survived without tools—he is too weak and puny a creature to fight nature with only his hands and teeth. The lion is stronger, the horse is faster, the giraffe can reach higher; tools serves as extensions of man's hand and amplifers of his muscle power, enabling him to adjust his hereditary organic equipment to an almost infinite number of operations in virtually any environment.

Mankind and Tools

It is not surprising, therefore, that some anthropologists defined the human species on the basis of tool-using and tool-making, or to be more exact, tool dependency. Modern physiology, psychology, evolutionary biology, and anthropology all combine to demonstrate to us that *homo sapiens* (man the thinker) cannot be distinguished from *homo faber* (man the maker). Indeed, we now realize that man could not have become a thinker had he not at the same time been a maker. Thus, we find that technology is perhaps one of the earliest and most basic of human cultural characteristics.

Mankind and Fire

The current generation of anthropologists is no longer so concerned about the relations between man's physiological evolution and his tools. Following the lead of the great French anthropologist Claude Lévi-Strauss, they seek to find the underlying structural unity between the myths and habits of mankind which link together the human species. But even the structuralist anthropologists do not ignore the role of technology in man's development. The title of one of Lévi-Strauss's most famous works, *The Raw and the Cooked*, shows that. In that book he finds that the dividing line between nature and culture is control of the technology of fire—with the auxiliary technique of cooking vessels and cooking implements. The primitive man eats his food raw; civilization begins when man cooks his food.

Mankind and Agriculture

Novel as this interpretation might be, it is not far removed from an older tradition associated with the late British anthropologist Gordon Childe, who equated the start of civlization with the development of settled communities. But that too rested upon a technological innovation—agriculture. Prior to that time, men had been hunters and food-gatherers; in a sense, they had been parasites upon nature. We do not know exactly how or when agriculture began, yet once men discovered that they could cooperate with nature by sowing seeds and waiting for nature to perform the miracle of growing crops, there arose the possibility of settled and civilized life. Unlike the hunter, the agriculturalist could not live in constant conflict, and had to cooperate not only with nature but with other human beings. If he spent all his time in fighting, he would have neither time nor energy for carrying on agricultural pursuits. Yet, if he ran away from his enemies, his crops would go untended and he would lose his livelihood. With the coming of agriculture, therefore, the beginnings of civilized society emerge, and along with it man's dawning awareness that he must live and work together with others if he is to survive.

Mankind and Environment

From savagery to barbarism to civilization, man's material progress throughout the ages has been bound up with his technology. The story of technology is a story of man's attempts to control his environment—in terms of materials, from stone to bronze, from bronze to iron, and iron to steel; in terms of energy, from human muscle power to animal, to wind and water, to steam and oil, to rockets and nuclear power; and in terms of machines, from hand tools to powered tools, to mass production lines, to computer-controlled factories.

TECHNOLOGY VS. HUMANITY

Despite the role of technology in the panorama of human history, we frequently come up against those who regard technology as something divorced from the essence of humanity. The humanists ask: What is human about a monkey wrench, a lathe, a computer? What repels them about technology is the very "inhumanity" of its objects, for example, the monster automatons which it has created and which threaten to make man expendable, or the "antihumanity" of its byproducts such as the pollution which threatens our environment.

It is true that we ordinarily think of technology as something mechanical: yet the fact is that all technical processes and products are the result of the human creative imagination and human skills. The story of how man has utilized technology in order to master his environment is part of the great drama of man fighting against the unknown. This is the continuation of the struggle of Prometheus, who stole fire from the gods in order to bring it to mankind. Ever since that time, man has sought to use nature to his own ends, to master the environment, to control nature. This is a very human activity—this is technology.

THE SIGNIFICANCE OF TECHNOLOGY

Furthermore, the significance of technology lies in its use by human beings, what it does. Take, for example, the telephone. If we regard it as a collection of wires through which current passes from a transmitter to a receiver, it would seem to have little interest, except for telephone technicians and repairmen. But the significance of the telephone lies in its use in transmitting messages. It is the communications function of the telephone that gives it importance. Similarly, the function of technology is its use by human beings.

Technology is inseparable from men and communities. The technologist is concerned with the application of science and other forms of knowledge to the needs of man in society. He is up to his neck in human problems whether he likes it or not. When our Peace Corps built a road in tropical Africa, that road was more than an exercise in civil engineering. It was, in fact, a major experiment in social anthropology, for it affected the primitive villages up-country and acted as a communication link which stimulated the acculturation of those people to modern Western society. Every technological activity has social consequences. Technology is thus of humanistic interest, not only because it is a product of the human mind, but also because it affects the course of human and societal development.

THE IMPACT OF TECHNOLOGY

It is, of course, this impact of technology upon the social ecology that makes many sensitive individuals fear that this very human activity—technology—has grown so large and has presented mankind with such awful byproducts that it threatens to engulf man, making life uninhabitable on this planet. Technology, it has been said, might be a good thing—but it might be too much of a good thing. Look at the inhuman uses to which it has been put. What about the devastation wrought by wars during the 20th century, and the present possibility of destroying the human race through nuclear warfare? What about the deterioration of the environment created by air and water pollution, the spoilation of the countryside, and the rot of our cities? Technology, it is claimed, has destroyed the ecological balance between man and nature. Not only that, but we are robbing future generations of their inheritance by plundering the earth of irreplaceable natural resources.

Yet, when this 20th century began, it was confidently anticipated that technology would put an end to ignorance and superstition, misery and hunger, domestic strife and international conflict. Things have not quite turned out that way. Not only have we not reached the material utopia promised by technology, but every step we take along the way makes it seem as if we were running madly on a gilded treadmill. As we approached the ideal of producing enough goods and services to provide middle-class comfort to all America, we discovered that we still had an enormous number of people living at or below the poverty level. Our astronauts can whiz around the earth in less than an hour, and anyone with a credit card and a reservation on the Concorde will be able to have breakfast in both New York City and London on the same morning; yet suburban commuters cannot get to work on time. Our cars are the most powerful and luxurious in the world, but we are running out of the gas to drive them. In less time than it takes to tell you about this—indeed, in the half second it would take for spilled water to fall from the table to the floor—today's large computers can examine the electrocardiograms of 100 patients, and score 150,000 answers on 3000 exams, and figure the payroll for a company of 1000 employees, and debit 2000 checks to 300 different bank accounts while at the same time driving people frantic as they try to straighten out the computer's mistakes in their bank accounts. Our Xerox machines multiply papers faster than the sorcerer's apprentice could dump them in a garbage can, and television and radio bring instant information to us; yet men feel less and less capable of understanding the world in which they live. Despite our vast communication network, people agonize as never before over "the failure to communicate."

In 17th-century London, John Donne preached that "no man is an island," and technology has made this a reality through knitting an infinite number of threads of global communications and international trades. Yet this very dependency on our neighbours fills us with fear: the breakdown of a distant electrical relay plunges us into darkness; strikes by delivery men, garbage collectors, policemen, teachers can deprive us of food, pile up mountains of refuse, and threaten our education and security: the vagaries of international power politics can cut off vital fuel and material resources; the actions of a few violent terrorists can destroy international transportation, disrupt the harmony of international sports, end the careers of gifted and talented leaders, and threaten the vital but fragile threads of our highly complex industrial society.

We find ourselves baffled by this series of paradoxes. Who would have thought that this nation of presumably unlimited resources, glorying in the enormous treasure that it extracted from the earth, would now be concerned by the profligate squandering of its treasury? Who would have thought that "America the beautiful" would be pockmarked by the rash of urban blight, the ugly scars of stripmining, and the varicose arteries of freeways? Who would have thought that our cities would become so large as to become virtually unmanageable, that our traffic would be so heavy we might find it quicker to walk than to drive, that new college buildings would resemble medieval fortresses rather than fake Gothic or Tudor castles?

SOCIAL PROBLEMS AND TECHNOLOGY

Is technology to blame for all the problems which beset us? Yes, say some of the most influential humanists of our time. Though willing to admit that technology has "raised the ceiling of human achievement," Lewis Mumford claims that modern technology has become authoritarian and is "transferring the attributes of life to the machine and the mechanical collective." Jacques Ellul, the French legal philosopher, has a similar apocalyptic view, feeling that technology has become the end of human life. Fusing ideas borrowed from both Freud and Marx, Herbert Marcuse, the guru of the new left, attacks industrial civilization on the grounds that it has flattened man into a "one-dimensional" being. Even admitting that more men may be happier today than ever before, their happiness, he claims, is "a state of anesthesia." Though technology has done away with scarcities, it forces men, says Marcuse, to "exhausting, stupefying, inhuman slavery," alienating the workers from each other, from their product, and from work itself. Mass society provides bread, circuses, and technology. But the material plenty, according to Marcuse, yields no spiritual gratification and leads to social oppression. Marcuse holds these principles to be self-evident in both capitalist and communist societies. They characterize industrial civilization no matter what the socio-political arrangements might be.

Is technology really responsible for all these problems? Social thinkers like Mumford, Marcuse, and Ellul think so, as do the adherents of the counter-culture, such as Theodore Roszak and the late Paul Goodman. Theirs is a naive technological determinism, the notion that technology determines the pattern for our lives, actions, values, and institutions. This is an old complaint. Over a century ago, Ralph Waldo Emerson made the same criticism when he said, "Things are in the saddle and ride mankind."

Not all scholars would accept this version of technological omnipotence. Lynn White, the great medievalist and recent president of the American Historical Association, has said: "Technology opens doors, it does not compel one to enter." In this view, technology is an enabling mechanism, a means which man is free to employ as he sees fit. But the question arises as to whether we can distinguish between ends and means, and if the means do not sometimes determine the ends. True, man is not compelled to enter Professor White's open door, but an open door is an invitation. Besides, who decides which doors to open—and once men have entered the room, are not their future actions and directions perhaps guided by the contours of the chambers and corridors into which they have stepped? More importantly, once we have crossed the threshold, can we turn back?

The Social Consequences of Technology

Frankly, we historians of technology do not know the answer to this question of technological determinism; ours is a new discipline, and we are still working on the problem. We might never reach agreement upon an answer. Yet, there are several things that we do know. One is that technology's interaction with the social ecology is such that technical developments frequently have social and human consequences which go far beyond the immediate purposes of the technical devices and practices themselves. I have already pointed out how a technological innovation, namely agriculture, made possible the development of settled communities and the start of civilization. I could show you how the introduction of gunpowder helped bring about the demise of feudalism, or how the Industrial Revolution changed the dwelling places, work activities, and thoughts of mankind. Rather than retracing all history, however, I shall focus on two developments in the 20th century which have had—and continue to have—profound repercussions upon American society and which manifest my statement that technology has social ramifications which do not necessarily seem inherent in the technology itself.

Women's Rights

Take women's rights, for instance. The emancipation of women is a recent phenomenon within Western society, dating back only some 50 years, and it can scarcely be said that engineers consciously sought to advance that cause. When we see old newsreels showing suffragettes marching down Fifth Avenue demanding equal rights for women, we assume that the pressures exerted by Susan B. Anthony and her followers were responsible for the freedom of women in the modern world. Don't you believe it!

Except at a few colleges, female emancipation never really moved forward until Charles Kettering came along with the self-starter for the automobile. Now the hand that rocked the cradle, but could not turn a crank, had only to turn a key. The cradle itself gave way to the kiddy-seat beside Mom, with both Mom and kiddy hellbent on seeing the world. The suffragettes won women the right to vote, but women were far too smart to use that club. Engineers were handing them much better ones. The true emancipation of women came when the auto gave the mobility and when technological devices within the home freed them from onerous household chores. When frozen foods came in, every woman became a great cook. Automated machines and detergents made her a great laundress. Supermarkets converted her into a canny purchasing agent. The pill allows her to have children, if, as, and when she wants them. The upshot is that, while we men still decide the really important questions of the world like stemming Communism in Angola, the women make all those unimportant decisions having to do with money, food, shelter, clothing, and the good life.

The Black Revolution

Let me take another example—the Black Revolution, manifested through violence, destruction, and death. We attribute this, perhaps rightly, to many different socio-cultural factors: the frustration of the Negro in white society; the intolerance, "racism," or just plain indifference of the white community; the lack of foresight and sensitivity on the part of the public officials; the failure of effective law enforcement; the presence of outside agitators, and so on and on. Yet the deeper causes of the black revolt have technological foundations. Let me show you.

Many historians claim that Negro slavery was languishing in this country near the close of the 18th century, when an ingenious innovation, Eli Whitney's cotton gin of 1793, lowered the price of upland cotton in the United States. The obvious economic advantages of this invention fastened the plantation system and the institution of slavery upon the American South. Within another half century, however, some historians claim, slave labor had proved itself so inefficient as to be technologically obsolete. This view has recently been challenged, but no matter. The movement toward its abolition cost us a civil war, and the South, being technologically underdeveloped, lost to the more industrialized North. Although the slaves were emancipated, they remained in economic poverty, social bondage, and political disenfranchisement for nearly another century. In the end, however, it was technological advance in the 20th century which finally delivered the Negro from southern to northern peonage.

In this connection, three advances are of special note. First was the industrialization of the South. Machines are color-blind: they do not know or care whether the hands which guide them are white, black, blue, yellow, or green. And for this, among other reasons, the South resisted machines far into the 20th century. It was a hopeless cause—just the kind our Southland has always fancied. Second was the development of new agricultural products. They ended the reign of King Cotton and the dependence of the South on one crop. Third was the development of the mechanical cotton-picker which deprived the Negro fieldhand of his livelihood. Only a quarter of a century ago it took a man with a mule 160 hours to plant, cultivate, and pick an acre of cotton. Now with the aid of machines, the same work is done with as little as 13½ hours of labor. In the Mississippi Delta, some 30,510 workers were employed in the back-breaking task of picking cotton in 1960; by 1967, the introduction of cotton-picking machinery had caused that number to dwindle to 7225 to pick the same amount of cotton.

There was, therefore, no longer a place or need for unskilled labor in the South. So, on or off relief, blacks either died of malnutrition—the polite American term for starvation—or moved out. More moved than died because another marvel of technology emerged. The automobile gave mobility to the entire population of the United States including even poverty-strickened blacks and whites in battered jalopies.

The urban industrial centers beckoned like promised lands flowing with machines and jobs which, unfortunately, were in the hands of the Canaanites and Philistines. Possessing mobility and no longer having even a miniscule economic stake in southern rural regions, a vast migration of Negroes flowed from southern farms to northern cities. Between 1940 and 1970, some 4.4 million blacks left the South, flooding the North with "unemployables," a term perhaps invented by those who wished to avoid doing anything about the problem.

This technologically powered flight of the rural poor to the big cities had far-reaching consequences, none of which was avoidable by the invention of more new words like "in-migrant," "underprivileged," and "disadvantaged." These new words, however, helped explain away the next cruel technological blow which fell upon the poor southern black migrating to the promised land, "where the streets are paved with gold." The black migrant could find no place in today's sophisticated technology. Perhaps for the first time in Western history, we saw in the 1950's and 1960's the spectacle of large-scale unemployment while thousands of jobs went begging. There was no match between the openings that required highly skilled workmen and the available labor supply.

Compounding the frustration was another miracle of modern technology, the television set. As a revolutionary instrument, TV must rank alongside the Declaration of Independence and the Rights of Man. Into the shabby living room of the Negro relief client were piped the dreams and sugarcoated realities of an affluent society. He was invited to spend hours watching how good life was for everybody else and hearing about products which everyone, except him, seemed able to buy. For the first time in history, the disinherited could see intimately the affluence of the wealthy with immediate perception of what they were missing. They wanted to partake of the great outpouring of goods which engineers had made possible, yet they were denied all legal and socially acceptable means to obtain them. And so we packaged wealth and privilege for television. We displayed it publicly to the poor and nonprivileged whom we barred from the system. Then we wondered why they rioted and looted.

This sad history, whose social repercussions are still working themselves out, illustrates the wide-ranging interactions of technology with many different elements of our social environment.

THE INTERACTION OF TECHNOLOGY AND SOCIETY

Yet we take our technology for granted. Our daily lives have been so inextricably bound up with our technology, that we fail to realize how unique and significant it is, until we are faced with altering the patterns of our lives by a sudden energy crisis. I am reminded of the ninety-year-old women who flinched from taking a ride in an airplane, one of the technical marvels of our time: "No siree," she said, "not for me. I'm going to sit right here at home and watch television just the way the good Lord intended I should."

Many of our technology-related problems arise because of the unforeseen consequences when apparently benign technologies are employed on a massive scale. Hence many technical applications which seemed a boon to mankind when first introduced became threats when their use became widespread. Let me take a case which is known to all of us: pollution of the atmosphere caused by automobiles. Yet, at the turn of the century, automobiles were regarded as a solution to the pollution, congestion, and safety problems created by horse-drawn transportation. That was a time when in New York City alone horses deposited some 2½ million pounds of manure and 60,000 gallons of urine in one day. The automobile promised cleaner, quieter, and more rapid transportation—but as we know, pollution, congestion, and safety problems have returned in altered form as a result of the large-scale use of the automobile.

So we angrily demand that Detroit change the engine design in order to do away with the dangerous exhaust and at the same time use less fuel. In other words, we rely upon the "technological fix," that is, remedying matters by the application of more or better technology rather than our own forbearance. We continue to drive our cars: it is always the other fellow who should walk or take public transportation, not us. We favor our own inhuman use of technology in order to suit our own comfort and convenience—and we ignore the health and safety of the community at large. If and when gasoline rationing comes upon us, we will blame the government, industry, oil potentates, and what have you—everybody but ourselves and our materialist values—and we will look to a technological fix to relieve us of the consequences of our own selfishness.

A PARABLE OF THE USE OF TECHNOLOGY

But the way in which our technical solutions give us only a temporary respite from further problems is illustrated by the old prospector in Mexico who ran out of food and water and was forced to wander over a barren desert until he finally staggered upon a convent located in the midst of the desolate landscape. When the nuns had fed and nursed him until he was able to get about again, they offered him a mule on which he could ride to the nearest town—some 35 miles away. Although he promised to repay their great generosity, the nuns assured him that he owed them nothing and that they would retrieve the mule on the next visit to the town some months thence.

Gratefully, our prospector got astride the mule, shouted "Giddap," and waited for it to get moving. But the mule would not move. The old prospector kicked the balky mule and cursed it with the traditional mule-skinner's strong language, but still it refused to budge.

Angered at the mule's display of stubborness, the prospector stalked from the stables to the Mother Superior, thanking her again for her generosity, but explaining that the mule refused to move. "Oh," said the Mother Superior, "I forgot to tell you that this mule does not respond to the language of mule drivers. This mule was brought up in our convent, so it obeys only religious commands."

"What do you mean?" asked the prospector.

"Well," said the Mother Superior, "If you want it to walk, you say 'Thank God'; if you want it to trot, you say, 'Thank God, thank God'; and if you want it to canter at full speed, you say, 'Thank God, thank God, thank God.' To get it to stop you simply say 'Amen.' You will find that the mule is quite obedient when you order it about in those terms, and it will do whatever you say."

The prospector thanked the Mother Superior, mounted the mule, and said, "Thank God," and, sure enough, the mule began to move. About a mile from the convent the terrain flattened out, and the prospector thought that he could make somewhat better time than the ambling pace of the mule. So he said, "Thank God, thank God," and, lo and behold, the mule began to trot. After a few minutes, the prospector urged the mule to even greater speed by saying, "Thank God, thank God, thank God." The mule responded, and our hero was galloping across the plateau with great satisfaction when he suddenly saw a precipice yawning about a hundred yards ahead of him.

He immediately shouted "Whoa!" to the mule, but it did not stop. He yelled, "Stop!" Then "Halt!" But the mule continued its mad gallop. For several seconds, as the mule continued its dash toward the edge of the precipice, our prospector was transfixed because he could not remember the command for stopping which the Mother Superior had given him.

Finally, when just a few yards away from the edge of the precipice, he remembered and shouted, "Amen!" The mule immediately came to a halt, just a step away from the edge of the precipice. Exhausted, but relieved at this narrow escape from a sure death, our prospector leaned back in the saddle and heaved a sigh of relief, saying, "Thank God."

Well, that is a parable of our use of technology. It sometimes obeys our commands, and when our successful technology brings problems in its wake, we ameliorate the situation by still more technology, which brings us again to catastrophe.

BETTER USE OF TECHNOLOGY

But catastrophe can be averted, if we use our technology with humane compassion and with some understanding and comprehension of the social ecology in which our technology will be employed. For one thing, we must understand that technology has provided us with goods and services, food, shelter, and clothing; it has quickened transportation, heightened communication, and so forth. But a better world has not yet come into being—largely because we have not had the social innovations to accompany our technological advancement. Instead, we have utilized our technical achievements in the service of values and institutions belonging to an earlier, more competitive society, one based on scarcity. We have allowed many of the benefits of an advancing technology to be appropriated by narrow interest groups, and we have ignored our own consciences and our own ideals. What can we do about it?

Within the past few years we have begun to develop mechanisms for directing our technology to serve the common weal rather than the interest of narrow interest groups. These new socio-political mechanisms go under the general heading of Technology Assessment, which attempts to evaluate the social and human consequences of the applications of science and technology—before these are applied. Man has always assessed the effectiveness of his technology, but his past assessments were confined to seeing if it would murder his enemies more effectively or bring him greater profits. Now we are trying to extend this assessment to second-order and third-order social and human consequences.

You are already familiar with some of the manifestations of this technology assessment movement. They go under the headings of environmentalism, consumerism, and accountability. For example, the public is beginning to demand that scientists, engineers, business corporations, and government officials be held accountable for the environmental, human, and social consequences of their actions. The passage of the Environmental Quality Control Act and the establishment of the Environmental Protection Agency are demonstrations of the public's concern, as was the institution of the Office of Technology Assessment by the Congress a couple of years ago. The simple fact is that the public has been aroused by the spectre of damage to the ecology and environment; the public is frightened by the possibility of lowering our current standard of living; and the public is furious about the shoddiness and inadequacy of the consumer products offered by American industry. As Betty Furness, former consumer adviser to President Lyndon B. Johnson and later chairman of the New York State Consumer Protection Board, said in Washington at a Congressional hearing—and I quote—"If American industry continues to sow contempt for the consumer, it will reap contempt from the consumer. And from Congress it will reap statutes. This could be the most spectacular case of statutory reap in history."

THE SOCIO-TECHNOLOGICAL INTERFACE

The point is that most of the urgent problems facing man today and tomorrow involve technology, human values, social organization, environmental concerns, economic resources, political decisions, and the like. These are "interface problems," that is, the interface between technology and society, and they can only be solved—if they can be resolved at all—by the application of scientific knowledge, technical expertise, social understanding, and humane compassion.

These interface problems have another feature in common: the engineers cannot solve them alone, yet they cannot be resolved without the aid of the engineers. The intrusion of social and environmental questions into the practice of engineering has blasted engineers out of their smug preoccupation with purely technical problems. Heretofore, the engineer achieved success in problem solving by the process of reductionism—that is, reducing the problem to manageable parts. He restricted himself to the technical components of problems—and rightly so, because he could not cope with the nontechnical elements.

But then engineers complained that they had little direct say in public decisions involving technological matters. The reason for their lack of influence is that the engineers, in their innocence believed that such matters as the SST debate were purely technical problems. But they were not—and are not—purely technical. They are economic and social and political problems, and in cases involving the environment, almost religious problems. True, technology may be the *major* component of such policy problems, but the *decisive* policy elements are value-laden political and economic factors. If you do not realize that, you simply do not understand Secretary of Transportation Coleman's recent decision allowing the Concorde to land in the United States.

By now we must come to the realization that all these various parts—technical, economic, social, political—must be dealt with in an holistic fashion, not separately. Call it a systems approach or an ecological approach, if you will. The point is that we cannot solve individual parts of problems when we are coping with mushy socio-technical issues; we must look at the whole in order to make the necessary and difficult trade-offs between technical possibilities and conflicting socio-economic pressures.

THE LESSON OF HISTORY

If we can educate society to an understanding of technology and engineers to an understanding of society, we might be able to cope with the interface problems of the future. I remain hopeful that we will. After all, this is the Bicentennial year, and in thinking back on American history, I remember that Archibald MacLeish once wrote "America was promises." I believe that it still is.

I believe that, for one reason, because I teach young engineering students, and I am impressed by their sense of calling. Sure, they want well-paid jobs, but they also have a deep sense of commitment to serve mankind and make America a better place for themselves and for their children—just the same sentiments which have inspired Americans for the past 200 years.

This sense of calling perhaps accounts for the increased enrollments in environmental engineering and in urban planning, for the growing interest of our students in solar energy, antipollution devices, energy-saving machines, and the like. It is this mixture of idealism and practicality which I find in full measure in our engineering students of today, and which reflects the spirit and attitude of those who founded and built our country.

The second reason for my belief that America is still "promises" is that I am an historian of technology, and America's technological history has been one of growth and achievement. Ever since the foundation of our Republic, history has been on the side of the technological optimists.

Yet the engineering future will not be easy. To meet the challenges of the future, our engineers must incorporate new social elements into their thinking and provide us with more socially aware technologies. Ours is a society of so-called "high" technology. We may be no more happy nor more secure than our ancestors, but ours is an exciting age in which to live, and we must prepare our young people to live in a world of continuous and rapid change. It is the business of the future, said Alfred North Whitehead, to be dangerous. But we can accept the risks with composure and confidence if we set ourselves to the task of educating our future citizens to understand the values underlying our technological system and the social forces accompanying scientific and technological change, and to play a meaningful part in making our technology contribute to the achievement of our human goals and the solution of the problems of our social and natural ecology.

If we professors do our jobs properly, I think our students will be able to do theirs. I think they can solve the socio-engineering problems of the future. I am sure that, in so doing, their success will be only partial. They will create more problems for their successors even as our generation has done for them. I think their successors will also solve those problems, and in turn create new ones, and so history will be made. And, as an historian of technology, let me conclude by reminding you that the lesson of that history is hope: H-O-P-E, hope.

Selected Additional Readings

A. Brandt, "Views," *The Atlantic*, July 1977, pp. 46–49.

P. L. Bereano, *Technology as a Social and Political Phenomenon*, J. Wiley, New York, 1976.

J. G. Burke and M. C. Eakin (editors), *Technology and Change*, Boyd and Fraser, San Francisco, 1979 (see especially the annotated bibliography in Appendix B).

S. C. Florman, "Technology's Minor Moments," *Harpers Magazine*, June 1978, pp. 16–19.

M. Kranzberg and W. H. Davenport (editors), *Technology and Culture: An Anthology*, Schocken Books, New York, 1972.

B. Mazlish, "The Fourth Discontinuity," *Technology and Culture*, January 1967, pp. 1–15.

E. G. Mesthene, *Technological Change: Its Impact on Man and Society*, Harvard University Press, Cambridge, Mass., 1970.

R. W. Peterson, "Impacts of Technology," *American Scientist*, January–February 1979, pp. 28–31.

C. Susskind, *Understanding Technology*, Johns Hopkins University Press, Baltimore, 1973.

A. Wildavsky, "No Risk Is the Highest Risk of All," *American Scientist*, January–February 1979, pp. 32–37.

J. Woodcock, "Disaster Thrillers: A Literary Mode of Technological Assessment," *Science, Technology and Human Values*, winter 1979, pp. 37–45.